ANCIENT TEXTILES SERIES VOL. 24

SPINNING FATES AND THE SONG OF THE LOOM

The Use of Textiles, Clothing and Cloth Production as Metaphor, Symbol and Narrative Device in Greek and Latin Literature

Edited by

Giovanni Fanfani, Mary Harlow and Marie-Louise Nosch

OXBOW | books
Oxford & Philadelphia

First published in 2016. Reprinted as a paperback in the United Kingdom in 2023 by
OXBOW BOOKS
The Old Music Hall, 106–108 Cowley Road, Oxford, OX4 1JE

and in the United States by
OXBOW BOOKS
1950 Lawrence Road, Havertown, PA 19083

Paperback Edition: ISBN 978-1-78925-986-5
Digital Edition: ISBN 978-1-78570-161-0

A CIP record for this book is available from the British Library

Library of Congress Cataloging-in-Publication Data

Names: Fanfani, Giovanni, editor. | Harlow, Mary, 1956 – editor. | Nosch, Marie-Louise, editor.
Title: Spinning fates and the song of the loom: the use of textiles, clothing and cloth production as metaphor, symbol and narrative device in Greek and Latin literature / edited by Giovanni Fanfani, Mary Harlow and Marie-Louise Nosch.
Other titles: Ancient textiles series; v. 24.
Description: Oxford; Philadelphia: Oxbow Books, 2016. | Series: Ancient textiles series ; vol. 24
Identifiers: LCCN 2016013176 (print) | LCCN 2016016704 (ebook) | ISBN 9781785701603 (hardback) | ISBN 9781785701610 (epub) | ISBN 9781785701627 (mobi) | ISBN 9781785701634 (pdf)
Subjects: LCSH: Weaving in literature. | Clothing and dress in literature. | Classical literature–History and criticism.
Classification: LCC PA3015.W48 S65 2016 (print) | LCC PA3015.W48 (ebook) | DDC 880.09–dc23
LC record available at https://lccn.loc.gov/2016013176

Printed in the United Kingdom by CMP Digital Print Solutions

For a complete list of Oxbow titles, please contact:

UNITED KINGDOM
Oxbow Books
Telephone (0)1226 734350
Email: oxbow@oxbowbooks.com
www.oxbowbooks.com

UNITED STATES OF AMERICA
Oxbow Books
Telephone (610) 853-9131, Fax (610) 853-9146
Email: queries@casemateacademic.com
www.casemateacademic.com/oxbow

Oxbow Books is part of the Casemate Group

Front cover: Detail of the Fates from the Krater of Kleitias and Ergotimos; Florence, Museo Archeologico Nazionale 4209 (so-called François Vase); drawing from Furtwängler and Reichhold 1904, pl. 1.

Contents

1. Of Metaphorical Matrices and their Networks: Generally Speaking, and in the Field of Textile Activities

*Nicole Guilleux**

Technical vocabulary has given rise to countless metaphors stemming from the fact that some human activities – for example tactics, wiles, and literary creation (which was first oral) – are difficult to express properly, as they are often intangible or hidden. When broaching such topics, speakers have to resort to concrete referents. It is, however, less commonly accepted that stable semantic associations exist beyond the restricted frame of etymological links, even though such associations are countless. No matter their name – 'conceptual metaphors' according to George Lakoff and Mark Johnson[1] or '*matrices métaphoriques*' ('metaphorical matrices') in Jean Taillardat's words[2] – these associations are expressed through a great variety of lexemes. They form a large field, further organized in networks, making it a hard task to spot and describe them all.

Firstly, my starting point is the philosophical scope of Lakoff and Johnson's *Metaphors we live by* (1980) and the practical frame the French linguist Jean Taillardat independently worked out in his 1977 paper. Secondly, I will focus on the network of '*matrices métaphoriques*' rooted in the vocabulary of textile crafts. My observations will be mostly based on classical languages, and I will not aim at painting a complete picture of the data but rather at outlining the key elements.

'Conceptual metaphors' (CM) and 'matrices métaphoriques' (MM)

An epistemological convergence

Lakoff and Johnson's theory of 'conceptual metaphors' (CM) and 'conceptual metaphorical mappings' (CMM)

Lakoff and Johnson's book was first published in 1980. It was widely distributed[3] and republished in 2003 with an afterword. The authors took this opportunity to assess their theory placing metaphors at the core of human experience. They also took a look at the different contributions their work triggered. The purpose of their book encompassed a very large field as they aspired to:

* I am very much indebted to Marie-Louise Nosch for her welcome at the CTR in June 2012, and it is a great pleasure for me to thank her and her whole team in Copenhagen. I owe very special thanks to Alice Kabanoff for her valuable help in translating my paper into English. However, all the remaining infelicities are mine. Greek terms are transliterated with their accent and length in this text.

[1] Lakoff & Johnson 1980.

[2] Taillardat 1977.

[3] The first edition was translated into French (1985) and into German (1998), the second edition into Russian (2004).

> Rethink some of the most fundamental ideas in the study of mind: meaning, truth, the nature of thought, and the role of the body in the shaping of mind. As a result it [has] far-reaching implications in field after field – not just linguistics, cognitive science, and philosophy but also literary studies, politics, law, clinical psychology, religion, and even mathematics and the philosophy of science.[4]

Metaphors extend beyond the traditional borders of philosophy and into the field of cognition. Indeed,

> A great deal of everyday, conventional language is metaphorical, and the metaphorical meanings are given by conceptual metaphorical mappings that ultimately arise from correlations in our embodied experience.
>
> In short, (…) conceptual metaphor is a natural part of human thought, and linguistic metaphor is a natural part of human language. Moreover, which metaphors we have and what they mean depend on the nature of our bodies, our interactions in the physical environment, and our social and cultural practices.[5]

It is crucial to admit that these conceptual metaphors are rooted in every aspects of human experience. As Lakoff and Johnson prompt every scholar interested in CM and CMM:

> What is needed is still more empirical research that seeks converging evidence and is gathered by using different empirical methods of inquiry.[6]

Lakoff and Johnson then mention[7] the many books and papers, from 1980 to 2003, in which conceptual metaphors have been identified. The fields concerned are as varied as they are numerous: literary analysis, politics, law, social issues, psychology, mathematics, philosophy, and of course cognitive linguistics, including (among others) Talmy's, Langacker's, and Fauconnier's works.

Jean Taillardat's 'metaphorical matrices' (MMs)

At the end of the 1970s, similar kinds of questioning arose among European linguists specialising in both historical and comparative linguistics. Aurélien Sauvageot, while reviewing a paper written by Jószef Erdödi[8] for the *Bulletin de la Société de Linguistique de Paris,* witnessed that Hungarian linguists had begun taking interest in the mental operations involved in metaphorical designations. The most successful contribution was however made, as far as I know, by Jean Taillardat when coining the concept of 'matrice métaphorique' ('metaphorical matrix') during a session of the Guillaume Budé Association.[9] Although Taillardat's resulting paper was not easy to access and thus to read, his work was nonetheless followed by several French scholars.[10]

For example, Jean-Louis Perpillou[11] highlighted the existence of the INSULT = CUT MM in most Indo-European languages and beyond. This MM, which should be more widely defined as VERBAL AGGRESSION = ATTACK, equates the utterance of aggressive words with various kinds of attacks such as wounds, cuts or bruises, words being assimilated to projectiles. Such an MM is similar to two English CMs noticed by Lakoff and Johnson: ARGUMENT IS WAR and IDEAS ARE CUTTING INSTRUMENTS.[12] Perpillou identified another MM that consistently and usually connects three

[4] Lakoff & Johnson 2003, 243.
[5] Lakoff & Johnson 2003, 247.
[6] Lakoff & Johnson 2003, 247.
[7] Lakoff & Johnson 2003, 267–272, and bibliography: 274–276.
[8] Sauvageot 1978, 380–383. For details, *cf. infra.*
[9] Taillardat 1977.
[10] This is why I take the liberty of quoting all the examples given by Taillardat.
[11] Perpillou 1996.
[12] Lakoff & Johnson 2003, 4, 48. Many English verbs (and most of their derivatives as well) pertain to this MM likewise

elements. Examples are found especially in Latin and Greek vocabularies in the designations of YOUNG HUMAN BEINGS, YOUNG ANIMALS, and YOUNG PLANTS.[13] I would personally add to this matrix the German word *Baumschule* 'tree nursery' (literally, 'school for trees') and the English *nursery* that can be used for children, as well as for plants and fingerlings. In the same line of thought, Marine Bretin-Chabrol's book *L'arbre et la lignée* further exploited this MM by highlighting the crucial function of plant metaphors in filiation and marriage in classical Latin.[14]

Genesis, definition, and uses of MMs

Genesis of the MMs

When he coined the phrase 'matrice métaphorique' in 1977, Taillardat was however not the first scholar to consider the leading role of metaphors in the structuring of vocabulary and in the creation of neologisms. Marcel Schwob and Georges Guieysse[15] had already identified coherent series among colloquial metaphors in their 1892 paper on French slang:

> On reconnaîtra (…) dans la série de mots qui signifient *tête* des métaphores très simples qui se rapportent à la forme. *Calebasse* (*Jargon de l'argot*), *coloquinte* (*ibid.*), † *poire* (*loirepoque*),[16] † *couatche* (*all. questche*), † *ciboulot* et *ciboulotte* (*ciboule*), † *citronnade* (*citron*), † *pomme*, † *balle*, † *boule*, † *bobine*, † *fiole*, † *cafetière*. (…) La métaphore joue ainsi un rôle dans l'argot, mais elle ne paraît pas y avoir un rôle plus grand que dans les autres langues.[17]

No matter the referent – plants or manufactured items – slang neologisms rest upon inherent semes shared by both the source-lexeme and the target-lexeme: /roundness/ or /cylindrical characteristic/, with /hollow/ as an additional seme for the manufactured items only.

Definition of MMs

Eight decades later, when Taillardat coined the phrase 'matrice métaphorique', as well as the tool for lexicological analysis, linguistics had become an established discipline. He had therefore access to books and papers written by linguists and philosophers such as Algirdas Julien Greimas, Bernard Pottier and Paul Ricœur.[18] Taillardat also called on his thorough knowledge of Aristophanes' plays and brought to light, in his seminal 1965 book, the numerous semantic constants within the

---

cut, hurt, injure, rack, skin, sting, tear.

[13] Perpillou 2004, 41–54, chapter 3, 'Marcotte, drageon, plant' [layer, sucker, sapling] et 'veau' [calf]: une métaphore.

[14] Bretin-Chabrol 2012, reworked version of her 2007 PhD thesis, *La naissance et l'origine: métaphores végétales de la filiation de Caton à Gaius.*

[15] Schwob & Guieysse 1892. Taillardat (1977, 350) quotes their examples. The sign † before a word indicates that it was collected from an oral source by the authors, whereas the other words come from a chapbook (Halbert d'Angers 1849). It should be noticed that some of these words are still in use.

[16] The slang word *loirepoque*, which is not used anymore – unlike *poire* – is to some extent similar to the British English back slang, *e.g. tekram* = *market*. Indeed, the source-lexeme *poire* is transformed as follows: the first phoneme *p-* is replaced by *l-*, *p-* is then moved to the end of the word, after *l-oire-*, and the suffix *-oque* is finally added. Such a linguistic process is typical of French slangs called *largonji* and *loucherbem*. *Cf.* Guiraud 1973, 66–69; Christol 2015.

[17] "Some very simple metaphors related to shape can be identified in a series of words meaning 'head'. *Calebasse* 'gourd' (*Jargon de l'argot*), *coloquinte* 'colocynth' (*ibid.*), † *poire* 'pear' (*loirepoque* [on which see n. 16 above]), † *couatche* 'damson plum' (German *questche*), † *ciboulot* and *ciboulotte* 'spring onion' (*ciboule*), † *citronnade* 'lemonade' (*citron* 'lemon'), † *pomme* 'apple', † *balle* 'ball', † *boule* 'bowl', † *bobine* 'reel', † *fiole* 'flask', † *cafetière* 'coffee pot'. (…) Metaphors do play a role in slang, but seemingly not more than in other languages." (author's translation) Schwob & Guieysse 1892, 49–50.

[18] Greimas 1966; Pottier 1964; Ricœur 1975, for instance.

metaphorical networks abounding in the language of comedy. He borrowed Aristotle's definition of the image and of the metaphor, and rephrased in structuralist terms the philosopher's idea that the metaphorizing and the metaphorized terms must share at least one common seme. In this spirit, Taillardat added:

> La matrice métaphorique, qui joue le rôle d'un moule, est la source et donne l'explication des images (métaphores et similitudes) qui peuvent, en foisonnant, créer des champs métaphoriques considérables. C'est en rassemblant tous les métaphorisants exprimant la même notion qu'on peut découvrir la matrice à l'origine du champ métaphorique.[19]

It is obvious that Taillardat's reasoning has mostly a practical aim and it has the merit of providing a global explanation. The point is to avoid describing semantic changes as catachreses, because that would imply to consider them as random phenomena contradicting the regular working of the lexicon. On the contrary, identifying the MMs results in a better understanding of the semantic links between the lexemes and reinforce the coherence of the lexical system. In doing so, Taillardat adopts the same position as Lakoff and Johnson, and his concerns are also similar to those of cognitive linguistics.[20]

Uses of MMs in classical philology: establishing a reading, providing a more relevant semantic option, and shedding new light on etymology

As a linguist specialising in classical languages and in comparative grammar, Taillardat pointed out several applications of MMs. First, identifying an MM works to the advantage of philology since it can help establishing the meaning of a passage whose transmission was disrupted. It is the case of line 326 of Aristophanes' *Knights*,[21] in which two readings compete:

> *Hêi sù pisteúōn amérgeis († amélgeis) tôn xénōn toùs karpímous*
>
> You, shameless, you pluck whatever you can from those wealthy strangers († you milk them, draining them off of all they have).[22]

Now, unlike French and English, which do possess the expression *vache à lait* or *cash cow*, ancient Greek is deprived of the MM STEAL = MILK. It does however feature the less obvious MM STEAL = BEAT DOWN PIECES OF FRUIT. The idea of 'plucking a profit' *karpoûsthaí ti* (= 'drain something from someone, exploit someone'), be it fraudulent or not, is very frequent in Greek.[23]

Other scenarios that could benefit from the identification of MMs involve reliable sources that cannot be satisfactorily understood.[24] The phrase *phōnēn empeprēménēs huós* "burning voice of a sow" appearing in *Wasps* 34 provides a good example. Some scholars understand it wrongly as the "voice of a sow in heat" or, worse, the "voice of a scalded pig". Taillardat's take on it is that two metaphors are combined in the line: FIRE = ANGER and ANGRY MAN = BOAR.

---

[19] "The metaphorical matrix as a mould gives rise to and explains the images (metaphors and similarities), which may proliferate and create huge metaphorical fields. Gathering all the metaphorizing elements is the key to discover the matrix at the root of the metaphorical field." (Translation by the author) Taillardat 1977, 349.

[20] On the developments of cognitive linguistics, *cf.* Lakoff & Johnson 2003, 270–271, and *MSL* 2010 as well.

[21] Taillardat 1977, 350–351. *Amélgeis* is the reading chosen by V. Coulon (Aristophane, *Cavaliers*, Paris, Belles-Lettres, 1923).

[22] All the translations from Greek and Latin are my own if not specified otherwise.

[23] *Cf.* Taillardat 1965, 420–421.

[24] Taillardat 1977, 351.

He thus reveals how complex the stylistic processes at work are and how refined and violent the onslaught against Cleon is, as he becomes the butt of a twofold joke about his effeminate manners:

> *moudókei/ dēmēgoreîn phálaina pandokeútria,/ ékhousa phōnèn empeprēménēs huós.*
>
> It seemed to me that a lecherous (ready to swallow anything) whale was holding forth to the people with the (shrill) voice of a sow in a towering rage.

Lastly, Taillardat demonstrates[25] that in the framework of an etymological approach identifying a MM sheds light on the lost coherence of a lexical group. A good illustration is provided by the family of the Greek verb *pseúdomai* 'lie, speak false', the origin of which can be traced back to the Indo-European root **bhes-*'blow'. Taillardat has the great merit to show that two isolated lexemes, the adjective *psudrós* 'lying' and the substantive *psúdrax* 'blister, pustule', should actually be linked to the many words based on the *pseud-* stem. Indeed, the LIE = BLOW MM is not only attested in Greek but in many other European languages as well.[26]

Likewise, Perpillou[27] tackled the Greek and Latin data pertaining to the INSULT = CUT MM: in Greek *kertoméō* 'taunt, sneer at'/*kértomos* 'mocking', *loidoréō* 'abuse, revile'/*loídoros* 'railing', and *kēkázō* 'abuse'/*káduros* 'hog', and in Latin *contemnō* 'despise'/Greek *témnō* 'cut'.[28] As a result, he managed to give a much more convincing etymologies to each of these lexical pairs.

Extension and relativity of MMs

Extension of MMs

Taillardat's investigations did not limit to ancient Greek as he was eager to show that MMs are transcending linguistic boundaries. Indeed, he noticed that a similar MM can be encountered in many different languages if the metaphorizing term is related to the human body and the way it is perceived, and/or if the MM implies cognitive processes that all human beings share.[29] For instance, Taillardat mentioned in his paper the GORGE, GULLY = THROAT MM (*cf.* French *gorge*, German *Schlund*, Russian *gorlo*, Serbian *gȑlo*, Latin *faucēs*)[30] and the LIE, TWADDLE = BLOW, BREATH OF AIR MM (*cf.* English *winds* 'empty talk', *windy* 'pompous', *blast*/*blister*, German *windige Worte* 'windy words', *Blase* 'bubble, blister'/*blasen* 'blow', Italian *buffa* 'wind, breath of air/twaddle', French *souffler* 'blow'/*boursoufler* 'puff up'…).[31]

As for the HEAD = PIECE OF FRUIT and HEAD = CONTAINER MMs,[32] Taillardat borrowed them from Schwob and Guieysse, but he added the following elements to their corpus:

> – The collocations taken from Greimas *se mettre quelque chose dans la tête* ('to put something into one's head'), *une tête bien pleine* = 'a head full of useful knowledge', *se creuser la tête* ('to cudgel one's brains').[33]

[25] Taillardat 1977, 352–354; *DELG*, 1241–1242.
[26] Examples are given *infra* under *Extension of MMs*.
[27] Perpillou1996, 118–124.
[28] Against Meillet's objection (*DELL*, 680), taken over by Chantraine (*DELG*, 1066).
[29] It would be presumptuous, in my opinion, to assert that an MM is universal since it is absolutely impossible to verify this, considering that there are countless languages in the world and speakers' creativity knows no bounds either.
[30] Taillardat 1977, 352.
[31] Taillardat 1977, 353.
[32] Taillardat 1977, 350.
[33] Greimas 1966, 48.

– The Latin *testa* 'skull', then 'head', etymon of the French *tête*, which has initially a concrete meaning (in particular 'earthenware vessel').[34]

– The German *Kopf* 'head', borrowing from the Latin *cuppa* 'vat, cup'.[35]

I would add the English *nut* and *mug* for 'head',[36] the French *boîte crânienne*, the American-English *brain-pan*, the German *Hirnschale* [= *brain-cup], as well as the phrases *a head stuffed with dates* = *une tête farcie de dates* and *to be empty-headed* = *avoir la tête vide* that both match the collocations mentioned by Greimas.

All these examples confirm that the stable element in the MMs is the semantic link between two referents, and consequently between two signified. The choice of signifiers is on the contrary totally free, which explains their wide range: speakers, whatever their language, can pick any semantically suitable word from the lexical stock in order to expand and renew MMs *ad libitum* with absolutely no concern for etymology.

Universality vs. relativity of MMs

Some MMs, such as HEAD = PIECE OF FRUIT, CONTAINER, or INSULT = CUT, are very widespread, but others not so much because they are rooted in a specific culture or a specific referential world. Taillardat gave a good example by mentioning the case of ancient Greek, in which ANGER is equated with an ACRID PLANT.[37] The prototype *drīmù blépein* 'to glance bitterly at' coexists with variants referring to different Mediterranean phytonyms (e.g. nose-smart, oregano, savory, and silphium), for example *nâpu blépein* – literally 'look bitterly (like mustard) at'. The metaphorizers chosen clearly depend on the plants found in the Greek speakers' surroundings and on the way they used them.

The nouns for the pupil of the eye present a similar case. I am referring here to the review Aurélien Sauvageot wrote of József Erdödi's paper.[38] The Hungarian scholar found some Finno-Ugric forms puzzling. Indeed, in dialectal Ostyak (*sem-kew*), in Cheremis (*šindža-kü*), and in Votyak (*śin-köl'i̯*), the PUPIL is called the STONE OF THE EYE, each word consisting in an identically shaped compound [eye+stone].[39] Erdödi hesitated, first suggesting the possibility of a russianism due to the double meaning of *kameń* 'stone' and 'stone (in a fruit)'.[40] Drawing a parallel with *ögon-sten* (Swedish) and *öie-sten* (Norwegian), he then leaned towards the idea of an areal phenomenon involving the northern Germanic languages as well. As for Sauvageot, he preferred the hypothesis of:

> Une convergence produite uniquement par les opérations mentales par le moyen desquelles l'esprit humain relie certains faits entre eux ou, plus précisément, 'motive' certaines acceptions à la faveur de raisonnements divers.[41]

---

[34] With a cross-reference to Benveniste 1966, 295–296. Taillardat, however, considers that the seme /*hardness*/ was determining in selecting *testa* as the substitute for the classical Latin *caput*.
[35] *DELL* 158–159, 688–689. Kluge 2002, 528.
[36] With a perfect match between the two slang phrases *he got a bang on the mug* and *il s'est pris un coup sur la cafetière* (coffee pot).
[37] Taillardat 1977, 350, with cross-references to Taillardat 1965, 216–218.
[38] Sauvageot 1978; Erdödi 1977.
[39] About 'micro-syntax' of nominal compounds, *cf.* Benveniste 1974, 145, 149–151.
[40] The polysemy of the English *stone* is an illustration of this MM.
[41] "A convergence produced only by the mental processes by means of which the human mind connects some facts together or, more precisely, justifies some acceptations under cover of various arguments".

He consequently proceeded to add the following forms to those gathered by Erdödi:[42]

– From the Finno-Ugric group, the dialectal Zyrian *śin-mol'* [eye+glass bead]; from the Turkic-Mongolic group, the Chuvash Turk *kuś šərśi* [eye+glass bead].

– From the Uralic-Altaic group, the Tungusic *jāsal ńaŋtan* [stone of the eye]; from the Polynesian group, the Marquesan *pu'u koku'u mata* [tree-fruit of the eye], to which the adjective *ke'eke'e* 'noir' [black] is sometimes added, making it a similar form to the French *prunelle* [sloe/pupil].

– From the Finno-Ugric group, the standard Votyak *śin-nuni̯* [eye+child], the Zyrian *śin-kaga* [*id.*], and the dialectal Zyrian *śin-akań* [eye+doll]; from the Turkic-Mongolic group, the Ottoman Turkish *göz bebeği* [eye+baby], the Mongolic Urdu *k'uxen kara nudu* [maiden/young animal+black+eye]; from the Indo-European group, the French *pupille* [ward/pupil of the eye], and the Spanish *niña* [little girl/pupil of the eye].

I suggest adding to the latter the Latin *pupula* 'girl/pupil of the eye' and *pupilla* 'ward, orphan girl/pupil of the eye', diminutives of *puppa* 'girl, doll', the ancient Greek *kórē* 'maiden/pupil',[43] and the Hebrew *ʾīšōn* 'pupil of the eye', diminutive of *ʾīš* 'man'.[44] To explain this comparison of the pupil of the eye with a small spherical body (pearl, little stone, stone [in a fruit], and berry) or with a scaled-down human being, Aurélien Sauvageot put forward the possible existence of "panlinguistic laws" in semantics. He referred there to the lectures by Zoltán Gombocz, the great Hungarian linguist of the interwar period. As a matter of fact, the different ways of saying 'pupil of the eye' are the result of cognitive experiences that could have occurred in different places independently from each other.

In short, two MMs are competing here: PUPIL = SCALED-DOWN HUMAN BEING and PUPIL = SMALL SPHERICAL BODY. The latter has four sub-varieties seemingly depending on the cultural and/or material context: PUPIL = GLASS BEAD, STONE, (SMALL DARK) STONE or BERRY.[45] Therefore no universal MM exists because speakers select various and *a priori* unexpected referential options, even within the same cultural areas.

Taillardat, as well as Lakoff and Johnson, have a similar standpoint: human experience controls the way speakers create the matrices of metaphors and organize their networks, whether they are labelled 'conceptual metaphors' or 'metaphorical matrices'. As a result, it is unnecessary for linguists to explore the paths of borrowing or calques in such cases. However, it should be noted that human experience might sometimes be influenced by environmental and/or cultural factors that may restrain the extension of these matrices. Taillardat's approach is also of great interest, and differs from that of Lakoff's and Johnson's, insofar as it favours linguistic data and is built on both the synchronic and diachronic observation of languages. As we will see now, a diachronic perspective is indeed especially useful in the field of textile activities.

[42] Sauvageot 1978, 381–382.
[43] *DELL*, 546 and *DELG*, 567–568.
[44] I owe the *ʾīšōn* example to Lyliane Sznajder.
[45] The designations for the pupil are not limited to these MMs. In ancient Greek, the pupil can also be defined metonymically by its shine: thus the noun *glḗnē* stems from the nominalization of the adjective **gl̥h₂s-no-* 'shiny' (*DELG*³, 227, 1283). Incidentally the PUPIL = SCALED-DOWN HUMAN BEING MM derives from this characteristic, since the surface of the pupil reflects the image of the observer.

The metaphorical matrices in the field of textile activities

In order to draw a picture of the numerous MMs pertaining to textile activities, I will refer to data that are generally known to classicists. Nevertheless, I owe special thanks to Marie-Louise Nosch, whose 2014 stimulating paper helped me to renew my thoughts on Greek textile crafts and their related social practices, and to deepen my reflection on what is the source of metaphors. I will thus focus my examination (but not exclusively) on the data provided by ancient Greek and Latin languages and cultures, which represent for me a privileged field of investigation. More precisely, my presentation will follow the main technical processes involved in textile crafting: first, making thread and winding it into a ball, then weaving, and finally other techniques of intertwining using either flexible fibres (plaiting) or rigid elements fitted together (carpentering). The fundamental semes at the root of these technical terms are also at the root of many MMs about time and fate, speech and poetry, lie and cunning.

Spinning and winding up: thread and time, thread and life, thread and speech, tangled threads

Spinning

After the raw fibre material has been prepared,[46] the next step is spinning – that is twisting, drafting, or splicing – thus creating a continuous thread. This continuity is precisely the reason why thread as a concrete item is well adapted to metaphorize the abstract notion of time.[47] This very equivalence explains the polysemy associating time and place in the prepositional organisation of several languages: English *before/after*, German *vor/nach*, French *avant/après*, Ancient Greek *pró/metá*, and Latin *ante/post*.[48]

This way thread is crucial for the creation of various MMs: thread of time or thread of life, thread of speech or entangled thread of lies and cunning.

The TIME = (TIGHT) THREAD *and* THE LIFE *or* FATE = THREAD *MMs*

Both French and English (though less frequently) include MMs characterized by the /continuity/ seme in phrases such as: *au fil du temps, des jours* 'as time goes by', 'with the passing days',[49] *le fil des événements* 'the chain of events', *le fil conducteur* (d'un roman) = *the unifying thread* (of a novel) or *le(s) fil(s) de la vie, de la destinée* = *the thread(s) of life*, *of fate*.

Connecting thread and fate refers back to the figure of the Moirai (ancient Greek *Moîrai*),[50] via the Roman *Parcae*.[51] The Moirai are actually said to govern human fate that Klotho spins, Lachesis measures by determining its duration, and Atropos inexorably cuts. Nevertheless, they are not the only *Klôthes* ('spinners') dealing with fate. Indeed, according to Pausanias (8.21.3), the poet Olen

[46] Barber 1991, 9–38.

[47] About concrete phrasing of abstract notions, *cf.* among others Sweetser 1990, 19–20 (mental states and speech acts) and Cairns 2012 (feelings: shamelessness, grief metaphorized by clothes). It is also possible to represent time as a line (*cf.* below), something easily embodied in a tight thread.

[48] Lakoff & Johnson (2003, 41–45) noticed that there were contradictions in phrases related to time: "While expressions like *ahead of us, I look forward,* and *before us* orient time with respect to people, expressions like *precede* or *follow* orient times with respect to times" (*ibid.* 43). To my mind, facts are less linked to the lexicon than to the question of whether or not the enunciator is taken into account in the utterance.

[49] In French, the passing of time is assimilated to the flow of water. As a result, an MM associating STREAM with THREAD developed in phrases such as *au fil de l'eau, au fil du courant* 'along the current, along the stream', and also explains the intransitive use of the verb *filer* 'fly' when applied to time in particular.

[50] Pirenne-Delforge & Pironti 2011.

[51] Dasen 2011.

labelled Ilithyia, the goddess of childbirth, a 'good spinner', *eúlinos* 'the one with a good yarn'. As a result, cases in which the Moirai and the Parcae are related to the vocabulary of spinning abound both in literary works and funerary epigrams.[52] Here are some examples:

> – In ancient Greek, the verbs *nḗthō* or *klṓthō* 'to spin' and their prefixed forms; the nouns *línon, mítos, nêma*, 'thread, yarn'; poetic collocations such as *klṓthō télos bíou* "to spin the end of life" (Q. S. 11.141), *pikrà(...) peprōména klṓthō* "to spin a bitter fate" (Nonn. *D.* 2.678).
>
> – In Latin, the phrases *natalis stamina* "threads of nativity" (Ov. *Tr.* 1.63), *subtegmina ducere* "to pull on the weft threads" (Cat. 64.427), *sororum fila(...) atra* "the dark threads spun by the (three) sisters" (Hor., *Od.* 2.3.16), *stamina fatalia* "warp threads of fate" (Tib., Ov., etc.), *rumpere fila* "to break the threads of fate" (*CE* 443.5, 1114.4…).[53]
>
> – In Old Norse, the compounds *ørlǫgþáttr* 'fate-strand' and *ørlǫgsíma* 'fate-thread', although there is no direct evidence of the Norns' spinning.[54]

All these occurrences show elements that derive from the FATE = THREAD MM without being necessarily linked to a triad of spinning deities.

The SPEECH = THREAD MM

Thread can also metaphorize speech since uttering words takes place in time and is based on the linear nature of the signifier as defined by Ferdinand de Saussure.[55] That is to say, in the speech stream, phonic segments – the so-called signifiers – are heard one after the other, without any possibility of either overlapping or going backwards. As Almuth Grésillon shows,[56] the abnormal case of portmanteau words, which are monstrous lexical creations, on the contrary confirms this statement. For example, in *motel* [məʊtel], *motor* and *hotel* overlap each other: the word starts with [məʊt], the beginning of [məʊtəʳ], and ends with [əʊtel], the final sequence of [həʊtel], which requires going backward in the utterance chain in order to understand the meaning at stake. The SPEECH = THREAD MM can be found in English as well as in French in *suivre, perdre, interrompre le fil d'un discours, de ses pensées = to follow, to lose or to interrupt the thread of a speech, of one's thoughts*.

The CUNNING, LIE = THREAD MM

Spinning is a process more complex than meets the eye and almost impossible to undo once finished.[57] This may appear mysterious to the uninitiated, who might in turn suspect a trick. The CUNNING, LIE = THREAD MM derives from it and is found in English in the figurative meaning of *spin* and *yarn* in the collocations to *spin a tale* or a *yarn* or just to *yarn*. Furthermore, a strand (or several) is sometimes spun twice, plyed, in order to achieve a strengthened thread. French has the adjective *retors* 'twisted [= *-tors*] twice [= *re-*]' which can be said literally of a thread and figuratively of a person (= 'wily'). The French *filou* 'swindler', a dialectal variant of *fileur* 'spinner', belongs to the same MM, as well as the English *to twist the truth, it is a twist*, and *a twister*.

---

[52] *Cf.* Giannakis 1988, 15–22.

[53] *Pace* Giannakis 1989, it is impossible to include in this MM the Indo-Iranian motif of the noose of death (*cf.* Andrés-Toledo 2009), since it pertains to the equivalence VIOLENT DEATH = HUNT, in which human beings are preys.

[54] Bek-Pedersen 2011, 123–132. About the weaving Valkyries, *cf. infra.*

[55] Saussure 1974, 70: "The signifier, being auditory, is unfolded solely in time from which it gets the following characteristics: (a) it represents a span, and (b) the span is measurable in a single dimension; it is a line".

[56] Grésillon 1984.

[57] Bek-Pedersen 2009, 32 n. 14.

Thread(s) can easily get entangled, figuratively muddling the thread of utterance and communication. Other collocations belong to the CUNNING = THREAD MM: for both French and English, one can mention *embrouiller quelqu'un* 'to confuse somebody', *s'embrouiller dans ses explications* = *to get into a tangle when trying to explain*, as well as the Latin hapax *migdilix* (Plaut. *Poen.* 1033) the meaning of which Bruno Rochette disentangled.[58] This Greek-Latin compound appears in a diatribe from Milphion against Hannon, a wily Punic character who masters both Latin and Punic languages. *Migdilix* is made up of a Greek adverb, *mígda* or *mígdēn* 'confusedly', and of the stem *līc-* of the Latin *līcium* 'heddle, thread'. *Migdilix* thus means literally 'with an entangled thread', but in context it is a way to charge Hannon of being 'someone who muddles people up'.

Winding and unwinding a ball of yarn

As soon as it is made, the thread is wound onto a reel before being folded into skeins or wound into *balls*. The semes /winding up/ and /winding off/ associated with these processes are at the root of two MMs, one concerning life, and another the stream of speech or thought.

The PASSING OF TIME, OF LIFE = WINDING INTO A BALL/ UNWINDING *MM*

Except for French in which the MM is common (just think of the *déroulement du temps, de la vie*),[59] few examples of this particular MM can be found. Ancient Greek however features the verb *tolupeúō* and its prefixed form *ektolupeúō*, both deriving from *tolúpē*.[60] This noun means literally 'roves of wool fastened to the distaff and ready for spinning' or 'ball of thread ready for weaving'. But it is sometimes used figuratively as well. Unlike the substantive *tolúpē*, the verbs above are mainly used metaphorically in poetry until the Byzantine period. *Tolupeúō*, literally 'to wind into a ball', mostly means 'to (make someone) endure' painful events (*álgea* 'pains', *polémous* 'wars', *pénthos* 'grief'), but can also signify 'to finish building (a house)' when associated with *dómon*. Winding metaphorically here leads to an accumulation as time goes by. Conversely, *ektolupeúō* 'to unwind a ball of thread' figuratively means 'unwinding the thread' of a *khalepòn pónon*, a 'difficult labour' (Hes. *Sc.* 44), namely 'bringing to an end'.

Disjecta membra of a related lexical family, preserved in Hesychius' lexicon, show the same MM as *tolúpē*. If *ordikón*: *tòn khitōnískon* ('*ordikón*: the short frock') and *órdēma*: *hē tolúpē tôn eríôn* ('*órdēma*: 'the ball of wool') belong to textile terminology,[61] the aorist *ōrduleusámēn* glossed *emókhthēsa* 'I was sore distressed' is semantically parallel to *álgea tolupeúō* 'to unwind the thread of pains'.

[58] Rochette 2000.

[59] *Cf. TLFi* s.v. *déroulement* 'course, sequence, passing' (of time, of life) and *se dérouler* 'take place, occur'.

[60] More detailed information can be found in Maurice 1991 and Guilleux 2009, in which the uses of *tolúpē* are investigated: it appears that two concurrent semes, */thread/* and */spherical mass/*, are at stake in this lexeme. *Cf. infra* about *Od.* 19.137.

[61] These lexemes are, in my opinion (*contra* Meillet *DELL*, 467), etymologically akin to the Latin *ōrdior* 'begin to weave a web, weave, spin; begin', *ōrdō* 'row, order', and *ōrnō* 'adorn', given that before actually weaving (running the weft thread through the yarn threads), the spinner has to put in order and arrange the threads, a long and rather complex process that will determine the technical and esthetical characteristics of the fabric in progress. These conclusions were drawn from life-size tests made at the Centre for Textile Research of the University of Copenhagen, and by Ellen Harlizius-Klück. She has all my thanks.

The UNFOLDING OF A NARRATION, OF MEMORIES = UNWINDING A THREAD *MM*

This MM appears in the English *to reel off verses*, *a list* and mostly in the French verbs *dérouler* 'unwind/go through' or *se dérouler* 'come unwound/develop'. They are figuratively used about a narrative, the action of a novel, as well as about memories, an investigation or a range of other events.[62] *Se dérouler* and its derivative *déroulement* are commonplace, yet the phrase *dérouler un fil* in one's mind, be it orally or in writing, pertains to a more formal level. Also, the use of the verbs *dévider* 'to reel off' and *débobiner* 'to wind off' belongs to a more archaic stage of French. It explains why many examples of the UNFOLDING OF A NARRATION, OF MEMORIES = WINDING A THREAD OFF MM are actually found in literary French.[63]

Weaving

The weaving stage transforms the thread much more radically than spinning transforms the fibre material. Intertwining threads is a more complex process, as are the many technical variants in twisting threads or choosing colours that intervene while weaving. Since weaving however follows spinning and uses thread, it is not surprising that the MMs related to this technique are partly similar to those associated with spinning.

The FATE = WEAVING *MM*

While the FATE = SPINNING MM is easy to understand, intertwining threads can also, in fewer cases yet, metaphorize fate. The motif exists in Old Norse sagas: as Bek-Pedersen explains,[64] the warp corresponds to the characteristics given at birth, whereas the woof represents the part that every individual builds on their own. The best example is found in the *Darraðarljód*, in which the Valkyries are weaving a bloody banner during a battle with the flesh and bones of combatants. It has parallels in Old English and Old Irish,[65] and even in Vedic and Lithuanian.[66] This MM remains rare elsewhere, although Latin poetry provides another testimony in a funerary epigram (*CE* 443.4): *sanguinea palla quae texit prodiga Clotho*, "the bloody garment which the wasteful Clotho has woven".[67]

THE SOCIAL LIFE = WEAVING *MM*

As for the Greek and Roman worlds, John Scheid and Jesper Svenbro[68] showed that political weaving and nuptial blanket rituals used to structure social life. Even though these rituals have faded away, modern languages retain some traces of them in their vocabulary, such as the equivalence between the French *tisser des liens (affectifs...)* and the English *to weave (emotional...) ties*.

The SPEECH = WEAVING, LIE, STUPIDITIES = FABRIC, *and* POETRY = WEAVING *MMs*

The equation of speech with weaving comes less from the fact that it was accompanied by singing than from the characteristics of speech itself. As a matter of fact, speech is organized along two

62 *Cf. TLFi* s.v.
63 *Cf.* for example the narrative by Irène Frain (2013, 33–34, 55, 61, 103).
64 Bek-Pedersen 2009, 34–36.
65 Bek-Pedersen 2011, 132–158.
66 Durante 1960, 238, n. 28.
67 Example and translation from Giannakis 1988, 16.
68 Scheid & Svenbro 1996, 7–49, 51–107.

intertwining axes: syntax on the horizontal axis and lexical choices on the vertical (*i.e.* paradigmatic) one. If the SPEECH = WEAVING MM is attested in Vedic,[69] it is quite rare in ancient Greek, in which it actually competes with the SPEECH = PLAITING MM. It however appears already in the *Iliad* (3.212) in the phrase *múthous kaì mḗdea húphainon* "they wove a web of words and ideas". Likewise, Pherecrates, in one of his comedies, coined the phrase *katà míton tà prágmat' eklogízesthai* "to present the facts in detail", namely "thread by thread" (fr. 146.7). In Latin several figurative uses are noteworthy in the lexical group of *texō* 'to weave': *textum* "fabric of style" (Quint. 9.4.17), *textūra* and *textus* (*-i*) 'content (of a speech)', *orationem contexō* "to weave a speech" (Quint. 10.6.2).[70]

Verbal utterance can also be conceived as a fabric, though generally in a depreciatory context: *tissu de mensonges = web, tissue of lies*; *tissu d'obscénités, d'inepties* 'jumble of obscenities, of stupidities', *tissu de contradictions, woven with contradictions*.

Based on formulaic poetry, traditional poetic diction combines a horizontal axis of narration with the vertical axis of items selected among the richness of existing formulas. The poets' skill consists then in mastering perfectly the combination of both constraints. The POETRY = WEAVING MM is typical of Greek choral lyric poetry, as illustrated by the verbal phrase *húmnous huphaínō* 'to weave hymns', but also of Old English poetry (*cf. wordcraeft wæf* 'I wove the poetic art') and Vedic hymns.[71]

The CUNNING = WEAVING MM

Arguing, or weaving a net around the one who needs convincing, and lying, or deceiving with words, are not so different from the ruse metaphorized by the preparation of the loom before weaving. It is a long and meticulous process[72] conditioning the final aspect of the fabric, which a layperson is unable to see. The Latin *ōrdior*, etymon of the French *ourdir* [to warp], is at the core of this MM since, besides its technical sense, it also means 'to begin' a speech, a talk. In the same lexical family, the noun *ōrdō* similarly means 'row' and 'order'. Once the loom is warped, the weft yarn is drawn through the warp yarns. In the same way, *on trame un complot* ('a plot is hatched'), *a web of intrigue is woven, on tisse un tissu d'intrigues*. And Odysseus fittingly says: *dólous kaì mêtin húphainon* "I wove wiles and craft" (*Od.* 9.422).[73]

There remains the particular case of Penelope, whose trick consisted in unravelling during the night the piece of fabric woven during the day. It explains the unique and seemingly contradictory formula *dólous (...) tolupeúō* "I wind my wiles into a ball" (*Od.* 19.137).

Other related techniques: plaiting and carpentering metaphorizing time, poetry, and cunning

Plaiting and carpentering can be compared to weaving, because all three techniques imply the intertwining of elements, whether they are flexible fibres for plaiting or rigid pieces of wood carved out, and then fitted together for carpentering. Beyond these specific processes, some ancient languages exploit a larger variety of techniques to create other MMs. Some even pick as metaphorizer the hyperonyms of this very semantic field (*technique* or *making*).

[69] Schmitt 1967, 298–300.
[70] *Cf.* Scheid & Svenbro 1996, 107–130.
[71] Schmitt 1967, 300; McIntosh Snyder 1981.
[72] *Cf.* Barber 1991, 118–119, 134–137, about the complexity of starting borders that are however absolutely necessary when weaving on a warp-weighted loom.
[73] *Cf.* Detienne & Vernant 1974, about *mêtis*.

The MMs based on plaiting

In ancient Greek, plaiting gives rise to several MMs. The PLAITING one's LIFE MM – *tòn bíon diaplékō*/ *tḕn zoḕn kataplékō*/ *khrónon toû zên plékō*[74] – is rather common. The SEXUAL INTERCOURSE = PLAITING MM uses the verb *sumplékō* and its derivative *symplokḗ* that denotes an embrace.[75] Nor is it surprising that SPEECH, POETRY, ARGUMENTATION or CUNNING as PLAITING exist in numerous and frequent MMs: *plékō* and its derivatives are thus linked with *rhêsin*, *rhḗmata*, and *lógous* 'speech/es', with *húmnous* 'hymns' and *thrênon* 'dirge', as well as with *dólous* and *mēkhanás* 'wiles'. As for the nominal derivatives from these verbs, *plokḗ* refers to the plot of a play or to a ruse, *periplokḗ lógōn* to 'circumlocutions', and *symplokḗ* to various combinations of speech elements (letters in words, words in sentences or a mix of various styles). Finally, poetical words can be added to this list: *doloplókos* 'plaiter of wiles' (Sapph.1.2 V., Tryph. 264) and its derivative *doloplokía* 'wiles plaiting, crafting' (Thgn. 226, Hp. Ep. 17), as well as the obscure phrase *epiklópos múthōn* (*Il.* 22.281) that Claire Le Feuvre suggested reading **epiplókos múthōn* 'the one who plaits lies' on the basis of very convincing arguments.[76]

MMs related to carpentering

Like plaiting, carpentering can replace weaving in the creation of MMs. Within the Indo-European languages, ancient Greek, as well as Vedic Sanskrit and Avestan, all feature the POETRY = CARPENTERING MM. This formulaic match, that was recognized as inherited a long time ago,[77] is based on the perfect identity of the Greek and Indo-Iranian lexemes. Its reconstructed form is *$wek^w esh_2$ tekt-* 'to carpenter words, poems', from the ancient Greek *épea tektaínomai* 'to carpenter words' and from the Sanskrit *vácāṃsi tákṣam* 'I will carpenter words'. As for Latin, the same formulaic phrase exists but the complement changed: *sermones texō*, 'to carpenter/weave words'.

Latin actually occupies a special place in this general picture, for it did not keep the **web^h-* inherited root. 'Weave' is said *texō*, which appears formally akin to **tekt-* 'to carpenter'. The explanation is quite simple:[78] the Latin *texō* could replace the inherited verb 'weave', since it also held the /intertwining/ seme. This hypothesis, owed to Darmesteter,[78] relies on the fact that Latin includes several examples of *texō* and of its derivatives, which have retained the etymological meaning 'to carpenter'. To the epic hapax *textrīnum* 'shipyard' (Enn. *Ann.* 468), Darmesteter added three occurrences of *texō* in which the verb is applied to the building of ships (Verg. *Aen.* 11.326, Cat. 64.10), and of a basilica (Cic. *Att.* 4.16.14). Lewis and Short[79] actually describe meaning in the *texō* group as a semantic *continuum*, from 'weave' (literally and figuratively), to 'plait' or 'intertwine', and to 'carpenter' and 'build' (by adjusting pieces of wood or stones). Thus they go from the more usual occurrences to the rare ones that are restricted to poetry. From a diachronic viewpoint, however, facts must be presented backwards, starting from the etymological meaning, i.e. 'to carpenter'. Taking into account the MMs network prevents putting forward a very vague common meaning (which Meillet rightly rejected), and sheds light on the etymology of *texō*.

[74] Taillardat 1965, 46–47, with other examples and references.

[75] Scheid & Svenbro 1994, 33–34.

[76] Le Feuvre 2015, 307–338. She demonstrates that *epiklópos múthōn* results from an erroneous reinterpretation of the genuine **epiplókos*.

[77] Darmesteter 1878.

[78] Darmesteter 1878.

[79] Lewis & Short 1879, *texō, textilis, textrīnum, textura* s.v.

Ancient Greek also has, which is not surprising at all, the LIES, WILES, INTRIGUES = CARPENTERING MMs. They are found especially in comedy,[80] whereas in Homeric poetry they have been renewed, FITTING replacing CARPENTERING. It results in the compound noun *artiepḗs* 'a glib tongue, spinner of yarns', or in the phrases *pseúdea artúnō* 'to arrange lies', and *dólon artúō* 'to plot wiles'.[81]

Beyond carpentering

It must be added that in Indo-European languages metaphorical networks have been extended to other techniques – fighting gear, forging, building, etc. – in order to express poetry, the organisation of speech and thought, as well as cunning.[82] Since poetry is "un fare artistico, cioè attività che presuppone un sapere tecnico",[83] the use of hyperonyms of technical activities can easily be understood. Whatever the lexeme used,[84] POETRY can be, as in Greek with the verb *poieō* 'make, create', a MAKING, without any reference to a specific TECHNIQUE. Likewise, CUNNING, as the mastering of technical processes unknown to laypeople, is at the basis of the CUNNING = TECHNIQUE MM. It is at work for example in the Greek *tékhnē* and the Latin *ars*, eventually mingling technical abilities with fraudulent artifice, or in the Greek *mēkhanḗ* 'contrivance'/plural 'wiles', borrowed then by Latin (*machina, machinari, machinatio*…) and passed down to Roman languages and English.

Conclusion

Although this examination can exhaust neither the examples nor the MMs, it allows us to appraise provisionally those pertaining to textile activities and to other fields beyond. The MMs related to the main two technical stages of spinning and weaving are tremendously varied. Both metaphorize various notions such as the passing of time and the duration of life, the thread of speech in all its forms, and literary creation, social ties and sexual intercourse, tricks and lies. Nevertheless, this metaphorical network is not limited to textile crafts: it extends beyond and to other techniques, such as plaiting and carpentering, in which the seme /intertwining/ is crucial too. This network further includes technique in itself as a hyperonym of every particular technical process, insofar as it is a concrete making. It is of interest to note that even when techniques have become obsolete, languages as a result of their being conservative retain phrases rooted in those MMs. Such phrases are often not understood anymore, but etymology provides a way to shed light on their original meaning. Finally, while MMs are generally growing and developing synchronically, some of them, based on etymologically related lexemes, form a common inheritance shared by a linguistic group. It is the case of the CARPENTERING OF WORDS, a MM defining poetry in several Indo-European languages.

All in all, Taillardat's approach to the *matrices métaphoriques* and their working can help broadening the conceptual perspective of Lakoff and Johnson, as it takes diachrony into account and is anchored in "la chair chaude des mots",[85] whether they pertain to everyday life or not.

[80] Taillardat 1965, 232–233, for examples.
[81] Le Feuvre 2015, 307–308, 324.
[82] Durante 1960, 236, Taillardat 1965, 445, Lakoff & Johnson 2003, 46.
[83] "An artistic making, that is an activity presupposing a technical knowledge" (Durante 1960, 234).
[84] Durante 1960, 234.
[85] "The warm flesh of words" (Raymond Queneau, in *Le chien à la mandoline*, Paris, 1958).

Bibliography

Andrés-Toledo, M. A. 2009 The Indo-Iranian noose of death. In E. Pirart, X. Tremblay (eds), *Zarathushtra entre l'Inde et l'Iran*. Wiesbaden, 1–16.

Barber, E. 1991 *Prehistoric Textiles*. Princeton NJ.

Bek-Pedersen, K. 2009 Fate and Weaving: Justification of a Metaphor. *VMS* 5, 23–29.

Bek-Pedersen, K. 2011 *The Norns in Old Norse Mythology*. Edinburgh.

Benveniste, E. 1966 Problèmes sémantiques de la reconstruction. In *Problèmes de linguistique générale*, vol. 1. Paris, 289–307.

Benveniste, E. 1974 Fondements syntaxiques de la composition nominale. In *Problèmes de linguistique générale*, vol. 2. Paris, 145–162.

Bretin-Chabrol, M. 2007 La naissance et l'origine: Métaphores végétales de la filiation de Caton à Gaius, PhD thesis, Ph. Moreau (supervisor), Université Paris XII.

Bretin-Chabrol, M. 2012 *L'arbre et la lignée: Métaphores végétales de la filiation et de l'alliance en latin classique*. Grenoble.

Cairns, D. L. 2012 Vêtu d'impudeur et enveloppé de chagrin. In F. Gherchanoc, V. Huet (eds) *Les vêtements antiques*. Paris, 149–162.

Christol, A. 2015 Le *largonji* des *loucherbem*. *Lalies* 35, 145–163.

Darmesteter, J. 1878 *Iranica* VI: Une métaphore grammaticale de la langue indo-européenne. *MSL* 3, 319–321.

Dasen, V. 2011 Le pouvoir des femmes: des Parques aux *Matres*. *Études de Lettres* 289, 115–146.

DELG = Chantraine, P. *Dictionnaire étymologique de la langue grecque*. Paris, 1968–1980¹, 1999² (with a *Supplement*), 2009³ (with a *Supplement*).

DELL – Ernout, A. & Meillet, A. *Dictionnaire étymologique de la langue latine*. Paris, 1959⁴.

Detienne, M. & Vernant, J.-P. 1974 *Les ruses de l'intelligence: la* mètis *des Grecs*. Paris (English translation, 1991, *Cunning intelligence in Greek culture and society*. Chicago IL).

Durante, M. 1960 Ricerche sulla preistoria della lingua poetica greca: la terminologia relativa alla creazione poetica. *Atti Acad. Lincei* (*Classe di Scienze Morali, storiche e filologiche)* 15, 231–249.

Erdödi, J. 1977 Ostj., wotj., tscher. 'Augenstein' = 'Augenstern'. *Finnisch-ugrische Forschungen*, 42, 195–196.

Frain, I. 2013 *Sorti de rien: Récit*. Paris.

Greimas, A. J. 1966 *Sémantique structurale*. Paris.

Giannakis, G. 1988–1989 The 'fate-as-spinner' motif: a study of the poetic and metaphorical language of Ancient Greek and Indo-European. *IF* 103, 1–27, and 104, 95–109.

Grésillon, A. 1984 Le mot-valise: un "monstre de langue"? In S. Auroux *et alii* (eds) *La linguistique fantastique*. Paris, 246–259.

Guilleux, N. 2009 Vieux pots et bonne soupe: l'emploi de *tolúpē* en LXX 4 *Règnes* 4.39. In D. Petit, N. Guilleux (eds), *A-TI-DO-RO* (= *Ktèma* 34), 73–87.

Guiraud, P. 1973 *L'argot*. Paris.

Halbert d'Angers, A. 1849 *Le nouveau dictionnaire complet du jargon de l'argot ou Le langage des voleurs dévoilé*. Paris.

Kluge, Fr. 2002 *Etymologisches Wörterbuch der deutschen Sprache*. Berlin–New York.

Lakoff, G. & Johnson, M. L. 1980 *Metaphors We Live By*. Chicago–London; 2003 2nd edition with an afterword.

Le Feuvre, C. 2015 Ὅμηρος δύσγνωστος. *Réinterprétations de termes archaïques en grec archaïque et classique*. Geneva.

Lewis, C. T. & Short, C. 1879 *A Latin Dictionary*, Oxford (www.perseus.tufts.edu/hopper/text?doc=Perseus:doc:1999.04.0059).

McIntosh Snyder, J. 1981 The Web of Song. *CJ* 76, 193–196.

Maurice, N. 1991 Τολύπη ou les écheveaux de l'étymologie. *RPh* 65, 161–164.

MSL 2010 *Grandes voies et chemins de traverse de la sémantique cognitive*. *MSL*, nouvelle série 18. Paris.

Nosch, M.-L. 2014 Voicing the loom: Women, Weaving, and Plotting. In D. Nakassis *et alii* (eds), *KE-RA-ME-JA: Studies presented to Cynthia W. Shelmerdine*. Philadelphia PA, 91–101.

Perpillou, J.-L., 1996 De 'couper' à 'insulter', in *Recherches lexicales en grec ancien*. Louvain, Paris, 113–124.

Perpillou, J.-L., 2004 *Essais de lexicographie en grec ancien*. Paris.
Pirenne-Delforge, V. & Pironti, G. 2011 Les Moires entre la naissance et la mort: de la représentation au culte. *Études de Lettres* 289, 93–113.
Pottier, B., 1964 Vers une sémantique moderne. *Tralili* 2, 107–137.
Ricœur, P., 1975 *La métaphore vive*. Paris.
Rochette, B. 2000 Latin *migdilix*. Études *Classiques*, 68/4, 371–374.
Saussure, F. de 1974 *Course in General Linguistics*. London (translation of Saussure, 1967 *Cours de linguistique générale*, 3rd edition. Paris).
Sauvageot, A. 1978 review of *Finnische-ugrische Forschungen*, 42, 1977. *BSL* 73/2, 380–383.
Scheid, J. & Svenbro, J. 1996 *The craft of Zeus*. Cambridge MA (translation of Scheid & Svenbro 1994 *Le métier de Zeus*. Paris).
Schmitt, R. 1967 *Dichtung und Dichtersprache in indogermanischer Zeit*. Wiesbaden.
Schwob, M. & Guieysse, G. 1892 Études sur l'argot français. *MSL* 7, 33–56.
Sweetser, E. 1990 *From Etymology to Pragmatics*. Cambridge.
Taillardat, J. 1965 *Les images d'Aristophane*. Paris.
Taillardat, J. 1977 Images et matrices métaphoriques. *BAGB*, *Lettres d'humanité* 36, 344–354.
TLFi Trésor de la langue Française informatisé, (site created by Jacques Dendien). CNRS/Université de Nancy-Lorraine (atilf.atilf.fr).

2. Indo-Iranian Weavers of Old and New Hymns

Miguel Ángel Andrés-Toledo

The Indo-European poets identified their compositions with handicrafts well-known to their audience and often used the technical terminology of a particular occupation as a poetic device.[1] Thus, the jargon of carpentry, plaiting and textile was metaphorically applied to poetic compositions in the oldest literary texts of every Indo-European linguistic subgroup, giving symbolic expression to their world's view. In Watkins' words, "virtually any technology can be exploited for such metaphorical purpose, such as weaving".[2]

Previous studies by eminent scholars have been devoted to the wide spread motif of poetry as textile in some Indo-European literatures,[3] of which those of Old Indian and Old Iranian offer good examples. Nevertheless, the related metaphor of the new or old thread, referring to the poet's attempt to innovate or to continue traditional patterns, has so far remained unnoticed as regards the Indo-Iranian milieu. In this chapter, I will explore the Vedic and Avestan texts containing this textile metaphor and its links to other related metaphors in the Indo-Iranian context.

In the oldest literary sources of the Indo-Iranian linguistic subgroup, the Indian *Ṛgveda* and the Iranian *Gāϑās*, belonging to the 2nd millennium BCE, the topic of the poet's originality is expressed by the same etymologically related word, namely Ved. *ápūrvya-*[4] and OAv. *apaouruuiia-* "not formerly, as never before",[5] going back to IIr. **a-pr̥Hu̯i̯a-*. The former is attested in RV 6.32.1, 8.66.11 and 10.23.6:[6]

RV 6.32.1

> *ápūrvyā purutámāny asmai mahé vīrā́ya taváse turā́ya*
>
> *virapśíne vajríṇe śáṁtamāni vácāṁsy āsā́ sthávirāya takṣam*
>
> As never before, the richest, the most beneficent words I carve with my mouth to this, to the great, the strong, the powerful hero, to the rich, the mighty mace-carrier.

[1] See the most usual examples in West 2007, 35–40.
[2] Watkins 1995, 14.
[3] Bartholomae 1904, 1346; Durante 1960, 238–244 and 1976, 173–179; Schmidt 1967, 298–301; Campanile 1977, 36–37; Scheid & Svenbro 1996, 111–155; Nünlist 1998, 110–118.
[4] Mayrhofer 1992–2001, 2.157.
[5] Bartholomae 1904, 79.
[6] Vedic texts in Anantanarayana 2000. For references to Ved. *ápūrvya-*, see Humbach & Elfenbein & Skjærvø 1991, 2.20.

RV 8.66.11

vayáṃ ghā te ápūrvyéndra bráhmāṇi vṛtrahan

purūtámāsaḥ puruhūta vajrivo bhṛtíṁ ná prá bharāmasi

We indeed offer you as offering, o Vṛtra-slayer Indra, formulas as never before, the richest ones, o much invoked mace-carrier.

RV 10.23.6

stómaṁ ta indra vimadā́ ajījanann ápūrvyam purutámaṁ sudā́nave

vidmā́ hy àsya bhójanam inásya yád ā́ paśúṁ ná gopā́ḥ karāmahe

The richest hymn as never before the Vimadás have created for you, o bountiful Indra, for we know what pleases him, the mighty one, namely that we drive it (= the hymn) together like herdsmen the cattle.

As we observe, the Vedic poets wrapped the topic of their originality either in a metaphor from carpentry, as the verb *takṣ-* "to carve" in RV 6.32.1 indicates, or simply used the Vedic adjective *ápūrvya-* with no metaphor at all, as in the other two passages already mentioned. In contrast, the Iranian poet, presumably Zaraϑuštra, claimed his originality by means of a textile metaphor, as the use of the verb *vaf-* "to weave" in Y 28.3 demonstrates:[7]

yə̄. vå̄. aṣ̌ā. ufiiānī. manascā. vohū. apaouruuīm.

mazdąmcā. ahurəm. yaēibiiō. xšaϑrəmcā. aγžaonuuamnəm.

varədaitī. ārmaitiš. ā.mōi. rafəδrāi. zauuə̄ṇg. jasatā.

I (am) who shall weave a hymn for you, o Aṣ̌a and Vohu Manah, as never before, / and for Ahura Mazdā, for (all of) whom Ārmaiti also increases unfading power.[8] Come to my calls for support!

The core of this metaphor in Y 28.3, part of the *Ahunauuaitī Gāϑā*, lies in the Old Avestan verb *ufiiānī* "I shall weave" and the adverb *apaouruuīm* "as never before." The former belongs to the verb *vaf-* "to weave",[9] exclusively applied in Avestan to the 'weaving' of poetic compositions, but never to textile craft, with the only exception of its related Young Avestan adjective *ubdaēni-* "woven", referring to textiles in V 7.15 and 8.23.[10] Conversely, the etymological Indo-European cognates of this verbal root, all of them going back to PIE. **(h$_2$)u̯eb^h-* "to weave", are commonly applied to textiles.[11] With regards to its morphology, it must be noticed that Old Avestan *ufiiānī*, a 1st sing. pres. subj. act., was chosen by the poet to emphasize his personality and his active role in the composition as well as to call the audience's attention to him.

As we have seen, the Indo-Iranian poets highlighted the originality of their compositions by means of the same word, but (at a first glance) only the Iranian used a textile metaphor. I say at

[7] Avestan text and English translation in Humbach & Elfenbein & Skjærvø 1991, 1.117.

[8] I have preferred to leave untranslated the proper names of the divinities Aṣ̌a (= Righteousness), Vohu Manah (= Good Thought), Ahura Mazdā (= Wise Lord) and Ārmaiti (= Right-Mindedness).

[9] Bartholomae 1904, 1346; Mayrhofer 1992–2001, 2.506. Conversely, Cheung 2007, 401–402 distinguished between PIr. **u̯ab/f^1-* "to call, utter" and **u̯ab/f^2-* "to weave", and assigned Av. *vaf-* to the first root, which in my opinion, founded on the basis of the parallels in other Indo-European texts, is unlikely.

[10] Bartholomae 1904, 401.

[11] Pokorny 1959–1969, 1114; Mayrhofer 1992–2001, 2.506; Adams & Barber 1997, 572; Cheung 2007, 402; Andrés-Toledo 2010b, 437.

a first glance, because Vedic parallels in which the adjective *pūrvyá-* "former, old", antonym of *ápūrvya-*, is linked with the noun *tántu-* "thread" might point to the existence of a textile metaphor related to that metaphorical domain in Vedic too.

The thread is endowed with a very rich symbolism in Indo-Iranian poetry and accordingly is part of various intertwined metaphors, of which the following are found in Vedic:[12]

1) The thread of Righteousness or Truth, a cosmic thread from which the entire universe is created, as we see for instance in AVŚ 2.1.5 and in its Old Iranian parallel in Y 34.10:

AVŚ 2.1.5:[13]

pári víśvā bhúvanāny āyam ṛtásya tántuṁ vítataṁ dṛśé kám

yátra devā́ amṛ́tam ānaśānā́ḥ samāné yónāv ádhy aírayanta

Around all beings I went in order to see the stretched thread of Righteousness,[14] / in the same womb where the gods, having attained immortality, were begotten.

Y 34.10:[15]

ahiiā. vaŋhə̄uš. manaŋhō. š́iiaoϑanā. vaocaṯ. gərəbąm. huxratuš.

spəṇtąmcā. ārmaitīm. dąmīm. vīduuā̊. hiϑąm. aṣ̌ahiiā.

tācā. vīspā. ahurā. ϑβahmī. mazdā. xšaϑrōi. ā. vōiiaϑrā

By means of the action of this Good Thought, the well-minded proclaims his understanding, / he who knows the Beneficent Ārmaiti, the creator, the thread of Righteousness, / and all the tools of the loom in her power, o Ahura Mazdā.

2) The threads of the sky and the threads of light, woven by the Night and the Day, for instance in AVŚ 10.7.42–44:[16]

42. *tantrám éke yuvatī́ vírūpe abhyākrā́maṁ vayataḥṣáṇmayūkham*

prā́nyā́ tántūṁs tiráte dhatté anyā́ nā́pa vṛñjāte ná gamāto ántam

Two young maids of different colours (= Dawn and Night) without equal weave the six-pegged loom grasping it strongly. / The one spreads out the threads, the other lays them; they do not deviate nor reach the end.

43. *táyor aháṁ parinṛ́tyantyor iva ná ví jānāmi yatarā́ parástāt*

púmān enad vayaty ⁺*út* ⁺*kṛṇatti*[17] *púmān enad ví jabhārā́dhi nā́ke*

Of these two, like two dancers dancing around, I cannot distinguish which one precedes. / A man weaves it, he finishes it, a man has stretched it across the sky.

[12] Andrés-Toledo 2010a, 33–92.

[13] Whitney & Lanman 1905, 39; Lüders 1959, 472; YV, VS 32.10; TĀ 10.1.4; MnārUp 2.5. AVŚ 2.1.5a = AVP 2.6.4a; AVŚ 2.1.5b = AVP 2.6.5b.

[14] *cf.* Ved. *ṛtásya tántur vítataḥ* in RV 9.73.9 and Ved. *ṛtásya tántuṁ* in AVŚ 13.3.19.

[15] Geldner 1886–1896; Skjærvø 2005, 272.

[16] See Whitney & Lanman 1905, 595; RV 6.9.1–3; TB 2.5.5.3.3–6.

[17] I emend *úd gṛṇatti* by ⁺*út* ⁺*kṛṇatti*, following Whitney & Lanman 1905, 595 on the basis of RV 10.130.2. In the electronic edition of the Titus Projekt, however, we find *úd gṛṇanti*. See Kuiper 1937, 129 and Mayrhofer 1992–2001, 1.505, *s.v. granth- (grath-)* too.

44. *imé mayū́khā úpa tastabhur dívaṁ sā́māni cakrus tásarāṇi vā́tave*

These pegs have buttressed up the sky, these chants have worked as loom weights for weaving.

3) The thread of life, from which the creatures are made, for instance in AVŚ 10.8.37–38:[18]

37. *yó vidyā́t sū́traṁ vítataṁ yásminn ótāḥ prajā́ imā́ḥ*

sū́traṁ sū́trasya yó vidyā́d sá vidyād brā́hmaṇaṁ mahát

Whoever knows the stretched thread in which all these creatures were woven, / whoever knows the thread of the thread, he will know the great *brā́hmaṇa-*.

38. *védāháṁ sū́traṁ vítataṁ yásminn ótāḥ prajā́ imā́ḥ*

sū́traṁ sū́trasyāháṁ vedā́tho yád brā́hmaṇaṁ mahád

I know the stretched thread in which all these creatures were woven, / the thread of the thread, so I know what the great *brā́hmaṇa-* is.

4) and finally, the thread of sacrifice, designating the milk as well as the *sóma-* juice poured in the ritual performance,[19] for instance in RV 2.3.6 and YV, VS 20.41:[20]

RV 2.3.6:

sādhv ápāṁsi sanátā na ukṣité uṣā́sānáktā vayyèva raṇvité

tántuṁ tatáṁ saṁváyantī samīcī́yajñásya péśaḥ sudúghe páyasvatī

Beneficent Dawn and Night, like two happy weavers, have stretched from old their works for us, / while they weave together the stretched thread, adornment of the sacrifice, yielders of rich milk, overflowing.[21]

YV, VS 8.61–62:[22]

61. *cátustriṁ̐śat tántavo yé vítatniré yá imáṁ yajñáṁ̐ svadháyā dádante*

téṣāṁ chinnáṁ̐ sám v etád dadhāmi svā́hā gharmó ápy etu devā́n

The thirty four threads that have been stretched,[23] that have been offered to this sacrifice according to the offering, / whatever of these shall be cut off, I join it together. *Svā́hā*, may the warm (drink) reach the gods.

62. *yajñásya dóho vítataḥ purutrā́ só aṣṭadhā́ dívam anvā́ tatāna*

sá yajña dhukṣva máhi me prajā́yāṁ̐ rāyáspóṣaṁ víśvam ā́yur aśīya svā́hā

[18] Whitney & Lanman 1905, 600–601; BĀUp 3.7.1.
[19] In Avestan, on the contrary, these metaphorical uses are underrepresented, probably because of the scanty Avestan texts that have survived. In fact, only the thread of Righteousness (Av. *hiϑąm. aṣ̌ahiiā*) is explicitly mentioned in Y 34.10, while the thread of life might be guessed in the Avestan compound *vīkərət̰.uštānəm* "cutting off the (thread of) life" in V 5.37 and 13.12–15.
[20] See Geldner 1951, 1.280; Skjærvø 2005, 270–271.
[21] Regarding the comparison between Dawn and Night and the cows, see Lüders 1959, 393.
[22] See Griffith 1987, 77; Shastri 1971.
[23] *cf.* RV 1.164.5, 10.52.4 and 10.124.1 *saptátantum* "having seven threads" referred to the sacrifice. See Geldner 1951, 1.228, 3.215, 3.353.

The sacrifice's milk has been spread in all directions; this has been spread in eight directions across the sky. / O sacrifice, yield fullness of milk to my offspring; may I obtain fullness of richness during my whole life, *svā́hā*.

A fifth thread, referring to poetic compositions, might exist in Vedic, but it has remained unnoticed until now, probably because of the ambiguity of the passages including words for 'thread'. Indeed, the metaphorical richness concerning the thread contrasts in Vedic with its lexical expression, insofar as the same word, be it Ved. *tántu-* or Ved. *sū́tra-* "thread", usually designates more than one of the aforementioned threads, so that in the absence of further details we cannot easily identify which of them is referred to in each Vedic passage. Because we are not as acquainted with the mythical and ritual context as the audience to which the Vedic poets recited their compositions, or simply because these poets deliberately preferred to maintain a metaphoric ambiguity, only the comparison between those passages and other parallels, and specially the identification of formulas, may help us to figure out which of these threads is implied.

A Vedic formula *tántu- tan- pūrvyá-* "to stretch the old thread", in which Ved. *tántu-* "thread" appears together with the adjective *pūrvyá-* "former, old", antonym of the already mentioned *ápūrvya-* "as never before", is found in the following two texts:

RV 1.142.1:

sámiddho agna ā́ vaha devā́m̐ adyá yatásruce

tántuṁ (the final ṁ does not have _ underneath) tanuṣva pūrvyáṁ sutásomāya dāśúṣe

Burning, o Agni, bring close the gods today to him with the sacrificial spoon; / stretch the old thread for the giver who has pressed the *sóma-*.

RV 8.13.14:

ā́ tū́ gahi prá tú drava mátsvā sutásya gómataḥ

tántuṁ tanuṣva pūrvyáṁ yáthā vidé

Come, hasten, get drunk with pressed (*sóma-*) mixed with milk; / stretch the old thread as it is known.

Apart from noticing the *figura etymologica* and the alliteration in *tántu- tan-*, we must try to understand what *tántu- pūrvyá-*, the "old thread", means and to answer the following question: if Ved. *ápūrvya-* "as never before" was clearly applied to the poet's originality, might its antonym *pūrvyá-* "former, old" on the contrary indicate the poet's attempt to continue traditional patterns? In the preceding two passages, the sacrificial context and the fact that *sóma-*juice and milk are explicitly mentioned suggest that the "old thread" designates the thread of sacrifice. In that case, 'to stretch the old thread' would mean to pour the sacrificial liquids according to the ancient ritual praxis. However, we must take into account that the Indian as well as the Iranian ritual performances were accompanied by the recitation of hymns and other sacred texts, which cannot be separate from each other. Accordingly, "to stretch the old thread" might not necessarily refer only to the ritual performance, but could also imply the poetic composition "woven" or recited according to traditional patterns, "as it is known" (*yáthā vidé* according to RV 8.13.14). If so, the Vedic composers of those texts would not focus on their originality, as did others by means of the adjective *ápūrvya-*, but on their knowledge of the poetic tradition by means of its antonym *pūrvyá-*, but would in any case express this topic by a textile metaphor, namely *tántu- tan- pūrvyá-* "to stretch the old thread". If we accept this interpretation, common to Indo-Iranian religious poetry would not only be the use of the same etymologically related word, but also of textile metaphors to express the topic of the poet's originality, or the lack of it.

Bibliography

Adams, D. Q. & Barber, E. J. W. 1997 Textile; Textile Production. In J. P. Mallory & D. Q. Adams (eds), *Encyclopaedia of Indo-European Culture*. London, 569–574.

Anantanarayana, H. S. *et al.* 2000 *R̥gveda-Saṁhitā*. E-text of the TITUS Project in http://titus.uni-frankfurt.de/texte/etcs/ind/aind/ved/rv/mt/rv.htm. Last update: 1.6.2000.

Andrés-Toledo, M. Á. 2010a *El hilo de la vida y el lazo de la muerte en la tradición indoirania*. Valencia.

Andrés-Toledo, M. Á. 2010b Some Considerations about Vedic, Avestan and Indoiranian Textile Terminology. In C. Michel & M.-L. Nosch (eds), *Textile Terminologies in the Ancient Near East and Mediterranean from the Third to the First Millennia BC*. Ancient Textile Series Vol. 8. Oxford, 430–444.

Bartholomae, C. 1904 *Altiranisches Wörterbuch*. Strasburg.

Campanile, E. 1977 *Ricerche di cultura poetica indoeuropea*. Pisa.

Cheung, J. 2007 *Etymological Dictionary of the Iranian verb*. Leiden.

Durante, M. 1960 La terminologia relative alla creazione poetica. *RAL* 15, 231–249.

Durante, M. 1976 *Sulla preistoria della tradizione poetica greca*. Rome.

Geldner, K. F. 1886–1896 *Avesta. The Sacred Books of the Parsis*. Stuttgart.

Geldner, K. F. 1951 *Der Rig-veda aus dem Sanskrit ins Deutsche übersetzt und mit einem laufendem Kommentar versehen*. Cambridge MA.

Griffith, R. T. H. 1987 *The texts of the White Yajurveda*. New Delhi.

Humbach, H., Elfenbein, J. & Skjærvø, P. O. 1991 *The Gāthās of Zarathuštra and the Other Old Avestan Texts*. Heidelberg.

Kuiper, F. B. J. 1937 *Die indogermanischen Nasalpräsentia. Ein Versuch zu einer morphologischen Analyse*. Amsterdam.

Lüders, H. 1959 *Varuna*. Göttingen.

Mayrhofer, M. 1992–2001 *Etymologisches Wörterbuch des Altindoraischen*. Heidelberg.

Nünlist, R. 1998 *Poetologische Bildersprache in der frühgriechischen Dichtung*, Stuttgart–Leipzig.

Pokorny, J. 1959–1969 *Indogermanisches etymologisches Wörterbuch*. Bern.

Scheid, J. & Svenbro, J. 1996 *The Craft of Zeus: Myths of Weaving and Fabric*. Cambridge MA.

Schmitt, R. 1967 *Dichtung und Dichtersprache in indogermanischer Zeit*. Wiesbaden.

Shastri, J. 1971 *Vājasaneyi-Mādhyandina-Śukla-Yajurveda-Saṁhitā*. New Delhi.

Skjærvø, P. O. 2005 Poetic and Cosmic Weaving in Ancient Iran. Reflections on Avestan *vahma* and Yasna 34.2. In D. Haug & E. Welo (eds), *Haptačahaptāitiš. Festschrift for Fridrik Thordarson*. Oslo, 267–279.

Watkins, C. 1995 *How to Kill a Dragon: Aspects of Indo-European Poetics*. Oxford–New York.

West, M. L. 2007 *Indoeuropean Poetry and Myth*. Oxford–New York.

Whitney, W. D. & Lanman, C. R. 1905 *Atharvaveda Saṁhitā*. Cambridge MA.

Abbreviations

Av.	Avestan
AVP	*Atharvaveda Saṁhitā, Paippalāda* recension
AVŚ	*Atharvaveda Saṁhitā, Śaunakīya* recension
BĀUp	*Bṛhad Āraṇyaka Upaniṣad*
PIE.	Proto-Indo-European
IIr.	Indo-Iranian
MnārUp	*Mahānārāyaṇa Upaniṣad*
OAv.	Old Avestan
PIr.	Proto-Iranian
RV	*Ṛgveda Saṁhitā*
TĀ	*Taittirīya Āraṇyaka*
TB	*Taittirīya Brāhmaṇa*
V	*Vīdēvdād*
Ved.	Vedic
Y	*Yasna*
YV, VS	*Yajurveda Vājasaneyi Saṁhitā*

3. Clothed in Shamelessness, Shrouded in Grief. The Role of 'Garment' Metaphors in Ancient Greek Concepts of Emotion

*D. L. Cairns**

This chapter deals with the role of garment metaphors in the conceptualization of emotion. Metaphor plays a central role in the formation of emotional concepts, as indeed in the formation of concepts in general. Typically, this is a matter of mapping the concrete onto the abstract and the physical onto the psychological.[1] A culture's emotional categories are experiential, based on the physical interaction of embodied human beings with their environment and extended *via* metonymies and metaphors that derive from such experience. The properties of these emotions are not objectively given, but felt; they depend not just on objective processes in the body, the brain, and the world, but on the representation of the phenomenology of such processes in the intersubjective system that is language. Emotional concepts are constrained by physical embodiment and a shared external reality, but they are also variable – from one individual to another and especially from one linguistic community to another. Garment metaphors for emotion are a case in point: they illustrate a universal tendency to derive abstract concepts from concrete physical experience, but they themselves are not universal. As elements of ancient Greek ways of conceptualizing emotion, they depend on the pragmatics and symbolism of Greek dress as an aspect of the body's interaction with the environment.

In what follows I set garment metaphors for emotion in two contexts: first that of emotional metaphor in general, and second that of garment metaphors in general (including the wider nexus of imagery in which garment metaphors belong). This will help establish precisely what garment metaphors do in the Greek conceptualization of emotion, but also remind us (first) that emotions may also be conceptualized in other (related or unrelated) terms; (next) that garments are good to think with in a variety of different ways; and (thirdly) that "emotion" is not a clearly defined

* I am grateful to the Leverhulme Trust, the Alexander von Humboldt-Stiftung, and the European Research Council (*via* the Oxford University Project, The Social and Cultural Construction of Emotions: The Greek Paradigm, directed by Professor A. Chaniotis) for funding the research project from which this paper derives. Many thanks also to audiences in Paris (*bis*), Marburg, Kyoto, Copenhagen, and Oxford for their comments on various preliminary and partial versions.

[1] See Lakoff and Johnson 1980; Lakoff 1987; Johnson 1987; Gibbs 2008. On the widespread acceptance (from Locke to the present) of the basic idea that abstract concepts derive from embodied human beings' interactions with the physical environment, see Jäkel 1999; *cf.* also Geary 2011, 137–138 on Giambattista Vico's *New Science* (1725). On emotion-metaphors, see Kövecses 2000; 2008; Theodoropoulou 2012.

and unified category, either for us or for the Greeks: the metaphors that contribute to the Greek conceptualization of what we regard as emotions also apply in a range of non-emotional concepts, and there is no absolute qualitative difference between emotion and other domains of experience in terms of the import or implications of such metaphors – neither in conceptual nor in phenomenological terms do they isolate features that are unique to the categories of *pathos* or "emotion".

It was not my initial intention in preparing this paper to concentrate on Homer. But my research showed that the relevant metaphors are (with a few significant exceptions) largely poetic, and that their main source is Homer and early hexameter. Thus we are not dealing precisely with the everyday conceptual metaphors of ordinary spoken Greek; but nor are we dealing with the products of individual artistic invention: Homeric metaphor is part of a shared and popular poetic idiom based in ordinary speech and fundamental to Greek thought and culture; even though some of the metaphors we shall examine are not common in everyday prosaic language, nonetheless they are rooted in Greek thought and in the practices and performances of Greek emotional, social, and religious life. They can often be regarded as bringing out conceptual schemata and associations that are latent not only in ordinary language, but also in the languages of dress, gesture, and ritual.[2]

All emotions which have garment metaphors are also conceptualized in terms of other metaphors, many of which are shared with other emotions. At the most basic level, emotions are conceptualized as entities: they 'come into being', they 'exist', they 'come'; one can 'have' them, one can 'lose or destroy' them, 'let them go', 'let go of' them, 'turn one's heart away from' them; and so on. At one end of the scale, there is an emotion such as *nemesis* (indignation), which one can "put in one's *phrenes*" (*Il.* 13.122), but which otherwise merely "is (or is not) there".[3] *Nemesis* is thus not an especially vehement passion. At the other extreme, there is *cholos* (anger): it grips, takes hold, or seizes a person (like a wild beast); it enters one, it gives one orders, it makes one's heart swell; one's own *cholos* can overcome one's enemies; one digests it, one's chest strives to contain it, one can try to wear it away or delay it, one can try to avoid it; one can put it in one's heart, put it out like a fire, cure it like a disease; and so on.[4] Despite its etymological link with *cholē* (bile) and the *cholades* (bowels), *cholos* is not primarily seen in terms of physiological processes; its conceptualization is fundamentally metaphorical, and exhibits a feature common to emotional concepts in a variety of cultures, the notion that emotions are forces of which we are passive objects.[5] The force can be internal, with the mind or the body as its container, or it can be external, when the emotion is an opponent in a struggle or a beast that attacks us. Thus at the heart of the conceptualization of *cholos* are expressions that are common to a number of other Homeric emotions: like *cholos*, *aidōs* (shame), *achos* (grief), *deos* (fear), *oiktos* (pity),

[2] For the relation between poetic and conceptual metaphor, see Lakoff and Turner 1989, esp. 57–139.

[3] "There is/is not *nemesis*", *Il.* 3.156, 14.80; *Od.* 1.350, 2.136, 20.330, 22.40; with causal/modal dat. *Il.* 6.335. On *nemesis* in the context of other Homeric anger-terms, see Cairns 2003.

[4] With *hairein*, *Il.* 4.23, 8.460 (NB *agrios*, savage), 18.322; *Od.* 8.304 (NB *agrios*); with *damazein* (subdue, with target, not patient as object), *Il.* 18.119; with *dyein* (enter), *Il.* 9.553, 19.16, 22.94; with *empiptein* (attack), *Il.* 9.436, 14.207, 306, 16.206; with *echein* (hold), *Il.* 9.675; with *keleuein* (order), *Il.* 20.255, with *lambanein* (take hold of), *Il.* 1.387, 16.30; heart swells with *cholos*, *Il.* 9.646; obj. of *aleeinein* (avoid), *Od.* 1.433; obj. of *aleisthai* (avoid), *Il.* 15.223; obj. of *ballesthai* (put [in the heart]), *Il.* 14.50; obj. of *diatribein* (wear away, delay), *Il.* 4.42; obj. of *entithesthai* (put in), *Il.* 6.326; *Od.* 24.248; obj. of *exakein* (heal), *Il.* 4.36; *Od.* 3.134; obj. of (*kata*)*pessein* (digest), *Il.* 1.81, 4.513, 9.565; obj. of *sbennynai* (extinguish), *Il.* 9.678; obj. of *chandanein* (contain), *Il.* 4.24, 8.461.

[5] See esp. Kövecses 2000.

pothos (longing), and *sebas* (awe) can all 'hold' (*echein*), 'take hold of' (*lambanein*), or 'seize' (*hairein, ainysthai*) a person.[6] Likewise, *aidōs* can be something a person 'yields' or 'gives in' to (*eikein*, *Il.* 10.238), *deos* can 'fall on' or 'attack' one (*empiptein*, *Il.* 17.625, *Od.* 14.88), and *achos* can 'constrain' (*bian*, *Il.* 10.145, 16.22), 'strike' (*typtein*, *Il.* 19.125; *cf.* struck by, *bebolēmenos*, *Il.* 9.9, 10.247), and 'enter' a person (*dyein*, *Il.* 19.367, *Od.* 18.348, 20.286), 'flow' from one's eyes like tears (*cheisthai*, *Il.* 20.282), 'carry one down to Hades' (*Il.* 22.425), be 'poured around' (*Od.* 4.716), 'go around', or surround a person (*Od.* 8.541),[7] or be something that another person 'gives' one to (*didonai*, *Od.* 19.167; *cf. prostrepesthai*, to give oneself over to, *Il.* 6.336). Yet, though such locutions are typical of concepts that we regard as emotions, they do not isolate emotions as a class: entities such as *skotos* (darkness), *thanatos* (death), and *atē* (delusion, disaster) also seize their patients, and one can give in to one's *thymos* (spirit) – an internal agent that is a source of emotion – as well as to one's *aidōs*.[8] The basic point, however, is clear: the metaphors that are typical of emotions tend to emphasize the phenomenological passivity that is one of our own folk criteria for emotion.

Covering and veiling

Before discussing garment metaphors for emotion, we need to consider garment metaphors in general. In general, these are a sub-species of a larger category of metaphors of covering or concealment, most commonly expressed using the verb *kalyptein*. The basic image-schema in all uses of this verb is of two substances or entities, one forming a layer that conceals the other. The location of the layer is normally above the thing concealed, though (in the case of objects such as shields) it may occasionally be in front of it. The covering may be a positive thing (as a shield protects the body) or a negative thing (as a cloud covers the sun). The entity which forms the cover is either an actual substance or a substance by ontological metaphor. In the two most basic constructions of *kalyptein* in Homer the substance itself is (in the first) the subject of the verb (*e.g.* the cloud covers the earth) and (in the second) the instrument (the agent covers the object with the substance).[9] The kinds of things that can be coverings (in Homer and after) include:[10] garments and armour; substances that can form piles or layers (leaves, sand, snow, earth, mud, dust, ashes, waves, mist, clouds); mountains that overshadow cities;[11] urns, graves, buildings, a city, structures such as the Trojan Horse;[12] the layer of fat that covers the thighbones and the paunch of the animal that covers the other parts in sacrifice (or similarly the fat that Achilles uses

[6] Taking, holding, seizing: with *aidōs* (*Il.* 15.657); *achos* (*Il.* 13.581, 14.475, 16.599; *Od.* 11.279, 21.299); *deos* (*Il.* 4.421, 5.812, 817, 7.479, 8.77, 13.224, 14.387, 15.657–658, 17.67; *Od.* 6.140, 11.43, 633, 12.243, 22.42, 24.450, 533); *oiktos* (*Od.* 2.81, 24.438); *phobos* (flight, rout but also, by metonymy, the impulse to flee at *Il.* 11.402, 13.470; hence the meaning "fear"); *pothos* (*Od.* 4.596, 14.144); *sebas* (*Od.* 3.123, 4.75, 4.142, 6.161, 8.384).

[7] *Cf. ponos* as mental disturbance at *Il.* 6.355.

[8] *Skotos* (*Il.* 5.47, 13.672, 16.607), *thanatos* (*Il.* 5.83, 16.334, 20.477), and *atē* (16.805) with verbs of taking/holding; *thymos* with *eikein*, *Il.* 9.109–110, 597, *Od.* 5.126 (*cf.* also phrases such as "restrain your *thymos*", *Il.* 1.192); and esp. *hypnos* (sleep), which very frequently takes, seizes, holds, releases, subdues, *etc.* (x16 *Il.*, x22 *Od.*).

[9] See Dyer 1964, esp. 29 (his constructions A and D, respectively).

[10] I give references only for the less common examples; the common cases, for which references are very numerous, are readily traced using (*e.g.*) the *TLG*.

[11] *E.g. Od.* 8.569, 13.152, 158, 177, 183.

[12] *E.g. Il.* 23.91 (urn); *Od.* 4.618 (building), 8.502–503 (wooden horse), 8.511 (city); S. *Ant.* 28; E. *El.* 1277–1279 (grave).

to cover Patroclus' corpse from head to foot);[13] a person's hand gripping a stone;[14] the eyelids over the eyes;[15] and the parts of the anatomy in relation to the internal organs.[16] Some of these may be slightly extended senses, but all involve physical objects, placed over, around, or in front of other physical objects. Clearly, garments are not the only form of covering to be used with this verb: though all the substances and entities that can *kalyptein* can be seen as forming layers of some sort, by no means all are garment- or blanket-like. If, therefore, not all coverings are garments, how is it possible to use *kalyptein* metaphors as evidence for the conceptualization of emotions as garments? First, there are signs that *kalyptein*, its compounds, and its derivatives bear a prototypical reference to dress: *kalyptein*, *kalyptesthai*, *etc*. are very frequent indeed with reference to clothing; and each of the two nominal derivations in early Greek, namely *kalyptrē* and *kalymma*, refers to a type of garment. Next, there is no hard and fast separation between garment and non-garment applications of the relevant terms, but rather a mutual attraction based on an underlying visual similarity (especially between coverings as things that one pours and as things that one wears).[17] Since garments are the prototypical example of things that *kalyptein*, metaphors of covering in general regularly generate or are transformed into garment metaphors. Even when clothing is not the primary motive for the metaphor, that notion often appears to be latent in metaphorical applications of *kalyptein etc*., to the extent that all the target domains in which the general source metaphor of covering is applied may also be expressed specifically by garment metaphors.

We see a good example of this in the most common image-complex in which *kalyptein etc*. occur, the use of verbs of covering and uncovering in the metaphorical sense of concealing or revealing the truth.[18] This is a wholly post-Homeric phenomenon, and is the only common application of metaphors of *kalyptein* (*etc*.) in prose literature. The everyday cognitive metaphor that objects of knowledge are somehow concealed, until their true nature is uncovered, is an aspect of the 'sight = knowledge' schema that is found in modern languages such as English, but which is especially firmly rooted in a language, such as Greek, in which the verb 'to know' is a perfect tense of the verb 'to see'. We need not translate every such instance of a relevant verb as 'clothe/unclothe', 'veil/unveil', or the like, but the fact that the prototypical idea behind such locutions is covering by means of an item of clothing is demonstrated by the way that this imagery so readily gives rise to full-blown garment metaphors, as in the case of the unveiling of

[13] *E.g. Il*. 1.460; *Od*. 17.241; Hes. *Th*. 539, 541; Hdt. 2.47.3; A. R. 1.434; for Achilles and the corpse of Patroclus, see *Il*. 23.167–168.

[14] *E.g. Il*. 16.735.

[15] *E.g. Il*. 17.136; Eustath. *Comm. Od*. i. 233.43–44.

[16] *E.g.* Hp. *Ep*. 23.14; *Medic*. 8; *de Arte* 10.

[17] Both metaphors, for example, are used of death, sleep, mist, *achos*, and *erōs* (see below). For the metaphorical pouring of literal garments, see *e.g. Il*. 5.734–735; A. *Ag*. 239; *cf. Od*. 16.47 (a fleece for sitting on). *Cf*. Eng. "flowing robes", the "shedding" of both garments and tears, *etc*. The claim is not that pouring and wearing are the same metaphor, merely that they overlap and interact.

[18] See *e.g.* E. *Hipp*. 191–192 (ἀλλ' ὅτι τοῦ ζῆν φίλτερον ἄλλο | σκότος ἀμπίσχων κρύπτει νεφέλαις, "whatever else is dearer than life darkness envelops in clouds and conceals"); *cf*. Pl. *Crat*. 395b (ἡ οὖν τοῦ ὀνόματος [sc. Atreus, derived from ἀτηρά] ἐπωνυμία σμικρὸν παρακλίνει καὶ ἐπικεκάλυπται, "The form of the name diverges a little and is covered over"); discovery/disclosure as *ekkalyptein* (*anakalyptein, apokalyptein*), see *e.g.* [A.] *PV* 193; S. fr. 918 Radt; Hdt. 1.112; E. *IA* 872; Crit. B 16 DK = E. fr. 591 Kannicht; Evenus fr. 5 W; Ctesias fr. 26. 9 Jacoby; Theopompus fr. 20a. 35 Jacoby; Plu. *Lys*. 30. 2 (*etc*.); falsehood = *kalyptein* (and cpds): Critias B 25.41 DK; *cf*. Call. fr. 75.39 Pfeiffer.

the Heliades as they make their journey towards Truth in the proem of Parmenides' *On Truth*.[19] It seems that whenever the metaphor of (un)covering that structures this conceptualization of truth is expanded in more specific metaphorical terms, it is by means of the image of putting on or taking off items of clothing that this happens. The same elaboration is evident in mystery cult, where the veiling and unveiling of the initiate enacts and visually symbolizes the passage from darkness to light, ignorance to knowledge.[20]

This image-complex intersects most closely with the theme of this chapter when the truth that is or should be concealed is something ignoble, for in that case, its metaphorical 'covering' becomes a metonymy for the emotion of *aidōs*, just as actual covering with a garment as an expression of *aidōs* can stand for the emotion itself (see further below). The association is apparent (for example) at Euripides, *Ion* 1521–1527: Ion recognizes that Creusa's story (that he is the result of her rape by Apollo) is the kind of thing that might be motivated by a desire to avoid disgrace, and so this is a suggestion that he makes by whispering into her ear, an attempt to "envelop the matter in darkness" (περικαλύψαι τοῖσι πράγμασι σκότον, 1522), his veiled language performing the same function as would a literal veil as a metonymy for *aidōs*.[21] The same basic image generates an elaborate veiling metaphor in Pindar's fifth *Nemean* (14–17): the truth of Peleus' and Telamon's murder of their half-brother, Phocus, is a disreputable one that the narrator himself feels *aidōs* to tell, a sentiment that is vividly conveyed by the image of a quasi-personified Alatheia covering, i.e. drawing a veil over, her face:[22]

> αἰδέομαι μέγα εἰπεῖν
> ἐν δίκαι τε μὴ κεκινδυνευμένον,
> πῶς δὴ λίπον εὐκλέα νᾶσον
> καὶ τίς ἄνδρας ἀλκίμους
> δαίμων ἀπ᾽ Οἰνώνας ἔλασεν.
> στάσομαι· οὔ τοι ἅπασα κερδίων
> φαίνοισα πρόσωπον ἀλάθει᾽ ἀτρεκές.

> I feel *aidōs* to speak of a mighty exploit, a risk not ventured with justice, how in fact they came to leave the famous island and what fortune drove the heroes from Oinona. I shall stop: to be sure, not every truth is better for showing her face precisely.

Darkness, death, and related phenomena

Returning to earlier literature, and specifically to hexameter poetry, we find that in Homer, Hesiod, and the Hymns the images that are relevant to our discussion are centred on a complex of intersecting metaphors based on various kinds of darkness, from actual night to phenomena such as sleep, old age,

---

[19] Parm. B 1.9–10 DK (Ἡλιάδες κοῦραι, προλιποῦσαι δώματα Νυκτός, | εἰς φάος, ὠσάμεναι κράτων ἄπο χερσὶ καλύπτρας, "The maiden daughters of Helios, leaving the halls of Night, into the light, removing their veils, *kalyptrai*, from their heads with their hands"). *Cf.* Pl. *Prot.* 352ab (as the doctor asks the patient to bare his chest, so Socrates asks Protagoras to bare his thoughts); *G.* 523cd (as clothes conceal the truth about bodies, so the body conceals the truth about the state of the soul); Philo *De vit. cont.* 78 (uncovering thoughts brings them naked into the light).

[20] *Cf.* Cairns 2009, 53–54 with refs. in nn. 29–30. On light and darkness in mystic initiation, *cf.* Patera 2010; Seaford 2010.

[21] *Cf.* Bias 34 DK; Aesop *Proverb* 104; Pi. *P.* 11.25–27; *N.* 9.7; E. *Hipp.* 712; S. *OC* 282–283; Pl. *Charm.* 169d; *Ninus P. Berol.* 6926, col. A, 3. 24 Stephens and Winkler = fr. A, 100 Zimmermann; Plb. 12.6b.2 (*etc.*).

[22] *Cf.* Cairns 2002, 78–79; also Eustath. *Comm. Il.* i. 572.7–8 (on 2.857): liars are like those who veil out of *aidōs*.

and death.[23] These categories bleed into each other, giving rise to further metaphors or metonymies (as when death is represented as a cloud or when night is used as a metonymy for death). The general notion of 'covering' in these metaphors is regularly particularized as covering by means of a garment; and the metaphors are regularly related to the use of real or figurative garments in life and in ritual.

The use of metaphors of 'covering' for aspects of the natural environment (such as night, darkness, and clouds) presents those phenomena as concrete physical entities with which humans interact and whose nature is thoroughly permeated by human symbolism. A case in point is the covering of Gaia by Uranus at *Th.* 126–127 (ἵνα μιν περὶ πάντα καλύπτοι, "in order to cover her completely"): Uranus is at once the starry sky, commensurate with Earth in extent, but also a divine being, Earth's offspring and consort; the basic image-schema of the sky as a layer that covers the earth encompasses not only the celestial, but also the sexual (the male covers the female) and the moral (Gaia is a female being whose nakedness *must* be covered). Similarly, at *Il.* 5.506–507 'night' is a substance with which Ares is able to "cover the battle around" much as Odysseus is able to cover his head with rags at *Od.* 14.349, and if a person is still out of doors at nightfall, night can be said to "cover him all around" (*Il.* 10.201).[24] Such a covering makes it difficult to see:[25] presumably, this is the image's basic experiential motivation. Hence night also covers the eyes as a metonymy for death and for fainting.[26] That this can be perceived in terms of covering by a garment is apparent in the seer Theoclymenus' vision at *Od.* 20.351–357:

> "ἆ δειλοί, τί κακὸν τόδε πάσχετε; νυκτὶ μὲν ὑμέων
> εἰλύαται κεφαλαί τε πρόσωπά τε νέρθε τε γοῦνα,
> οἰμωγὴ δὲ δέδηε, δεδάκρυνται δὲ παρειαί,
> αἵματι δ' ἐρράδαται τοῖχοι καλαί τε μεσόδμαι·
> εἰδώλων δὲ πλέον πρόθυρον, πλείη δὲ καὶ αὐλή,
> ἱεμένων Ἔρεβόσδε ὑπὸ ζόφον· ἠέλιος δὲ
> οὐρανοῦ ἐξαπόλωλε, κακὴ δ' ἐπιδέδρομεν ἀχλύς."

> "Ah, wretched men, what evil is this that you suffer? Clothed in night are your heads and your faces and your knees beneath you; kindled is the sound of wailing, bathed in tears are your cheeks, and sprinkled with blood are the walls and the fair rafters. Full of phantoms is the porch, full too the courtyard, of phantoms that hasten down to Erebus beneath the darkness. The sun has perished out of heaven and an evil mist has run over all."

[23] *Cf.* Onians 1954, 422–431; Dyer 1974. Dyer (p. 31) denies that metaphor is involved, but it seems to me undeniable that the experience of covering with something garment-like is the source from which these conceptualizations (or 'conceptual metaphors') derive. For Near Eastern parallels, *cf.* West 1997, 212–213, 349, 363, 579–580. *Cf.* Katz 2000 for the argument that Greek *hesperos* and Latin *vesper* derive from an Indo-European root meaning "to be clothed". I am grateful to Alex Purves (UCLA) for bringing this article to my attention.

[24] *Cf.* personified Night covered in cloud at *Th.* 757 (also the house of Night at 744–745). For references to or depictions of Night veiled or enveloped in a garment in Greek and other languages, *cf.* Katz 2000, 77–81.

[25] At *Il.* 5.23, however, the night with which Ares covers Idaeus makes him invisible, unseen by others (*cf.* rescue in mist or a garment, p. 33 below). For 'covering' as a salutary rather than an ominous occurrence, *cf.* (more remotely) the armed men who 'cover' (*kalyptein*) a baggage-train, so that it does not appear "naked" to the enemy, at Xen. *Cyr.* 5.4.45.

[26] 'Night covered' (νὺξ ἐκάλυψε etc.) as metonymy for death: *Il.* 5.310, 659, 11.356, 13.424–425, 580; *Cf.* Moreux 1967. The same metonymy is used for the swoon: *Il.* 14.438–439, 22.466. On the analogy between fainting and death, see Clarke 1999, 139–143. I refer to these locutions as metonymy rather than metaphor because, though the use of the term 'night' is metaphorical, the loss of vision is an aspect, symptom, or effect of the overall experience of death or fainting. The locution in fact involves both metaphor and metonymy, and illustrates the close interaction between the two. On metaphor, metonymy, and their affinities and differences, see Lakoff and Johnson 1980, 35–40; Lakoff and Turner 1989, 100–106; Theodoropoulou 2012, esp. 450–460; *cf.* Silk 2003. See also my remarks at Cairns 2013, 239 and n. 12.

Theoclymenus sees the suitors' heads, faces, and knees clothed in night (the primary reference of the verb εἰλύειν is to dress). This metaphorical and metaphysical clothing represents at once their own deaths (their phantoms, *eidōla*, go to the darkness of Erebus, 355–356) and their lamentation at their own deaths (353), and thus unites death and mourning.[27]

In Homer, the metaphor of darkness (*skotos*) as a covering is not applied to simple environmental phenomena, but is exclusively a metonymy for death.[28] Here again the general 'covering' metaphor can be specified in terms of dress, and thus the darkness of death specifically represented as a garment, as it is at Sophocles, *OC* 1701, where Antigone addresses her dead father as "wearing darkness under the earth" (ὦ τὸν ἀεὶ κατὰ γᾶς σκότον εἱμένος).

Just as images of night and darkness as covering layers or garments are used as metonymies for death, so death is often conceived as a dark cloud, either in the phrase "dark cloud of death" or simply as a dark cloud which stands metonymously for death.[29] But death itself can also serve as a covering, either in the simple phrase θάνατος δέ μιν ἀμφεκάλυψε ("death enveloped him") or using the periphrasis τέλος θανάτοιο κάλυψε(ν) ("the end that is death covered") with a direct object.[30] Death is not the personified agent in such locutions, but the covering itself.[31] This is an image of the experience of death as something that comes over or envelops a person, just as darkness, night, and clouds do. Thus it lends itself to elaborations which bring out the latent notion of dress: the dark cloud that covers the dying in Homer becomes something that the dead have willingly "put on" (ἀμφεβάλοντο) in Simonides' epigram for the Spartan dead at Plataea (121 D = *AP* 7.251.2); and the Nurse in Euripides' *Hippolytus* activates the notion of death as a garment when she draws a link between her literal covering of Phaedra's head and the death whose covering will represent her own final release from trouble (250–251): κρύπτω· τὸ δ' ἐμὸν πότε δὴ θάνατος σῶμα καλύψει; ("I hide [your head]; but when will death cover *my* body?").

This notion of death as a covering or garment connects with historical social and religious practices in a number of ways, because (of course) the dead *are* literally covered. Bodies or cremated ashes are buried in the earth, and in Homer and later "the (poured) earth covers" (in phrases such

[27] On the association of veiling with death, grief, and mourning, see Cairns 2009, and *cf.* below, nn. 34, 53.

[28] *Skotos* covers the eyes (as metonymy for death): *Il.* 4.461, 503, 526, 6.11, 13.575, 14.519, 15.578, 16.316, 325, 20.393, 471, 21.181; *h.Ap.* 370; *cf.* Létoublon 2010, 168. Contrast the environmental application at Pi. fr. 108b.3–5, where "black-clouded darkness" covers the sun in an eclipse.

[29] "Dark cloud of death": *Il.* 16.350; Thgn. 707; B. 13.63–64; Sim. 121 D = *AP* 7.251.2; for the dark cloud (*simpliciter*) as a metonymy for death, see *Il.* 20.417. *Cf.* mist (*achlys*) poured over the eyes as a metonymy for death: *Il.* 5.696, 16.344; Od. 22.88. A version of the same phrase is used of the *grief* of Hector at the death of a brother at *Il.* 20.421; *cf. achos* at *Il.* 20.282; *Od.* 4.716 (above, p. 27; *cf.* also n. 48 below). *Cf.* also E. *Pho.* 1310–1312: the city is surrounded by νέφος τοιοῦτον ὥστε δι' Ἀχέροντος ἰέναι, "a cloud such as to send it over Acheron", with an allusion to veiling as a mark of the passage between life and death (n. 34 below).

[30] See *Il.* 5.68, 553 (τέλος θανάτοιο, "the end that is death"), 16.502 (ditto), 855 (ditto), 22.361 (ditto); *cf. Il.* 13.544, 16.414, 580 ἀμφὶ δέ μιν [or οἱ, 13.544] θάνατος χύτο ("around him death was poured"); E. *Tro.* 1315 (black death); A. R. 3.1130 (πάρος θάνατόν γε μεμορμένον ἀμφικαλύψαι, "until fated death envelops us"). On τέλος θανάτοιο ("the end that is death") in particular and *telos* in general, see Onians 1954, 426–466.

[31] *I.e.* construction A rather than construction D, in Dyer's terms (1964, 29); *cf.* n. 9 above. At *Il.* 12.116–117 (πρόσθεν γάρ μιν μοῖρα δυσώνυμος ἀμφεκάλυψεν | ἔγχεϊ Ἰδομενῆος ἀγαυοῦ Δευκαλίδαο, "For before that ill-named fate had enveloped him by the sword of Idomeneus, brave son of Deucalion") the subject of the verb is *moira*, fate; for Dyer (1964, 36) *moira* is here a personified agent, rather than the 'substance' that covers. This might be supported by Q. S. 1.763; but A. *Pers.* 915–917 and A. R. 2.892–893 suggest that these authors took *moira* in the Iliadic original as an enveloping substance, a metonymous equivalent to *thanatos*, "the end that is death", or "the cloud of death" (so Leaf 1900 on 12.116; Onians 1954, 425).

as "the poured earth covers", χυτὴ κατὰ γαῖα καλύπτει) is a regular metonymy for death.[32] The notion of earth as a layer of stuff that covers the dead may be more or less literal, but it is also used as a symbol for the abstract notion of death, and the latent idea of a garment in such phrases is activated in references to 'wearing earth' as a variant of the same metonymy.[33] By the same token, the use of actual garments, whether by the dying to cover their faces or by the bereaved to cover the corpse or the cremated remains, serves as a physical embodiment of the metaphor of death as a cover or garment.[34]

Death is the ultimate impairment of sensory experience; the swoon, which can be presented in the same terms (n. 26 above), is a temporary analogue. Sleep is similar,[35] and is likewise something that covers a person, the eyes, or the eyelids.[36] That there is an analogy between sleep as something that covers a person and the way that the eyelids cover the eyes and the bedclothes cover the body is not something we need to labour. The connexion between literal and figurative coverings is brought out in passages such as *Od.* 5.491–493:[37]

> ὣς Ὀδυσεὺς φύλλοισι καλύψατο. τῶι δ' ἄρ' Ἀθήνη
> ὕπνον ἐπ' ὄμμασι χεῦ', ἵνα μιν παύσειε τάχιστα
> δυσπονέος καμάτοιο, φίλα βλέφαρ' ἀμφικαλύψας.
>
> So Odysseus covered himself with leaves. Athena poured sleep over his eyes to relieve him at once of troublesome weariness, enveloping his dear eyelids.

Again, this can be particularized in the specific image of sleep as a garment, as at Euripides, *Bacchae* 384–385, where the mixing bowl is said to "dress men in sleep" (ἀνδράσι κρατὴρ ὕπνον ἀμφιβάλλῃι).[38]

Clouds and garments

This brings us to clouds. Though the Homeric poems do have some literal language for clouds as environmental phenomena, it is striking how little of it there is. They are high above us (*Il.* 5.867,

[32] *Il.* 6.464, 14.114; Hes. *Op.* 121, 140, 156; Thgn. 973; Pi. *N.* 8.38; A. *Pers.* 645–646; [A.] *PV* 582; Lyc. *Leocr.* 89.

[33] For "wearing earth" as a metonymy for death, see Alc. 129. 17 L-P; Pi. *N.* 11.16; Xen. *Cyr.* 6.4.6; A. R. 1.691; *cf.* Vernant 1983, 154; Sourvinou-Inwood 1995, 123–127; *cf.* also the riddling allusion to the same notion at Sim. 25. 4 W.

[34] For the covering of the face on the part of those who are about to die, see E. *Hipp.* 1457–1458; Pl. *Phd.* 118a; Plu. *Per.* 16.8; *Cic.* 47.9–10; *Caes.* 66.12. For the covering of the corpse or the remains, see *Il.* 18.352–353, 23.254, 24.587–588, 795–796, with further evidence (from *e.g.* funerary legislation, artistic depictions, *etc.*) in Cairns 2009, 50.

[35] *Cf.* other forms of temporary impairment of consciousness, *e.g.* dizziness at A. R. 2.203–204 (κάρος δέ μιν ἀμφεκάλυψεν | πορφύρεος, "purple dizziness enveloped him"); *cf.* old age as a form of mental impairment that envelops the eyes and the mind at Mimn. 5.8 W (βλάπτει δ' ὀφθαλμοὺς καὶ νόον ἀμφιχυθέν, "it pours round the eyes and the mind and impairs them"). Old age is a 'covering' that will envelop Anchises also at *h.Ven.* 244 (γῆρας(…)ἀμφικαλύψει), an eventuality that will leave his lover, Aphrodite, enveloped in a sympathetic and analogous covering of *achos* (243). Mental impairment is perhaps the implication here too (one reason why old age is "cruel, accursed, and wearisome", 245–246). On clouds, mist, and other coverings as forms of sensory impairment, *cf.* Onians 1954, 420–431; *cf.* in general Moreux 1967, 238–240, 254–255.

[36] Sleep is something that is poured at *Il.* 2.19, 14.253, 23.63, 24.445; *Od.* 2.395, 5.492, 7.286, 11.245, 12.338, 18.188, 20.54, 19.590; something that is both poured and covers at *Od.* 5.491–493; and something that simply covers at *Il.* 14.359; *Od.* 18.201, 20.85–86, 23.16–17; *Hes. Th.* 797–798; A. R. 3.748. *Cf.* sleep as a metaphor for death at [Hes.] fr. 278.6.

[37] *Cf. Hes. Th.* 797–798, where the divine perjuror "lies breathless and voiceless on a covered bed, and an evil sleep (*kōma*) covers him" (κεῖται ἀνάπνευστος καὶ ἄναυδος | στρωτοῖς ἐν λεχέεσσι, κακὸν δ' ἐπὶ κῶμα καλύπτει).

[38] *Cf.* the use of the verb *epiballein* at *Od.* 1.363, 16.450, 19.603, 21.357.

13.523, 15.20, 170, 192, 23.874; *Od.* 8.374, 16.264, 20.113–114), and birds move through them (*Il.* 22.309). They are scattered by the wind (*Il.* 5.525–526; *cf.* 11.305, 12.157, 16.374–375, 23.213) and 'appear' over land and mountains (*Il.* 17.372–373). Their upward movement in the air can be used as an image for a rising cloud of dust (*Il.* 23.366). But even literal clouds are regularly symbolic: they are signs of bad weather to come, and so lend themselves to allegory: at *Il.* 4.275–279, for example, the shepherd sees a cloud, shivers, and takes shelter, in a simile which is used to illustrate the ominous approach of the 'cloud of infantry' in 274.[39] From there it is a short step to metaphor, as in the image of the frown as a cloud that crosses the face, a clear connexion between cloud-imagery (though not in this case garment-imagery) and emotion.[40]

Clouds can cover the sea or the landscape in a simple application of the *kalyptein* image-schema,[41] but much more often cloud cover is the result of deliberate, sometimes hostile divine action.[42] This conception is to the fore when clouds are used by the gods as a means of supernatural concealment.[43] Such clouds regularly function as garments, as in Aeneas' rescue by Apollo in *Iliad* 5 (344–345), a direct analogue to Aphrodite's use of her robe for the same purpose earlier in the same book (315). In a similar way, the cloud that Apollo takes from the heavens and uses to cover Hector's exposed body (*Il.* 23.188–190) serves as the shroud that would be used in regular funerary ritual.[44] We might compare the golden cloud with which Zeus promises he will envelop himself and Hera in *Iliad* 14 (341–345):

[39] *Cf.* the simile at *Od.* 22.3–4 (birds cower before clouds). For allegorical clouds, see Archil. fr. 105 W (as reported by Hclt. *Alleg. Hom.* 5.2) and *cf.* Sol. 13.24 W. For the cloud of war or warriors, see *Il.* 4.274 (of infantry), 16.66 (of Trojans, with *amphibainein*), 17.243 (of war itself, which covers everything all around), 23.133 (infantry); Pi. *N.* 10.9 (of war); E. *Hec.* 908 (of Greeks, a cloud that conceals, κρύπτει, the city of Troy); Hdt. 8.109 (a rare instance in prose); Ar. *Pax* 1090 (evoking Homer). The basic notion here (and in other metaphors such as the cloud of jackdaws and starlings in the simile at *Il.* 17.755–757, and the cloud of wealth at Pi. *Enc.* fr. 119.4 S-M) seems to be one of *density*, and so we might see a relation between clouds as dense coverings and the use of *pykazein* as a synonym of *kalyptein* in Homer (below nn. 43, 48).

[40] See A. *Th.* 227–229 (*amēchania* as a cloud over the eyes). For the frown or scowl as a cloud, see S. *Ant.* 528–529; E. *Hipp.* 172; *El.* 1078; *Pho.* 1308; *cf.* [Arist.] *Physiog.* 809b22 (the lion's forehead hangs over its nose and brows like a cloud), 811b34–812a2 (the adjectives *galēnes/synnephes*, calm and cloudy, used of facial expressions); with these metaphors, which describe physiognomy in terms of the natural environment, *cf.* and contrast the 'smiling' of earth at Hom. *Il.* 19.362–363 and *h.Cer.* 14, where the environment is described in terms of physiognomy. I am more inclined than is Clarke (2005, 39–40) to see the latter as metaphors.

[41] The cloud itself covers the sky/landscape at *Od.* 11.15 (*i.e.* there is no divine agent; *cf.* Xen. *Anab.* 3.4.8, where the cloud, having covered the sun, προκαλύψασα, makes it disappear). *Cf.* Hes. *Op.* 555 (a storm cloud covers a person caught outdoors).

[42] See *Il.* 5.522–523 (Zeus places clouds on the mountain tops), 16.298 (Zeus moves a cloud from a mountain top), 17.594 (agent = Zeus); *Od.* 5.291–294 (agent = Poseidon, who gathers clouds and hides both land and sea), 9.68–69 (agent = Zeus), 12.314–315 (agent = Zeus), 12.405–406 = 14.303–304 (Zeus places ominous cloud over ship; storm = divine hostility). For the anthropomorphism, *cf.* *Il.* 12.280–282: Zeus pours snow to cover the peaks, having put the winds to sleep. See also *Il.* 5.750–751 = 8.394–395, where the Horai are charged with opening and closing the dense cloud of heavens (ἠμὲν ἀνακλῖναι πυκινὸν νέφος ἠδ' ἐπιθεῖναι), like a door: *cf.* 8.395; *Od.* 11.525, 22.156. For other metaphors, see *Od.* 12.74 (*amphibainein*; also used of the cloud of warriors at *Il.* 16.66; with *ponos* at *Il.* 6.355, *achos* at *Od.* 8.541, *thrasos* at E. *Supp.* 609, and *moira* at *Andr.* 1081–1082); *Il.* 11.62–63, 16. 364–365 (other verbs of motion); 15.625 (the cloud nourishes the winds); 11.27–8 (cloud as 'frame' for the rainbow); *Od.* 9.144–145 (cloud holds the moon).

[43] As well as the examples discussed in the text, *cf.* 16.668–669, 17.551 (with *pykazein*); *Od.* 8.562 (together with *aēr*); A. R. 4.1361–1362 (together with *achlys*); *cf.* *aēr* at *Il.* 3.381, 11.752, 16.790, 21.549, 597; Hes. *Th.* 9; E. *Hel.* 44–6; *cf.* the cloud that covers Apollo's shoulders at *h.Merc.* 217. Similarly mist (*aēr, achlys*) can be poured as a form of supernatural concealment (*Il.* 20.321; *Od.* 7.15, 140, 143).

[44] *Cf.* the covering of the body with the aegis at *Il.* 24.20–21 and the actual burial practice referred to at *Il.* 18.352, 23.254, 24.796.

τὴν δ' ἀπαμειβόμενος προσέφη νεφεληγερέτα Ζεύς·
"Ἥρη μήτε θεῶν τό γε δείδιθι μήτέ τιν' ἀνδρῶν
ὄψεσθαι· τοῖόν τοι ἐγὼ νέφος ἀμφικαλύψω
χρύσεον· οὐδ' ἂν νῶϊ διαδράκοι Ἠέλιός περ,
οὗ τε καὶ ὀξύτατον πέλεται φάος εἰσοράασθαι."

Zeus the cloud-gatherer addressed her in reply: "Hera, have no fear of god or man on *that* score: in such a golden cloud shall I wrap us that not even Helios could see through it, and his eyesight is keenest of all."

This cloud clearly functions as a blanket. But since the point of Zeus' assurance is to assuage Hera's ostensibly modest reluctance to make love in the open air (330–340), this figurative cloud-blanket has an obvious symbolic and affective significance in serving, as literal garments often do, as a metonym for *aidōs*.[45] Accordingly, when the divine couple lie down together, they dress themselves (ἕσσαντο) in the beautiful golden cloud (ἐπὶ δὲ νεφέλην ἕσσαντο | καλὴν χρυσείην, 350–351). Clouds as supernatural forms of concealment are not only *like* blankets or garments, they are explicitly said (using the verbs *eilyein, hennynai*, and *stephanousthai*) to be worn *as* garments (or garlands).[46] As we noted above, the same is true of metaphorical clouds: just as the concealment of a black cloud can be a metonymy for death (νεφέλη δέ μιν ἀμφεκάλυψε | κυανέη, "the dark cloud of death enveloped him", *Il.* 20.417–418), so the dark cloud of death can be something that one is said to "put on" or "dress oneself in" (κυάνεον θανάτου ἀμφεβάλοντο νέφος, Simonides 121.2 D = *AP* 7.251.2).[47]

Garments, clouds, and grief

This brings us, finally, to metaphorical clouds as images of emotion and to garment metaphors for emotion in general. The Homeric term *achos* is normally translated "grief", but is in fact used more generally of a number of different forms of mental distress. In hexameter, *achos* is something that covers or envelops a person or a person's *phrenes*, both as such and as the 'cloud of *achos*': the image may be conveyed by the verbs *amphikalyptein*, *pykazein*, or *amphibainein*.[48] The intersection of this

[45] On *aidōs* as a garment, see below, pp. 36–37.

[46] See *Il.* 5.186 (νεφέλῃ εἰλυμένος ὤμους, "his shoulders clothed in cloud"); hence *cf. h.Merc.* 217, cited in n. 43 above); *Il.* 14.350–351 (quoted in text), 15.308 (εἱμένος ὤμοιιν νεφέλην, "wearing a cloud on his shoulders"), 20.150; *cf.* ἠέρα ἑσσαμένω ("dressed in *aēr*") at *Il.* 14.282. *Cf.* the cloud by which Zeus is crowned at *Il.* 15.153; that with which Athena crowns Achilles at 18.205–206 (*cf.* the aegis that she puts on his shoulders in 204); those with which Zeus crowns the sky at *Od.* 5.303–304. For the synonymy of *eilyein* and *kalyptein* in the sense "dress (someone else)", see *Il.* 21.318–321; for metaphorical uses of *eilyein*, see *Il.* 12.286 (NB *kalyptein* in 281), 16.639–640; *Od.* 5.403, 14.136, 20.352; *h.Merc.* 245); the verb is used of literal clothing at *Il.* 17.492, 18.522; *Od.* 14.479; *h Merc.* 151; *cf. h.Ap.* 451 (of a hairstyle, worn like a garment).

[47] For *amphiballesthai* of metaphorical garlands, see Pi. *O.* 1.8 (*hymnos* as a garland for the minds of the wise), *P.* 5.31 (*geras*). The verb is used of a (metaphorical) yoke at A. *Pers.* 50 and E. *Andr.* 110; *cf.* its use of sleep at E. *Ba.* 385, p. 32 above, and of *phonos*, slaughter, at E. *Andr.* 797–798.

[48] For *achos* (as such) as the enveloping garment, see *Il.* 8.124, 316, 17.83 (all three with *pykazein*). *Pykazein* is used of literal garments at *Il.* 10.271, 24.581, and *Od.* 22.488, and possibly also at *Il.* 2.777; *cf.* its use of chariot frames covered with gold and tin at 23.503; of cloud cover at *Il.* 17.551. In fact, only 2 of 12 Homeric cases of the verb refer to density, rather than to literal or figurative garments. For the same image in other terms, see *h.Ven.* 243 ("envelop the mind", φρενὰς ἀμφικαλύπτοι), *Od.* 8.541 ("surrounded the mind", φρενὰς ἀμφιβέβηκεν); *cf.* Q. S. 4.471. *Cf.* also penthos, sorrow (which covers the eyes, ἐκάλυψε, at *Il.* 11.249–250). For the 'cloud of *achos*' see *Il.* 17.591, 18.22; *Od.* 24.315 (all three with the formula "the black cloud of *achos* covered him", τὸν δ' ἄχεος νεφέλη ἐκάλυψε μέλαινα). Similarly, mist (poured over the eyes) is a metaphor for grief at *Il.* 20.421 (n. 29 above); *cf.* the pouring of *achos* itself at *Il.* 20.282; *Od.* 4.716. On all the above, *cf.* Onians 1954, 421–422.

imagery with the metaphorical construction of night and darkness and with the literal use of actual garments is also apparent in a number of tragic and similar contexts,[49] and especially in the *exodos* of Euripides' *Heracles*. Heracles initially reacts to the realization that he has killed his wife and children by referring to the cloud of lamentation that envelops him (αἰαῖ· στεναγμῶν γάρ με περιβάλλει νέφος, 1140), but the urge to kill himself that ensues (from line 1146) encompasses the desire to escape not only his suffering, but also his disgrace (1152), and with the arrival of Theseus (1153) his sense of shame becomes acute: at 1157–1162 he envelops his head in darkness, out of shame, but also out of a concern not to transmit *miasma* (the "pollution" that arises from bloodshed) by visual contact:

> οἴμοι, τί δράσω; ποῖ κακῶν ἐρημίαν
> εὕρω, πτερωτὸς ἢ κατὰ χθονὸς μολών;
> φέρ', ἀμφὶ κρατὶ περιβάλω σκότον (...).[50]
> αἰσχύνομαι γὰρ τοῖς δεδραμένοις κακοῖς,
> καὶ τῷδε προστρόπαιον αἷμα προσβαλὼν
> οὐδὲν κακῶσαι τοὺς ἀναιτίους θέλω.
>
> Ah me, what am I to do? Where am I to find escape from evils: by taking wing or by descending into the earth? Come, let me wrap my head in darkness [of robes]. For I am ashamed of the wrongs I have done, and I have no wish to harm the innocent by afflicting this man here with the taint of shed blood.

Both Theseus and Amphitryon then urge Heracles to uncover himself, Theseus observing that "no darkness has a cloud dark enough to hide the disaster of [his] misfortunes" (1215–1217). The primary motives for Heracles' veiling are his shame and his concern for pollution (as Amphitryon explains to Theseus at 1199–1201, corroborating Heracles' own explanation at 1160–1162), but grief is also part of his reaction, and there is a direct link between the cloud of lamentation that represents the initial shock of grief and the literal covering of Heracles' head, described by Theseus as an attempt to hide his misfortunes in cloud and darkness.[51]

It is no accident that the metaphors for the subjective experience of grief are so similar to those which occur in connexion with death. The homology between grief and dying as processes is mirrored also in the use of dress in external ritual expression: the dying cover their faces and the corpse is covered during the funeral,[52] while veiling is both a spontaneous expression of grief and an element in mourning ritual.[53] The Homeric metaphor of grief as a dark and enveloping substance belongs to a complex of images typically used of states that are either negative and disruptive or at least irruptive and overwhelming, especially of states that disturb conscious functioning.

---

[49] See A. *Cho.* 51–53 (darkness, *dnophoi*, covers the house as a metaphor for mourning, lines that are sung by a veiled chorus, 81–83); S. *OT* 1313–1314 ("cloud of darkness", σκότου νέφος, referring to self-blinding as an expression of emotion); E. *Med.* 106–108 ("cloud of lamentation", νέφος οἰμωγῆς, as prelude to a thunderstorm of emotion). *Cf.* relief as the removal of mental cloud and darkness at Chariton 3.9.11 ("Dionysius, as it were, lifted a veil of cloud or darkness from his *psychē*", καθάπερ οὖν νέφος ἢ σκότος ἀπεκάλυψε τῆς ψυχῆς Διονύσιος).

[50] Pflugk (*apud* Diggle, OCT app. crit.) suggests σκότον <πέπλων> in 1159; Diggle himself has a slight preference for πέπλων σκότον, comparing *Or.* 1457.

[51] On the link between clouds and veils (encapsulated in the Latin *nubes, nubere) cf.* Onians 1954, 421.

[52] See n. 34 above.

[53] For veiling as expression of/metonymy for grief and mourning, see Cairns 2002, 75–76, 82–83; 2009. The non-verbal expression of grief and mourning gives rise to a similar kind of metaphor when the verb *amyssein*, used especially of the tearing of one's clothes or flesh (*e.g. Il.* 19.284–285), is applied to the psychological pain of grief itself, as at A. *Pers.* 115, 161; B. 17.18–19, 18.11; Call. *Aetia* fr. 75.10–11 Pfeiffer; [Aeschin.] *Ep.* 12.10; *cf. Il.* 1.243–244: Agamemnon will tear his *thymos* in anger that he paid no honour to the best of the Achaeans.

The *kalyptein* metaphors for night, darkness, clouds, and death are not merely visual or descriptive images, but have a strong affective charge. They represent a generalized, intersubjective version of a first-person perspective, an attempt to get inside what it feels like to die or to fall asleep. They are, in the strict sense, phenomenological. Accordingly, when directly applied to an emotion such as ἄχος, the image of the cloud of grief or of grief itself as an enveloping garment amplifies the notion of the emotion as something that comes from outside the normatively functioning self and emphasizes it as a phenomenologically passive, irruptive mental event.

The garment of *aidōs*

In Euripides, Heracles first experiences grief as a cloud that descends upon him, but then uses a literal garment (described both as a cloud and as darkness) as a way of coping with the emotions (principally shame) to which his grief gives rise. This has a very close analogue in a passage of the same poet's *Orestes* (459–469), where the arrival of Tyndareus occasions *aidōs* in Orestes, and this is manifested in the desire to hide his face in his garments (figured as 'darkness' and 'cloud') and thus avoid the old man's gaze:

> Τυνδάρεως ὅδε
> στείχει πρὸς ἡμᾶς, οὗ μάλιστ' αἰδώς μ' ἔχει 460
> ἐς ὄμματ' ἐλθεῖν τοῖσιν ἐξειργασμένοις.
> [...]
> τίνα σκότον 467
> λάβω προσώπωι; ποῖον ἐπίπροσθεν νέφος
> θῶμαι, γέροντος ὀμμάτων φεύγων κόρας;

Here comes Tyndareus towards us; because of what I've done I'm especially ashamed to meet his gaze.
(…)
What darkness can I find for my face? What sort of cloud can I put in front of it, to avoid the old man's eyes?

The veil or the mantle, of course, is a typical symbol of *aidōs*, used both as a concrete physical expression of the emotion and as a metonymy for the emotion itself.[54] The notion of *aidōs* as itself a metaphorical garment is perhaps clearest at Herodotus 1.8.3 (and its reflexions in Plutarch and Diogenes Laertius), where Gyges reacts with horror to Candaules' suggestion that he should contrive to see the queen, Candaules' wife, naked – ἅμα δὲ κιθῶνι ἐκδυομένωι συνεκδύεται καὶ τὴν αἰδῶ γυνή ("When a woman takes off her *chitōn*, she simultaneously takes off her *aidōs* along with it").[55]

[54] For veiling as expression of *aidōs*: see Cairns 1996a; 1996b; 2002; Llewellyn-Jones 2003. For its function as symbol, see *e.g.* the explicit comments of Σ E. *Pho.* 1485–1486 (*cf.* Σ *Od.* 1.334). That the gesture is symbolic of the emotion is shown by its use as a purely linguistic metonymy (*Sprachgebärde*), as at Plu. *Mor.* 987C = *Brut. an. rat. uti* 4 (where οὐκ ἐγκαλύπτηι (…) ἀποκαλούμενος means "you are not ashamed to be called").

[55] With Hdt. 1.8.3, *cf.* Plu. *Mor.* 139C = *Conj. praec.* 10 ("No: the modest woman puts on her *aidōs* in its place", ἡ σώφρων ἀντενδύεται τὴν αἰδῶ); *Mor.* 37D = *De aud.* 1 (some boys are like Gyges' wife in removing their sense of shame and fear, τὸ αἰδεῖσθαι καὶ φοβεῖσθαι, when they take off their clothes); D. L. 8.43 (Theano "advised a woman about to make her way to her own husband to take off her shame, *aischynē*, along with her garments, and to put it back on again along with them when she leaves", τῆι δὲ πρὸς τὸν ἴδιον ἄνδρα μελλούσηι πορεύεσθαι παρήινει ἅμα τοῖς ἐνδύμασι καὶ τὴν αἰσχύνην ἀποτίθεσθαι, ἀνισταμένην τε πάλιν ἅμ' αὐτοῖσιν ἀναλαμβάνειν). For *aidōs* as a metaphorical garment, *cf.* Pi. *P.* 9.12 (Aphrodite "cast lovely *aidōs* on the sweet love-making" of Apollo and Cyrene, ἐπὶ γλυκεραῖς εὐναῖς ἐρατὰν βάλεν αἰδῶ) and *cf.* the cloud as blanket as expression of *aidōs* at *Il.* 14.343, 350–351, pp. 33–34 above. See further Cairns 1996a; Ferrari 1990; 1997, 6; 2002, 54–56.

The physical clothing is the outward manifestation of a woman's sense of propriety and of her claim to the respect of others. According to Gyges in Herodotus and Theano in Diogenes, when a woman takes off her clothes, so she takes off all these social and moral accoutrements. For Plutarch (*Mor.* 139C), on the other hand, when she takes off her clothes, she maintains her social and moral role by putting on an invisible garment of *aidōs* in their place. In either case, an actual garment stands for an emotional disposition itself conceived as a garment. Actual garments may serve as visible, external symbols of the disposition, but the disposition-as-garment and the actual garment can also be dissociated, as they are by Plutarch, and also by Plato's Socrates, when he insists that naked female athletes can remain clothed in an invisible garment of *aretē* (virtue).[56]

The notion of *aidōs* as a garment is clearly presupposed in Homer. There, garments not only express *aidōs* (as in the case of Odysseus' veiling at *Od.* 8.83–86); they can also be said to cover one's *aidōs*, a locution in which the noun stands – by a standard form of metonymy – not for the emotion, but for its object (in this case, the genitalia, *ta aidoia*).[57] In hexameter poetry, however, the idea that *aidōs* can be represented by a concealing garment emerges as a metaphor only in the phrase "clothed in shamelessness (*anaideiē*)", used twice of Agamemnon in the *Iliad* and once of Hermes in the *Hymn*.[58] This phrase paradoxically presents *anaideiē* not as the absence of the garment of *aidōs*, but as the presence of a garment of a completely opposite nature. The only other psychological quality to which the verb *heimai* (or a compound) is applied in Homer is *alkē*, martial courage.[59] In these locutions, there is a link (but also a disjunction) between the physical armour that one needs to put on in order to enter battle and the right emotional attitude that is the more important form of armour. This confirms that garment metaphors for *aidōs*, *alkē*, and *aretē* are different from similar metaphors for grief (or dizziness, or sleep, or old age, or death). The notion of grief as a cloud-like covering emphasizes the occurrent emotion, the sudden onset of an irruptive mental state. This fits with the non-garment metaphors for *achos* as an external force, as an opponent in a struggle, *etc.* Garment metaphors for *aidōs*, on the other hand, align that emotion with dispositions such as courage and virtue; these do not suddenly cover a person, but are things that one can deliberately put on.[60] *Aidōs* certainly can be an occurrent emotion, a sudden and involuntary reaction; we saw that it can be something that 'holds a person back' or to which a person can 'give in'. But *aidōs* is also the name of a disposition, the sense of honour, shame, or modesty;[61] the distinction between occurrence and disposition is neatly encapsulated in the fact that *achos* simply 'covers' (a simple event in the aorist), whereas *anaideiē* (and so, by extension, *aidōs*) is *worn* (perfect with present sense).[62]

---

[56] Pl. *R.* 457a: "then the guardians' wives must strip for exercise, since they will put on *aretē* in place of their clothes", ἀποδυτέον δὴ ταῖς τῶν φυλάκων γυναιξίν, ἐπείπερ ἀρετὴν ἀντὶ ἱματίων ἀμφιέσονται.

[57] *Il.* 2.262 (Thersites' clothes cover his *aidōs*); for the metonymous shift from the subjective to the objective field, *cf.* 22.7. Here too, then, the clothes themselves stand for the subjective sense of *aidōs*.

[58] *Il.* 1.149, 9.372; *h.Merc.* 156.

[59] *Il.* 7.164, 8.262, 18.157, 20.381 (NB esp. "wearing *alkē* in his mind", φρεσὶν εἱμένος ἀλκήν); *Od.* 9.214, 514; *cf. Il.* 9.231, 19.36 for "putting on" (*dyesthai*) *alkē*; see further Collins 1998, esp. 13, 63. For Semitic parallels, see West 1997, 238–239.

[60] *Cf.* Onians 1954, 420.

[61] See in detail Cairns 1993.

[62] Similar distinctions are apparent also in the imagery of death: as an event or passive experience, death 'covers' (typically in the aorist: nn. 26, 28–29 above); in reference to the permanent state of those who have died, however, verbs of 'wearing' can be used (*e.g.* S. *OC* 1701; Pi. *N.* 11.16; *cf.* above p. 31 and n. 33); and the death of voluntary self-sacrifice can be something that one willingly 'puts on' (Sim. 121 D = *AP* 7.251.2, above, p. 34).

Conclusions, and a brief look at *erōs*

Dress, even in its concrete, literal form, is laden with symbolism. It extends, prosthetically, the body's capacity to express thought and emotion. These physical expressions of emotion are then available as symbols for the emotional concepts that they express. All of the emotions for which garment metaphors exist can also be expressed or referred to metonymously by means of dress. An apparent exception is *erōs*, which (in two passages of the *Iliad*) is said to envelop one's *phrenes* (but which can also be "poured out around the *thymos*" and "subdue" it).[63] Ancient scholars were divided on the meaning of this imagery, whether it represents a further example of the 'emotion as cloud' schema or is a metaphor from hunting, a frequent source of erotic metaphors.[64] Either a cloud of *erōs* or a net of *erōs* would be virtual garment metaphors, given that metaphorical clouds can be worn like garments and that actual garments can be metaphorical hunting nets.[65] But the former is more likely, since *amphikalyptein* is not used of literal (fishing) nets until Oppian, and the suggestion seems to be an inference first from the regular use of hunt-imagery for sex and from the analogy of the relatively common word *amphiblêstron*, a fishing net.[66] The primary force of the image is that of the mental impairment caused by a particularly irruptive and disruptive affective state. At *Il.* 14.316 the enveloping movement takes place within the chest (ἐνὶ στήθεσσι περιπροχυθείς), as the cloud of *eros/erôs* subdues the *thymos*, which suggests the general Homeric analogy between external environmental phenomena and the physical underpinnings of mental events.[67] Yet, as we have already had occasion to note, *Iliad* 14 is replete with associations between clouds and garments as images and instruments of sleep, seduction, and sexuality,[68] and it is likely that the images of *erôs* as a type of covering are related also to the actual use of textiles as covers for love-making, and in particular to the use of the blanket as a metonym for sexual activity.[69]

However that may be, the presentation of *erōs* in these passages certainly belongs with that of other phenomenologically passive experiences. In contrast to this group stands a second, in which imagery of putting on, taking off, and wearing is used to represent emotion as what we might call *habitus*, both dress and disposition. In the case of both groups, the actual use of garments in emotional and ritual behaviour underlies the metaphor: garment metaphors recur in the conceptualization of

---

[63] *Il.* 3.442 ("Never yet has *erōs* so enveloped my *phrenes*", οὐ γάρ πώ ποτέ μ' ὧδέ γ' ἔρως φρένας ἀμφεκάλυψεν), 14.294 ("As soon as he saw her, *erōs* tightly enveloped his *phrenes*", ὡς δ' ἴδεν, ὥς μιν ἔρως πυκινὰς φρένας ἀμφεκάλυψεν); *cf.* 14.315–316 ("never before has *eros* [*sic*] for a goddess or a woman so subdued me, streaming around the *thymos* in my chest", οὐ γάρ πώ ποτέ μ' ὧδε θεᾶς ἔρος οὐδὲ γυναικὸς | θυμὸν ἐνὶ στήθεσσι περιπροχυθεὶς ἐδάμασσεν).

[64] See Eustath. i. 681.11–13, i. 682.9–10 on *Il.* 3.442, iii. 647.14–22 on 14.294; bT ad locc.

[65] For *amphiblēstron* (commonly a kind of fishing net) as a garment, see A. *Ag.* 1382; *Cho.* 492; S. *Tr.* 1052; E. *Hel.* 1079; *Telephus* fr. 697 Kannicht.

[66] In favour of 'cloud' over 'net', see Onians 1954, 420–421; Janko 1992, 198 on *Il.* 14.294.

[67] *Thymos* is itself often conceived as a substance that can be breathed in or out: *Il.* 4.522–524, 13.653–655, 16.468–469, 20.403, 21.416–417; *cf.* Onians 1954, 44–50, 421; Clarke 1999, 61–126. But if *eros* (*sic*) in *Il.* 14.316 is a cloudy substance within the breast, so is it also an agent that exerts power over an opponent (ἐδάμασσεν); the ontological metaphor of emotion as cloud is extended by its personification as internal agent.

[68] See esp. 14.282 (Hera and Hypnos dressed in air, above n. 46), 342–343 (Zeus will cover himself and Hera with a cloud, above, pp. 33–34), 350–351 (they put on the golden cloud, above, p. 34), 359 (Hypnos has covered Zeus with *kōma*, above n. 36); *cf.* the pouring of sleep over the eyes at 165–166, around the mind at 252–253, Hera's dressing at 178–186 (including κρηδέμνῳ δ' ἐφύπερθε καλύψατο δῖα θεάων, "the bright goddess veiled herself from the head down with a head-dress", at 184).

[69] For 'covered by a blanket' as a metonymy for sex, see *h.Ven.* 158; Archil. 196a 45; S. *Tr.* 539–540 (with Davies 1991, 152); fr. 483.2 Radt (with Pearson 1917, ii. 138); E. fr. 603.4 Kannicht; Theocr. *Id.* 18.19; Asclepiades *AP* 5.169.3–4; Longus 3.24; see further Arrigoni 1983.

those emotions that are most associated with the use of garments in their physical expression. But there are some emotions for which garments are common as expressions and metonymies, but which seem not to generate garment metaphors. There are, as far as I can see, no garment metaphors for anger, fear, or despair, though all of these emotions can be expressed and symbolized by the veiling of the head.[70] In these cases, it seems, covering the head is felt to be merely one of many possible expressions, while in the case of (*e.g.*) grief or *aidōs* it is a characteristic and prototypical expression, and thus a much clearer symbol of the emotion. This is a strong indication that in the latter cases the visual expression of the emotion by means of dress has become emblematic of the emotion itself, and thus shows how the role of the body and its extensions in the phenomenology and expression of emotional experience can inform the language and thought of emotion itself.

[70] Anger: *e.g.* E. *Med.* 1147 (with Cairns 2001); fear: *e.g. h.Ven.* 182–183; Arist. *Insomn.* 462a13–15; despair/*amēchania*: *Od.* 10.53–54; S. *Aj.* 1145–1146; A. R. 4.1294. *Pace* Dyer 1974, 32, *Il.* 16.200–202 (Achilles exhorts the Myrmidons not to forget the threats they made towards the Trojans πάνθ' ὑπὸ μηνιθμόν, *i.e.* during the whole period of his anger) is very unlikely to presuppose the notion of anger as a garment.

Bibliography

Arrigoni, G. 1983 Amore sotto il manto e iniziazione nuziale. *QUCC* 15, 7–56.
Cairns, D. L. 1993 *Aidōs: The Psychology and Ethics of Honour and Shame in Ancient Greek Literature.* Oxford.
Cairns, D. L. 1996a Off with her Αἰδώς: Herodotus 1. 8. 3–4. *CQ* 46, 78–83.
Cairns, D. L. 1996b Veiling, Αἰδώς, and a Red-Figure Amphora by Phintias. *JHS* 116, 152–158.
Cairns, D. L. 2001 Anger and the Veil in Ancient Greek Culture. *G&R* 48, 18–32
Cairns, D. L. 2002 The Meaning of the Veil in Ancient Greek Culture. In L. Llewellyn-Jones (ed.), *Women's Dress in Ancient Greece.* Swansea and London, 73–93.
Cairns, D. L. 2003 Ethics, Ethology, Terminology: Iliadic Anger and the Cross-cultural Study of Emotion. In S. Braund and G. W. Most (eds), *Ancient Anger: Perspectives from Homer to Galen,* Yale Classical Studies 32. Cambridge, 11–49.
Cairns, D. L. 2009 Weeping and Veiling: Grief, Display, and Concealment in Ancient Greek Culture. In T. Fögen (ed.), *Tears in the Graeco-Roman World.* Berlin, 37–57.
Cairns, D. L. 2013 The Imagery of *Erôs* in Plato's *Phaedrus*. In E. Sanders *et al.* (eds), *Erôs in Ancient Greece.* Oxford, 233–250.
Christopoulos, M., Karakantza, E. D., & Levaniouk, O. (eds) 2010 *Light and Darkness in Ancient Greek Myth and Religion.* Lanham MD.
Clarke, M. J. 1999 *Flesh and Spirit in the Songs of Homer: A Study of Words and Myths.* Oxford.
Clarke, M. J. 2005 On the Semantics of Ancient Greek Smiles. In D. L. Cairns (ed.), *Body Language in the Greek and Roman Worlds.* Swansea, 37–53.
Collins, D. 1998 *Immortal Armor: The Concept of Alkē in Archaic Greek Poetry.* Lanham MD.
Davies, M. (ed.) 1991 *Sophocles: Trachiniae.* Oxford.
Dyer, R. R. 1964 The Use of Καλύπτω in Homer. *Glotta* 42, 29–38.
Dyer, R. R. 1974 The Coming of Night in Homer. *Glotta* 52, 31–36.
Ferrari, G. 1990 Figures of Speech: The Picture of Αἰδώς. *Mètis* 5, 185–204.
Ferrari, G. 1997 Figures in the Text: Metaphors and Riddles in the *Agamemnon. CP* 92, 1–45.
Ferrari, G. 2002 *Figures of Speech: Men and Maidens in Ancient Greece.* Chicago IL.
Geary, J. 2011 *I is Another: The Secret Life of Metaphor and How it Shapes the Way We See the World.* New York.
Gibbs, R. W. (ed.) 2008 *The Cambridge Handbook of Metaphor and Thought*. Cambridge.
Jäkel, O. 1999 Kant, Blumenberg, Weinrich: Some Forgotten Contributions to the Cognitive Theory of Metaphor. In R. W. Gibbs & G. J. Steen (eds), *Metaphor in Cognitive Linguistics.* Amsterdam, 9–27.
Janko, R. 1992 *The Iliad: A Commentary* vol. iv. Cambridge.
Johnson, M. 1987 *The Body in the Mind.* Chicago IL.
Katz, J. 2000 Evening Dress: The Metaphorical Background of Latin *uesper* and Greek ἕσπερος. In K. Jones-Bley, M. E. Huld, & A. della Volpe (eds), *Proceedings of the eleventh Annual UCLA Indo-European conference. Los Angeles, June 4–5, 1999.* Washington, DC: Institute for the Study of Man = *Journal of Indo-European Studies* Monograph Series 35, 69–93.
Kövecses, Z. 2000 *Metaphor and Emotion.* Cambridge.
Kövecses, Z. 2008 Metaphor and Emotion. In Gibbs 2008, 380–396.
Lakoff, G. 1987 *Women, Fire, and Dangerous Things: What Categories Reveal about the Mind.* Chicago IL.
Lakoff, G. and Johnson, M. 1980 *Metaphors We Live By.* Chicago IL.
Lakoff, G. and Turner, M. 1989 *More Than Cool Reason: A Field Guide to Poetic Metaphor.* Chicago IL.
Leaf, W. (ed.) 1900 *The Iliad* vol. i. London.
Létoublon, F. 2010 To See or Not to See: Blind People and Blindness in Ancient Greek Myths. In Christopoulos *et al.* 2010, 167–180.
Llewellyn-Jones, L. 2003 *Aphrodite's Tortoise: The Veiled Woman of Ancient Greece.* Swansea.
Moreux, B. 1967 La Nuit, l'ombre, et la mort chez Homère. *Phoenix* 21, 237–272.
Onians, R. B. 1954 *The Origins of European Thought about the Body, the Mind, the Soul, the World, Time, and Fate.* 2nd ed. Cambridge.
Patera, I. 2010 Light and Lighting Equipment in the Eleusinian Mysteries. In Christopoulos *et al.* 2010, 261–275.
Pearson, A. C. (ed.) 1917 *The Fragments of Sophocles*, 3 vols. Cambridge.

Seaford, R. 2010 Mystic Light and Near-Death Experience. In Christopoulos *et al.* 2010, 201–206.

Silk, M. S. 2003 Metaphor and Metonymy: Aristotle, Jakobson, Ricoeur, and Others. In G. Boys-Stones (ed.), *Metaphor, Allegory, and the Classical Tradition: Ancient Thought and Modern Revisions.* Oxford, 115–147.

Sourvinou-Inwood, C. 1995 *"Reading" Greek Death to the End of the Classical Period.* Oxford.

Theodoropoulou, M. 2012 The Emotion Seeks to be Expressed: Thoughts from a Linguist's Point of View. In A. Chaniotis (ed.), *Unveiling Emotions: Sources and Methods for the Study of Emotions in the Greek World.* Stuttgart, 433–468.

West, M. L. 1997 *The East Face of Helicon: West Asiatic Elements in Greek Poetry and Myth.* Oxford.

Vernant, J.-P. 1983 Hestia-Hermes: The Religious Expression of Space and Movement in Ancient Greece. In *Myth and Thought among the Greeks.* London, 127–175.

4. The Robes of Alcman's and Pindar's *Parthenoi*

*Thomas R. P. Coward**

Testimonia on *partheneia* (songs performed by a chorus of maidens) are scant: there are however discernible, though tentative, traits in the fragments of Alcman and Pindar.[1] *Partheneia* were performed both publically and privately to mark the initiation of those who performed a *partheneion* on the threshold of womanhood, and formed part of wider social practices which included the *hymenaios* (a wedding song), and the *paian* (originally a sacred hymn addressed to a divinity called *Paian/ Paieōn*, who later corresponds to Apollo, which was sung and danced traditionally by young men,[2] usually to the accompaniment of the *kithara* and *aulos*).[3] The *partheneia* of Alcman and Pindar are noted for their focus on the gender and status of the performers. There are ten named girls in Alcman 1 *PMGF*, Astymeloisa in 3 *PMGF*, and Damaina and Andaisistrota in Pindar (94b S-M). One way this focus is achieved is through the clothing and adornments of the *parthenoi*, which can also act as poetological imagery. Greek lyric poetry interacts with its audience and communicates between a real or fantastical event involving ritual practice and poetic creativity. Poetological imagery express the ways and means of composition, portray the poet and/or performers as literary and social figures, and convey the effect and quality of the song situated beside cultic/non-cultic practices.

This chapter examines examples of such imagery in Alcman (1.60–76, 46, 53, 117 *PMGF* = 3.60–76, 114, 119, 154 Calame) and Pindar (94b S-M).[4] Parallels from Sappho, Attic comedy and tragedy, and depictions on vases will also be considered for comparison and comment. This

* I am grateful to the organisers of and to the participants at the Copenhagen conference at the Danish National Research Foundation's Centre for Textile Research at the Saxo Institute, Copenhagen University for their helpful suggestions and comments on the oral version of this paper. All translations are adapted from the relevant Loeb editions.

[1] There are also references to *partheneia* by Anacreon, Simonides and Bacchylides but there are no surviving fragments. There is a probable example from Anacreon (501 *PMG* = 190 Gentili), but scholars are uncomfortable to attribute a *partheneion* to Anacreon. See Ps.-Plut. *De mus*. 14.1136a; Dion. Hal. *Dem*. 39; Σ Pind. *Pyth*. 3.139a Drachmann; Σ Theocr. 2.10b Wendel. References to choruses of *parthenoi*: Sapph. 17.13–15, 44.14–15 Voigt; Bacch. 3.58, 11.110; Pind. *Pyth*. 10.38, 112 S-M; Eur. *Hel*. 1313.

[2] With the exception of the Delian women in the *Homeric Hymn to Apollo* (158–61); the *paian* is uniquely associated with women in tragedy in opposition to the male characterization of *paian* in ritual practice, see Swift 2010, 64–66 on this matter. A unanimous and singular definition of the *paian* is difficult to propose, see Rutherford 2001, 21–23, Ford 2006, 292, and Rodighiero 2012, 124.

[3] See *Suda* π 661 Adler; Hutchinson 2001, 77–8; Swift 2010, 175 on the *partheneion*. Calame 1977 thinks private performance only.

[4] Alcm. 162 fr. 2 *PMGF* = º240 Calame is of unsure authorship, but does contain several interesting references to 'hairbands', colour, perfume, and young women. The fragment may be describing a chorus of *parthenoi* or a divine chorus.

chapter demonstrates how the clothing and adornments of the *parthenoi* can interact with the ritual background and performance of a *partheneion*, in order to show how these pieces of descriptive imagery and their symbolic associations evokes a set of heterogeneous concepts of what this particular cultic song should or could contain.

The *Partheneia* of Alcman and Pindar

Alcman's first *partheneion* or Louvre *Partheneion* (1 *PMGF* = 3 Calame) is the longest and most studied extant fragment of a *partheneion*. This monostrophic poem of fourteen line stanzas was at least one hundred and one lines long and was performed by a chorus of young girls (1.90, 99 *PMGF*), presumably at or in the environs of Sparta. The first extant part (1.1–35 *PMGF*) told a mythical narrative about the conflict between Heracles and the Hippocoontidae to restore Tyndareus to the throne of Sparta, and of the activities of the Dioscouri, in particular their conflict with the Apharetidae.[5] Two gnomic statements punctuate the narrative (1.16–17, 34–36 *PMGF*). The remaining parts of the poem are self-praise and self-description of the performers (36–105) where they refer to named individuals, their attire and actions, some sort of cultic dedication, and the roles of the individual performers. Examples of clothing imagery and references to clothing from this *partheneion* include:

ταὶ Πεληάδες γὰρ ἇμιν
Ὀρθαίαι[6] **φᾶρος** φεροίσαις
Νύκτα δι᾽ἀμβροσίαν ἄτε σήριον
ἄστρον ἀυηρομέναι μάχονται.

Alcm. 1.60–63 *PMGF* (adapted)

For the Pleiades soaring up through the ambrosial night fight against us,
so we carry the **mantle**, like the Sirius star, to Orthia.

The passage explains why the chorus previously stress the outstanding qualities of Hagesichora and Agido, the two prominent named figures named in this *partheneion*, (39–59). Priestly (2007) has made strong arguments that Sosiphanes' gloss (ΣA Alcm. 1.61 *PMGF*) of φᾶρος as a 'plough' is highly doubtful on papyrological, practical, and literary grounds.[7] This φᾶρος as a 'mantle' or 'wide cloak' (*cf.* Hom. *Il.* 2.43, *Od.* 5.258; Hes. *Op.* 198; Soph. *Aj.* 916), a piece of cloth which wraps around the body, is like the Dog Star, the brightest in the night sky, which chases away the

[5] Alcm. S1 *PMGF* = 1 Calame is the invocation from the beginning of this *partheneion*.

[6] The suggestion of Davison 1968, 157–158, followed by Bowie 2011, 60. It is an early attested spelling of Orthia/Ortheia, see Davison 1968, 169–72 for evidence. The transmitted text is ΟΡΘΡΙΑΙ. Bergk 1882, IV 41 prints Ὀρθίαι as a dative singular, which is based on the reading of Aristophanes (ΣA Alcm. *sub col. ad* 1.60ff. *PMGF*). Calame 1983, 30, Hutchinson 2001, 6 and Neri-Tosi 2009, 13 have Ὀρθρίαι, which was inspired by Aelius Herodianus' *De prosodia catholica* (III.1 392–93 Lentz) and Περὶ μονήρους λέξεως (III.2 942 Lentz). *PMGF* prints ὀρθρίαι, which was proposed in Blass 1870, 184. If Ὀρθρίαι/ὀρθρίαι are nominative plural adjectives, they would agree with ταὶ Πεληάδες, i.e. 'the Peleades of the early morning.' If it is in the dative singular, it is qualifying an implied σιᾷ = θεᾷ (*cf.* Alcm. 1.87 *PMGF*) or it is a proper noun itself. See Calame 1983, 37 for other conjectures, and Davison 1968, 154–157 and Priestley 2007, 179 n.22 for a summary of interpretations to this passage of Alcman. D'Alessio 2013, 646 notes that the discussion on this matter is not fully resolved.

[7] Turner 1987, 44–5 and McNamee 2007, 162 show that the suggested gloss of Sosiphanes was crossed out on the papyrus.

Pleiades.[8] In this *partheneion*, there are a series of contrasting pairs; Agido and Hagesichora are the sun and moon, a Venetian and Kolaxian horse respectively. The Sirens are posited as rival singers to the song of the *parthenoi*.

ἁ[9] δὲ τᾶν Σηρηνίδων
ἀοιδοτέρα μ[ὲν οὐχί,

Alcm. 1.96–8 *PMGF*

And she [Hagesichora?] is naturally not more melodious than the Sirens

The song without Hagesichora is compared to the owl's hoot (Alcm. 1.85–7 *PMGF*) and the Sirens (1.96–7), whereas under her control the song is like a swan (1.100–101 *PMGF*).[10]

Pindar composed a *daphnephorikon*, a form of *partheneion*, for Agasikles of Thebes (94b S-M).[11] Proclus (*ap.* Phot. *Bibl.* 321ab Henry) noted that the *partheneion* was a mixed song-type as it praises both man and god.[12] Alcman (10b *PMGF* = 82c Calame) also refers to a beardless male chorus-leader who leads a *partheneion*. Pindar's *daphnephorikon-partheneion* overlaps in several places with Alcman's *partheneia* such as the comparative use of the siren.

(...) ὑμνήσω στεφάνοισι θάλ-
λοισα παρθένιον κάρα,
σειρῆνα δὲ κόμπον
αὐλίσκων ὑπὸ λωτίνων
μιμήσομ' ἀοιδαῖς (...)

Pind. 94b.11–15 S–M

(. . .) I will sing with my maidenly head flourishing with garlands, and I shall imitate in my songs, to the accompaniment of lotus pipes that siren's loud song (...)

The Sirens are associated with the power of music over man and nature (Hom. *Od.* 12.39–54, 158–200; Hes. 28 M-W; Apoll. Rhod. 4.891–919), and Alcman (30 *PMGF* = 86 Calame) further equates the Muse with a Siren.[13] The Sirens in both Alcman and Pindar thus echo a performance and ritual context as well as thematic trope of the *parthenoi*'s ability and quality to sing.

A parallel to the passage of Alcman (1.60–63 *PMGF*) above where a garment is described as an astronomical object is in the *Homeric Hymn to Aphrodite* (*cf.* also Hom. *Il.* 5.315):

στῆ δ' αὐτοῦ προπάροιθε Διὸς θυγάτηρ Ἀφροδίτη
παρθένῳ ἀδμήτῃ μέγεθος καὶ εἶδος ὁμοίη,
μή μιν ταρβήσειεν ἐν ὀφθαλμοῖσι νοήσας.

---

[8] Anglophone scholarship traditionally translates φᾶρος as a 'robe' on this passage, however it is better understood as a 'large piece of cloth, specifically a wide cloak drawn over the head to veil by either sex, also used as a shroud or bedspread,' as stated in Cleland, Davies and Llewellyn-Jones 2007, 147.

[9] See Hutchinson 2001, 100–101 on the problems of identifying what ἁ is referring to.

[10] *Cf.* also Alcm. S2 *SLG* = °231 Calame = Terp. 6 *SLG* = 1 Gostoli = Ar. 590.20–29 *PCG*[3.2].

[11] The *Daphnephoria* was an octennial Boiotian festival in honour of Apollo Ismenios or Galaxios (Paus. 9.10.4; Proclus *Chrest. ap.* Phot. *Bibl.* 321ab Henry). *Parthenoi* carried laurel branches, processed to a temple of Apollo and sang songs in his honour. A boy of good family led the procession dressed in special clothes and carried laurel branches (the *daphnephoros*). At his side a relative of his held a staff decorated with laurel, flowers and emblems. The chorus followed. See Wilamowitz-Moellendorff 1922, 433; Calame 1997, 59; Moggi and Osana 2010, 277–79.

[12] First observed in Severyns 1938, II 211, 229. *Cf.* also D'Alessio 2000, 259–60.

[13] *Cf.* 'Sirens' 1093–1104 *LIMC*[7.1]; Eur. *Hel.* 167–78.

Ἀγχίσης δ᾽ ὁρόων ἐφράζετο θαύμαινέν τε
εἶδός τε μέγεθος καὶ εἵματα σιγαλόεντα.
πέπλον μὲν γὰρ ἕεστο φαεινότερον πυρὸς αὐγῆς,
εἶχε δ᾽ ἐπιγναμπτὰς ἕλικας κάλυκάς τε φαεινάς,
ὅρμοι δ᾽ ἀμφ᾽ ἁπαλῇ δειρῇ περικαλλέες ἦσαν
καλοὶ χρύσειοι παμποίκιλοι· ὡς δὲ σελήνη
στήθεσιν ἀμφ᾽ ἁπαλοῖσιν ἐλάμπετο, θαῦμα ἰδέσθαι.

Hymn. Hom. Ven. 81–90

And Aphrodite the daughter of Zeus stood before him, being like a pure maiden in height and appearance, that he should not be frightened when he noticed her with his eyes. Now when Anchises saw her, he marked her well and wondered at her appearance and height and shining garments. For she was clad in a *peplos* out-shining the brightness of fire, and she had twisted bracelets and shining ear buds. Round her tender neck were beautiful necklaces of gold, enriched with all manner of rich and varied work, which shimmered like the moon over her tender breasts, a marvel to behold.

In the *Homeric Hymns* (*Hymn. Hom. Cer.* 188–90, 275–80; *Hymn. Hom. Sol.* 31.13–14; *Hymn. Hom. Lun.* 32.8), the radiance of clothing acts as part of an epiphany and in Sappho (96.6–11 Voigt), the beauty of a girl is like the moon amongst the stars.[14]

φᾶρος ('mantle' or 'wide cloak') not only marks the performers as *parthenoi* and acts as an offering to elicit the favour and protection of 'Orthia': it represents the song itself. Pindar also shows this when he describes his song and its music as a Lydian fillet:

πόλιός θ᾽ ὑπὲρ φίλας
ἀστῶν θ᾽ ὑπὲρ τῶνδ᾽ ἅπτομαι φέρων
Λυδίαν μίτραν καναχηδὰ πεποικιλμέναν

Pind. *Nem.* 8.14–15

And on behalf of his beloved city and of these citizens
I am bringing a Lydian fillet embellished with ringing notes.

Poetic composition is often described as weaving in Indo-European literature.[15] A Pindaric fragment states:

ὑφαίνω δ᾽ Ἀμυθαονίδαισιν ποικίλον
ἄνδημα

Pind. 179 S–M

I am weaving an elaborate headband for the sons of Amythaon

Pindar uses weaving verbs to refer to poetic composition less than Bacchylides does, but this fragment implies that the song is formed from a number of different ingredients into one variegated

[14] On clothing and adorning goddesses see Richardson 2010, 230–31, 234.

[15] e.g. Pind. *Ol.* 6.85–87, Sapph. 188 Voigt; Bacch. 5.9–14. See Nünlist 1998, 110–24; West 2007, 36–8 for further examples. It has been suggested that there is an etymological association between ὕμνος and ὑφαίνω, however this has been rejected on phonological grounds, see West 2007, 37.

piece of fabric.[16] ἄνδημα as a poetic headband is only seen here, indicated by ποικίλος, which is used with construction, clothing and song.[17]

In Alcman, the description of the attire of the performers or attendees continues:

> οὔτε γάρ τι πορφύρας
> τόσσος κόρος ὥστ᾽ ἀμύναι,
> οὔτε ποικίλος δράκων
> παγχρύσιος, οὐδὲ μίτρα
> Λυδία, νεανίδων
> ἰανογ[λ]εφάρων ἄγαλμα,
> οὐδὲ ταὶ Ναννῶς κόμαι,
> ἀλλ᾽ οὐ[δ᾽] Ἀρέτα σιειδής,
> οὐδὲ Σύλακίς τε καὶ Κλεησισήρα,
> οὐδ᾽ ἐς Αἰνησιμβρ[ό]τας ἐνθοῖσα φασεῖς·
> "Ἀσταφίς [τ]έ μοι γένοιτο
> καὶ ποτιγλέποι Φίλυλλα
> Δαμαρ[έ]τα τ᾽ ἐρατά τε ϝιανθεμίς·
> ἀλλ᾽ Ἁγησιχόρα με τείρει.

Alcm. 1.64–77 *PMGF*

For in no way is our purple sufficient so that it protects us, nor the intricate snake of solid gold, no nor a Lydian headband, pride of dark eyed girls, nor the hair of Nanno, nor again godlike Areta, nor Thylacis and Cleësithera; nor will you go to Aenesimbrota's and say, "If only Astaphis were mine, if only Phulylla were to look my way and Damareta and lovely Ianthemis"; no Hagesichora protects me.

The whole stanza shows a gradual shift from the luxury and beauty of material objects (64–69) to the performers (70–77), the turning point is lines 68–69. The γάρ (64) joins this description to the previous section of this *partheneion* analyzed. The Sirens threaten the nobility of the *parthenoi* and the threat of Pleiades causes the *parthenoi* alarm. The passage describes how their own unique qualities are insufficient to protect them. All the singers in this chorus are beautiful and have particular characteristics, but Hagesichora and Agido are superior. Calame thinks that these descriptions of the girls do not mark them out as *parthenoi*.[18] It however marks them out precisely for their nubile beauty. Their status is illustrated by their adornments and gender, thereby marking the song as a *partheneion*. πορφύρας ("purple") is a description of the girl's dresses and χρυσέος ("golden") is an image of imperishable supremacy.[19] A variant is ἁλιπόρφυρος ("sea-purple"), which is an Odyssean adjective for the colour of the wool on the distaff or a woven garment (Hom. *Od.* 6.53, 6.306, 13.108). It is used in Anacreon for a dye (447 *PMG* = 129 Gentili), in some lyric hexameters of Alcman for the colour of a bird (26 *PMGF* = 90 Calame), and for the waves of the sea in a hymn to Poseidon spoken under the *persona* of Arion (939.19 *PMG*).[20]

[16] See Pfeijffer 1999, 22 for examples. *Cf.* also Sapph. 188 Voigt; 955 *PMG*; Call. 26.5 Pfeiffer.

[17] Ibyc. S257(a) fr. 27.3 *PMGF*; Pind. *Ol.* 4.2, 6.87, *Pyth.* 9.77–8, *Nem.* 4.14, 5.42, 179, 194.1–3 S-M; Prat. 708.5 *PMG*. *Cf.* ΣT Hom. *Il.* 16.222 Erbse; Pind. *Ol.* 1.29; Philox. 461.2 Theodoridis; Apoll. *Lex. Hom.* p56.7 Bekker; [Ael. Herod.] *Part.* p19.4 Boissondale; Hesych. δ 42, 47, 49–51 Latte.

[18] Calame 1977: II 87 n.78, 98.

[19] **πορφύρας**: Hom. *Il.* 3.126, 8.221, 24.796; Hom. *Od.* 4.115, 4.154, 19.242; Sapph. 44.8–9, 54.1, 92.8 Voigt; Eur. *Or.* 1457a; Hor. *Carm.* II.18.7. **χρυσέος**: Theogn. 449–552; Sim. 256 Poltera = 541.4 *PMG*; Pind. *Ol.* 6.1–4; Bacch. 3.85–7. Wealth: Pind. *Ol.* 1.1–10; 3.42.

[20] See West 1982, 5–9 on Arion.

πορφύρας and ἀλουργής frequently occur in temple inventories lists from Greek sanctuaries for garments presented to a temple as part of the dress code of instructions of what not to wear.[21] On the μίτρα Λυδία, which is νεανίδων ἰανογ[λ]εφάρων ἄγαλμα ("pride of dark-eyed girls"), there are parallels in Sappho (98a.10–11 Voigt), Pindar (*Nem.* 8.15), the chastisements of Xenophanes (3.1 W = 3.5 D-K), and in the fragments of a *krateriskos*, where a woman with her head back and long hair bound with purple fillet.[22]

This passage (Alcm. 1.64–77 *PMGF*) draws attention to the totems of the performer's attire, whether real or imagined, as symbolic adornments of initiation and show that the song is formed from a number of different ingredients that is brought together by one leader. It places them within a cultic context in that not only are these items required but a *choragus* ('leader of the chorus') too (*cf.* 1.88–89), performed at the proper place and time. The φᾶρος marks the song; the clothing and jewellery mark the performers.

Some of the imagery in the Pindaric *partheneia* strongly mark out the *partheneion*, while others are drawn from a wider tradition of imagery. Poetry as craft is seen in Pindar's *daphnephorikon-partheneion* (94b S-M), where the choral leader states:

πολ]λὰ μὲν [τ]ὰ πάροιθ' [⏑ - × - ⏑ –
  δαιδάλλοισ'[23] ἔπεσιν, τὰ δ' α [× - ⏑ -
Ζεὺς οἶδ'(...)

Pind. 94b.31-33 S-M[24]

Many things that came before (...)adorning them with my verses(...)things(...)Zeus knows(...)

δαίδαλος describes craftsmanship and wrought or fabricated objects; δαιδάλλω is only used metaphorically in Pindar. δαιδαλ- words describe the lyre in a noun-epithet pair (Hom. *Il.* 9.187; Pind. *Pyth.* 4.296) and illustrate poetry only in Pindar (*Ol.* 1.29, 105; *Nem.* 11.18). Gerber suggested a parallel usage of δαίδαλος with πτύχη (Pind. *Ol.* 1.105), a 'fold' or something 'bent over/around', like clothing, to refer to poetic style.[25]

κλυταῖσι δαιδαλωσέμεν ὕμνων πτυχαῖς.

Pind. *Ol.* 1.105

to embellish in famous folds of songs

Craft/construction metaphors are not exclusively Pindaric, however the embellishment of poetry by a cognate of δαίδαλος is.[26] In 94b S-M there are words for clothing, ζωσαμένα τε πέπλον (6),

[21] See *IG* II² 1514–1518, *ID* 1473, *SEG* 43.212 for examples.

[22] Hamilton 1989, pl. 83; Krateriskos fragments, Piraeus Museum Kk3; Palaiokrassa 1983, pl. 44.

[23] Ferrari 1991, 393 also suggests that δαιδάλλοισ' is a third-person indicative plural, though cf. Kurke 2007, 87 -8. See now Spelman (2014) on this section of the *partheneion*.

[24] **Armour**: Hom. *Il.* 6.418; 8.195; 13.331, 719; 18.479, 482, 612; 19.13,19, 380; 22.314; [Hes.] *Sc.* 137. **Female clothing**: Hes. *Theog.* 574–578; Hom. *Il.* 14.179; Aesch. *Eum.* 635; Eur. *Hec.* 470. **Chariots**: Hom. *Il.* 17.448; Pind. *Pyth.* 5.36, 106.7 S-M. **Thrones**: Hom. *Od.* 1.131, 10.315, 367, 17.32. **Jewelry**: Asius 13.6 *PEG*; Hom. *Il.* 18.400; Hom. *Od.* 19.227. **Wooden chest**: Hom. *Il.* 16.222; Sim. 271.2 Poltera = 543.2 *PMG*; Bacch. 5.140. **Handles**: Hom. *Il.* 18.379. **Belt**: Hom. *Il.* 4.135. **Bed**: Hom. *Od.* 23.200. **Altar**: Bacch. fr. 4.64 M. **Colonnade**: Apoll. Rhod. *Argon.* 3.237. **Art**: Theocr. *id* 1.32; Luc. *Am.* 13; Nonn. *D.* 37.127. **Craftsman**: Hom. *Il.* 5.60. *Cf.* Gerber (1982): 64; Lehnus 1984, 75; Morris 1992, 36–53.

[25] Gerber 1982, 159. *Cf.* Hom. *Il.* 18.479–482; Nonn. *D.* 43.209; Hesych. π 4249 Latte.

[26] See Steiner 1986, 52–65 and Nünlist 1998, 83–125 for examples.

which δαίδαλος describes. A clear statement of intent follows δαιδάλλοισ' ἔπεσιν that configures itself within the strong poetic *persona* of a *partheneion* and its performers:

ἐμὲ δὲ πρέπει
παρθενήϊα μὲν φρονεῖν
γλῶσσᾳ τε λέγεσθαι·
ἀνδρὸς δ' οὔτε γυναικός, ὧν θάλεσσιν ἔγ-
κειμαι, χρή μ[ε] λαθεῖν ἀοιδὰν πρόσφορον.

Pind. 94b.33–7 S-M

But it is fitting for me to think maidenly thoughts and to utter them with my tongue. Neither for a man nor woman, to whose offspring I am devoted, it is necessary for me to forget a fitting song.

This passage is within the Alcmanic tradition of *partheneia* with its strong feminine choral narrator, however this *partheneion* also speaks of the addressee's victories akin to the epinician narrator (*cf.* Pind. 94b.36–41 S-M), thereby showing a Pindaric take on things.[27] The craft meta-literary metaphor δαιδάλλοισ' ἔπεσιν ("to adorn things in fine words") also shows this Pindaric take with the adaption of an adjective used with clothing that chimes with the self-description of the female narrator in Pindar's *daphnephorikon-partheneion*.

Some *Incerta* fragments of Alcman

The second part of this chapter examines three fragments of Alcman, which cannot be firmly assigned to *partheneia*, that show further clothing imagery. The first two (46, 53 *PMGF* = 114, 119 Calame) describes the clothing of goddesses; the last fragment (117 *PMGF* = 154 Calame) probably describes the clothing of a chorus-leader.

ἔκατον μὲν Διὸς υἱὸν τάδε Μῶσαι κροκόπεπλοι

Alcm. 46 *PMGF* = 114 Calame

The Muses in saffron-*peploi* (taught/led?) these things to the far-shooting son of Zeus (Apollo)

This citation is from Hephaestion (*Ench.* 12.2 Consbruch) and was used as an example of the ionic metre (⏑⏑ - - ⏑⏑ - - ⏑⏑ - - ⏑⏑ - -). Johannes Tzetzes (Tzetz. in Hom. *Il.* 1.384 Lolos) quotes it as a parallel of Apollo's epithet (ἕκατος). A poet invokes the Muses without Apollo at the beginning of a song or section of a poem, however both the Muses and Apollo, who is portrayed as the leader of the Muses and is understood as the origin of the Muses' knowledge, could define the poet.[28]

It is unusual for Apollo to be a 'companion' or 'follower' of the Muses, though the two may dance, sing and play instruments together or separately at the same venue.[29] They perhaps perform an amoebaean song; perhaps short lyric pieces, at a feast in the *Iliad* (1.595–604). This banquet

[27] Hutchinson 2001, 367.

[28] **Apollo and Muses**: Hes. *Theog.* 94–5; *Hymn. Hom. Ap.* 186–206; *Hymn. Hom.* 25.2–3; Sapph. 208 Voigt; Call. 203.1 Pfeiffer. **Leader of Muses**: Hom. *Il.* 1.603–4; Pind. 94c.1 S-M; 941 *PMG*; Pl. *Leg.* 653d. **Source of knowledge**: Hom. *Od.* 8.479; Aesch. *Eum.* 1–8; Apoll. Rhod. 1.1–2. *Cf.* Nagy 1999, 292–306.

[29] Hom. *Il.* 1.604; Hes. *Theog.* 93ff; *Hymn. Hom. Ap.* 186–206; *Hymn. Hom.* 25; Pind. *Pyth.* 1.1–15.

of the gods describes obliquely the rituals of the symposium, where the gods feast in the palace of Zeus with Hephaestus as cupbearer and Apollo and the Muses singing alternatively.[30] Here it is not obvious if this fragment is an invocation to the Muses, but it is a description of them doing something with Apollo as an object of whatever they are doing. The Muses in 46 *PMGF* are described by their clothing, which is a very rare description of the goddesses.

Muses in invocations are usually unornamented, though they can be described as λιγύς 'clear-voiced' (Hom. *Od.* 24.62; *Hymn. Hom. Matr.* (14).2; *Hymn. Hom. Dioscur.* (17).1; *Hymn. Hom. Vulc.* (20).1; Pind. *Pae.* 14.32 S-M = S3.32 Rutherford; 1045 *PMG*). λιγύς can also describe the Sirens (Hes. 150.33 M-W), who can be equated with the Muses (*cf. Seirenes* 1102 *LIMC*).[31] An extended invocation is found in Alcman:

> Μῶσ' ἄγε Καλλιόπα θύγατερ Διὸς
> ἄρχ' ἐρατῶν ϝεπέων, ἐπὶ δ' ἵμερον
> ὕμνωι καὶ χαρίεντα τίθη χορόν.
>
> Alcm. 27 *PMGF* = 84 Calame
>
> Come, Muse Calliope, daughter of Zeus, begin the lovely verses;
> set desire for the song and make the choral dance graceful.

Muses are described either with regard to their eyes (Bacch. 9.3; *Hymn. Hom. Dioscur.* (33).1), and quite often to their hair (Sapph. 128 Voigt; Sim. 264 Poltera = 577a *PMG*; Pind. *Ol.* 6.90–1, *Pyth.* 1.3, *Isth.* 7.23, 215b col.i.8 S-M; Bacch. 29d.9 M? *cf.* also Ibyc. 303 *PMGF* (Cassandra); Pind. *Pyth.* 4.136 (Tyro)); 1001 *PMG*). There are also the rare descriptions of jewelry in their hair ([ἰ]οστεφάνων Μοισᾶν (Bacch. 5.4)), their seating (Μοίσαις γὰρ ἀγλαοθρόνοις (Pind. *Ol.* 13.96)), and the unique Μοῖσαι ἀργύρεαι ([Pind.] 287 S-M). The Graces can also act as Muses (Hes. *Theog.* 63; *Hymn. Hom. Dian.* 27.15; Sapph. 53, 103.5 Voigt; Theogn. 15; Pind. *Nem.* 10.1; Bacch. 9.1), though their function is not to act as sources of inspiration but to embellish the composition with divine grace. Their epithets are similar to the Muses above such as their hair (Stes. 173 Finglass = 212 *PMGF*; Ibyc. 288.1–2 *PMGF*; Pind. *Pyth.* 5.45), their eyes (Bacch. 19.5–6), their holiness or regality (Sapph. 53, 81b.3, 103.8, 386 Voigt; Pind. *Ol.* 2.50, 14.3–4), and their jewelry (Bacch. 5.9, 9.1).

Compound-epithets for female clothing are derived from the *peplos* or the *chiton*.[32] The most common are for the *peplos*. The only other colour-compounds of *peplos* are prefixed by **κυανό-**: *Hymn. Hom. Cer.* 319, 360, 374, 442 (for Demeter); Hes. *Theog.* 406; Orph. *Hymn.* 35.1 (Leto),[33] and **λευκο-** : Corin. 655b.3 *PMG* (women of Tanagra, addressees); Hippon. 47 *IEG* = 51 Degani (personified day); Eur. *Hec.* 933 (chorus at fall of Troy *cf.* also Eur. *Hel.* 1088). Black clothing is usually associated with mourning (μέλας for mortals, κυάνεος for deities).[34] A golden *peplos* is naturally used for female

[30] Kirk (1990), 114; Murray (2008), 164. Other epic amoebean songs: Hom. *Od.* 24.58–64; *Hymn. Hom. Ap.* 186–206; [Hes.] *Scut.* 201–6.

[31] λιγύς describes the *phorminx* at line-end in early Greek hexameter poetry: Hom. *Od.* 8.67, 105, 254, 261, 537; 22.332; 23.133; *Hymn. Hom. Ap.* 3.

[32] **λευκοχίτων**: 929d.5 *PMG; Batrachmyomachia* 37; Nonn. *D.* 20.124; **χρυσοχίτων**: Pind. 195 S-M; *Anth. Pal.* 6.102.6; **ἑλκεχίτωνες**: Hom. *Il.* 13.685; *Hymn. Hom. Ap.* 147. εἷμα has no compound forms, but does have noun + adjective pairs.

[33] n.b. cult of Leto *Nuxia* or *Muxia* (Plut. *Περὶ τῶν ἐν Πλαταιαῖς Δαιδάλων/ 'On the Festival of Images at Plataea'* §3 = 157 Sandbach).

[34] Hom. *Il.* 24.94; *Hymn. Hom. Dem.* 183, 319; Aesch. *Choe.* 11; Eur. *Alc.* 427, 819, *Phoen.* 372; Dion. Hal. 2.19. It is also associated with Night (Bacch. 3.13; Eur. *Ion* 1150; Alexis 93 *PCG*[2]), Death (Eur. *Alc.* 843), and the Erinyes (Aesch. *Eum.* 370).

divinities (Anacr. 418 *PMG* = 74 Gentili (Aphrodite); Pind. *Isth.* 6.75 (Mnemosyne), 94b.1 S-M (perhaps Artemis?); Bacch. 19.22 (Hera). Pindar also coins the compound καλλιπέπλος for Koronis (Pind. *Pyth.* 3.25, *cf.* also ironic use in Eur. *Tr.* 338).[35] Other epithets focus on the design or cut of the *peplos*:[36]

> **ἑλκεσι-** 'of the trailing robe'[37]: Hom. *Il.* 6.442, 7.297, 22.105 (Trojan women); [Hes.] 193.2 M-W = 90 Hirschberger (Daugthers of Kadmus); Alc. 130b.18 Voigt (?); Nonn. *D* 1.103.
>
> **τανύ-** 'with flowing robe': Hom. *Il.* 3.228; *Od.* 4.305, 15.171 (Helen); Hom. *Il.* 18.385, 18.424 (Thetis); Hom. *Od.* 12.375 (Lampetia); Hom. *Od.* 15.363 (Ktimene); [Hes.] *Scut.* 83 (Henioche); [Hes.] 251a.4 M-W = *Meg. Ēh.* F12 Hirschberger (Aristaechme); 291.3 M-W (Eudore); Stes. 183.7 Finglass = 222.7 *PMGF* (?); Bacch. fr. 20a.15 M (Marpessa).
>
> **εὐ-** :Hom. *Il.* 5.424; *Od.* 21.160 (Achaean women); *Od.* 6.49 (Nausicaa); Hom. *Il.* 6.378, 383, 24.769 (attendants);[38] Bacch. 15.49 (Graces); Bacch. 9.61 (Sinope, Cleone, Corcyra?); Bacch. 11.42 (daughters of Proteus perform at sacrifice).

There is a similar distribution of epithets for the ornamentation of hair on divinities, humans or animals, which are derived from θρίξ, κομή or ἔθειρα. Compounds of κομή, prefixed usually by ἠϋ- or καλλι-, are the usual epic adjectives for the hair of female divinities or mortals.[39] There are the less frequent prefixes for divinities: χρυσο- and ξανθο-.[40] θρίξ compounds are rare in lyric (Bacch. 5.28, 11.105; 929d.6, 1007.1 *PMG*). εὐ- and καλλι- compounds describe horses in epic.[41] ἔθειρα as a compound-epithet does not occur in epic for metrical reasons, but it does in lyric and elegy.[42] The opening to Alcman's first *partheneion* (S1 *PMGF*) likewise contains a similar invocation to a Muse as χρυσοκόμα φιλόμολπε.[43] The closest parallel to 46 *PMGF* in terms of vocabulary is in Sappho:

> π̣ε̣[
> κρ[.........]περ[
> **πεπλον**[...]π̣υ̣σ̣χ[
> καικλ̣ε̣[..]σαω[
> **κροκοεντα**[

[35] There are also noun and adjective pairs of *peplos*: Aesch. *Ag.* 1580, *Eum.* 635; Eur. *Alc.* 923, *Hel.* 1088, *Med.* 1159, *Androm.* 148, *Or.* 1457a; Lycoph. *Alex.* 652.

[36] On the implications of these epithets on the spatial dimensions and production of these garments, see van Wees 2005, 45–46.

[37] The translation of Graziosi and Haubold 2010, 206; *contra LfGrE*'s 'drawing up their robe'.

[38] Hom. *Il.* 6.372 for the attendant of Andromache. A variant on ἀμφιπόλοισιν ἐϋπλοκάμοις (Hom. *Il.* 22.442; *Od.* 6.198).

[39] **ἠϋκόμ-** : Hom. *Il.*1.36, 2.689, 3.329, 4.512, 6.92, 273, 303, 7.335, 8.82, 9.339, 10.5, 11.369, 505, 13.766, 16.860, 19.413, 24.466, 602; *Od.* 8.452, 11.318, 12.389; *Hymn. Hom. Cer.* 1; Hes. *Theog.* 241, 267, 625, 634; *Op.* 165; 10a.32, 25.17, 37.18, 185.8, 195.2, 199.2, 200.2, 200.11, 204.33, 204.55, 280.20, 343.4 M-W; *Cypria* 9.2 *PEG*; Pind. *Ol.* 6.91, *Pyth.* 5.45. **καλλι-** : Hom. *Il.* 9.449; *Od.* 15.58; Hes. *Theog.* 915, *Op.* 75; 141.10 M-W; Sapph. 128 Voigt; Stes. 173.1-2 Finglass = 212.1-2 *PMGF*; Ibyc. 288.2 *PMGF*; Anacr. 390 *PMG* = 92 Gentili; Sim. 264 Poltera = 577a.2 *PMG*; Pind. *Pyth.* 9.106, *Nem.* 10.10, *Pae.* 5.41 = D5.41 Rutherford.

[40] **χρυσο-** : Hes. *Theog.* 947; Alc. 327 Voigt; Anacr. 358.3 *PMG* = 13 Gentili; Sim. 115 Poltera = 519 fr. 84.6 *PMG*; *CEG* I.308; Pind. *Ol.* 6.41, *Ol.* 7.32; Bacch. 4.2. **ξανθο-** : Hes. 25.5 M-W; Pind. *Nem.* 9.17.

[41] **εὐ-** : Hom. *Il.* 23.13, 301, 351. **καλλι-** : Hom. *Il.* 5.323, 8.348, 433, 503, 10.491, 11.280, 11.531, 17.504, 624, 18.223, 23.525, 530; *Od.* 3.475, 5.380, 9.336, 469, 15.215; *Hymn. Hom. Lun.* 32.10; [Hes.] *Scut.* 372.

[42] **εὐ-** : Stes. 222b.243 *PMGF*; Anacr. 418 *PMG* = 74 Gentili; Bacch. fr. 20a.26 M. **χρυσο-** : Archil. 323 *IEG*; Ibyc. S151.9 *PMGF*. **πυρι-** : Bacch. 17.56.

[43] The scansion of this assigned beginning indicates that it is the second line of the strophe as it is one syllable too long for strophe line one.

πεπλον πορφυ[ρ̣.........]δ̣εξω̣[.]
χλαιναι περσ̣[
στεφανοι περ[
καλ[.]ρ̣σ̣σ̣α̣μ̣[
φρυ[γ?
πορφ[υρ
τ̣α̣π̣α̣[

Sapph. 92 Voigt

(…)robe(…)saffron(…)garment(…)purple
peplos(…)cloaks(…)garlands(…)
[beauty?](…)[Phrygian?](…)purple(…)

Unfortunately, only the line beginnings are preserved, though there are other instances of coloured clothing in Sappho such as a purple girdle (ἰόκολπος) for Aphrodite(?) (21.13, 103.3 Voigt), or for a bride (30.4–5 Voigt), a purple sash (κἄμματα / πορφλυρ[α], 44.8–9 Voigt), Eros' purple short mantle (54 Voigt, πορφυρίαν χλάμυν), and possibly a purple headband (π̣ορφύρῳ̣(…) *vel* στέφει) is ideal for a dark-haired maiden (98a.4 Voigt, the supplement of Vogliano 1941, 11–12).

Pindar has an εἶμα ('shirt') of Jason as κρόκεος 'saffron-dyed' (*Pyth.* 4.232), Aristophanes' Kinsman, in the costume of Euripides' maiden Andromeda, dresses in a saffron coloured frock (Ar. *Thesm.* 1044), and Antigone wears one too (Eur. *Phoen.* 1491). A scrappy piece of Tyrtaeus describes a woman exulting in a saffron-coloured dress(?), though the context and identity of the woman (high-status or goddess) is lost (Tyrt. 18.2 *IEG*). The κρόκος flower (*crocus sativus*) itself is present in a number of myths of abduction and/or seduction:

Part of the bed of Zeus and Hera on Mount Ida (Hom. *Il.* 14.348)

One of the flowers, which Persephone was picking when Hades abducted her (*Hymn. Hom. Cer.* 6, 426, 428)

One of the flowers, which Europa was picking when Zeus abducted her (Mosch. *Eur.* 68, *cf.* also ΣD *Il.* 12.397 Ersbe; [Hes.] 140 M-W)

The yellow saffron colour is associated with festivity, luxury or femininity. In three of four instances in Homer (Hom. *Il.* 8.1, 19.1, 24.695) it describes the attire of Dawn as she spreads light over the earth or sea whilst also representing the light, on the other occasion (*Il.* 23.227) it is a substitute for Dawn's usual epithet 'rosy-fingered'. In Hesiod, it unusually describes Enyo (Hes. *Theog.* 273) and the Oceanid Telesto (358). A κροκωτός was a floor length yellow dress (Ar. *Lys.* 44, 51, *Thesm.* 941, *Ecc.* 332, 879) and it was also associated with Dionysus (Ar. *Ran.* 46; Crat. 40 *PCG*^{4}; Dionysus 84, 87, 111 *LIMC*$^{3.1}$). In Euripides (*Hec.* 468), a slave girl offers a saffron-dyed *peplos* (κροκέωι πέπλωι) to Athena.[44] There are also examples of clothing items with κροκο- as the suffix (Pind. *Ol.* 6.39 (*cf.* also Hom. *Od.* 11.124); Eur. *Hec.* 471). Darius' shoe is a κροκόβαπτος (Aesch. *Pers.* 660), Iphigeneia's robe at her sacrifice is κρόκου βαφὰς (Aesch. *Ag.* 239), and Heracles' swaddling clothes are κροκωτὸν σπάργανον (Pind. *Nem.* 1.38).

Purple, like saffron, is used for divinities or high status women. A useful parallel are φοινικό- adjectives, which describe a cloak in Homer (Hom. *Il.* 10.133; *Od.* 14.500, 21.118). In the *Homeric Hymn to Hermes*, when Apollo rifles through the possessions of Maia, her clothing in the storerooms is described as containing φοινικόεντα καὶ ἄργυφα εἵματα ("purple and white

[44] This is probably a reference to Athena's new *peplos* in the occasion of the Panathenaic festival.

garments" *Hymn. Hom. Merc.* 250).[45] An interesting use of a φοινικό- compound is in Pindar where Hecate is described as:

> ἄγγελλε δὲ φοινικόπεζα λόγον παρθένος
> εὐμενὴς Ἑκάτα
> τὸν ἐθέλοντα γενέσθαι.

Pind. *Pae.* 2.77–79 = D2.77–79 Rutherford

> And Hekate, the maiden with the purple feet, was graciously
> announcing her prophecy eager for fulfillment.

Demeter is also φοινικόπεζα (Pind. *Ol.* 6.94), though this has a non-cultic significance and is designed to balance the colour and ornate section of this ode and is a variant on ἀργυρόπεζα.

In 46 *PMGF* where the saffron robed Muses (teach?/lead?) Apollo, there is an inversion of the usual placement of Apollo and the Muses. This is probably acting as a mythological paradigm to either the origin of the *partheneion* or to evoke a particular performative context. It does not look like an invocation to the Muses, but perhaps part of a mythological narrative involving the Muses and Apollo. The choice of colour is interesting and the Muses in Alcman and elsewhere are also *parthenoi* or κοῦραι/κόραι Διός.[46] κροκόπεπλοι is also used for Dawn in line formulae and Iphigeneia in *Agamemnon* was sacrificed as a *parthenos* wearing a saffron robe, which evokes the shedding of the *krokotos* by the Brauronian *arktoi*.[47] Spartan *partheneia* were probably performed at dawn or an immediately post-dawn (Alcm. 1.40–43, 87 *PMGF*) and the Muses, acting as *parthenoi*, visualise the first light ceremony with their clothing.

Another description of another goddess, in this case Artemis, also focuses on her clothing:

> ϝεσσαμέναι πέρι δέρματα θηρῶν

Alcm. 53 *PMGF* = 119 Calame

> She who is clad in the skins of wild animals

This is a description of Artemis according to the Homeric *scholia* (Σ Hom. *Il.* 21.485b1 (5.238) Erbse) which discusses how ἐναίρειν means 'despoil.' It comments on a speech of Hera where she rebukes Artemis (Hom. *Il.* 21.481–88), saying that Artemis cannot match her, even though Zeus made Artemis better against women, as she is better suited to hunting than fighting in battle. The anonymous ancient critics in this particular case use Alcman to further clarify ἐναίρειν since Artemis περιάπτεται γὰρ νεβρίδας ("wears fawn-skins").

Descriptions of Artemis naturally focus on her hunting and its noisy activities.[48] Artemis' clothing, like the clothing of the Muses, is rarely described (Call. *Hymn. Dian.* 3.11–12, 16). Pindar

---

[45] Alcm. 162 fr. 2.1 *PMGF* = °240 Calame refers to something as φοινικέα.

[46] ***Parthenos*:** Stes. 90 Finglass = 193 *PMGF*; Emped. 3.3 D-K; Cor. 692.2.1 *PMG*; Cat. 1.9. ***Parthenoi*:** Alcm. 59b.1–2 *PMGF* = 149 Calame; Bacch. 1.1–2; Pind. *Isth.* 8.57, *Pae.* 6.54 = D6.54 Rutherford; Ar. *Ran.* 875–6; Euphorion 416.2 *SH*; Cat. 65.2; Prop. 2.30.33. **κοῦραι/κόραι Διός**: Hom. *Il.* 2.598; Hes. *Theog.* 25, 52, 966, 1022; Theogn. 15; Sapph. 53 Voigt; Pind. *Ol.* 10.96.

[47] *Cf.* Sourvinou (1971); Stinton (1976).

[48] Πότνια θηρῶν, / Ἄρτεμις ἀγοτέρη (Hom. *Il.* 21.470–1, *cf.* Anacr. 348.3 *PMG* = 1.3 Gentili); Ἄρτεμι θηροφόνη (Theogn. 11); Ἀρτέμις χρυσηλάκατος κελαδεινὴ (Hom. *Il.* 16.183; Hes. 23a.17 M-W; *Hymn. Hom. Ven.* 16, 118; *Hymn. Hom. Dian.* (27).1); Ἄρτεμις χρυσηλακάτῳ εἴκυῖα (Hom. *Od.* 4.122); Ἄρτεμις ἰοχέαιρα (Hes. *Theog.* 14, 918; *Hymn. Hom. Ap.* 15, 159, 199; *Hymn. Hom. Cer.* 424); κυνηγέτις (Cornut. 34). Though *cf.* Bacch. 5.98–9; Sim. 100 Poltera = 519 fr. 35b.7–8 *PMG*.

perhaps describes Artemis as χρυσοπ[επλ (94b.1 S-M) at the beginning of the extant fragment of the 'second' *Partheneion*. Claudian (*Panegyricus dictus Probino et Olybrio consulibius* 1.183–91) mentions that Leto gave purple robes to the infant Apollo and Artemis. Iconography tends to focus on her dressed in an decorated knee-length dress and sandals, armed with spears or a curved bow, at times with hunting dogs (Artemis 87, 88abcd, 109a–12, 1135 1325, 1346, 1362 *LIMC*$^{2.1}$). Artemis is also occasionally depicted as wearing animal skins in art in the archaic and classical period (e.g. cult statue in Arcadia (Paus. 8.37.1); Lydos *dinos* (*ca.* 550 B.C.) 107.1, 684 *ABV* = Athens, N.M. Acr. i, 607.+).[49] An anonymous ancient critic (Σ Ar. *Lysist*. 645 Hangard) notes that at the sanctuary of Artemis at Brauron the young female initiates aged 5–10 would dress up as bears and wear saffron coloured robes at the Brauronia festival. Although Brauronia was an Attic festival, it was an important cult centre of Artemis, though there is little evidence for the formal training of female choruses in Athens.[50]

The fragment is understood as an address to a goddess, though it is really just a description of female clothing. This does not necessarily mark the fragment out as a *partheneion*, as it could be from a hymn or another cultic song, though given Artemis' association with *partheneia*, it is possible that this was a description designed to mark the attire of the goddess or of a performer dressed as the addressee.

The final example possibly describes a chorus-leader in a *partheneion*.

λᾶδος ἡμένα καλόν

Alcm. 117 *PMGF* = 154 Calame

wearing a beautiful simple dress

This is a quotation from Eustathius (*ap*. Eust. ad *Il*. 18.352 (IV.190 (1147)) Van der Valk), who quotes this piece of Alcman for the purpose of commenting upon ἐν λεχέεσσι δὲ θέντες ἑανῷ λιτὶ κάλυψαν / ἐς πόδας ἐκ κεφαλῆς ("and they (Achilles' companions) covered him with a soft linen cloth placing it from head to foot" Hom. *Il*. 18.352–353). In this part of the *Iliad* the body of Patroclus is prepared for burial, his body has been washed and wrapped in a soft linen cloth and then a white robe placed on top. Eustathius first discusses the accentuation of λίς ('linen cloth') in the dative, offering the Attic variants (λῆδος, λήδιον 'a cheap, common dress') and then the Doric alternatives with Alcman as an example.[51] Hesychius (λ 122 Latte) and Aelius Herodianus (*ca*. 185–250 AD) gloss (*Peri Orthographias* 3.2 542 Lentz) this word as meaning a Τριβώνιον, which was a short, cheap, woolen cloak worn by poor Athenians, ascetics and by Spartans.[52]

In the Louvre *Partheneion* (1.64) and here, the beauty of the dress is emphasised. A contrasting description is of Electra in Euripides (*El*. 175–85 (lyr.), *cf*. also Sapph. 57 Voigt) where she describes her own tattered clothing and wretched state as un-*parthenos*. Vases tend to have the dresses of *parthenoi* as either having a criss-crossed pattern, which matches other iconographical remains of

[49] Heracles in animal skins: Theocr. *Id*. 25.63, 175; Call. 268c, 597 Pfeiffer; perhaps Panyassis 4–5 *PEG* = 1–2 *EGF* = 6–7 *GEF*.

[50] Parker 2005, 182–3.

[51] Van Wees, 2005, 45 notes that in epic poetry, women while alive wore linen garments, while the bodies or ashes of dead men were covered with linen.

[52] Ar. *Vesp*. 33, *Lys*. 278, *Ek*. 850, *Plut*. 714, 842–6, 882, 897, 935; Eupol. 298.6 *PCG*^{5}; Men. *Georg*. Fr. 1.4 Sandbach; Antiphan. 76 *PCG*^{2}; Dem. 54.34, Lys. 32.16. Lexicographers later simply viewed it as an ancient cloak (Phot. τ 422 Theodoridis, *Suda* τ 958 Adler).

Artemis Orthia or with painted white folds on them.[53] This fragment could be a description of the chorus leader in a *partheneion*, as with Hagesichora and Agido, the *choragus* (Pind. 94b.6–12), or as Aristophanes notes at the end of *Lysistrata* (1315), in a Spartan *partheneion*, ἁγνὰ χοραγὸς εὐπρεπής ("the pure and comely chief of the chorus").[54]

Conclusions

A variety of traditional imagery abounds in the cultic songs, several have parallels in Greek and Indo-European poetry and some are unique to a particular poet. Generally these images and their symbolic and poetological associations reveal the identity of a thing only with the application of some intellectual effort and present a challenge to an audience or hearer, which may be understood with difficulty through familiarity with the style.[55] Poetological imagery express the ways and means of composition, portray the poet or performers as literary and social representatives of a community, and convey the effect and quality of the song. There are of course some passages where due to the fragmentary nature of the texts it is difficult to reconstruct their cultic function, but can be seen as part of a wider tradition of images. Literary criticism essentially involves the consideration of a work as a whole, which it is difficult to apply to fragments. An almost complete poem is still incomplete. Nevertheless as Colin Macleod noted 'all literature is nourished from two sources: life and other literature.'[56]

Alcman's *partheneia* simultaneously place themselves within their probable local Spartan cultic context and anticipates their Panhellenic reception: Pindar likewise does the same for Thebes in the *daphnephorikon-partheneion*. Pindar's *partheneia* show a close affinity to his epinician *persona* and add to the strong focus on the gender and characteristics of the performers. Alcman's *partheneia* convey a sense of distance between the narrator and the poet, whereas Sappho's poems formally mark the two as one.[57] Pindar demonstrates a perception of Sparta's poetic output and identity in an unassigned fragment in dactylo-epitrites:

ἔνθα βουλαὶ γερόντων
καὶ νέων ἀνδρῶν ἀριστεύοισιν αἰχμαί,
καὶ χοροὶ καὶ Μοῖσα καὶ Ἀγλαΐα

Pind. 199 S–M[58]

There (in Sparta) the counsels of elders and the young men's spears
prevail, and the choruses, the Muse, and Splendour.

In the first half there are the activities of Spartan men, in the second the activities of men and women.

The clothing of *parthenoi* in *partheneia* are descriptions of cultic attire or attire and ornaments designed to attract attention, mark them out as performers by their luxury and superiority, and as eligible *parthenoi*. The Louvre *partheneion* can be understood to merge a cultic offering to the goddess, a woven object, to represent the song itself, while the appearance of the performers

53 See Sourvinou-Inwood 1988, 1990; Hamilton 1989 for images.
54 See Bierl 2009, 2011 on Alcman and Aristophanes.
55 West 2007, 77.
56 Macleod 1983, 15.
57 See the comments on the non-dramatic lyric narrator in Hutchinson 2001, 435–436.
58 This is from Plutarch (*Lycurg*. 21.3–4) and is paired with a quotation assigned to Terpander (5 Gostoli) to demonstrate a connexion between music and military virtue, (*cf.* also the poems of Tyrtaeus). *Cf.* Kienzle (1936), 74–6; D'Alessio (2009), 139. Pindar (*Ol*. 13.22–3) has a similar description of Corinth.

can mark them out as *parthenoi*. The *Incerta* fragments of Alcman show the means to visualise divinities and performers by their clothing. Pindar's *daphnephorikon-partheneion* shows that while the *partheneion* requires a different style and tone from Pindar's other songs, it shows a similar sort of 'nobility and seriousness.'[59] The clothing imagery is not as prominent as it is in Alcman, but the embellishment of the poetry by other craft imagery sequences with the self-description of female narrator in Pindar's poem. Divinities can be given splendid clothing, while mortals are given approximations of these or the divinities act as archetypes or Ur-performers for a particular song-type by their voice, dance, and attire. The clothing and trinkets of the *parthenoi* can interact with and be part of the ritual background and performative context of a *partheneion*. These pieces of descriptive imagery and their symbolic associations evoke notions of what this particular cultic song should or could contain. The examples of clothing imagery from Alcman and Pindar above demonstrate the poetry and splendour of the *partheneia* and its performers.

[59] Dion. Hal. *De Dem.* 39.41–46.

Bibliography

Adler, A. 1967–71 Suidas. *Lexicographi Graeci* I, 5 vols, Lexicographi Graeci I. Leipzig.

Bekker, I. 1833 (ed.) *Apollonii Sophistae lexicon Homericum*. Berlin.

Bergk, Th. 1882 *Poetae Lyrici Graeci*. Leipzig.

Bierl, A. 2009 L'uso intertestuale di Alcmane nell finale dell Lisistrata di Aristophane. In F. Perusino (ed.), *Dalla lirica corale alla poesia drammatica: forme e funzioni del canto corale nella tragedia e nella commedia greca*. Pisa, 259–290.

Bierl, A. 2011 Alcman at the end of Aristophanes' Lysistrata: ritual interchorality. In L. Athanassaki & E. Bowie (eds), *Archaic and Classical Choral Song: Performance, Politics and Dissemination*. Berlin–New York, 415–436.

Blass, F. 1870 Zu Alkman, II. Das aegyyptische Fragment des Alkman, *RhM* 25, 177–201.

Boissonade, J.F. 1819 *Herodiani partitiones*. London.

Bowie, E. 2011 Alcman's first Partheneion and the song the Sirens sang. In L. Athanassaki & E. Bowie (eds), *Archaic and Classical Choral Song: Performance, Politics and Dissemination*. Berlin–New York, 33–66.

Calame, C. 1977 *Les choeurs de jeunes filles en Grèce archaïque*. Rome.

Calame, C. 1983 *Alcman*. Lyricorum Graecorum quae exstant 6. Rome.

Calame, C. 1997 *Choruses of Young Women in Ancient Greece: Their Morphology, Religious Role, and Social Function* (trans. D. Collins and J. Orion. Lanham). London.

Carey, C. 2009 Genre, occasion and performance. In F. Budelmann (ed.), *The Cambridge Companion to Greek Lyric Poetry*. Cambridge 21–38.

Cingano, E. 2003 Entre skolion et enkomion: Réflexions sur le "genre" et la performance de la lyrique chorale grecque. In J. Jouanna and J. Leclant (eds), *La poésie grecque antique*, Cahiers de la Villa 'Kerylos', 14. Paris 17–45.

Cleland, L., Davies, G. & Llewellyn-Jones, L. 2007 *Greek and Roman dress from A to Z*. London–New York.

D'Alessio, G. B. 2000 "Tra gli dèi ad Apollo, e tra gli uomini ad Echecrate" P. Louvre E7734+7733 (Pind.fr. dub. 333 S.-M.). In M. Cannatà Fera & S. Grandolini, (eds) *Poesia e religione in Grecia. Studi in onore di G. Aurelio Privitera*, I. Naples, 233–262.

D'Alessio, G. B. 2009 Defining local identities in Greek lyric poetry. In R. Hunter & R. B. Rutherford (eds) *Wandering Poets: Travel and Locality in Greek Poetic Culture*. Cambridge, 137–167.

D'Alessio, G. B. 2013 Parallela Graeco-Latina: ΦΑΡΟΣ (Antimachus, fr. 154 Matthews) and Other Glosses in an Unpublished Lexicographical Excerpt. *CQ* 63.2, 633–50.

Davies, M. and Finglass, P. J. 2014 *Stesichorus. The Poems*. Cambridge Classical Texts and Commentaries 54. Cambridge. [References to the fragments of Stesichorus are referred to as 'Finglass' only]

Davison, J. A. 1968 *From Archilochus to Pindar*. London–New York.

Dover, K. J. 1964 The Poetry of Archilochus. In O. Reverdin (ed.), *Archiloque*, Entretiens Hardt 10. Vandoeuvres–Geneva, 183–212.

Drachmann, A. B. 1903–27 *Scholia Vetera in Pindari Carmina*. 3 vols. Leipzig.

Erbse, H. 1950 *Untersuchungen zu den attizistischen Lexika*. Berlin.

Erbse, H. 1969–88 *Scholia Graeca in Homeri Iliadem*. 7 vols. Berlin.

Ferrari, F. 1991 Tre papyri Pindarici. *RFIC* 119, 385–407.

Ford, A. 2002 *The Origins of Criticism: Literary Culture and Poetic Theory in Classical Greece*. Princeton NJ.

Ford, A. 2006 The Genre of Genres: Paeans and Paian in Early Greek Poetry. *Poetica* 38, 277–295.

Fowler, R. L. 1987 *The Nature of Early Greek Lyric: Three Preliminary Studies*. Toronto.

Gentili, B. 1958 *Anacreonte: Introduzione, testo critico, traduzione, commentario*. Naples.

Gentili, B. 1984 *Poesia e pubblico nella Grecia antica da Omero al V secolo*. Rome–Bari.

Gerber, D. E. 1982 *Pindar's Olympian One: a Commentary*. Phoenix Suppl. 15.

Graziosi, B. & Haubold, J. 2010 *Homer: Iliad VI*. Cambridge.

Gostoli, A. 1990 *Terpander: veterum testimonia, collegit, fragmenta edidit*. Lyricorum Graecorum quae exstant 8. Rome.

Hamilton, R. 1989 Alkman and the Athenian Arkteia, *Hesperia* 58, 449–472.

Hangard, J. 1996 *Scholia in Aristophanis Lysistratam. Scholia in Aristophanem II.4*. Groningen.

Hirschberger, M. 2004 *Gynaikōn katalogos und Megalai ēhoiai: ein Kommentar zu den Fragmenten zweier hesiodeischer Epen, Beiträge zur Altertumskunde*. Bd. 198. Munich.

Hutchinson, G. 2001 *Greek Lyric Poetry. A Selected Commentary on larger pieces*. Oxford.

Kienzle, E. 1936 *Der Lobpreis von Städten und Ländern in der älteren griechischen Dichtung*. Basel.

Kirk, G. S. 1990 *The Iliad: A Commentary*, vol. 1, books 1–4. Cambridge.

Kurke, L. 2007 Visualising the Choral: Epichoric Poetry, Ritual, and Elite Negotiation in Fifth-Century Thebes. In C.W. Kraus, S. Goldhill, H.P. Foley & J. Elsner (eds), *Visualizing the tragic: drama, myth, and ritual in Greek art and literature: essays in honour of Froma Zeitlin*. Oxford, 63–104.

Latte, K. 1953–66 (ed.) *Hesychii Alexandrini lexicon*, 2 vols. Copenhagen.

Lehnus, L. 1984 Pindaro: Il Dafneiforico per Agasicle (Fr. 94b Sn.-M.). *BICS* 31, 61–92.

Lentz, A. 1870 *Grammatici Graeci*, vol. 3.2. Leipzig.

Lolos, A. C. 1981 *Der unbekannte Teil der Ilias-Exegesis des Ioannes Tzetzes (A 97–609)*. Beiträge zur klassischen Philologie 130.

Macleod, C. M. 1983 *Collected Essays*. Oxford.

McNamee, K. 2007 *Annotations in Greek Literary Papyri*. American Studies in Papyrology. Oxford.

Moggi, M. & Osanna, M. 2010 *Pausania. Guida della Grecia. Libro IX. La Beozia*. Milan.

Morris, S. P. 1992 *Daidalos and the origins of Greek Art*. Princeton NJ.

Murray, O. 2008 The Odyssey as Performance Poetry. In M. Revermann & P. Wilson (eds), *Performance, Iconography, Reception. Studies in Honour of Oliver Taplin*. Oxford 161–176.

Nagy, G. 1990 *Pindar's Homer: The Lyric Possession of an Epic Past*. Baltimore MD–London.

Nagy, G. 1999 *The Best of the Achaeans. Concepts of the Hero in Archaic Greek Poetry*. Baltimore MD–London.

Neri, C. & Tosi, R. 2009 Un seminario sul papiro di Alcmane (PMGF 1). *Eikasmos* 20, 11–18.

Nünlist, R. 1998 *Poetologische Bildersprache in der frühgriechischen Dichtung*. Stuttgart.

Palaiokrassa, L. 1983 *To hiero tēs Artemidos Mounichas*. Thessaloniki.

Parker, R. C. T. 2005 *Polytheism and Society at Athens*. Oxford.

Pfeijffer, I. L. 1999 *Three Aeginetan Odes of Pindar: A Commentary on Nemean V, Nemean III, & Pythian VIII*. Mnem. Suppl. 197.

Poltera, O. 2008 *Simonides Lyricus, Testimonia und Fragmente: Einleitung, kritische Ausgabe, Übersetzung und Kommentar*. Basel.

Priestley, J. M. 2007 The φαρος of Alcman's Partheneion 1. *Mnemosyne* 60.2, 175–95.

Privitera, G. A. 1974 *La rete di Afrodite: Studi su Saffo*. Palermo.

Richardson, N. J. 2010 *Three Homeric Hymns: To Apollo, Hermes, and Aphrodite. Hymns 3, 4, and 5*. Cambridge.

Rodighiero, A. 2012 *Generi lirico-corali nella produzione drammatica di Sofocle*. Tübingen.

Sandbach, F. H. 1969 *Plutarch's* Moralia *XV*. Cambridge MA.

Sandbach, F. H. 1972 *Menandri Reliquiae Selectae*. Oxford.

Severyns, A. 1938 *Recherches sur la Chrestomathie de Proclos*. 2 vols. Liège.

Sourvinou-Inwood, C. 1988 *Studies in Girls' Transitions. Aspects of the Arkteia and Age Representation in Attic Iconography*. Athens.

Sourvinou-Inwood, C. 1990 Ancient rites and modern constructs: on the Brauronian Bears, again. *BICS* 37, 1–14.

Spelman, H. 2014 Zeus and the Maidens: Pindar 94b.31-7. *ZPE* 192, 31–3.

Stinton, T. C. W. 1976 Iphigeneia and the Bears at Brauron. *CQ* 26, 11–13.

Swift, L. A. 2010 *The Hidden Chorus: Echoes of Genre in Tragic Lyric*. Oxford.

Theodoridis, C. 1976 (ed.) *Die Fragmente des Grammatikers Philoxenos*. Sammlung griechischer und lateinischer Grammatiker 2. Berlin–New York.

Turner, E. G. 1987 Greek Manuscripts of the Ancient World 2. rev. P. Parsons. *BICS* Suppl. 46.

Valk, M. H. A. L. H. van der 1971–87 *Eustathii archiepiscopi Thessalonicensis commentarii ad Homeri Iliadem pertinentes*. 4 vols. Leiden.

Vogliano, A. 1941 *Una nuova ode della poetessa*. Milan.

Wees, H. van 2005 Trailing Tunices and Sheepskin Coats: Dress and status in early Greece. In L. Cleland, M. Harlow & L. Llewellyn-Jones (eds), *The clothed body in the ancient world*. Oxford, 44–51.

Wendel, C. 1914 *Scholia in Theocritum vetera*. Berlin.

West, M. L. 1982 Metrical Analyses: Timotheos and others. *ZPE* 45, 1–13.

West, M. L. 2007 *Indo-European Poetry and Myth*. Oxford.

Wilamowitz-Moellendorff, E. F. W. U. von 1922 *Pindaros*. Berlin.

Abbreviations

ABV	J.D. Beazley, *Attic Black-figure Vase-painters* (Oxford 1956).
CEG	P.A. Hansen (ed.), *Carmina Epigraphica Graeca*, 2 vols. (Berlin 1983–89).
D-K	H. Diels (ed.), *Die Fragmente der Vorsokratiker*[6] 3 vols., rev. W. Kranz, (Berlin 1951–2).
EGF	M. Davies (ed.), *Epicorum Graecorum Fragmenta* (Göttingen 1988).
FGrHist	F. Jacoby *et al.* (ed.), *Die Fragmente der griechischen Historiker* 15 vols. (Berlin: Leiden 1923–58).
GEF	M.L. West (ed.), *Greek Epic Fragments. From the Seventh to the Fifth Centuries B.C.*, (London and Cambridge, Mass. 2003).
ID	*Inscriptions de Delos*. 7 vols. Paris 1926–1972. Vol. 6 [1], Nos. 1–88, ed. André Plassart (1950); vol. 7 [2], nos. 89–104[33], ed. Jacques Coupry (1972); vol. 1 [3], nos. 290–371, ed. Félix Durrbach (1926); vol. 2 [4], nos. 372–509, ed. Félix Durrbach (1929); vol. 3 [5], nos. 1400–1496, ed. Félix Durrbach and Pierre Roussel (1935); vol. 4 [6], nos. 1497–2219, eds Pierre Roussel and Marcel Launey (1937); vol. 5 [7], nos. 2220–2879, eds Pierre Roussel and Marcel Launey (1937).
IEG	M.L. West (ed.), *Iambi et Elegi ante Alexandrum cantati*[2] (Oxford 1989–92).
IG	*Inscriptiones Graecae* (Berlin 1873-). [Roman numerals indicate the volume; index figures the edition, Arabic numerals the number of the inscription. Thus *IG* II[3] 558 is inscription no. 558 in the third edition of volume II.]
LfgrE	B. Snell, H. Erbse *et al.* (eds), *Lexicon des frühgriechischen Epos.* 24 Bande (Göttingen 1955–2010).
LIMC	*Lexicon Iconographicum Mythologiae Classicae* (8 vols. in 16; Zurich, Munich, and Düsseldorf, 1981–97).
M-W	R. Merkelbach and M.L. West (eds), *Fragmenta Hesiodea* (Oxford 1967).
PCG	R. Kassel and C.F.L. Austin (eds), *Poetae Comici Graeci* (8 vols.; Berlin and New York 1983–1998). 2 (1991); 3.2 (1984); 4 (1983); 5 (1986); 6.2 (1998); 7 (1989); 8 (1995). [Volume as indicated]
PEG	A. Bernabé (ed.), *Poetae Epici Graeci: Testimonia et Fragmenta* (2 parts, part 2 in three fasc.). Leipzig (Pars I) and Munich (Pars II). (1996–2007).
Pfeiffer	R. Pfeiffer (ed.), *Callimachus* (2 vols.; Oxford 1949–53).
PMG	D. L. Page (ed.), *Poetae Melici Graeci* (Oxford 1962).
PMGF	M. Davies (ed.), *Poetarum Melicorum Graecorum Fragmenta* Vol. 1 (Oxford 1991).
SEG	J. J. E. Hondius *et al.* (eds), *Supplementum Epigraphicum Graecum* (58 vols. to date; Leiden 1923–present day).
SH	P. H. J. Lloyd-Jones and P.J. Parsons (eds), *Supplementum Hellenisticum* (Texte und Kommentare 11; Berlin and New York 1983).
SLG	D. L. Page (ed), *Supplementum Lyricis Grecis* (Oxford 1974).

S-M	B. Snell, rev. H. Maehler (eds), *Pindari Carmina cum Fragmentis*[8] 2 vols. (Leipzig 1987–9).
TrGF	*Tragicorum Graecorum Fragmenta*. Vol. 1 *Didascaliae Tragicae, atalogi Tragicorum et Tragoediarum, Testimonia et Fragmenta Tragicorum Minorum* (ed. B. Snell; Göttingen 1971[1], 1986[2]); vol. 2 *Fragmenta Adespota* (edd. R. Kannicht and B. Snell; 1981); vol. 3 *Aeschylus* (ed. S.L. Radt; 1977); vol. 4 *Sophocles* (ed. S.L. Radt; 1985[1], 1999[2]); vol. 5 *Euripides* (ed. R. Kannicht; 2 parts; 2004).
Voigt	E. –M. Voigt (ed.), *Sappho et Alcaeus. Fragmenta* (Amsterdam 1971).

5. (B)orders in Ancient Weaving and Archaic Greek Poetry

*Ellen Harlizius-Klück and Giovanni Fanfani**

In this chapter we offer an investigation of textile terms where they are used for describing intellectual production or knowledge (*epistēmē*) in ancient texts. We focus on archaic Greek poetry as well as on selected philosophical works and experience from weaving experiments.

The choice of the passages and the material to be discussed as well as the way the argument is built up reflect two different but – we believe – complementary approaches to the topic we explore throughout this chapter: one is guided by an interest in ancient mathematics and philosophy and the other draws on literary criticism. Both have a role to play when tracing and contextualizing the (somewhat elusive) technological significance of textile imagery for archaic poetry and prose as well as for philosophy. This study arises from a common interest in the question of how ancient textile production and particularly weaving might have affected the very early discourse on poetry-making and the question of how poetic composition or the composition of texts as a *technē* might resemble the order of nature. We therefore lay emphasis on the connection of art, technology, and nature that is encompassed by the ancient term *technē* and embraces weaving as well as forging, writing a poem or a philosophical dialogue. In all such cases, we see a concept of *technē* at work that refers to a notion of the genesis of cosmic order that has never been explicated. In some instances it is best explained by weaving. Despite the obvious difference in the subjects of our interest, we build our arguments on this shared foundation.

The passages we refer to are commonly perceived as metaphorical. We focus instead on the German philosopher Hans Blumenberg's linking of metaphors and technology. Here, metaphors indicate a technological language prior to the era of proper technology.[1] As the technological aspects of Blumenberg's philosophy and history of science are not very well known even in Germany, we will provide a short introduction.

* We would like to express our gratitude to foundations and institutions supporting the research whose results are presented here. Ellen Harlizius-Klück's investigations were supported by a Marie Curie Research Fellowship (COFUND) of the Gerda Henkel Foundation which she conducted at the Danish National Research Foundation's Centre for Textile Research (DNRF 64). Giovanni Fanfani would like to thank the Danish Council for Independent Research and FP7 Marie Curie Actions – COFUND (DFF – 1321-00158); he is especially grateful to Douglas Cairns for helpful comments on an earlier draft of this chapter.

[1] The term 'proper technology' indicates that most scholars use the term technology as applying to the result of industrialization in the 19th century.

Blumenberg's itinerary from technology to metaphor and back again

In the 1950s, Blumenberg published a series of articles concerning technology, investigating its influence on our image of man (*Bild des Menschen*) and the autonomy or powerlessness (*Ohnmacht*) of human beings facing the intrinsic logic of the world of machines.[2] Blumenberg thus scrutinises the history of science in its changes avoiding to subscribe a history of inventions and inventors or technological problems. To Blumenberg, it is necessary to verify in every case the relation and exchange between idea and matter, or: mind and technology.[3] This project he later called "Geistesgeschichte der Technik",[4] which might be translated as "history of technological ideas" and thus reacts ironically to the usual way of writing histories of technology as well as to new sociological explanations of this history.[5]

Blumenberg claims that an analysis of the backdrop for the origins of technology in terms of the historical development of ideas is lacking.[6] He starts outlining this development by investigating technology or *technē* in ancient Greece. The result has been summarised thus: "For the Greeks, the notion of *techne* summed up those abilities and skills which follow nature on its own course of becoming and perfection, and, in following nature, offer it help. *Techne* promotes and sustains *physis*."[7] But *technē* also figures as the very beginning of things when the godly creator is called *demiourgos* (artificer, craftsman, *Timaeus*, 28a),[8] or even *phytourgos* (begetter, father, lit. gardener; *Republic* 597d) by Plato.[9] *Technē* imitates *physis* since *physis* is made by *technē*.[10]

To Blumenberg, it were the sophists that introduced the idea of *thesis*: the postulation of something not already given.[11] They demonstrate how *technē* can be isolated and independent from nature as being and truth. They offered a training in tricks of speech without respect to justice, and trained

[2] *Cf.* Blumenberg 2015; *cf.* the epilogue of the editors Schmitz and Stiegler 2015, 281–298.

[3] *Cf.* Schmitz and Stiegler 2015, 289.

[4] Blumenberg 2009, published posthumously. The manuscripts in the archive in Marbach are dated to 1966/67. *Cf.* Blumenberg 2015, 278. If not stated otherwise, translations are by the authors of this chapter. In some cases, we give the German original text in the notes.

[5] In fact Blumenberg himself states that "the term 'history of ideas' does not sound good any more." Blumenberg 2009, 51. Taking into account that the manuscripts were finished 1966 after the publication of "The Copernicanic Turn" (*Die kopernikanische Wende*, 1965) and at the same time as "The Legitimacy of the Modern Age" (*Die Legitimität der Neuzeit*, 1966), this might refer to the works of Thomas S. Kuhn who also investigated the Copernicanic revolution and the role of metaphors and paradigms in science (Kuhn 1957, 1962). For Blumenberg on Kuhn *cf.* Blumenberg 1996, 540–541; Blumenberg 1983, 467; Ingram 1993, esp. 23.

[6] *Cf.* Blumenberg 2015, 23.

[7] Campe 2000, 114.

[8] On technological imagery drawn from arts and crafts in the *Timaeus* in the context of a comprehensive investigation of the Greek conception of the *kosmos* as an artefact, see Lloyd 1966, 277–285, who notes that "πλέκειν (plait) and ὑφαίνειν (weave) and their compounds are used to describe the joining together of soul and body (36e2), the interlacing of the veins (77e1) and so on".

[9] *Cf.* Blumenberg 2015, 99.

[10] An excellent treatment of the archaic and early classical conception of *technē*, especially in relation to the emergence of the idea of poem/song "as verbal craftsmanship" in early 5th century choral lyric is Ford 2002, 93–157 (quote from p. 93), with further bibliography. See also Durante 1976, 170–179. On the interesting etymological connection between the root of Greek *tektōn* 'carpenter' and *technē*, and Latin *texō* ('weave, join, carpenter') see Nagy 1996, 74, who concludes that "[T]hese and other such facts lead to the general conclusion that the metaphor of carpentry as songmaking in Indo-European languages in parallel to the metaphor of weaving."

[11] Blumenberg 2015, 97. On the archaic poetic conception of *thesis* in the sense of *poiesis*, *i.e.* of poetry as 'composition' or 'ordered structure of words' see Gentili 1990, 50 and Pindar *O.* 3.8 (ἐπέων θέσις). On *logos* as *technē* in Gorgias in the context of the Sophists' 'materialistic poetics' see Ford 2002, 172–187.

know-how instead of know-what. It is the birth of the *rhetorikē technē* and the framework for the category that is later called – metaphor.[12] From that moment onwards, philosophy takes as its task to separate itself from this type of speech that is not concerned with truth but with persuasion: "Plato's rejection of the sophists' technique had the implication of *banning technology from the intellectual legitimitacy of the European tradition*."[13]

But this ban was not characteristic of antiquity. The final dissection of nature and technology, according to Blumenberg, is a result of the Christian concept of creation and might be exemplified by the case of weaving coloured garments: "A polemic against some textile refinements can be found with some patristic authors reasoning god would have created coloured sheep if he had wished men to wear coloured clothes."[14]

In the early modern, the foundation of pure reason followed this dissection of nature and technology and masked the historical and practical condition of theory. Blumenberg refers here to a most influential example: "Descartes obviously disavowed the historical background that provided him with decisive suggestions for the new idea of science, in order to establish the myth of an absolute beginning through reason assuring itself."[15]

At the famous Jesuit school *La Flèche,* Descartes not only learned mathematics and philosophy but also fencing and tennis. On fencing, he later wrote a geometrical treatise; tennis provided him with metaphors for his theory of light.[16] The value of such body technologies for science and philosophy remains however marginal, because, when they appear on paper they are perceived as rhetorical tropes and then belong to the realm of literary style. According to Blumenberg, Descartes aimed at a terminology where the presence and precision of given facts is captured by defined terms/concepts. Metaphors instead only have a meaning in functional transition; in metaphor, the human mind hurries ahead the execution of reasoning.[17] To Descartes, only arithmetic and geometry provide reliable and evident knowledge.

Above, we quoted from a publication that does not deal with technology at first sight. In 1960, Blumenberg published a volume in the German series *Archive for the history of concepts*[18] entitled "Paradigms of metaphorology" dealing with so-called 'absolute metaphors' as devices for the development of concepts in the philosophical tradition.[19] Absolute metaphors[20] stand in for missing concepts in cases of logical awkwardness (*logische 'Verlegenheit'*)[21] and they resist attempts of finally making concepts out of them. They answer to questions like "What is truth?", "What is man?" "What is God?" "How is the world made?". To Blumenberg such metaphors force us to re-think

[12] Metaphor is meant here in the sense of early rhetorics. To Blumenberg, there cannot be something like a definition of a metaphor, but only historical cases where we can find certain functions of language that are called metaphorical.

[13] "(…) Platons Abweisung der Sophistik [implizierte] die *Ausschließung der Technik aus der geistigen Legitimität der europäischen Tradition*" (Blumenberg 2015, 197, emphasis by Blumenberg).

[14] *Cf.* Blumenberg 2015, 106.

[15] Blumenberg 2015, 235: "*Descartes* hat den geschichtlichen Hintergrund erkennbar verleugnet, aus dem ihm entscheidende anregungen für die neue Wissenschaftsidee zugekommen waren, um den Mythos vom absoluten Anfang durch die sich selcbst vergewissernde Vernunft etablieren zu können". (Emphasis by Blumenberg).

[16] *Cf.* Specht 1996, XVI.

[17] Blumenberg 1960, 7.

[18] Original title *Archiv für Begriffsgeschichte*.

[19] Blumenberg 1960; *cf.* Campe 2000, 105.

[20] Examples for such metaphors are "Macht der Wahrheit" (power of truth), *cf.* Blumenberg 1960, 12–18, "Wahrscheinlichkeit" (probability), *cf.* Blumenberg 1960, 88–105, "Wirklichkeit" (reality), *cf.* Blumenberg 1981. Some of these lose their metaphorical character when translated into English.

[21] Blumenberg 1960, 9.

the relation of imagination and *logos*. They show that imagination is not a reservoir of images that, element by element, dissolve into proper concepts in the course of history. Rather it is a catalystic domain that enriches the world of concepts without its basic stock ever being consumed.[22]

Also metaphorology is a historical project. Absolute metaphors change in the course of history and thus reflect changes in the approach to fundamental questions. Metaphors appear to be a technology to situate such questions. Campe, who develops in detail the connection of technique and metaphor in Blumenberg's work, concludes: "What is at stake in the exploration of metaphors in scientific texts is the impact of 'the technical' (*das Technische*) in science and in the text of philosophy."[23]

Blumenberg states: "Man *is* a technical being" ("Der Mensch *ist* ein technisches Wesen").[24] Until the 1960s, he claims man's natural deficiency in environmental adaption as reason for this statement. Later Blumenberg withdraw from this reasoning and took 'the technical' as a human condition independent from the idea of deficiency. He then stressed the role of 'life-world' (a concept inherited from Husserl's phenomenology) for the history of technology and explored the relation of life-world and technology in several case studies. Life-world here refers to the seemingly self-evident character of the reality at hand, leading to assumptions such as nature being opposed to technology or technology being applied science or, in our case, the easy availability and cheapness – and thus worthlessness – of textiles.[25] As we already heard, for the ancient Greeks the term *technē* being at the root of our word technology, shows no opposition against nature[26] and against the second assumption Blumenberg argues: "Technology can only be applied science because this science already originates in a technical understanding of being and truth."[27] To question the third assumption by objecting to the tradition that treats weaving as mythical or metaphorical subject will be part of our chapter.

Textiles within the framework of technization[28]

The worthlessness of textiles may be illustrated by the way in which Campe introduces a metaphor for Blumenberg's notion of metaphorology.[29] The integration of metaphorology and technology is underlined by a term Blumenberg used in explaining why a theory of metaphor could not result in a proper philosophical theory: *Halbzeug*. This evokes Heidegger's *Zeug* and his philosophy of being at the beginning of the paradigms for a metaphorology: "What I will present here is a mere

[22] Blumenberg 1960, 10: "Der Aufweis absoluter Metaphern müßte uns wohl überhaupt veranlassen, das Verhältnis von Phantasie und Logos neu zu durchdenken, und zwar in dem Sinne, den Bereich der Phantasie nicht nur als Substrat für Transformationen ins Begriffliche zu nehmen — wobei sozusagen Element für Element aufgearbeitet und umgewandelt werden könnte bis zum Aufbrauch des Bildervorrats —; sondern als eine katalysatorische Sphäre, an der sich zwar ständig die Begriffswelt bereichert, aber ohne diesen fundierenden Bestand dabei umzuwandeln und aufzuzehren."

[23] "In der Untersuchung der Metaphern im Text der Wissenschaft geht es um die Wirkung des Technischen in der Wissenschaft und im Text der Philosophie." Campe 2009, 284, translation by the authors. The German article is an extended version of the former English one, published in 2000. *Cf.* Mende 2009, 28, n. 120.

[24] Blumenberg 2015, 18.

[25] See footnote 55 below on the 'Primark-effect'.

[26] *Cf.* Blumenberg 2015, 19.

[27] Blumenberg 2015, 45.

[28] 'Technization' translates the German word *Technisierung* (*cf.* the chapter *Lebenswelt und Technisierung unter Aspekten der Phänomenologie*, in Blumenberg 2015, 163–202; *cf.* also Campe 2000, 109). Where technology (German: *Technik*) refers to the product, technization refers to the process of becoming.

[29] Blumenberg never gave a definition of what a metaphor or a metaphorology is.

half-stuff [*Halbzeug*] and the perfection and seamlessness with which one can speak about 'Being' is in this discursive field, the field of philosophical metaphors, utterly unreachable."[30]

According to Campe, Halbzeug "is a *terminus technicus* used exclusively in industrial technology" referring to things created from raw materials which are later made into finished products: "It renders invisible the raw material out of which it is made without evoking the technical application, which only the finished product will demonstrate."[31]

Campe's argument suggests that this was going on in Blumenberg's mind. However, a footnote surprisingly concedes that half-stuff as well as *Halbzeug* was and is still used in paper manufacture and the origin of the word lies there. In fact the technical encyclopaedias until roughly the turn of the 20th century only know half-stuff as a term in paper production, be it manufactured or industrialised.[32] The term denotes the disintegrated rags, the worthless textiles, from which the paper is later made. It denotes the textile part of the genesis of the paper on which all later philosophy will be written and printed.

Campe's claim thus follows the notion of a decisive cut induced by the invention of industrial machines like the mechanical loom in 1825, which is exactly what Blumenberg rejects in his work. To him, the potential of industrial utilization is a late and secondary trait. "The characteristic of natural being, namely carrying the principle of its design and function in itself, is transposted consequentially to the field of the technical work."[33] In the word 'transposted' we may recognize the 'transfer' of the metaphor.

For Campe's claim, that Blumenberg sees metaphor as a question of technology, it does not matter if we see half-stuff as a textile semi-paper-product belonging to a long history of paper manufacture or as an exclusively industrial term. But it matters for the approach presented here that takes textile terms seriously. Usually, textiles are taken for granted and not considered important for the way we make things today not to mention the way we think about this making. Some further cases might show how this affects our reading of textiles and textile production.

Outsourcing textiles from intellectual history

Pherecydes of Syros (6th century BCE) was one of the first to write down his ideas of the generation of the cosmos.[34] In his account (fr. 68 Schibli = DK 7 B2 = KRS 53), the god Zas (Zeus) weaves fabric representing the cosmos for Chthonie as a *geras*, a matrimonal gift.[35] Pherecydes, though writing his

---

[30] Blumenberg 1960, 29, here quoted after Campe 2000, 106.

[31] Campe 2000, 107.

[32] For example *cf.* Krünitz 1807, 590–595 and 625–626, also online at http://www.kruenitz1.uni-trier.de/ where the encyclopedia is called one of the most important sources on the history of science in the time of change to industrialization. Similarly later Hartmann 1838, columns 975–985 and 1017–1023. Also Karmarsch & Heeren 1843, under the entry "Papierfabrikation", 549–552, and 556–558. The English reference of this work is Ure 1840, where half-stuff is mentioned under "Paper, manufacture of", pp. 926 and 927. For the beginning of industrialized paper-production *cf.* Müller 1862, with several editions of which the ones from 1855, 1862 and 1877 since 2013 are available as Springer ebooks in the category of materials science.

[33] Blumenberg 2015, 29. In the original text: "Die Charakteristik des natürlichen Seins, daß es das Prinzip seiner Gestaltung und seiner Funktion in sich trägt, wird folgerichtig in den Bereich des technischen Werkes transponiert."

[34] For the tradition which makes Pherekydes the first Greek prose author see Schibli 1990, 2–4, who discusses the ancient sources (Theopompos *ap.* D.L. 1.116 and the *Suda*) also in relation to Anaximander's prose book. Previous account of divine geneaologies (*e.g.* Hesiod's *Theogony*) were poetic ones.

[35] We cannot deal at length with this example here. *Cf.* Schibli 1990, Harlizius-Klück 2004, 155–161; Scheid & Svenbro 1996, 63–67. Still today we refer to marriage as establishing a sort of fabric: 'the holy bond of matrimony'.

cosmology in prose, is not counted among the early cosmologists because weaving is taken as mythical content. Where the column drum of Anaximandros[36] might be addressed as a rational idea and the first scientific explanation of the cosmos, the loom of Zeus is taken as a myth or not mentioned at all.[37]

However in Pherecydes, weaving refers to the production of the cosmos and might well provide the technological backdrop of the question of how the cosmos is made and what its inner construction is like. Pherecydes, Anaximandros, the Pythagoreans and the atomists answer the same question, only the technologies are different: weaving, architecture, arithmetic and a behaviour of tiny particles (atoms). In this chapter, we object to the tradition that considers weaving as a mythical or metaphorical answer to the question of the order of cosmic composition.

The Pythagorean answer to the inner structure of the cosmos, namely to call it a harmony of numbers, might illustrate another outsourcing strategy: the distinction between pure and applied science. This can be traced back to a quote from Plato's *Statesman*, where knowledge is divided into disciplines such as arithmetic, which are "free from actions" (*psilai tōn praxeōn*) and "only provide insights" on one hand and, on the other hand, crafts such as carpentry, where the knowledge belongs to the action and together with it creates "bodily things" "that have not been there before" (258e). Unfortunately the young man named Socrates, to whom these words are directed in the dialogue, is not especially proficient in mathematics and meets his limits already when classifying squares according to the commensurability of the diagonal (266a). Therefore the interlocutor has to take a detour in order to exemplify the ideal of political knowledge to the young man: the paradigm of weaving. The statesman has to connect the brave and the tempered in fitting marriages, like the weaver connects threads to form a harmonious whole fabric.

Plato does not miss the chance to downsize the value of weaving. He says that one might as well, if nothing else is at hand, take "a quite minor example" (279a) like weaving, in order to correctly present what is to be recognized. He repeats this precaution doing his analysis of the presented example, saying: "No one in his right mind would ever consider the explanation of weaving for its own sake" (285d).

Descartes too, as already mentioned, is concerned with the distinction of proper reasoning and practical experience. Rule 10 of his famous rules for the direction of the mind states: "In order to acquire discernment we should exercise our intelligence by investigating what others have already discovered, and methodically survey even the most insignificant products of human skill, especially those which display or presuppose order." Descartes prides himself of being born with an *ingenium* that finds evidence by its own means and needs no such detours. But he has to confess: "Still, since not all minds have such natural disposition to puzzle things out by their own exertions, the message of this Rule is that he must not take up the more difficult and arduous issues immediately, but must first tackle the simplest and least exalted arts, and especially those in which order prevails – such as weaving and carpet-making, or the more feminine arts of embroidery, in which threads are interwoven in an infinitely varied pattern. Number-games and any games involving arithmetic, and the like, belong here. It is surprising how much all these activities exercise our minds, provided of course we discovered them from ourselves and not from others."[38]

Scheid and Svenbro cannot be accused of having outsourced weaving from the scholarly discussion in *The craft of Zeus*.[39] On the contrary: their impact on a serious investigation

[36] *Cf.* Kirk, Raven & Schofield 1983, 122.
[37] See for example Algra 2001.
[38] Descartes 1985, 35.
[39] Scheid and Svenbro 1996. The title of the French edition is *Le métier de Zeus* - with the double meaning of *métier*: craft and loom. Thus the title alludes to the fragment of Pherecydes mentioned earlier in this chapter.

of weaving allusions in ancient texts cannot be underestimated. They call Plato's weaving paradigm "the philosophical exegesis of ritual weaving"[40] that is accompanied by or based on a myth expressed in a philosophical context.[41] Myth here means a shared metaphor, a common knowledge, a figure of thought used by an entire civilization "without ever becoming fixed or dead".[42] Such a myth is thus generating stories and images, rituals and exegesis.[43] Scheid and Svenbro conclude that "the fundamental gesture of weaving is this interlacing of the warp and the woof of which Plato spoke in *The Statesman* – an interweaving signifying the union of opposites".[44]

Scheid and Svenbro ground their investigation on "this clear gesture": "To weave is to unite, to interlace, to bind: the act is so straightforward that it requires no explanation." But already the statement that weaving thus is unlike sacrifice in which "to divide is to unite"[45] shows the main problem of such an approach. Was it not Plato who called the shuttle an instrument of separation?[46] And is there not a huge discussion about the translation of *kerkis* as shuttle in this passage, because it is not at all clear to us what using a *kerkis (kerkizein)* means?[47]

We think that Plato actually far exceeds a mythographical exegesis and does not interpret the metaphor of a woven state. Instead, he describes a technological approach that he later addresses to the king as a valuable and necessary knowledge. Plato is interested in the genesis of the textile whereas Scheid and Svenbro concentrate on the product.

The guiding questions of our investigation follow the ideas outlined in this methodological introduction. To what question does the weaving allusion give an answer or provide a context or backdrop? Which are the characteristics of the answer, not in terms of a common knowledge but in terms of the coeval technological knowledge? What happens to the allusion to weaving when we interpret it in terms of the principles of ancient weaving technology? And what is the idea behind intellectual production when it is compared with textile production?

Differences in weaving technology

A consequence of our diagnosis of textiles and weaving being taken for granted is to unsettle this assumption first and provide an overview of the specific technological and structural features of ancient weaving.[48]

We need to avoid discussing ancient sources with modern concepts and terms of textile technology in mind. The construction of a weave, at least since medieval times, differs considerably from that of antiquity. Furthermore, the general understanding of the development of weaving technology leads to the misconception that fabrics made on primitive devices must have been primitive themselves. However, the advantage of industrial looms is mainly an increase of speed and automation, and not necessarily an increase in complexity of the fabric produced.

[40] Scheid & Svenbro 1996, 31.
[41] Scheid & Svenbro 1996, 22.
[42] Scheid & Svenbro 1996, 2.
[43] Scheid & Svenbro 1996, 4.
[44] Scheid & Svenbro 1996, 5.
[45] Scheid & Svenbro 1996, 10.
[46] *Cratylus* 436b.
[47] For a discussion of the passage see Ademollo 2011, 107–117 and Barber 1991, 273–274.
[48] But we not only point to the differences. Still nowadays weaving has properties in common with the ancient technology, especially the dualistic character to which Scheid and Svenbro refer as opposites.

Fig. 5.1 Telemachus and Penelope in front of a warp-weighted loom, drawing from Furtwängler and Reichhold, courtesy Museum for Plaster Casts Munich.

Blümner, in his comprehensive overview of ancient crafts, argued that the weave shown on the loom behind Penelope on the Chiusi skyphos (fig. 5.1)[49] could never have been woven on the depicted warp-weighted loom, which is far too primitive for such an elaborate work. The fabric must therefore have been embroidery.[50] Such an explanation has a long tradition, in which patterned textiles, *poikilia/poikilma*, that are so frequently mentioned in ancient sources or depicted on vases, are conceived as embroidery. This makes them a superficial decoration. A deeper structural analysis is then obsolete and cannot contribute to the meaning of weaving metaphors.

However, in 1948, Wace presented a study that explains why in almost all cases where the verb *poikillō* is used to specify a patterned textile, it must be translated, not by embroidery but weaving.[51] The fact that needle and thread belong to the tools of the *poikiltai* (the persons who produce patterned fabrics)[52] is no argument against weaving that is often done with needles or pointed instruments, especially tapestry.[53]

[49] Skyphos, ca. 440 BCE, Chiusi, Museo Archaeologico Nazionale 1831. Fig. 5.1 shows a drawing from Furtwängler and Reichhold 1921, pl.142.

[50] *Cf.* Blümner 1912, 158.

[51] *Cf.* Wace 1948. See also Barber 1991, 359–365.

[52] *Cf.* Aeschines, *Against Timarchos* 97.

[53] For a comprehensive discussion on this question *cf.* Droß-Krüpe and Paetz gen. Schieck 2015.

Gregory Nagy, when juxtaposing weaving and sewing imagery in archaic Greek poetry, states that "the second of the two is more complex than the first". Therefore, in the language of poets, many and various fabrics of songs, each one already woven, become a single new continuous fabric "by being sewn together".[54] In fact, we will show that weaving in archaic times was far more complex than sewing and much better suited to integrate different types of fabric into a unified whole (see also fig. 5.8).

Weaving terms are usually approached on the basis of the garments we wear, the way we wear them and the way they are made today. Cloth is associated with a textile produced and traded in considerable length, then cut and sewn into shape. Such textiles have a neat outer side and rarely seen inner side. Their cut follows the shape of our body. We enter our clothes by putting head, legs and arms into holes and tubes. We buy our clothes already made to shape and do not see them being constructed. We also do not see the technology behind the production of the fabrics. It is common for modern consumers to see textile production as a simple, lowbrow, undemanding, insipid activity we do not have to do ourselves. The current trend is to buy lots of cheap clothes and discard them quickly.[55] Acknowledging the complexity and effort of cloth production would give us an insight into the unfairness of the prices we usually pay.

In antiquity, garments were produced in almost every household; they were highly valuable and rarely cut and sewn. The most valuable ones were decorated while on the loom, woven into shape, and draped around the body. They had no inner and outer side and therefore designs could often be reversible. This might be the reason why painting and printing on fabrics is uncommon in antiquity, as preferred were those techniques with inverse, complementary or neat patterns on both sides. The shape of the garments does not follow the shape of the human body. It is not conceived as a second skin, and in draping such garments there are no holes or tubes to enter. In ancient times, textile production could be a task for highborn women and queens[56] and even stand symbolically for the fabric of a whole *polis* like in the case of the *peplos* woven and offered to Athena Polias on the occasion of the Great Panathenaia of Athens.[57]

However, the most important difference between modern and ancient weaving is the way in which weaving begins. Fabrics from the warp-weighted loom in use in ancient Greece are made by first weaving a starting border: a band that carries the warp threads for the fabric and remains attached to the fabric. In Roman times, this way of making a warp was called *ordior*, and all words for order stem from this root.

[54] *Cf.* Nagy 1996, 65–66, with special reference to the poetics of *rhapsōidein* in the performance context of the rhapsodic contest. Nagy focuses on the rhapsodic *agōn* at the Panathenaic festival in Athens: there the *rhapsōidoi* recited Homeric epics in a fixed sequence, one taking up the narrative where the previous had left it (see also Nagy 2002, 42–46 with references to epigraphic and scholiographic sources, esp. Σ Pind. *N.* 2.2).

[55] The so-called 'Primark-effect' proves this: where the proportion of household waste sent to landfill in the UK decreased nearly by a quarter within five years up to 2008, in that time textile waste has risen from 7% to 30% of total waste. *Cf.* House of Commons 2010, Vol I, 20; Vol II, Ev154 and n. 8.

[56] Helen is weaving in the *Iliad* (3.125–27), Penelope in the *Odyssey* (2.94–95; 19.136–40). Girls from noble families start the weaving of the *peplos* for Athena at the *Chalkeia* in Athens (Burkert 1990, 40–59).

[57] *Cf.* Scheid & Svenbro 1996, 18–21.

Fig. 5.2 Reconstruction of a weave like the one on the Chiusi skyphos. The starting border shows winged figures and was made by tablet weaving. The reverse side shows the same motif in reverse colour (photo and reconstruction by Ellen Harlizius-Klück).

Description of ancient weaving on a warp-weighted loom

In principle, weaving is the interlacing of two sets of yarn: a tensioned set of threads, set up first, constitutes the warp, while the threads introduced by interlacing constitute the weft. The device that provides the tension is the loom.[58] The two principal possibilities of tensioning the warp used in antiquity are to stretch the threads by weight or to stretch them between two beams. In the latter case the beams might be placed parallel to the ground, which has advantages for implements like heddles. The preferred loom type used in Greece was the warp-weighted loom, the one we see on archaic vase paintings.[59]

[58] *Cf.* Barber 1991, 5.

[59] *Cf.* Skyphos from Chiusi (fig.5.1), Plate from Cyprus, ca. 850–750 BCE, Collection of Antiquities, University of Bonn (Aspris 1996), Corinthian aryballos, ca. 600 BCE, Corinth Museum CP2038, and further examples collected in Hoffmann 1974, 297–307.

Before threads are attached to a loom to make a garment, many decisions have to be made to achieve the desired result: on yarn thickness, on quality and length, but also on details like the twist direction of yarn, the distribution of coloured threads, the number of threads necessary to achieve certain patterns and complete pattern repetitions. The warp-weighted loom is set up with a defining feature that also allows and affords most constructive decisions to be made beforehand: the starting border. This is a band or cord that collects, orders, and distributes the later warp threads, which are fixed in the cord or band as a weft (figs 5.2 and 5.6).

The starting border

The starting border is a band with weft threads prolonged on one side and secured and ordered by chaining them in groups.[60] Techniques for producing the band can vary. Sometimes the threads for the later warp are inserted into cords that later surround the fabric. Many archaeological textile finds have tablet-woven borders. The fact that these borders often shared the structure of the overall weave and were not sewn onto the fabric started the investigation of this technique by textile scholars at the turn of the 20th century.[61] They concluded that the preparation tasks of sorting the wool, spinning threads of the right quality and spin direction, planning the integration of the tablet woven borders, and setting up the warp threads by producing a tablet woven starting border might have taken much more time than the weaving of the fabric itself.

Setting up the loom

The warp-weighted loom has two uprights with an upper horizontal cloth beam and a lower horizontal rod used for shedding. Once the starting border has the length that is required for the width of the later fabric, it will be attached to the upper beam of the loom. The weft threads of the band now serve as warp threads and are distributed and attached to the *kairos* and the *kanōn(es):* the heddle rods.

In a tabby, one set of threads is laid in front of the lower rod (or *kairos*), the other half behind. Attaching loom weights provides the warp tension and leaning the loom against a wall at a slight angle produces a natural shed: the interval where the weft is introduced. A heddle rod to which the other half of the warp threads is tied, provides the countershed. In this form (with one heddle-rod) the loom is prepared for weaving tabby, but additional heddles make it possible to weave twills and satins.

Weaving

The range of possible techniques on the loom is wide, from simple tabby and twill to damask twills and other compound weaves like taqueté, but also tablet weaving, tapestry, flying needle, and brocades. In a tabby weave, the weft threads go over and under every second warp thread and the opposite way in the next row. It is the simplest form of a weave where every other warp thread is either attached to the single heddle rod (*kanōn*) or hanging in front of the bottom rod (*kairos*). Tabby therefore requires the threads to divide into odd and even numbered ones. The differentiation of odd and even, the division into back and front – this dualistic or dyadic principle of distinction is at the heart of the weaving process. To construct the weave, a weft thread is inserted into the shed that is formed by the distance of the *kairos* to the warp threads hanging vertically.

---

[60] For the setup and weaving on a warp-weighted loom *cf.* Barber 1991, 116–118, 129, 271–273, and the rich discussion in Edmunds 2012.

[61] *Cf.* Schinnerer 1895 as one of the first who published on sprang, tablet weaving, and nålebinding in archaeological finds.

Then the vertical threads are lifted towards the weaver by pulling the *kanōn* far enough to form the countershed where the next weft is inserted. Only this technique of alternation can produce a fabric, and the alternation of lifting and not-lifting the warp threads is as constitutive for the fabric as the alternation of thesis and arsis for poetry.

Regular alternation - + - + - + - +, a structure that gives tabby, is the basic but not the only method of generating fabrics. Several variations are possible like twills + + - + + - + + - + + - where the rule for the next row shifts by one thread: - + + - + + - + + - + +.[62] Twill weave has weft threads going over more than one warp thread with a translation of one or more threads in every row. Twill can also be made in several number combinations by changing the direction of translation like in diamond twill.

Weaves can be open, with a considerable and balanced distance of threads in warp and weft direction, or have threads packed closely with only warp or weft threads visible (warp- or weft-faced fabrics). Tapestry is a weft-faced technique where weft threads do not reach over the whole width of the fabric but only as far as the colour is needed for the motif.

Patterning a fabric

Patterns on a fabric can already result from how threads cross, even if these threads do not differ in colour. Damask weaves use the change of two twill weave structures to achieve images that are visible because of different shades and shadows created by the change of structure. Furthermore, carefully planned weave structures can be combined with different colour distributions and intertwine to create a visual pattern that is difficult for laymen to predict and appears like a wonder to see.[63]

Figured motifs on bands can be made by double-face tablet weaving (fig. 5.2). One of the less demanding techniques for inserting pictures in a weave on the loom is supplementary weft patterning[64] which uses an additional coloured weft thread that only shows where the design is meant to be but otherwise runs on the reverse side of the fabric. The fabric then shows complementary pictures on both sides.

Tapestry remains are rare among textile finds in Greece, but the few we have demonstrate a highly developed craft.[65] Literary sources often mention figurative weaves[66] and vase paintings show figured garments (see figs. 5.1, 5.3, 5.4 and 5.7), the depiction of Penelope's web on the *skyphos* from Chiusi is among the most famous. In Homer's works, patterned cloaks called *diplax*[67] are mentioned, a word that is sometimes translated as 'double weave'. Double weave is made out of two warp and weft systems on one loom, producing two fabrics that exchange in order to produce a pattern (see fig. 5.2). It also has then the reverse pattern on the reverse side of the fabric as with supplementary weft patterning and double-faced tablet weave, a reversion that seems to be a favoured style in archaic times.[68]

---

[62] For the hypothesis of a similarity between weaving structures and meter *cf.* Tuch 2006 and Nosch 2014, 94.

[63] *Cf.* Harlizius-Klück 2016 on the meander pattern depicted on a funeral plate by Exekias, figs. 45.5–45.9.

[64] *Cf.* Barber 1991, 139–40.

[65] For example the extraordinary tapestry from the tombs of Vergina, dated to the 4th century BCE, *cf.* Spantidaki 2014, 37–38; or the trousers from Sampula, *cf.* Wagner *et al.* 2009.

[66] Cf. Hom. *Il.* 3.125-127; Ov. *Met.* 6.53-82.

[67] *Il.* 3.126; 10.133; 22.441; *Od.* 19.226; 19.242.

[68] The effect of reverse design that comes naturally with this weaving technique also came up in bilingual vase painting where the black figured and the red figured technology meet on one vase. For these vases the question has been posed how it was possible to design the same motif in reverse on both sides as the artists can never see both sides at once (*cf.*

Fig. 5.3 The Fates from the Krater of Kleitias and Ergotimos, Florence Museo Archeologico Nazionale, 4209 (so-called François-Vase), drawing from Furtwängler and Reichhold 1904, pl. 1, courtesy The Royal Library, Copenhagen.

Composing a fabric

The warp-weighted loom is a very flexible device. The weaves that are produced can be rearranged when in progress: it is possible to detach and rearrange the weights at any time, to change the direction of the warp, to change the heddles and the rhythm of warp-lifts, and to combine different types of weave.[69] Such a combination of different bands within one weave is unusual to us today. However, it may explain a literary epigram where several girls weave hems for a fabric - and these are probably starting borders or vertical continuations on the selvedge. The poem, by Antipater of Sidon (*AP* 6.287), reads:

> Artemis, fairest of virgins, sovereign lady of women
> We three wove (*hyphēnametha*) this border (*pezan*) for thee.
> Bitie wrought (*kame*) the dancing girls

Cohen 2006, 29). This pottery technique is difficult to master and only occurs within a limited time span. In weaving, however, the reverse double depiction is an immediate result of the weaving technology and produced in one go (*cf.* Harlizius-Klück forthcoming).

[69] This was crucial especially for Ellen Harlizius-Klück's reconstruction for the Chiusi fabric (*cf.* fig. 5.2), where the weft threads of the short and narrow starting band are also the warp threads of the figured band, but in respect to the whole weave, they run again in weft direction. *Cf.* Harlizius-Klück 2016.

Fig. 5.4 Part of dancing scene with Theseus and Ariadne holding the ball with the thread in the right hand. Drawing from Furtwängler and Reichhold 1904, pl. 13, courtesy The Royal Library, Copenhagen.

> And the crooked stream of winding Maeander.
> Blonde Antianira devised the decoration (*kosmos*)
> That lies on the left side of the river,
> And Bittion that on the right,
> Measuring a span and a palm.[70]

Principles of ancient weaving

From the description of weaving on a warp-weighted loom we wish to stress three main concepts: 1) the ordered beginning or ordering border, 2) the dyadic structure and 3) the (ordered) composition of elements.

The ruling border or (b)order as beginning

The starting band or border is the most distinct feature of this type of weaving and probably evoked analogies of weaving with plotting and planning.[71] It provides the preliminary operation of ordering a textile by a band that predicts the measure, and often the quality, density, and the numbers of warp threads for the whole weave. Consideration for fitting patterns therefore begins with the starting border. At this stage two main choices must be correct: the distribution of colours and the divisibility of the total number of warp threads. If the latter were a prime number, no pattern repeat would ever fit. Knowledge of divisibility rules and number features such as odd, even, prime, relatively prime and knowledge of least common multiples and greatest common divisors was necessary to weave the elaborate patterns we see in the visual representations.

Dyadic structure

Another concept implied in the starting border is the dyadic structure that not only applies to the difference of warp and weft, but also of odd and even numbered warp threads.[72] This enables an

[70] Text and translation: W.R. Paton 1916 (Loeb).
[71] *Cf.* Nosch 2014, 93 and Guilleux in this volume.
[72] *Cf.* Tegle find in Rosenfeld 1958, 12; also Hoffmann 1974, 153 fig. 69.

easy distribution of the threads on the heddle rods – the *kairos* and *kanōn*. In the case of complex textiles, this pre-order by the band might involve arithmetical classifications that are much more complex.[73]

The dyadic arithmetic that is grounded on just these features of odd and even and handed down to us in Euclid's *Elements*[74] is traced back to the Pythagoreans, who thought that the universe was fitted together (*arariskō*) by harmonies of integers.[75] Becker, in his investigations of Pythagorean arithmetic, repeatedly stressed that the concept of number in ancient Greece resembled a pattern, a structure of elements, a texture, and that this textural aspect was the basis of the Pythagorean idea of the relationship or even kinship of things and numbers.

> What is called 'number' (*arithmos*) here obviously is a 'counted number', i.e. something counted or countable, to be precise: a multitude or texture of discrete (not continuous) sort that may be described by one or more numbers. (…) One should remember the etymology of the word *'arithmos'* that stems from *'ararisko',* 'I join', and indeed in its original sense means 'fugue' ['Gefüge']. A sense for this tangible patterned character, for this structure of the primordial number that is not applicable to anything and everything, but sticks to certain objects, is still alive in Aristotle. How much more with the Pythagoreans and also Plato! Beginning with this tangible, primordial, if you like primitive number concept, the Pythagorean thesis is much more understandable. 'Number' denotes something like a particular arithmetically describable structure inherent in things and constituting their proper nature. However, this 'structure' according to the Pythagoreans is not an arrangement of something else which is a proper reality, but an autonomous framework that carries the things or, so to say, keeps them together from inside.[76]

And Becker goes on saying that "mathematical structures thus only become particularly mathematical ones insofar as this structure of things is prescinded, isolated, and detached from its original context as soon as the observing person is mathematician; a detachment that is explicitly called ἀφαίρεσις, *abstractio*".[77]

Dyadic arithmetic was not part of the common everyday practice of architecture, astronomy, nor mercantile counting. The concrete value of integers is absent, and arguments concerning geometrical magnitudes "are expressed in the language of ratio and proportion, without assuming any numerical values".[78] To distinguish numbers as odd or even and explore the properties that are generated by combining them[79] is not a very demanding task for a mathematician, but finding numbers of threads that satisfy such required conditions is exactly what the weaver does when s/he plans the pattern. Textile craft is hardly ever considered in the works of historians of mathematics[80] and other crafts that require such a type of arithmetic are not at hand. Therefore historians instead argue that dyadic arithmetic was an invention serving as argument in philosophy and providing a logical classification method.[81]

[73] *Cf.* Rosenfeld 1958, 149–171.
[74] Books VII and IX of the *Elements* contain parts of an ancient arithmetic that is called theory of even and odd. *Cf.* Heath 1908; see also Harlizius-Klück 2004, 47–64.
[75] *Cf.* Kirk, Raven & Schofield 1983, 427, 428.
[76] Becker 1959, 9; *cf.* Becker 1963, 136. Translation by Ellen Harlizius-Klück.
[77] *Cf.* Becker 1963, 137, translation Ellen Harlizius-Klück.
[78] Saito 2009, 804.
[79] *Cf.* Heath 1908, *Elements* book 7 and 9.
[80] This is changing since ethnomathematics established itself and the analysis of Peruvian *khipus* showed that, at least in some non-European societies, textile structures are strongly connected to mathematics. See Urton 1997 and Brezine 2009.
[81] See Harlizius-Klück 2004, 54.

Mathematically it is possible and satisfying to attribute such abstract features as odd or even to a "concern with the generality of proof".[82] However, we propose to relate this way of thinking about numbers to a practice that is itself concerned with generality, though not with the generality of mathematics, but rather the practice of setting up a loom for weaving patterns. Some mathematical terms still show that threads and loom parts were connected to practices of measuring and ruling. The word 'line' shows its origin from the 'linen' thread via Greek *linon* and Latin *linum*. *Radius* was the Latin word for the shuttle, later also denoting the compasses. The weaver is constructing figures with *radius* and *regula*, just like the geometrician with compass and straightedge. *Regula*, the ruler or straightedge, *kanōn* in Greek, denotes the heddle-rod that controls the regular repeat of the pattern.

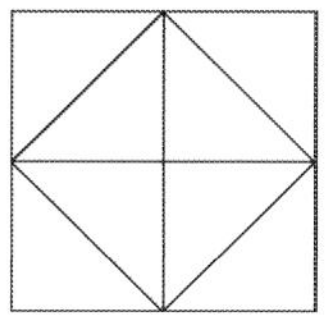

Fig. 5.5 Diagram with doubled square.

Ordered composition of elements

Woven ornaments are not just a question of filling frames, but a question of ordering the elements or atoms of the weave, namely the threads. Every geometric motif has to be transformed into a ratio of thread numbers that are positive integers. The tools for the construction of repeats fitting into the frame provided by the starting border are the least common multiple, the greatest common divisor, and especially all theorems on divisibility that we find in dyadic arithmetic. The so-called Euclidean algorithm is the way in which weavers would test the repeatability of a pattern in a fabric to weave. Because of the structure of weaves, geometrical problems have to be solved by arithmetical consideration, and the separation of proportion theory into a geometrical and an arithmetical one might have its origin here. Woven geometry is, so to speak, a tacit visual algebra.

Figured friezes and doubled squares

When we see complex fabrics like the ones on the François Vase (fig 5.3, 5.4 and 5.7) we now know that the woven parts need not be sewn together but can be integrated on the loom.[83] The krater shows not only plain fabrics and figured friezes as fabric decoration, but also patterns based on a repeating diagram with a square in a square (fig. 5.4).[84] The basic outline of this diagram (fig. 5.5) can be easily explained as a depiction of broken diamond twill, a pattern that is not difficult to weave and the frequency of its occurrence on vase paintings would testify to this.[85]

What makes this diagram interesting for our discussion is not the question how it was made, but the fact that it poses special problems for the practices of counting and measuring and hence for the

[82] Saito 2009, 805.

[83] Barber suggested that figured friezes were made in tapestry technique and this indicates that the whole fabric is tapestry. *Cf.* Barber 1991, 360–65 for a detailed discussion of figured cloth in ancient Greece. Ellen Harlizius-Klück's experience, moreover, is that combination with tablet weaving and other band weaving techniques is possible and often also suitable for figurative weaves.

[84] See the peplos of the dancing girl on the left of fig. 5.4.

[85] Another famous example is the *peplos* of Athena on a bilingual amphora by Andokides, ca. 520–510 BCE, Collection of antiquities, Munich, 2301.

development of ancient mathematics. It is the diagram which Plato used in his famous 'lesson on geometry'[86] and which underlies the important proof of the incommensurability of side and diagonal in a square[87] (a proof which is deemed to be at the beginning of deductive mathematics); the diagram that Plato/Socrates employs when he wants to test his interlocutor's ability of philosophical thinking, for example in the *Statesman* dialogue,[88] well-known for the central role played by the weaving paradigm. Incommensurability in this diagram means that the relation between the length of the diagonal in the square and the length of its side cannot be expressed by integers (whole positive numbers). But if weaving means to construct a diagram from a countable number of elements, namely the threads, why can such a diagram be woven?

The way we have to look at such elaborately depicted patterns is determined by Socrates in the Republic as follows:

> Then, said I, we must use the blazonry of the heavens as patterns (τῇ περὶ τὸν οὐρανὸν ποικιλίᾳ παραδείγμασι χρηστέον) to aid in the study of those realities, just as one would do who chanced upon diagrams drawn with special care and elaboration by Daedalus or some other craftsman or painter. For anyone acquainted with geometry who saw such designs would admit the beauty of the workmanship, but would think it absurd to examine them seriously in the expectation of finding in them the absolute truth with regard to equals or doubles or any other ratio.[89]

We think it is misleading to take this quote at face value. Plato's *Statesman*, if we investigate the dialogue carefully, shows that the argument leading to the determination of the statesman's knowledge builds upon the substitution of mathematics by weaving. In that dialogue Plato (respectively the Stranger as leader of the conversation) divides all sciences (*epistēmai*) into two classes: pure and applied (258c). To him, the arithmetic of odd and even numbers or dyadic arithmetic is pure, and craft is applied science (258d–e). In the dialogue, a nameless stranger from Elea is talking with a young man called Socrates and tries to describe the *epistēmē*, the science, knowledge, or art of a true statesman. It turns out that the statesman has to compose a state by 'royal weaving' (310e) that connects the citizens harmoniously.

It is hardly ever noticed that weaving serves here as a replacement for a pure science that goes beyond the understanding of the young man who is involved in the dialectical conversation: Socrates (not the philosopher, who instead listens to this conversation without interrupting) is asked to classify living beings by comparing the number of their feet (two or four) to the ratio of diagonal and area of a two-foot-square or a four-foot-square (266a-b). The crucial point is to test Young Socrates' knowledge of incommensurable quantities and the way in which they can be related. Doubled squares are the best examples for incommensurable relations: the ratio of side and diagonal cannot be expressed by a ratio of integers (which is incommensurability by definition). The Greeks called this ratio *alogos*, 'unspeakable.'[90] The only possibility to describe it was to give a proportional equation: an *analogia*, which is the Greek mathematical term for such equations of ratios. The unspeakability of incommensurability can only be discussed by analogies and this mathematical knowledge on incommensurable quantities is what the young interlocutor fails to master.

---

86 *Meno* 82a–85b.
87 *Cf.* Heath 1908, Vol. III, Introductory note on Book X, p. 2.
88 *Statesman* 266a–b.
89 *Cf. Republic* 529d–530a (quoted from *Plato in Twelve Volumes*, Vols. 5 and 6 translated by P. Shorey. Cambridge 1969).
90 *Cf.* Fritz 2004, 228–229.

The doubled square is used in the mathematical proof of incommensurability, which was handed down to us in the famous *Elements* of Euclid, written around 350 BCE.[91] It is a proof that is based on the definition of odd and even numbers in dyadic arithmetic, Plato's example of pure science. Hence, the diagram of the doubled square we see so frequently depicted as a fabric pattern is at the heart of the revolution in Greek mathematics,[92] however no diagrams are presented in the ancient texts of Plato and probably also not in Euclid.[93] The first mathematical diagrams we know are from Aristotle.[94]

The diagram depicted on the aforementioned plate from Cyprus showing a patterned fabric on a warp-weighted loom is dated to ca. 850–750 BCE. As woven patterns, the depicted geometrical shapes are constructed by ratios of countable elements (the threads). If we take the date of the plate as *terminus ante quem*,[95] this construction predates the arithmetical and logical reasoning and proofing of the geometrical properties by almost 500 years. For such reasons weaving can truly be addressed as an intellectual activity prefiguring concepts of mathematical order.

Technology into words

Turning into matters of early Greek literature, we set up to trace the ordering concept of the starting border as 'embedded' in key terms of archaic poetics. When looking at the semantics of such terms, their roots in textile technology seem to be reflected by two orders of factors: a) etymological derivation, with cases of popular etymology providing good evidence for the spread of certain conceptual associations, and b) the function they fulfil within the poem as both structural and performative principles.

In the scholarly debate to date, weaving imagery in ancient cosmogonies and the poetic exploitation of textile crafts as metaphors for the representation of man's fate, human cunning intelligence,[96] and the process of song-making in archaic Greek literature, have been studied and interpreted either through a historical-anthropological approach, or with the tools of comparative linguistics. In a reference study on the topic, textile imagery has been traced back to the pervasiveness of the 'myth of weaving' as interlacement of opposites in Greek (and then Roman) culture.[97] With a different focus, it has been considered a linguistic and literary inheritance from the wider Indo-European substrate, where, within a rich repertoire of craftsmanship metaphors, it seems possible to account for a special significance of weaving and spinning imagery to describe poetic composition and the making of fate by the gods.[98] In a number of cases, in fact, a more attentive look at the

[91] According to Heiberg, this proof is older and Euclid is not its author. It therefore does not occur in the English edition of Euclid's text, but only in the introduction to book X by Heath; *cf.* Heath 1908, Vol. III, 2.
[92] *Cf.* Szabó 1978, 94.
[93] The oldest extant diagram stems from a manuscript found at Oxyrhynchus and was dated to 75–125 CE, *cf.* Turner 1987. Today the papyrus fragment is in the Museum of Archaeology and Anthropology, University of Pennsylvania, E2748.
[94] *Cf.* Harlizius-Klück 2004, 66.
[95] *Cf.* Aspris 1996, 4.
[96] *Cf.* Nicole Guilleux's chapter in this volume. See as well Berggren 2008, 1–5 on the pre-Platonic tradition of transformative intelligence or *mētis* that is "fundamental to every *technē*" and connects the female art of weaving to language and the construction of truth. On the weaving of *mētis* as a narrative device in the Odyssey, see Slaktin 1996, 234–238. For the particular associations of *mētis*, weaving and feminine poetics in the *Odyssey* see Clayton 2004, 21–35.
[97] See the methodological remarks in the introductory chapter in Scheid & Svenbro 1996, 1–5 on their use of the concept of 'myth' (rather than 'metaphor') as a "figure of thought" (p. 2).
[98] Durante 1976, 173–179 gives a systematic account of weaving and spinning terminology and imagery for poetic creation in Indo-European languages. More recently see West 2007, 35–38. Onians 1951, 303–351 exhaustively treats Greek evidences for the image of the spinning, weaving and binding of human fate. On Indo-Iranic cosmic weaving and on the origin of the imagery of the thread of life see Andrés-Toledo 2010, 33–58, 65–75.

Fig. 5.6 Starting border spread out and fastened on a warp-weighted loom with the weft loops then distributed alternately in front or behind the shed-bar. The warp-ends behind the shed-bar are attached to a heddle-rod in order to cross the warp and make a counter-shed for weaving (courtesy Museum for Plaster Casts, Munich).

material and technological facts at the root of textile metaphors featuring in archaic Greek literature provides satisfactorily explanations to obscure images,[99] and it can help explaining the semantic stretch which generates figurative, abstract meanings from words originally designating realities

[99] Onians 1951, 303–309 is able to explain the phrase, recurrent in Homer, that the fate of man (and many other kinds of issues), 'lies on the knees of the gods' (θεῶν ἐν γούνασι κεῖται) by looking at the ancient method of spinning where part of the process was done on the knees while sitting; the phrase occurs at *Il.* 17.514, 20.435 and *Od.* 1.267, 400, 16.29.

and implements of weaving.[100] As an instance of this, a number of adverbial expressions built on textile terms seem to point to the relevance of concepts of ancient textile technology in shaping and producing ideas of order and spatial/temporal relations.[101]

Onians' rich discussion of πεῖραρ (plur. πείρατα) – which in LSJ[9] has the primary abstract meaning of 'end', 'limit', and a secondary material meaning of 'tackle', 'rope' – has outlined a major semantic value of the term, which is grounded on the materiality of textile crafts (binding, plaiting, spinning and weaving): the literal and concrete sense of 'bond' – and, depending on the context, 'band', 'knot', 'rope', 'loop', 'boundary', 'edge', 'zone', and even 'weft threads' – clearly emerges from a variety of passages that invite a literal, rather than an abstract or metaphorical, reading.[102] The meaning of 'weft threads' is by Onians attributed to *peirata* in a passage whose literal meaning has troubled interpreters. In Pindar's first *Pythian* (81–82), a gnomic sentence on epinician poetics opens the fifth strophe:

> καιρὸν εἰ φθέγξαιο, πολλῶν πείρατα συντανύσαις
> ἐν βραχεῖ, μείων ἕπεται μῶμος ἀνθρώ-
> πων·
>
> If you should speak according to (the function of) the καῖρος, keeping many weft-threads
> in tension in a short length (of warp) [or (fig.) 'with the texture of your speech tightly-woven'],
> less criticism follows from men.

The occurrence of both *kairos* and *peirata*, and the fact that the latter is the object of the *hapax* συντανύω ("to stretch so as to join together"),[103] which points allusively to the Homeric τανύω (in turn used in a description of weaving in *Il.* 23.761), invites a literal and 'textile-based' interpretation of these lines. This adds a subtler layer to the obvious figurative meaning of the *gnomē* ("if you should speak to the point by combining the strands of many things in brief").[104]

---

[100] In a book-length discussion of syllepsis in Pindar, Gallet 1990 has argued for the original identity of the abstract term καιρός ('due measure', 'critical time', 'opportunity') and the weaving implement καῖρος (which he identifies with the 'chained spacing-cord', according to the definition by Hoffman 1974, a device which keeps separated the even and odd threads of the warp and is located on the lower part of the warp-weighted loom), and applied this insight to a number of passages in epinician odes where καιρός conveys important statements of poetics: it emerges that the abstract term, which derives its conceptual domain from the function performed by the καῖρος, expresses the orderly interlacing of themes within the structure of the Pindaric poem. The semantic stretch from a material, textile related meaning to an abstract one for the term καῖρος/καιρός was first proposed by Onians 1951, 343–348, who identified καῖρος with the opening ('shed') produced by the heddles in the warp.

[101] While expressions such as κατὰ μίτον (lit. 'thread by thread', fig. 'in due order') point both to the subtlety of an argument and to the unbroken sequence of facts in a historical narration (*cf. e.g.* Pherecr. 146.7 and Plb. 3.32.3, on which see Durante 1976, 174), the case of adverbial phrases built on *kairos* (*e.g.* ἐς καιρόν, κατὰ καιρόν, πρός καιρόν 'at the right time', see LSJ[9] s.v. III b.) draws on the term's traditional usage in the sense of 'critical time, opportunity', developing from the spatial sense of 'due measure, proportion' (see Ford 2002, 16–22 on *kairos* as a key concept of archaic poetics and aesthetics, indicating appropriateness).

[102] *Cf.* Onians 1951, 310–17. In a monograph entirely devoted to the term, Bergren 1975, 153 proposes as the concrete meaning of πείρατα 'boundary lines', and 'determinants' as the abstract; for a detailed criticism of Onians' investigation of πεῖραρ see Bergren 1975, 170–180.

[103] This meaning of the verb is suggested by Bergren 1975, 153.

[104] Translation in brackets: Race 1997 (Loeb) modified. Onians 1951, 338–340 discusses briefly the passage. Bergren 1975, 148–162 and Gallet 1990, 103–115 offer rich and thorough discussions of this *gnome* and of the implication of Pindar's statement of poetics within the structure of *Pythian* 1. Interestingly, the weaving connotation of πολλῶν πείρατα συντανύσαις ἐν βραχεῖ had already been note by the scholiast (157b Drachmann): ἡ δὲ μεταφορὰ ἀπὸ τῶν δικτύων ("the metaphor is taken from the nets").

When considering the attestations of the term, we propose to think of *peirar* as a starting border spanned out at the upper beam of a warp-weighted loom with the loops of warp threads integrated and the warp threads distributed alternately in front or behind the shed bar or *kairos* (fig. 5.6).[105] This would in turn help explaining the similar semantic stretch of Latin *ora* as both 'cord' and 'edge'.[106] The starting border can be just such a cord that in the end of the weaving process surrounds the whole weave. In between, using the heddles in order to provide the shed for the weave, the warp ends that are part of the band as weft loops cross over again and again. The two systems of threads spread out from, and are alternately ordered and tensioned by, the starting band. (fig. 5.6).

In the remainder of this chapter we will argue (first) that the structural pattern of borders as an ordering principle invests early Greek views of an ordered *kosmos* surrounded by encircling bands/bonds; (second) that the poetic artefact is similarly conceived of as an ordered *kosmos* of words (though not necessarily in terms of textile imagery: cross-craft metaphors seem to play a role here); (third) that the poem, as mirror image of the physical *kosmos*, is in turn framed by borders, which take the shape of ring-structures; finally, that the semantics of a few key terms of archaic poetics, and their phenomenology and function are grounded in the technological principle of the starting border of the weave.

From cosmology to poetics: earth-encircling bands and the 'kosmos of words'

To investigate the role that weaving technology seems to play in providing a model for cosmological structure we need to recall πεῖραρ/πεῖρας.[107] In particular, we shall refer to occurrences of the term that attest to the significance of the idea of cosmological (and therefore physical) bonds and bands encircling the universe;[108] to the same semantic constellation of πεῖραρ/πεῖρας we would refer in turn the conception of the 'binding' of the gods.[109]

[105] That the meaning of 'starting border' can encompass occurrences, like the passage from the first *Pythian*, when πεῖραρ/πείρατα designate 'weft threads' is a consequence of an important technological principle of the starting border: this, being a border/band, always includes the weft threads.

[106] *Cf.* Onians 1951, 317.

[107] The relationship between the two forms of the term is generally considered a diachronic one, πεῖραρ being the older (mainly Homeric) form.

[108] An instance of such a view emerges in the widely-spread image of Okeanos as a river surrounding the earth (evidences from Homer, Hesiod, the Orphic hymns and later authors are collected in Onians 1951, 315–317 and Kirk, Raven & Schofield 1983, 10–17). See *e.g.* Hom. *Il.* 14.200–201 εἶμι γὰρ ὀψομένη πολυφόρβου πείρατα γαίης / Ὠκεανόν τε, θεῶν γένεσιν, καὶ μητέρα Τηθύν ("For I am going to visit the bonds of the all-nurturing earth, and Oceanus, from whom the gods are sprung, and mother Tethys"), where Okeanos and his wife Tethys "are clearly conceived of not as parts of the earth but as distinct beings encircling it" (Onians 1951, 316). Text and translation of Homer (here modified as to account for the material meaning of πείρατα) used in this chapter are taken from the Loeb edition: for the *Iliad*, translation by A.T. Murray, revised by W.F. Wyatt (1924 *Il.* 1–12, 1925 *Il.* 13–24); for the *Odyssey*, translation by A.T. Murray, revised by G.E. Dimock (1919 *Od.* 1–12, 1919 *Od.* 13–24).

[109] Which Onians 1951, 331 sees as "a literal description of an actual process, their mode of imposing fate upon mortals, a religious belief and not a metaphor": this seems to force the evidence, and we would rather endorse the more balanced view offered by Lloyd 1966, 192–193, who remarks that "[T]he spinning, weaving and binding of the fates is a fiction or a myth in the sense that it is imagery, even if it was anything but fictitious in the sense of false for the ancient Greeks who were convinced, or assumed, that this was indeed the way in which fate worked." Onians proposes for the formulaic phrase (τινι) ὀλέθρου πείρατ' ἐφέπται (usually interpreted as "the ending/consummation of death is hung over someone") a reading of *peirata* as 'bonds' or 'bands' of fate fastened upon someone, just like in the case of the *peirata* that bind Odysseus to the mast in *Od.* 12.51 (here the verb is another compound of ἅπτω, ἀνάπτω). While πεῖραρ/πείρατα is always qualified by a genitive often indicating 'ruin' or 'misery', these very words (ὀλέθρος and ὀιζύς) appear as objects in the formulaic Homeric phrase for the gods' spinning of men's fate (θεοὶ ἐπλέκωσαν): Onians points this out so as to suggest that πεῖραρ itself indicates a product of spinning, and that "the same fortunes which the gods spin for men, they also fasten upon them in the form of a cord or bond" (p. 335).

When looking at different cosmological conceptions such as, among the Presocratic philosophers, Parmenides' and the Atomists' (Leucippus and Democritus),[110] what strikes us is the recurrent image of something encircling and enveloping the cosmos – in a fashion which seems to suggest an analogy with the function of the ordering border in weaving. In the poem of Parmenides (28 B 8.26–31 DK = 298 KRS) we encounter again πεῖραρ, occurring twice in association with δεσμοί ('bonds') in a passage describing the conditions by which ἐόν (lit. 'what is') is unchangeable: it is said to be "immovable in πείρατα of huge bonds" (ἀκίνητον μεγάλων ἐν πείρασι δεσμῶν, 26) and held by "strong Necessity in bonds of the πεῖραρ which encloses it around" (κρατερὴ γὰρ Ἀνάγκη πείρατος ἐν δεσμοῖσιν ἔχει, τό μιν ἀμφὶς ἐέργει, 30–31).[111] The conception of the supreme power of Necessity, common to the Orphics and the Pythagoreans as well, is here associated with the idea of *peirata* binding the *kosmos*. It is thus tempting to establish a further relation, on the ground of textile terminology as a vehicle of cosmological realities, with Plato's image of the spindle of Necessity (Ἀνάγκης ἄτρακτος, *Rep.* 616c).[112]

In two different accounts of the Atomists' conception of the formation of worlds through atomic conglomerates[113] – where the isolated atoms form a vortex, become entangled, and make a spherical structure – it is reported that a membrane (ὑμήν) or, alternatively, a circular χιτών ('cloak') encloses the atomic congregation forming a *kosmos*.

> Λεύκιππος καὶ Δημόκριτος 'χιτῶνα' κύκλῳ καὶ 'ὑμένα'
> περιτείνουσι τῷ κόσμῳ διὰ τῶν ἀγκιστροειδῶν ἀτόμων συμπεπλεγμένον.
>
> Leucippus and Democritus envelop the *kosmos* in a circular *chitōn*
> or membrane, which was formed by the hooked atoms becoming entangled.[114] (Aetius 2.7.2 = 564 KRS)

The idea of *kosmos* as a structure ordered by encircling textiles finds application not only in regards to the complex and harmonious arrangement of elements that constitutes the physical world. Poetry is similarly conceived of as a *kosmos epeōn* ('*kosmos* of words'), an elaborated verbal construction that is the result of highly-developed technical skills and that displays in its inner structure unity, coherence and proportion. A view of poetic 'composition-and-performance' as craft has been positioned within the archaic Greek view of poetic activity as *mimesis*: poetry-making is seen as an imitation of nature, human life and previous poetic models.[115] A brief survey of the relevant

[110] For a recent overview on the question of the relative chronology of Leucippus and Democritus, and their respective contribution to the atomic theory (in both ancient doxographic accounts and modern scholarship) see Graham 2008, 333–352.

[111] Translation Onians 1951, 332 modified.

[112] Here we are drawing on Onians 1951, 332, who does not include the Atomists in his discussion. Interestingly, the πείρατα of Parmenides' 28 B 8.26–31 DK are taken as metaphors by Kirk, Raven & Schofield 1983, 253–254, who translate the term as 'limit' and note that "[T]he notion of *limit* Parmenides is employing here is obscure". Tarán 1965, 115–118 discusses the occurrences of πεῖρας/πεῖρατα in fragment 8 (13–15, 26–27, 30–31, 37–38, 42–43, 49) as being metaphorical rather than referring to a material reality. Mourelatos 1970, 26–29 sees in Parmenides' use of πεῖρας, which he interprets (after Onians) as referring to the ideas of 'bond', 'band', 'bound(s)' usually associated with the Fate and Ἀνάγκη, a peculiar characteristic of "Parmenides' polymorph deity", and he concludes that (p. 28) "[I]t is probably correct to visualize the polymorph deity as engaged in "binding" with every occurrence of the word πεῖρας in the text."

[113] Diogenes Laertius 9.31 = 67 A1 DK = 563 KRS; Aetius 2.7.2 = 564 KRS.

[114] The English translation of the fragments of the Presocratics are adapted from Kirk, Raven & Schofield 1983. It is interesting to note that the final participle συμπεπλεγμένον has a clear textile connotation, deriving from συμπλέκω which means 'to plait, interlace, weave together'.

[115] The 'poetics of *mimēsis*' as reflecting the concrete ways and processes of archaic poetic composition and performance has been theorized by Gentili 1990, 50–60.

passages and their contexts will help ground our discussion of a further aspect conveyed by the term and concept of *kosmos* in both cosmological accounts and archaic poetics.[116]

Given that it is among 5th century BCE thinkers that we find the conception of *kosmos* as an orderly structured world,[117] it is probably no coincidence that Parmenides and Democritus are among our sources for the *kosmos epeōn* phrase.

When the unnamed goddess of Parmenides' poem has completed her discourse on truth (ἐν τῷ σοι παύω πιστὸν λόγον ἠδὲ νόημα / ἀμφὶς ἀληθείης 28 B 8.51 DK),[118] she introduced her subsequent cosmological account as informed by δόξα βροτεῖα ("mortal opinion").

> (…) δόξας δ' ἀπὸ τοῦδε βροτείας
> μάνθανε κόσμον ἐμῶν ἐπέων ἀπατηλὸν ἀκούων.
>
> henceforth learn the beliefs of mortal men, listening to the
> deceitful *kosmos* of my words (28 B 8.51–52 DK)

The analogy between this *kosmos*, that we interpret as referring to the ordered structure of Parmenides' cosmological hexameters,[119] and *kosmos* as the ordering of the physical world, is marked in the poem by the compound term *diakosmos* (l. 60). This word moreover indicates the balanced arrangement of the two fundamental cosmic forms (μορφαί l. 53), 'light' and 'night', from which all other things derive.

The verbal richness and unity of Homer's poetry are referred to as *kosmos epeōn* in a statement by Democritus:

> Ὅμηρος φύσεως λαχὼν θεαζούσης ἐπέων
> κόσμον ἐτεκτήνατο παντοίων
>
> Homer, having been endowed with a nature that was sensitive
> to divine influence, built up a harmonious construction
> (*kosmos*) of words of every kind.[120] (68 B 21 DK = Dio Chrysost.
> *Orat.* 13.1)

The use of the verb τεκταίνομαι ('to build'), a clear reference to the craft of carpenters, suggests that Democritus' pronouncement on the orderly structure of Homeric poetry is in line with a further instance of the *kosmos epeōn*, one in which for the first time the difference between poetry and prose is explicitly stated.

Solon's elegy *Salamis* (fr. 1–3 W.^{2}) opens with a bold statement by the poet in the first person, who addresses his Athenian audience:

---

[116] In surveying the few extant occurrences of *kosmos epeōn*, it will be apparent that the phrase aligns itself with formulations of archaic poetics that describe and conceptualize song in terms of craftsmanship; in this regard, no special association with weaving imagery emerges in the texts: in fact, it is the transferring of the idea of an ordered *kosmos* to the song/poem that interests us here.

[117] See Empedocles B 134 DK, Anaxagoras B 8 DK, Diogenes of Apollonia B 2 DK. *Cf.* Ford 2002, 169 n. 35, who hints at the possible Pythagorean origin of such an idea.

[118] "Here I end my trustworthy discourse and thought concerning truth".

[119] For Parmenides' *kosmos epeōn* being 'deceitful'– *qua* a product of mortal opinion – see Kirk, Raven and Schofield 1983, 254; *cf.* also Mourelatos 1970, 226–227. Coxon 1986, 218 observes that "[T]he word (*sc.* κόσμος) is chosen for its aptness in relation to the διάκοσμος or 'system' which the 'composition' is to expound".

[120] English translation: Brancacci 2007, 201, who places Democritus' conception of the κόσμος ἐπέων in dialogue with Homer, Pindar, Parmenides and Solon (p. 203). Ford 2002, 169–170 has a fine discussion of this fragment, which he positions within what he calls Democritus' 'atomic poetics', arguing that the reference to *kosmos epeōn* points here to the ability of Homer as "word-constructor, not poem-maker" (170); this interpretation works well in the context of Democritus' interests in matters of language, such as the construction of words through the combinations of syllables, or the explanation of Homeric 'glosses' (68 B 20a DK). However, for the present purposes we see Democritus' *kosmos epeōn* in continuity with other instances of the phrase indicating the ordered structure of a poem.

Αὐτὸς κῆρυξ ἦλθον ἀφ' ἱμερτῆς Σαλαμῖνος
 κόσμον ἐπέων ᾠδὴν ἀντ' ἀγορῆς θέμενος.

I have come in person as a herald from lovely Salamis,
composing a song, a *kosmos epeōn*, instead of a speech.[121]

It has been suggested that with κόσμος ἐπέων Solon emphasizes the ordered and regular sequence of metrical units that characterizes the elegiac metre (as well as other stichic recited verses like hexameters).[122] Because it points to facts of poetic structure in terms of regularity of orderly elements and metrical patterns, this explanation of the poetic '*kosmos* (made) of words' has interesting implications for the argument that we are trying to build in this chapter. By pursuing the model of weaving in early accounts of the structure of both physical and poetic *kosmos*, we look for concepts and patterns of ancient textile technology that are reflected at the level of structural features of achaic Greek poetry (arrangements of thematic sections, metrical patterns, performative sequences).

A fragment by Pindar (194 S–M), typically featuring cross-craft imagery, offers an interesting variation on the *kosmos epeōn* phrase;[123] the poet describes the metapoetic erection of a 'building of song':

κεκρότηται χρυσέα κρηπὶς ἱεραῖσιν ἀοιδαῖς·
εἶα τειχίζωμεν ἤδη ποικίλον
 κόσμον αὐδάεντα λόγων

A golden foundation has been wrought for holy songs.
Come, let us now construct a variegated (*poikilon*)
kosmos of words (*logōn*) that is endowed with speech.[124]

We have here a complex texture of images evoking architecture (τειχίζωμεν, κρηπίς), the art of smith (κεκρότηται), sound (αὐδάεντα) and the technique designated by the adjective *poikilos*, which is often applied to the textile craft in Pindar.[125] The condition for composing a *kosmos*

[121] The English translation of Solon is the most recent Loeb, by D.E. Gerber (1999).

[122] Noussia 2001, 229–231 and Noussia-Fantuzzi 2010, 211–212, who provides a rich commentary to these lines and discusses at length the features of Solon's *kosmos epeōn* which she understands as an "ordered sequence of words" (p. 212). Noussia points out that already in Homer we find *kosmos* in relation to the ordered structure of themes in a poetic performance: the adverbial phrase λίην κατὰ κόσμον "in exceedingly good order" (transl. Garvie 1994 *ad l.*) is referred to Demodocus' song (and its complex arrangement of narratives) in *Odyssey* 8.489–491. Demodocus' proto-kitharodic performance presents many interesting structural features: see the rich discussion in Nagy 2009, 313–349. In a chapter entitled 'A *kosmos* of words', Lewis 2006, 60–73 links κόσμος ἐπέων in Sol. 1–3 W.[2] to the ring structure of fr. 4 W.[2] in terms of patterns of thought.

[123] A possible reason for the absence of *epeōn* here could be the fact that Pindar's song is not made of ἔπεα (*i.e.* stichic verses), but rather of lyric unities and patterns.

[124] The English translation: Race 1997 (Loeb) adapted. In particular, in the interpretation of the last line we diverge from Race as for the syntactical construction, and take the adjective αὐδήεις as qualifying the kosmos of words as endowed with speech (for this meaning of the word see the Homeric passages collected in LSJ[9] *s.v.*). We thank Douglas Cairns for suggesting this interpretation of the line to us. For a metaperformative reading of this fragment, as a self'referential utterance on the part of the chorus to their own performance, see Power 2011, 111: "[N]ot only the song, but the chorus and its performance itself should be understood as the κόσμος 'adornment' that is in the making before our eyes as well as ears. In other words, the chorus is notionally constructing *itself* through its very performance as an intricately crafted thing of beauty, a ποικίλος κόσμος, that yet has voice, an αὐδά".

[125] See *e.g.* another fragment, 179 S–M: ὑφαίνω δ' Ἀμυθαονίδαισιν ποικίλον / ἄνδημα ("I am weaving an elaborate (pattern woven) headband for the sons of Amythaon"), where ὑφαίνω seems in fact to replace πλέκω ('plait'), another favourite textile craft in Pindar's repertoire of metaliterary metaphors (see Gallet 1990, 77–82): we would like to thank Douglas Cairns for pointing this out to us. Nagy 2009, 554–555 argues for systematic references to pattern weaving in the case of the series ποικίλος/ποικίλλειν/ποικίλμα. On *poikilos* see the remarks in Barber 1991, 359, and Wagner-Hasel 2000, 146–147, 162, 376.

rests on the construction of the songs' foundations, much like in the famous opening of *Olympian* 6:[126] the beginning as a starting front of the poem bears a special relevance to archaic Greek poetics, in which it is possible to outline the traits of a 'proemial function' investing both the performative realization and the compositional structure of the poem/song.[127] In this pattern of an ordering proem in poetry we recognize a declension of the weaving paradigm: at the level of ancient weaving technology, the beginning of the weave is the structural pattern ordering the whole of the fabric-to-be.

Poem as mirror-image of the *kosmos*: borders and ring-composition

Before attempting to explore further the textile technological matrix of some structural patterns of archaic Greek poetry and their functioning in context, we shall return for a moment to the correspondence of world and poem to which the concept of *kosmos* seems to point. Early hexametric poetry in particular offers an excellent opportunity for investigating how formal patterns in the arrangement of the epic material are thought to reflect the cosmic order.[128] The poem discovers and reveals the *kosmos*, and the structure of the *kosmos* as it is described in the poem corresponds to the articulation of structural patterns framing and shaping the poem.[129] A remarkable instance of this set of correspondences can be appreciated in the description of the river Okeanos in the *Shield of Heracles*, a poem which was ascribed to the Hesiodic corpus by ancient editors but was most probably not composed by the poet. Just as the river-god was thought to circumscribe the earth as an encircling bond – and as, in a similar fashion, the streams of Meander, woven in the starting border, surround the fabric in Antipater's epigram[130] – so the physical and civic *kosmos* depicted on Heracles' shield in the poem is held together by the streams of Okeanos (*Aspis* 314–15):

> ἀμφὶ δ' ἴτυν ῥέεν Ὠκεανὸς πλήθοντι ἐοικώς,
> πᾶν δὲ συνεῖχε σάκος πολυδαίδαλον·

[126] "Let us set up golden columns to support the strong-walled porch of our abode and construct, as it were, a splendid palace; for when a work is begun (ἀρχομένου δ' ἔργου), it is necessary to make its front shine from afar" (ll. 1–4). On this opening, and in particular on the genitive absolute ἀρχομένου δ' ἔργου see Hutchinson 2001, 377: "[T]he actual link is not with the order in which the builder would work but with the front as the beginning when one 'reads' or approaches the house". *Cf.* Ford 2002, 124–125, who discusses both the incipit of Olympian 6 and fr. 194 S–M. *Cf.* Nagy 1990, 145 n. 45 on the semantics of *kosmos*; Nagy interprets Pindar fr. 194 S–M as depicting a chorus of Thebans that "is represented as if they were rebuilding the walls of Thebes, in that they are metaphorically 'building the walls' (τειχίζωμεν) of the kosmos 'arrangement' of the words of their song (lines 2–3)" (p. 145).

[127] See Aloni 1992 for a study of the 'proemial function' in archaic Greek poetry. Aloni notes that the main function of the archaic *prooimion* is to contextualize the song, making it suitable to the (cultic, sympotic, agonistic) occasion in which it was performed.

[128] The following discussion builds on the important insights that Thalmann 1983 provided into the function of formal features of early hexametric poetry: the use of parataxis and the juxtaposition through polarity and analogy, typical features of the archaic style, are reinforced at the level of structural features through the use of patterns of ring composition, *hysteron proteron*, chiasmus, the use of framings as means for the 'composer-in-performance' to organize his thoughts and control his material as he creates and expand it through patterns. Such formal devices serve to help the audience to follow the poem throughout the performance; this is made possible by the fact that the structural patterns of hexametric poetry are not just artistic convention: rather, "they are particular manifestations of an entire way of thinking" that the poet shared with his audience (Thalmann 1983, 31). Thalmann provides structural analyses of single passages and whole poems (*Iliad, Odyssey, Theogony, Works and Days,* the *Homeric Hymns*, the *Shield of Heracles*) as built around recognizable patterns: these serve as the vehicle for important thematic concepts and motifs.

[129] See Thalmann 1983, 32: "the sense of form exemplified by the shape of the physical world as the poetry depicts it is the same as that displayed by individual passages and indeed whole poems. There is a perfect correspondence between form and content".

[130] *AP* 6.287.

> Around the rim ran Ocean, and it looked as though it were in
> full flood. It held together the whole richly worked shield.[131]

As it has been pointed out, the verb συνεῖχε ("held together") emphasizes the fact that Okeanos provides a border to the scenes represented on the shield, and it alludes to the framing function (through ring composition) performed by the description of the river within the structure of the poem as a whole. This double layer of borders suggests that "there is a correspondence between the poetic pattern and the physical object described, for the whole account of the shield is enclosed by ring-composition".[132] Here we would like to propose that the borders framing both the physical *kosmos* and the poetic 'composition-and-performance' (and, we shall argue, also geometric motifs in pottery) are part of a broader pattern of thought which we see reflected (if not rooted) in ancient weaving technology.[133]

From 'bond' to 'thread' through 'rope' in archaic Greek poetics

The serpent-shaped Okeanos is represented by early sources as a bond or band encircling the earth, with the phrase πείρατα γαίης occurring twice in Homer in reference to the river-god and his consort Tethys.[134] As it has emerged for the semantic spectrum of πεῖραρ/πείρατα,[135] the notions of 'cord', 'bond', 'band', 'rope' and 'thread', have material, functional and conceptual affinities: like threads, ropes and cords wind up, unroll and intertwine.[136] In addition to designating cosmological realities through the images of the earth-encircling bands and the binding of man's fate, this semantic field bears importance as source material for some key terms of archaic poetics.[137] The meaning of οἶμος and οἴμη – the first commonly assumed to mean 'pathway' (but *cf.* Hom. *Il.* 11.24 where οἶμοι is 'strips'), the second 'heroic song'– can be traced back to a common etymology and to

[131] The English translation is taken from the most recent Loeb edition, G. Most 2007.

[132] Thalmann 1983, 10, who comments on the pattern of thought that gave rise to both the cosmological view of Okeanos encircling the earth, and the invention and use of ring composition as a means of creating functional borders for the poem (or sections of it): "[S]urely the idea that the earth was surrounded by the stream of Ocean, ever turning back upon itself in a wide ring, sprang from the same source as the impulse to round off a poetic passage by recurring to the beginning at the end – a feeling for satisfying form". A further picture of the *kosmos* on a shield is notoriously the narrative of the Shield of Achilles in *Iliad* 18, where we similarly have the river's stream circumscribing and delimiting the earth and the narrative itself: see Nagy 2009, 269–270. Slater 1983 makes a different and strong case for the structure of 'lyric narrative' (in both epics and choral lyric) as complex forms of folk narrative that were then embedded into epic and lyric genres, and responded to functional and structural principles of verbal narration.

[133] In general, the methodology of Thalmann's book is to test the relevance of structural patterns from smaller to larger scale in hexametric poetry. The opposite move, from appreciation of structural patterns functioning for the whole poem to the testing of those same patterns (often, again, ring patterns) in smaller structures of the text, down to groups of two verses, or even a single verse, has been successfully applied to the analysis of choral lyric composition. Cairns 1997 has shown in a paper on form and meaning in Bacchylides 5 the union of tripartition and bipolarity in the poem; the functioning of ring patterns on both large and small scale throughout the poem contributes as well to an appreciation of its meaning. Curiously, Bacchylides 5 is indeed the ode where we have the metaphor of 'weaving a song' (ὑφάνας ὕμνον 9–10; see also ode 19.8 for the same association), which might suggest the poet's awareness of his analogical model.

[134] *Il.* 14.200–201 = 14.301–302.

[135] Onians 1951, 310–342.

[136] See Durante 1976, 175–176 for an exhaustive discussion of evidences in Indo-European languages.

[137] To the repertoire discussed here could be added the term *humēn* (ὑμήν 'membrane'), that we have encountered in Democritus' account of atomic conglomerates, and whose etymology has been proposed as a possible explanation for the key-concept of archaic poetics *hymnos* (ὕμνος).

a constellation of notions that encompasses 'trail', 'bond', 'rope' and, through a connection that emerges at the level of poetic usage, the notion of 'thread'.[138]

This interpretation of οἴμη as related to the conceptual and technological domain of textile craft seems in turn to offer a good semantic and functional explanation for the notion of the compound term *prooimion* (προοίμιον 'that which comes before the *oimē*'), a fundamental concept in archaic Greek poetics and a structural and cross-generic component of poetic performance whose features and phenomenology invite comparison with the crucial starting border of weaving.

The semantics of another key term of early epic poetics, *rhapsōidos* ('he who sews/stitches together the song'), is similarly grounded in the materiality of the craft of *rhaptein*, which encompassed a broader range of operations than just sewing with a thread: it involved the use of leather laces, sinews, reeds to interlace wickers or repair shoes.[139] Drawing on the notions of ancient weaving technology outlined in the first part of this chapter, we suggest the possibility that *rhaptō* (and thus the poetic *rhapsōdein*) might point to the technology of pattern-connection within the (material and poetic) fabric rather than indicating the tailoring of different pieces of woven fabric to produce a whole. The relevant technological idea at work here seems to be the condition of orderly structure and continuity of the threads/laces/ropes that enables such operations.

A further instance of the contiguity, in textile technology as in the early Greek mind, of the notions and crafts of weaving and sewing is offered by the term *hymnos* (ὕμνος 'song in praise of the gods, hymn'). The etymology of the word is dubious,[140] a couple of stems related to the idea of weaving (making ὕμνος derive from the same verbal root ὑφ- from which ὑφαίνω originates)[141] and sewing (from the verbal root **syuH* that produces ὑμήν 'membrane')[142] respectively have been proposed. While the latter hypothesis is not ruled out by etymologists,[143] there seems to be phonological problems related to the first reconstruction (*hymnos* linked to *hyphainō*).[144] This leaves us with facts of poetic

[138] Pagliaro 1952, 34–40 is still a fundamental contribution to the history of the two words and their semantic development from the PIE root **sēi-*, whose meaning can be traced back to the idea of 'binding' and which seems to produce both οἴμη ('bond, trail') and οἶμος ('strip' as in *Il.* 11.24, and hence 'pathway'); Durante 1976, 176–179 adds further evidences to Pagliaro's insights (among which it is worth mentioning here the reconstructed root *ἵμη to be found in ἱμαῖον (μέλος) 'the song of those who draw water with a rope' in Call. *Hec.* 66) and proposes for οἴμη the meaning of 'knotting','plot' ("intreccio"). Interestingly, a comparable semantic extension from the concrete ('bond', 'rope', and – through 'trail' – 'thread') to the abstract has been postulated for πεῖραρ/πείρατα and for οἶμος/οἴμη ('end', 'limit' in the case of *peirata*; 'pathway (of song)' in the case of *oimos/oimē*). For a recent discussion on *oimos/oimē* which draws on Pagliaro and Durante, see Nagy 2009, 230–232. For a different view on both the etymology and semantics of *oimos/oimē* in the context of a reassessment of the function and history of the term *prooimion* see Maslov 2012, 198–203: the interesting insights provided and the questions raised by this article would deserve a more detailed discussion which is not within the scope of this chapter.

[139] See Durante 1976, 178–179, who proposes to interpret the semantics of *rhapsōideín* as complementary to the notion of οἴμη and indicating "the operation that produces the condition of continuity of epic discourse", namely "assembling the poetic product by arranging the threads of narration". Drawing on the historian Philochorus (*FGrHist* 328 F 212), who interprets the origin of the name *rhapsōidoi* as a reference to their technique of composition, Gentili 1990, 7 discusses the semantics and meaning of the term *rhapsōidos* tracing it back to the materiality of textile technology: "[I]t is obvious that *rháptein* is a concrete metaphor for the process of composition, describing the operation by which the strands or web of discourse are woven together".

[140] See Chantraine 1968–1980, 1156 ("L'étymologie en est obscure") and Beekes 2010, 1531 ("No certain etymology").

[141] Through the sequence ὑφ-μνος › ὕμ-μνος › ὕμνος.

[142] See Beekes 2010, 1531.

[143] Both Chantraine and Beekes present this etymological derivation as possible, and point to cases like λιμήν – λίμνη and ποιμήν – ποίμνη.

[144] The most exhaustive discussion is again Durante 1976, 159–162: he discards both the etymologies, for phonological (ὕμνος from ὑφ-μνος) and semantic (ὕμνος linked to ὑμήν) reasons.

usage, which should be regarded as no less significant than etymology for our purpose of tracing and understanding the role of textile craft in shaping ideas of song-making. Bacchylides hints twice at the 'popular' etymology which sees *hymnos* as connected to *hyphainō*: ὑφάνας ὕμνον 'having woven a song of praise (an epinician ode)' in 5.9–10, while in 19.8 the *figura etymologica* is alluded to by the juxtaposing of the two words, which are syntactically separated by punctuation (ὕμνοισιν· ὕφαινε). Whatever the etymology of the term, when we look back at what is possibly the first attestation of ὕμνος in Greek literature, namely the *iunctura* ἀοιδῆς ὕμνος '*hymnos* of song' (*Od.* 8.429, referred to the proto-citharodic performance by the bard Demodocos), the 'materiality' conveyed by the term seems to us to emerge strongly and to suggest the meaning of 'fabric (made) of song'.[145]

The textile imagery that pervades the semantics and usage of terms like *hymnos*, *rhapsōidos* and *prooimion* is especially explicit in passages where these words occur in association with each other, in contexts where the poetics and performance practice of epic recitation is at stake. These texts will ground our discussion of the ways in which textile imagery can reflect weaving technology in the next and final section of this chapter.

In the opening lines of his second *Nemean* ode, Pindar describes the performance sequence of the rhapsodic recitation with an emphasis on its beginning, characterized by the introductory function of the *prooimion*:

> Ὅθεν περ καὶ Ὁμερίδαι
> ῥαπτῶν ἐπέων τὰ πόλλ᾽ ἀοιδοὶ
> ἄρχονται, Διὸς ἐκ προοιμίου (...)
>
> just from the point where the Homeridai,
> singers of verses sewn-together, most often
> begin, from the *prooimion* of Zeus (...)

The passage is a precious source for our knowledge of rhapsodic practice, and it features an array of technical terms for epic performance. The '*prooimion* of Zeus' seems to refer to the genre of rhapsodic *prooimion*, the hexametric poem that introduced the recitation of *epos* by the epic singer and opened the sequence of heroic lays performed by competing rhapsodes in poetic contests. The corpus of the preserved *Homeric Hymns* (which we find referred to as *prooimia* in ancient sources) attest to this very tradition.[146] The genitive Διός is both a generic and a functional marker: together with the verb ἄρχονται (v. 3) it situates the *prooimion* within a literary tradition of formulaic openings,[147] and at the same time anchors the introductory poem to the occasion of the performance (a festival in honour of a particular god); the formulaic phrase σεῦ δ᾽ ἐγὼ ἀρξάμενος "having started from you" is addressed to the god in the closing line of several *Homeric Hymns* (5, 9, 18, 31), where it effects the transition from the *prooimion* to the *oimē* (*i.e.* the epic recitation) through a further formulation, μεταβήσομαι ἐς ἄλλον ὕμνον ("I will switch to another *hymnos*/to the rest of the *hymnos*").[148]

[145] See Garvie 1994, 322 ("a very odd phrase"). Interestingly, ὕμνος is *varia lectio* for οἶμος in *h.Herm.* 451 ἀγλαὸς οἶμος ἀοιδῆς. On ἀοιδῆς ὕμνος see Càssola 1975, x and Nagy 2002, 70.

[146] See Aloni 1980 for a study of the history of the term in reference to the *Homeric Hymns*, their structure and performative function. The 5th century usage of *prooimion* and *phroimion* (its Attic form) in drama and prose is investigated by Maslow 2012.

[147] ἄρχομαι + the genitive of the god appears in *incipit* in *e.g. h.Hom.* 25 and Hesiod's *Theogony* (in Alcman *PMG* 29 we find ἐκ Διὸς ἀρχομένα); in a number of the shorter *Homeric Hymns* (11, 13, 16, 22, 26, 28, 31) and in *h.Cer.* ἄρχομαι introduce the infinitive ἀείδειν ('to sing') + the accusative of the god celebrated. On the different scope of the two types of opening see Càssola, who considers the shorter *Homeric Hymns* as extracts from longer *prooimia*.

[148] The transitional formula μεταβήσομαι ἐς ἄλλον ὕμνον could be considered to belong in the domain of motion metaphor for poetic creation – a category in which *oimē-oimos*, when interpreted as instance of road-imagery ('pathway') may as well fall: see Thalmann 1984, 124.

Pindar's periphrastic definition of *rhapsōidoi* (ῥαπτῶν ἐπέων ἀοιδοί) became the object of a vast array of interpretations in ancient time as regards the nature of the 'sewing together' of Homeric *epē* on the part of the *Homeridai* (and in general the performers of epic poetry).[149] The scholia associate rhapsodic poetics with the sewing in a sequence (εἱρμῷ καὶ ῥαφῇ) of scattered parts of Homeric poetry (Σ Pind. *N.* 2.1d), or with the practice of reciting parts of the Homeric poems in an ordered sequence, so as to perform the whole poem (Σ Pind. *N.* 2.1d). The scholiast then goes on to offer a poetic fragment, attributed to Hesiod by the Attidographer Philochorus (Hes. fr. 357 M.-W. = 297 Most):

> ἐν Δήλῳ τότε πρῶτον ἐγὼ καὶ Ὅμηρος ἀοιδοὶ
> μέλπομεν, ἐν νεαροῖς ὕμνοις ῥάψαντες ἀοιδήν,
> Φοῖβον Ἀπόλλωνα χρυσάορον, ὅν τέκε Λητώ
>
> In Delos then for the first time Homer and I, singers,
> sang, sewing our song in new *hymnoi*,
> of Phebo Apollo with his golden sword, whom Leto bore.

Here again, the close association between the craft of poetic *rhaptein* and *hymnoi* suggests taking ὕμνοι in the sense of 'fabrics' of song(s), or even 'pattern connexions (of verses)',[150] the poetic product that Homer and Hesiod compose ῥάψαντες ('sewing/weaving together the song').

Starting borders in archaic Greek poetry: *hymnos* and *prooimion* in context

In what follows we propose to see the pattern of the ordering beginning as transferred from weaving to poetic composition and performance, and we explore how this pattern works in context. This in turn poses the question whether, in some specific cases, we are dealing with poetic metaphors, or textile terms are rather employed as technical terms.

The most consistent and sustained theory of a metaphorics of textile crafts (weaving, threading, sewing/stitching) that invests the poetics of early Greek epic has been elaborated upon and developed by Gregory Nagy throughout a number of scholarly works:[151] his is the first attempt to see the concept of the starting border in weaving as reflected (metaphorically, in his view) in the function and phenomenology of *hymnos* and *prooimion* in the case where the two terms are coextensive, *i.e.* in the rhapsodic performance, where the introductory *hymnos/prooimion* "refers to a notionally perfect beginning".[152] The character of 'right' or 'absolute' beginning of the hymnic performance as we have it in the *Homeric Hymns* is effected by two markers: a) the use of the verb ἄρχεσθαι, "which signals the invocation of the god or goddess who presides over the occasion of performance", and b) the transition to the rest of the performance, which the transitional formula ('metabasis' in Nagy's words) μεταβήσομαι ἄλλον ἐς ὕμνον marks metonymically as a *hymnos* ("the *hymnos* can make reference to its own hymnic consequent").[153] The *hymnos* is thus a performative continuum which encompasses the *prooimion* as its start, and whose wholeness is marked by its beginnig: "the *hymnos* is not just a proem that introduces epic but also the sequencing principle that connects with epic, then extends into epic, and then finally becomes the same thing as epic itself".[154]

[149] A detailed discussion of the ancient scholarly interpretation of Pindar N.2.2 is offered by Nagy 1996, 66–69.
[150] Thus drawing on the etymological derivation of ὕμνος from the same verb-root as ὑμήν.
[151] Nagy 1990, 1996, 2002, 2009.
[152] Nagy 2009, 232, from which the next citations in the text are also taken.
[153] See Nagy 2009, 241 "[I]n other words, the concept of *hymnos* is the concept of maintaining the song as the notionally same song by way of successfully executing a metabasis form the initial subject to the next subject."
[154] Nagy 2002, 70.

The idea of a beginning which makes a whole out of what follows from it, is the feature that the poetics of rhapsodic performance receives from weaving technology, and in particular from the mechanism of the starting border:[155] this intuition is grounded in the semantics and performance reality of the words *hymnos*, *prooimion*, *oimē* and *rhapsōidos* (all of which are rooted in the craft of fabric-making), and tested in the context of the poetics of the Panathenaic festival in Athens, where the rhapsodic recitation of Homeric poetry and the gigantic *peplos* woven for the statue of Athena are connected through the idea of *hymnos* and its divinely authorized beginning.[156]

It is worth quoting the core of Nagy's argument in his own words:

> (…) the general process of weaving as a metaphor is applied by epic to the specific activity of performng epic. The prime example I cited was the word *oimē*, which refers metaphorically to the 'story-thread' that begins the epic performance of the singer Demodoos in *Odyssey* viii 74. Such a beginning of epic, as we saw from the wording of Pindar's *Nemean* 2 (line 3), is a *prooimion*, which is metaphorically the starting point of the threading, of the *oimē*. Comparable to this Greek *prooimion* is the Latin *exordium* 'proemium'. Both words are applicable to the beginning of a song, a poem, or a speech. (…) the specific meanings of the Greek nouns *oimē* and *prooimion* are related to the general meaning of the noun *humnos*, which I interpreted etymologically as the overall process of weaving as expressed by the verb *huphainein* 'weave'. So the question is, can we say that the performing of epic in the age of Pheidias is visualized metaphorically as the work of male weavers in particular? In formulating the answer, I start with the metaphor inherent in the technical poetic term *prooimion*. In terms of this metaphor, a performance started by a singer is like a web started by a weaver. Performing the *prooimion* is like weaving the *exastis* which is a technical term for the initial phase of the weaving.[157]

Nagy grounds the semantics of the compound *rhapsōidos* in the performance context of the rhapsodic contest in the occasion of the Panathenaic festival in Athens; there, the competing bards recited pieces of Homeric epic in an orderly sequence, one picking up where the other had stopped.[158] From the point of view of the performance structure of such a rhapsodic contest, the *hymnos/prooimion* represented the first rhapsody in the sequence of successive recitations of Homeric *epē*.[159] Pursuing Nagy's argument, one might go further and suggest that the threads that unfolded from the *hymnos/prooimion*, just like the warp which is governed by the starting border, developed different patterns and narrative motifs throughout the rhapsodies performed by the competing singers.[160] A characteristic feature of the rhapsodic *prooimion* as we have it preserved in the *Homeric Hymns* is its function of linking the recitation of epic poetry to the occasion of the performance. This implied to pay poetic omage to the god in whose honour the festival hosting the rhapsodic contest was celebrated: as a consequence, the thematic relation between the content of the *prooimion* (a hymnic address to a god) and the heroic lays narrated in the *oimē* was often very loose.

---

[155] See Nagy 2002, 79 "[H]ere I return to my thesis, that the idea of *weaving*, just like the idea of *humnos*, is connected with the idea of *beginnings*. The essential point is the *point of departure*. Wherever you begin, you must have a continuum that follows (…) The wholeness of the performance is authorized by the beginning. To repeat, *arkhē* is both beginning and authorization."

[156] The ritual implications of the starting of the weaving (the making of the starting border) for the *peplos* of Athena at the *Khalkeia* festival in Athens (on which Arist. *Ath.* 60.1–3) are discussed in Nagy 2002, 86–88 and 2009, 550, 568.

[157] Nagy 2009, 568–569.

[158] On the so-called 'Panathenaic Rule', *i.e.* the regulation that imposed the recitation of Homeric *epē* in sequence at the rhapsodic contest during the Panathenaic festival see Nagy 2002, 9–20.

[159] See Càssola 1975, xiii–xxxi, who (xix) points out that the term προοίμιον could be used to denote both the beginning of a poem (which Càssola calls 'esordio') and a poem having an introductory function (the 'proper' *prooimion*, that featured an 'esordio' of its own – expressed by Greek ἀναβολή: see the ἀμβολαὶ προοιμίων in Pind. P. 1.4).

[160] In such a sequence of rhapsodies, we see the technology of *rhaptein* as encompassing the notion of 'pattern connections' within a pattern-woven fabric, rather than pointing to the craft of tayloring. For the hypothesis that rhapsodic contests featured 'closing *hymnoi*' see Càssola 1975, xxi–xxii.

Looking for a sample of epic/rhapsodic poetry that might reflect in its structure the function of the starting border – *i.e.* providing the warp threads for the weave, as well as the sequencing principles for the patterns to be woven into the fabric – we would rather turn to the only preserved sequence *prooimion-oimē* in hexametric poetry, namely Hesiod's *Theogony*.[161] Here the threads of the poem's warp, *i.e.* the themes of the divine genealogies which unfold throughout the *oimē*, are outlined as a poetic 'program' and embedded in the structure of the *prooimion*, a 'hymn to the Muses' (ll. 1–115) which bears strong formal resemblances with the corpus of the extant *Homeric Hymns*.[162] The structure of the *prooimion* is complex. We have a first shorter *hymnos* to the Heliconian Muses, introduced by the formulaic beginning Μουσάων Ἑλικωνιάδων ἀρχώμεθ᾽ ἀειδεῖν ("Let us begin to sing from the Heliconian Muses")[163] and describing the Muses singing and dancing on Mount Helicon (1–21) and performing a *hymnos* for the gods (ὑμνεῦσαι 11). This is followed by Hesiod's poetic investiture by the Muses (22–34), who command the poet to sing a *hymnos* (ὑμνεῖν 33) on the subject of "the race of the blessed ones" (μακάρων γένος 33). After a transitional line (35), a new *hymnos* begins much like the first (Μουσάων ἀρχώμεθα 36), and ends with a formulaic leave-off at l. 104 where the poets bids the Muses farewell (χαῖρετε, τέκνα Διός, δότε δ᾽ ἱμερόεσσαν ἀοιδήν "hail children of Zeus, and give me lovely song"). The second part of the *prooimion*, a *hymnos* to the Olympian Muses (36–103), contains a description of the Muses singing on Olympus about the *genos* of the gods (but beginning and ending with Zeus, ll. 47–49), then an account of the *Muses*' birth, their journey to Olympus, and their gifts to kings and poets. In a long transitional passage which introduces to the *oimē* (104–115), Hesiod commands the Muses to sing of divine genealogies, providing a sequence of three main topics which will then be followed in the central rhapsody of the poem.[164] It is beyond the scope of this chapter to discuss in detail the structure of this *prooimion*, where correspondences between the first, shortest *hymnos* (1–35) and the second one (36–103) not only regard formal features and recurring motifs, but invest broader thematic concerns of the poem as a whole.[165] Relevant for the present discussion are three observations that arise from a reading of the proemial '*hymnos* to the Muses' in Hesiod's *Theogony*: a) the emphasis on beginning (this *prooimion* contains three beginnings),[166] and the related fact that the poet feels the need to start from

[161] See Hamilton 1989, 11 "[T]he *Theogony* in fact is our best evidence for the function of the Homeric hymn as proem to a longer song." For the 'rhapsodic' nature of the *Theogony* see Càssola 1975, xxi, who recognizes in the structure of the poem (as we have it) several 'stitches' indicating successive rhapsodies (*e.g.* the leave-off at ll. 963–964 'Farewell now to you (…)' is followed by a new 'esordio' at ll. 965–968 "And now, sweet-voiced Olympian Muses, daughters of aegis-holding Zeus, sing of the tribe [φῦλον] of goddesses…"). Hamilton 1989, 15 further observes that the third invocation to the Muses (ll. 1021–1022), with which the *Theogony* concludes in medieval manuscripts, is the opening of the *Catalogue of Women* (see fr. 1 Most).

[162] On the importance of this feature for the overall structure of the *Theogony* see Hamilton 1989, 3 "[T]he crucial fact about the form of the *Theogony*, one virtually ignored by critics, is that the poem is articulated by the outline of its content given at the end of the hymn with which it begins."

[163] Hesiod seems here to conflate two different rhapsodic formulaic beginnings: the type ἄρχομαι + genitive of the god(s), and the type ἄρχομαι + ἀείδειν + accusative of the god(s); as West 1966, 151 observes, in *Theog.* 1 "the genitive is governed by ἀρχώμεθα (ἀείδειν being complementary)". On Hesiod's use of the first person plural here as to include the Muses in his exortation, see Strauss Clay 2003, 51–52, who also remarks how the construction with the genitive "reveals that the Muses constitute the appropriate or even necessary starting-point for Hesiod's song, but that they are not its object" (p. 53). The English translation of Hesiod's *Theogony* is Most 2007 (Loeb) adapted.

[164] The three themes are (quoting from Hamilton 1989, 14–15): "(1) the children of Gaia and Ouranos, Night and Pontos (106–107); (2) their children (111); (3) how they divided healt and *timai* and first got Olympus"; Hamilton shows how these three topics are consistently developed in the structure of the *Theogony*.

[165] See, among others, the rich discussion in Thalmann 1983, 134–152.

[166] Strauss Clay 2003, 52–53 aptly remarks how the motif of beginning is especially in place in a poem which sets up

the Muses again when he turns to the 'second' *hymnos* after the account of his poetic initiation (at line 36);[167] b) the metaliterary motif of the choral dancing of the Muses as divine paradigm for human songs: much like the god providing the *archē* ('beginning' and 'authorization') in the *Homeric Hymns*, this is their function of providing 'divine authorization' in the proem of the *Theogony*, thus enabling Hesiod to *hymnein* ('celebrate by means of a *hymnos*') the divine genealogies; c) the transition from the *prooimion* to the *oimē* is effected by means of an accurate description of the sequence of 'figurative patterns' that the starting border 'governs', and that constitute the condition for the continuity of the fabric (and of the main body of the *Theogony*).

In the first half of this chapter, it has been argued that the starting border is a crucial *condition* for weaving on a war-weighted loom: in a similar fashion, the cross-generic *prooimion* fulfils a fundamental function in archaic Greek poetics, connecting the song to the occasion of its performance[168] and providing a necessary (musical and poetic) introduction to the rhapsody (epic), the *nomos* (citharody)[169] or the choral song (melic).[170]

The pattern of 'embedded' *prooimia* is still recognizable in a sample of choral lyric which features the motif of the 'song that starts another song' or describes the practical modalities of the performance:[171] Alcman 14 *PMG* consists in an invocation to the Muse to "begin a new song (μέλος νεοχμόν) for girls to sing";[172] Pindar's first *Pythian* ode begins with a proemial invocation to the lyre and the description of a performance by Apollo and the Muses, where the god strikes up with his lyre "instrumental preludes to chorus-leading *prooimia*."[173] The beginning of the second *Nemean*, as we have shown in the previous section, is a precious evidence for the reconstruction of the performance of Homeric poetry enacted by the rhapsodic guilds: the ode is indeed unique within the corpus of Pindar's epinician odes in that it seems to show features typical of a *prooimion*: it is unusually short, and its closing lines, with the chorus inviting to lead-off (*exarkhein*) a song in honour of Zeus, resembles the transitional passages that close *prooimia* and lead into the 'proper' song.

The procession of gods painted on the main frieze of the François Vase (fig. 5.7) accompanies and celebrates the wedding of Peleus and Tethys. We choose here to draw attention to the *prooimion* that the Muses sing and dance on the occasion of this divine wedding in Pindar's *Nemean* 5.22–26:

---

to provide an account of cosmological beginnings: "[T]he proem of the song of beginnings, thus even as it makes its beginning, attempts to give and account of its beginning."

[167] See the remarks by West 1966, 151. For the possible juxtaposition of two successive 'starting bands' in weaving, see above p. 73.

[168] Aloni 1990, 119 observes that "the *prooimion* is not, and could not be, a literary genre with pre-determined formal and content-related features, that recur regularly. We are rather dealing with a function, or a necessity of the performance; as a consequence, all the elements characterizing it are to be placed at the level of pragmatics" (our translation).

[169] Power 2010, 185–215 provide an excellent and exhaustive treatment of the terminology and phenomenology of the citharodic *prooimion* and its connections with the rhapsodic *prooimion*.

[170] On the *prooimion* as introductory feature of any act of speech, and as an established practice in oratory speech see Aloni 1980, Maslov 2012. On Plato's recurrent puns on the citharodic sequence *prooimion-nomos* (where *nomos* stands for both 'song genre' and 'law') see Power 2010, 189 who aptly quotes *Timaeus* 29d and Socrates' address to Timaeus, who has started lecturing on cosmological theories: "[W]e have welcomed, then, your *prooimion* with wondrous admiration, but, following the sequence, begin to perform for us the *nomos*."

[171] See the remarks in Aloni 1990, 117–121; in the remainder of his essay (pp. 121–129), Aloni tackles the complex problem of the phenomenology of the *prooimion* and the 'proemial function' in Pindar's epinician odes.

[172] The translation of Alcman is by D. A. Campbell (Loeb).

[173] Translation: Powell 2010, 202.

Fig. 5.7 Krater of Kleitias and Ergotimos, Florence Museo Archeologico Nazionale, 4209 (so-called François-Vase), drawing from Furtwängler and Reichhold 1904, pl. 3, courtesy The Royal Library, Copenhagen.

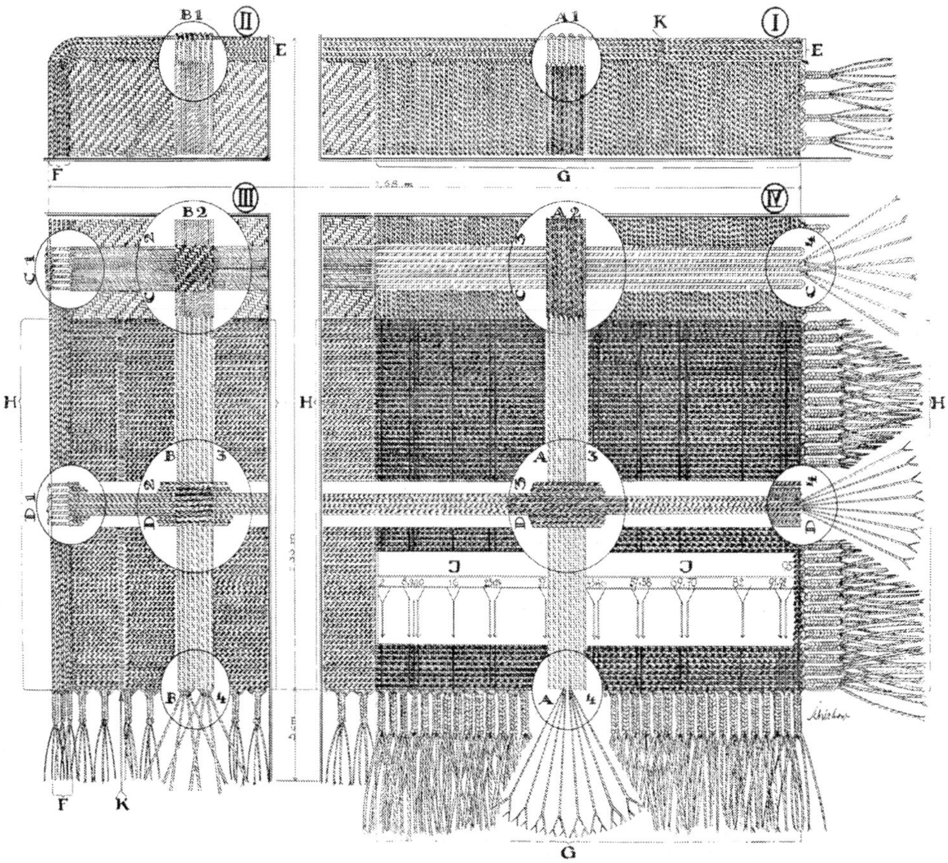

Fig. 5.8 Scheme of the construction and composition of the Thorsberg Mantle. Drawing from Schlabow 1956, 61, courtesy of the University Library, Düsseldorf.

Πρόφρων δὲ καὶ κείνοις ἄειδ᾽ ἐν Παλίῳ
Μοισᾶν ὁ κάλλιστος χορός, ἐν δὲ μέσαις
φόρμιγγ᾽ Ἀπόλλων ἑπτάγλωσσον
  χρυσέῳ πλάκτρῳ διώκων
ἀγεῖτο παντοίων νόμων· αἱ δὲ πρώτιστον μὲν ὕμνη-
  σαν Διὸς ἀρχόμεναι σεμνὰν Θέτιν
Πηλέα θ᾽·

Gladly did that fairest chorus of the Muses sing for those men on Pelion, while in their midst Apollo swept his seven-tongued lyre with a golden plectrum, and led them in tunes of all kind. And, after a prelude to Zeus, they first sang of august Thetis and Peleus (…)

Apollo and the Muses are here mirror images for the citharodos accompanying the epinician chorus: Apollo plays a prelude on the lyre, and the Muses move to sing a *hymnos* beginning from Zeus (ὕμνησαν ἀρχόμεναι Διός l. 25), a *prooimion* introducing the narrative of Peleus' heroic deeds. Much like in the *prooimion* of Hesiod's *Theogony*, we have here a vivid representation of divinely authorized and authorizing singing.

As a last instance of textile-like patterns in poetry, the mirror image of the Muses singing and dancing within a poem or a song, and replicating their performance *ad libitum*, reminds us of the pattern-woven cloth that both the Muses and the Moira wear in the François Vase (fig. 5.3). The figured motif in the Moira's cloth reduplicates the frieze's picture in a virtually endless Chinese box. As the Muses provide the epic singer with οἴμας παντοίας ("songs/threads of every sort") and with poetic authority through their divine performance, so the Moirai/Fates provide men with the thread(s) of one's destiny (fig. 5.7). We have tried to show that there is a convergence in the terminology, imagery and conceptual representation of these two sets of divinely given threads.

Conclusion

What we tried to show in our technological part is that weaving in antiquity is not a clear gesture nor a simple unification of opposites but a complex composition of woven and plaited parts integrated into a finished product by repeated gestures of turning the piece around or upside down. The scheme Karl Schlabow once made when reconstructing the Thorsberg mantle might demonstrate this complexity (fig. 5.8) and give an example of what archaic Greek poets had in mind and view when using weaving terms to describe poetic composition principles that not yet had their own terminology.

We assume that, starting with Pherecydes, weaving gives a possible answer to the question of how the order of the *kosmos* was and is produced and which generative principles are at work in the *kosmos* and thus in every *technē*. We have argued that specific features of ancient weaving like the starting-border or the repeated integration of bands into textiles (*cf.* fig. 5.8) can be recognised in the use of textile terms in ancient texts. Referring to textiles indicates a mode of invention or creation that differs considerably from the way in which poetry is perceived today, namely as a form of art including the concept of the artist creator who designs his work according to an idea in his mind. Such modern concept has influenced the understanding of both the history of Greek art as well as the history of technology.[174] It is especially the discussion of formulae or prototypes in archaic poetry and geometric art where this still affects the argumentation that often starts to become polemic as soon as textiles are claimed to be the the medium of memory and transfer throughout the Dark Ages.[175]

Homeric poetry has been consistently compared with Geometric pottery (especially Dipylon vases) on the ground of structural and compositional features, among which the framing function of rings bears a special relevance: "there is a circular composition also of scenes themselves (*sc.* in Homer), scenes framing scenes in concentric rings around centrepieces, exactly as central motifs are heavily framed by borders in Geometric paintings".[176] In turn, pictorial friezes and meander motifs

[174] The origins of modern art and modern technology have in common that they refer to the concept of invention, called *disegno* in the discourse of renaissance, *cf.* Blumenberg 2015, 27.

[175] For example in the reaction on Bensons book on the origins of Greek art (1970) by Coldstream (1974) and Hoffman (1997). Hoffman's book on imports from the Near East to Iron Age Crete deals with bronzes, faience, glass, gold, ivory, lead, pottery, scarabs, and even stone. However textiles are only mentioned in a footnote referring to the thesis of a textile continuity behind pottery and not saying anything about possible imports. *Cf.* note 3 on page 155.

[176] Whitman 1958, 98.

of Geometric vases have been traced back to figured textiles with in-woven mytho-histories.[177] Discussing the idea of friezes in Greek geometric art, Benson argues that "the impetus to depict a procession is more likely to have come at least partly from specific contemporary interest such as epic poetry (...). At that point, it would have been recalled that such vases and perhaps such wall paintings once existed".[178] Benson does not refer to textiles here as there are only secondary sources (texts) but no objects available. However he repeatedly points out the possibility of the artists being influenced by patterned textiles and the frieze lends itself to the constructive principle of weaving progressing in rows.

The principles of weaving are recognizable in the structure of the archaic Greek poem through the ring-sequence of frames and narrative sections. The sequence of geometric borders and pictorial friezes on the François Vase and on the clothes of both the Moira and the Muses represents a further instance of such a pervasive pattern stemming from weaving technology. The ordering and connecting *prooimion* in poetry reflects in both its terminology and function the pattern of the starting border in weaving. The 'proemial function' has been shown to represent a key performative and compositional pattern of archaic song-making: a remarkable cross-generic trait, it creates the condition for the rhapsodic and citharodic performance and is embedded in the structure of choral lyric songs.

We end our chapter with a passage from Euripides *Ion* (ll. 507–508), where translators are at pains to make sense of the juxtaposition of *kerkis* ("shuttle") and *logos* ("word") on the same level and, applying the results of our study to our translation, propose to accept that *kerkis* and *logos* might have the same value.[179]

> οὔτ᾽ ἐπὶ κερκίσιν οὔτε λόγοις[180] φάτιν
> ἄιον εὐτυχίας μετέχειν θεόθεν τέκνα θνατοῖς
>
> Not by shuttles nor by words has it ever been rumoured that godly children meant a share of blessing for mortals.

[177] Drawing on textual, iconographical and archaeological evidences, enriched by her own expertise as a weaver (see p. 365 "[T]o a weaver, early Greek vases look like textiles from the start"), Barber 1991, 358–372 rounds off a very convincing argument to ground her hypothesis that "(...) less-destructible arts – vases, terracottas, and later sculptured reliefs – were deriving their forms and concepts" from figured textiles.
[178] Benson 1970, 55–56.
[179] The two nouns are generally interpreted as pointing to the practice of singing while weaving, and making reference to 'songs to the loom' which accompanied textile activity. In the case of the expression ἐπὶ κερκίσιν, the plural of κερκίς is taken as a reference (via metonymy) to the loom and the expression translated as 'stories at the/my loom' (Kovacs, Tuck 2009) or simply 'at the loom' (Lee 1997), as paralled to λόγοις, interpreted as 'common talk' (Tuck), 'songs' (Kovacs), or 'telling of stories' (Lee): drawing on other passages in Euripdes that refer to in-woven mythical stories (Eur. *Ion* 196–200, *Hec.* 466–474, *IT* 221–224) Tuck argues for the possibility that "the retelling of such myths in the service of patterned textile production may have sometimes even related to the design motifs themselves".
[180] We print the text of L: Diggle (OCT) brings into his text Badham's λόγων, while Kovacs (Loeb) puts λόγοις between *cruces*.

Bibliography

Algra, K. 2001 Die Anfänge der Kosmologie. In A. A. Long (ed.), *Handbuch Frühe Griechische Philosophie*. Stuttgart and Weimar, 42–60.

Aloni, A. 1980 "Prooimia", "Hymnoi", Elio Aristide e i cugini bastardy. *QUCC* 4, 23–40.

Aloni, A. 1992 Proemio e funzione proemiale nella poesia greca arcaica. *AION(filol)* 12, 99–130.

Ademollo, F. 2011 *The Cratylus of Plato: A Commentary*. New York.

Andrés Toledo, M. A. 2010 *El hilo de la vida y el lazo de la muerte en la tradición indoirania*, Valencia.

Aspris, M. 1996 Ein zyprischer Teller mit der Darstellung eines Webstuhls. *Bonner Jahrbücher* 196, 1–10.

Barber, E. J. W. 1991 *Prehistoric Textiles. The Development of Cloth in the Neolithic and Bronze Ages with Special Reference to the Aegean*. Princeton NJ.

Becker, O. 1959 *Größe und Grenze der mathematischen Denkweise*. Freiburg.

Becker, O. 1963 Die Aktualität des pythagoreischen Gedankens. In O. Becker, *Dasein und Dawesen: gesammelte philosophische Aufsätze*. Pfullingen, 127–156.

Beekes, R. 2010 *Etymological Dictionary of Greek*. Leiden–Boston MA.

Benson, J. L. 1970 *Horse, Bird and Man: The Origins of Greek Painting*. Amherst.

Berggren, A. 2008 *Weaving Truth. Essays on Language and the Female in Greek Thought*. Cambridge.

Bergren, A. L. T. 1975 *The Etymology and Usage of πεῖραρ in Early Greek poetry*. University Park PA.

Bernadete, S. 2006 *The Being of the Beautiful. Plato's Theaetetus, Sophist, and Statesman*. Chicago–London.

Blumenberg, H. 1960 *Paradigmen zu einer Metaphorologie*. Bonn.

Blumenberg, H. 1981 *Wirklichkeiten in denen wir leben*. Stuttgart.

Blumenberg, H. 1983 The *Legitimacy of the Modern Age*, trans. R. M. Wallace. Cambridge MA.

Blumenberg, H. 1996 *Die Legitimität der Neuzeit*, rev. ed. Frankfurt.

Blumenberg, H. 2015 *Schriften zur Technik* (ed. A. Schmitz & B. Stiegler). Frankfurt.

Blumenberg, H. 2009 *Geistesgeschichte der Technik*. Frankfurt.

Blümner, H. 1912 *Technologie und Terminologie der Gewerbe und Künste bei Griechen und Römern*. Leipzig.

Brancacci, A. 2007 Democritus' *Mousika*. In A. Brancacci & P.-M. Morel (eds), *Democritus: Science, the Arts, and the Care of the Soul*. Leiden and Boston, 181–206.

Brezine C. 2009 *Algorithms and automation: mathematics and weaving*. In Robson & Stedall 2009, 468–492.

Burkert, W. 1990 *Wilder Ursprung. Opferritual und Mythos bei den Griechen*. Berlin.

Cairns, D. 1997 Form and meaning in Bacchylides' fifth ode. *Scholia* 6, 34–48.

Campe, R. 2000 From the theory of technology to the technique of metaphor: Blumenberg's opening move. *Qui Parle* 12, 105–126.

Campe, R. 2009 Von der Theorie der Technik zur Technik der Metapher: Blumenbergs systematische Eröffnung. In A. Haverkamp & D. Mende (eds), *Metaphorologie: Zur Praxis von Theorie*. Frankfurt, 283–315.

Càssola, F. 1975 *Inni Omerici*. Milan.

Chantraine, P. 1968–1980 Dictionnaire *Étymologique de la Langue Greque* I, IV-2. Paris.

Cohen, B. 2006 *The Colors of Clay: Special Techniques in Athenian Vases*. Los Angeles CA.

Clayton B. 2004 A *Penelopean Poetics. Reweaving the Feminine in Homer's Odyssey*. Lanham MD–Boulder CO–New York–Toronto–Oxford.

Coldstream, J. N. 1974 Benson, Horse, Bird and Man (Review). *Gnomon* 346, vol. 3, 273–278.

Coxon, A. H. 2009 *The Fragments of Parmenides. A Critical Text with Introduction and Translation, the Ancient Testimonia and a Commentary*. Assen–Maastricht.

Descartes, R. 1985 *Rules for the Direction of the Mind*, trans. J. Cottingham, R. Stoothoff & D. Murdoch. Cambridge, 9–77.

Droß-Krüpe, K. & Paetz gen. Schieck, A. 2015 Unravelling the Tangled Threads of Ancient Embroidery: a compilation of written sources and archaeologically preserved textiles. In Harlow & Nosch 2015, 207–235.

Durante, M. 1976 *Sulla Preistoria della Tradizione Poetica Greca*. Rome.

Edmunds, S. T. 2012 Picturing Homeric Weaving. In *Donum natalicum digitaliter confectum Gregorio Nagy septuagenario a discipulis collegis familiaribus oblatum*. http://chs.harvard.edu/CHS/article/display/4365. Center for Hellenic Studies.

Ford, A. 2002 *The Origins of Criticism. Literary Culture and Poetic Theory in Classical Greece*. Princeton–Oxford.
Fritz, K. von 2004 The Discovery of Incommensurability by Hippasus of Metapontum. In *Annals of Mathematics* 46 (1954), 242–264. Reprinted in J. Christianidis, (ed.) 2004, *Classics in the History of Greek Mathematics*. Dordrecht, 211–232.
Furtwängler, A. & Reichhold, K. 1904 *Griechische Vasenmalerei*. Serie I. Tafeln. Munich.
Garvie, A. 1994 *Homer. Odyssey VI-VIII*. Cambridge.
Gentili, B. 1990 *Poetry and its Public in Ancient Greece*, Baltimore MD–London.
Gleba, M. 2008 *Textile Production in Pre-Roman Italy.* Oxford.
Graham, D. W. 2008 Leucippus's atomism. In P. Curd & D. W. Graham (eds), *The Oxford Hanbook of Presocratic Philosophy*. Oxford, 333–352.
Hamilton, R. 1989 *The Architecture of Hesiodic Poetry*. Baltimore MD–London
Harlizius-Klück, E. 2004 *Weberei als episteme und die Genese der deduktiven Mathematik: in vier Umschweifen entwickelt aus Platons Dialog Politikos*. Berlin.
Harlizius-Klück, E. 2014 The importance of beginnings: gender and reproduction in mathematics and weaving. In Harlow & Nosch 2014, 46–59.
Harlizius-Klück, E. 2015 Denkmuster in der antiken Weberei? Eine Spurensuche. In H. Harich-Schwarzbauer (ed.), *Weben und Gewebe in der Antike: Materialität – Repräsentation – Episteme – Metapoetik. / Texts and textiles in the ancient world: Materiality – Representation – Episteme – Metapoetics*. Oxford.
Harlizius-Klück, E. 2016 Textile Technology. In G. L. Irby, *A Companion to Science, Technology, and Medicine in Ancient Greece and Rome*, vol. II. Oxford, 745–767.
Harlizius-Klück, E. (forthcoming) *Weberei und Vasenmalerei: eine Beziehungsgeschichte.*
Harlow, M. & Nosch, M.-L. (eds) 2014 *Greek and Roman Textiles and Dress: an Interdisciplinary Anthology.* Ancient Textile Series 19. Oxford.
Hartmann, C. 1838 *Encyclopädisches Handbuch des Maschinen- und Fabrikenwesens für Kameralisten, Architekten, Künstler, Fabrikanten und Gewerbetreibende jeder Art,* 2. Theil, 2. Abtheilung. Leipzig–Darmstadt.
Heath, S. T. 1908 *The Thirteen Books of Euclid's Elements*. Cambridge.
Hoffman, G. L. 1997 *Imports and Immigrants: Near Eastern Contacts with Iron Age Crete*. Michigan.
Hoffmann, M. 1974 *The Warp-Weighted Loom: Studies in the History and Technology of an Ancient Implement*, Oslo.
House of Commons, 2010 Environment, Food and Rural Affairs Committee: *Waste strategy for England 2007. Third Report of Session 2009–2010*, vol. I–II. London.
Hutchinson, G. O. 2001 *Greek Lyric Poetry*. Oxford.
Ingram, D. 1993 The Copernican Revolution revisited: paradigm, metaphor and incommensurability in the history of science-Blumenberg's response to Kuhn and Davidson. In *History of the Human Sciences* 6.4, 11–35.
Karmarsch K. & Heeren, F. 1843 *Technisches Wörterbuch oder Handbuch der Gewerbskunde. Bearbeitet nach Dr. Andrew Ure's Dictionary of Arts, Manufacturers and Mines,* Zweiter Band. Prague.
Kirk, G. S., Raven J. E. & Schofield, M. 1983[2] *The Presocratic Philosophers*. Cambridge.
Krünitz, J. G. 1807 *Ökonomisch-technologische Encyclopädie, oder allgemeines System der Staats-, Stadt-, Haus- und Landwirthschaft, und der Kunstgeschichte, in alphabetischer Ordnung,* Teil 106. Berlin.
Kuhn, Th. S. 1957 *The Copernican Revolution: Planetary Astronomy in the Development of Western Thought.* Cambridge MA.
Kuhn, Th. S. 1962 *The Structure of Scientific Revolutions.* Chicago.
Lee, K. H. 1997 *Euripides Ion*. Wiltshire.
Lewis, J. 2006 *Solon the Thinker. Political Thought in Archaic Athens*. London.
Lloyd, G. E. R. 1966 *Polarity and Analogy. Two Types of argumentation in Early Greek Thought*. Cambridge.
Maslov, B. 2012 The Real Life of the Genre of Prooimion. *CP* 107, 191–205.
Mende, D. 2009 Vorwort: Begriffsgeschichte, Metaphorologie, Unbegrifflichkeit. In A. Haverkamp & D. Mende (eds) *Metaphorologie: Zur Praxis von Theorie*. Frankfurt.
Mourelatos, A. P. D. 1970 *The Route of Parmenides. A Study of Word, Image, and Argument in the Fragments.* New Haven CT–London.
Müller, L. 1862 *Die Fabrikation des Papiers in Sonderheit des auf der Maschine gefertigten nebst gründlicher Auseinandersetzung der in ihr vorkommenden chemischen Prozesse und Anweisung zur Prüfung der angewandten Materialien.* Berlin–Heidelberg.

Nagy, G. 1990 *Pindar's Homer. The Lyric Possession of an Epic Past*. Baltimore MD.
Nagy, G. 1996 *Poetry as Performance: Homer and Beyond*. Cambridge.
Nagy, G. 2002 *Plato's Rhapsody and Homer's Music*. Cambridge.
Nagy, G. 2009 *Homer the Classic*. Cambridge MA–London.
Nosch, M.-L. 2014 Voicing the Loom: Women, Weaving, and Plotting. In D. Nakassis, J. Gulizio, & S. A. James (eds) *KE-RA-ME-JA. Studies Presented to Cynthia W. Shelmerdine*. Philadelphia PA, 91–101.
Noussia M. 2001 *Solone. Frammenti dell'opera poetica*. Milan.
Noussia-Fantuzzi, M. 2010 *Solon the Athenian, the Poetic Fragments*. Boston MA–Leiden.
Onians, R. B. 1951 *The Origins of European Thought about the Body, the Mind, the Soul, the World, Time and Fate*. Cambridge.
Pagliaro A., 1952 *Saggi di critica semantica*. Florence.
Power, T. 2010 *The Culture of Kitharôidia*. Washington DC.
Power, T. 2011 Cyberchorus: Pindar's Κηληδόνες and the aura of the artificial. In L. Athanassaki & E. Bowie (eds), *Archaic and Classical Choral Song: Performance, Politics and Dissemination*. Berlin–Boston MA.
Rosenfeld, H.-F. 1958 *Wort- und Sachstudien: Untersuchungen zur Terminologie des Aufzugs, zu Webstuhl und Schermethode der germanischen Bronze- und Eisenzeit und zur Frauentracht der Bronzezeit sowie der Frage ihres Fortlebens in der Volkstracht*. Berlin.
Robson, E. & Stedall, J. (eds) 2009 *The Oxford Handbook of the History of Mathematics*. New York.
Saito, K. 2009 Reading ancient Greek mathematics. In Robson & Stedall 2009, 801–826.
Scheid, J. & Svenbro, J. 1996 *The Craft of Zeus. Myths of Weaving and Fabric*. Cambridge MA–London.
Schibli, H. S. 1990 *Pherekydes of Syros*. Oxford.
Schinnerer, L. 1895 *Antike Handarbeiten*. Vienna.
Schlabow, K. 1965 *Der Thorsberger Prachtmantel: Schlüssel zum altgermanischen Webstuhl*. Neumünster.
Schmitz, A. & Stiegler, B. 2015 Nachwort. In Blumenberg 2015, 281–298.
Slaktin, L. M. 1996 Composition by theme and the *metis* of the Odyssey. In S. Schein (ed.), *Reading the Odyssey*. Princeton NJ, 223–238.
Slater, W. J. 1983 Lyric narrative: structure and principle. *ClAnt* 2, 117–132.
Spantidaki, S. 2014 Embellishment Techniques of Classical Greek Textiles. In Harlow & Nosch 2014, 34–45.
Specht, R. 1996 René Descartes (1596–1650). In Specht, R. (ed.) 1996 *René Descartes: Philosophische Schriften*, XV–XL.
Strauss Clay, J. 2003 *Hesiod's Cosmos*. Cambridge.
Szabó, A. 1978 *The Beginnings of Greek Mathematics*, Dordrecht–Boston MA.
Tarán, L. 1965 *Parmenides. A Text with Translation, Commentary, and Critical Essays*. Princeton NJ.
Thalmann, W. G. 1984 *Conventions of Form and Thought in Early Greek Poetry*. Baltimore MD–London.
Tuck, A. 2006 Singing the rug. Patterned Textiles and the Origins of Indo-European Metrical Poetry. *AJA* 110, 539–550.
Tuck, A. 2009 Stories at the loom. Patterned textiles and the recitation of myth in Euripides. *Arethusa* 42.2, 151–159.
Turner, E. 1987 *The Mathematics of Plato's* Academy. *A New Reconstruction*. Oxford.
Ure, A. 1840 *A Dictionary of Arts, Manufacturers and Mines: containing a clear exposition of their principles and practice*, 2nd ed. London.
Urton, G. and P. Nina Llanos 1997 *The social life of numbers: a Quechua Ontology of Numbers and Philosophy of Arithmetic*. Austin TX.
Wace, A. J. B. 1948 Weaving or Embroidery? *AJA* 52.1, 51–55.
Wagner, M. et al. 2009 The ornamental trousers from Sampula (Xinjiang, China): their origins and biography. *Antiquity* 83, 1065–1075.
Wagner-Hasel, B. 2000 *Der Stoff der Gaben. Kultur und Politik des Schenkens und Tauschens im archaischen Griechenland,* Frankfurt–New York.
West, M. L. 1966 *Hesiod. Theogony*. Oxford.
West, M. L. 2007 *Indo-European Poetry and Myth*. Oxford.
Whitman, C. H. 1958 *Homer and the Heroic Tradition*. Cambridge MA–London.

6. The Curse as a Garment in Greek Tragedy

Judith Fletcher

Textiles and garments generate some of the most common yet complex image systems in Greek tragedy. More than any other symbol they unite the world of metaphor with the physical realities of the theatre. Among its semiotic cargo, clothing in tragedy bears the meaning of death and disfigurement, a distortion of its normative function of protecting and adorning the human body. The swaddling robe that Clytemnestra uses to bind Agamemnon, pictured on the famous Boston krater, is the first surviving example of destructive clothing in tragedy (fig. 6.1).[1] Euripides' *Medea* and Sophocles' *Trachiniae* subsequently feature more magical flesh eating garments. In her study of the *peplos* in tragedy, Lee suggests that Aeschylus "invented the motif of the deadly *peploi*."[2] She argues that the senior tragedian manipulated and inverted epic and lyric connotations of the *peplos*, which include marriage, femininity and luxury, when he featured the deadly garment, and that Sophocles and Euripides appropriated this new symbol to signify distorted marriage and troubled gender dynamics.

Lee is certainly correct to see a similarity between the three deadly *peploi* wielded by Clytemnestra, Medea and Deianeira, and she offers important insights into the history and semantics of the tragic *peplos*, but I would like to suggest that these dramatic works also exploit a deeply ancient connotation of toxic textiles that predates Clytemnestra's deadly robe. I submit that the tragedians adapted the idea of the deadly robe from a long-established analogy that dates back to Hittite, Babylonian, Assyrian, and Old Testament magic and ritual. I categorize these different examples under the generalized rubric 'Near Eastern,' with the cautionary proviso that this term covers a range of different cultures over a period of several millennia in Western Asia and the Eastern Mediterranean Basin. The 'Ancient Near East' is by no means a cultural monolith, but it is the very diversity of the different versions of the curse as a garment image that leads me to suggest that the concept was diffused over a cultural area that included ancient Greece. My hypothesis is based on a group of ancient texts and inscriptions that employ the analogy of clothing for curses, oaths and spells. For instance a Hittite text from around 1200 BCE known as the 'Ritual for Infernal

[1] The Kalyx-krater by the Dokimasia Painter (c. 460 BCE, Boston 63.1246).

[2] Lee 2004, 262. While the first mention of a poisoned robe in Greek literature (a *chiton*) actually occurs in a Hesiodic fragment describing Deianeira's murder of Heracles (fr. 25.17–25) discussed below, Lee's focus is on the term *peplos*, which only occurs in tragedy in 5th century literature. All three of the deadly garments are called *peplos*, at some point, although other terms such as *phāros* and *chitōn* are also used.

Fig. 6.1 The Boston krater shows the swaddling robe that Clytemnestra uses to bind Agamemnon – the first surviving example of destructive clothing in tragedy.

Deities' reveals that spells can be cast by throwing a cloak over a person, or that one can get out of an oath or break a spell by removing their cloak.[3]

Garments and curses in ancient Near Eastern cultures

The analogy of the curse as a garment or as a skin disease that covers the body like a garment was widespread in the ancient world.[4] One good example occurs in Psalm 109, the so-called curse psalm.

> He clothed himself with curses as his cloak,
> And so it entered his inward parts like water,
> Like oil into his bones.

[3] See Bachvarova 2007, 260 n. 42. who cites KUB VII 41 obv. 13, 'A Ritual for the Infernal Deities.' The idea of a garment with supernatural powers, as Lee demonstrates, is evident in Homer. Aphrodite uses her peplos (5.311–17) to shield her son Aeneas in battle.

[4] See Kitz 2004, 315–321; Bachvarova 2007, 183.

> May it be like a garment that he wraps around himself,
> Like a belt that he wears everyday.
>
> Psalm 109. 16–17[5]

The garment simile of this psalm is typical of a range of Semitic curses, oaths and contracts: the curse is represented as a type of poisoned cloak that can inflict a skin disease on its victim. The idea may originate in the use of cloth to apply poultices to infected areas: by reverse analogy the medicinal properties of the poultice become poisonous in the curse as a garment simile. An early example of this curse simile appears in a Hittite oath ritual that compares the act of coating a person with oil to smearing oath–curses onto the body:

> [Just as] you rub yourself down with oil, [thus also] let these oath-curses be rubbed down onto [you]. Just as you put on a garment, so also put on these oath curses.[6]

In this case the oath is guaranteed with a self-curse that would be activated by perjury. Further examples and amplifications can be found in curses, many of them guaranteeing oaths, from Akkadian and Babylonian inscriptions, and a Neo-Assyrian loyalty oath. An Akkadian treaty from the 7th century BCE contains this curse against whoever breaks the treaty:

> May Sin-Nanna, the god (…) cover them with leprosy like a cloak.[7]

Why is the imagery of the curse as a garment so effective and wide spread, especially as conditional curses attached to oaths? One pragmatic explanation might be that clothing can actually transmit diseases. The Old Testament makes reference to leprous garments (e.g. *Leviticus* 13. 47–59): *Kings* 2.5 tells the brief tale of Gehazi who, along with his descendants, was cursed with leprosy after deceiving his master Elisha and accepting two cloaks and bags of silver as forbidden gifts from the former leper Namaan. But there are deeper sociological explanations for the curse as a garment simile as well. In her discussion of the oath curse, Bachvarova observes that clothing is a symbol of its wearer "and marks [the oath taker's] socially defined role (…) the state of his body (…) and the social structure which makes the oath powerful and is assisted by the oath."[8] I would add that the oath and its conditional curse are so deeply inscribed on the body of the oath taker that they cannot be shaken off. The idea of wearing an oath might explain why Pylades tells Iphigenia that she has "thrown an oath" around him in Euripides' *Iphigenia among the Taurians* (ὦ ῥᾳδίοις ὅρκοισι περιβαλοῦσά με, "having encircled me with an easy oath [to keep]" *I.T.* 788) as if she has cast a cloak over him – indeed the verb is used exactly that way in other authors.[9] By offering an oath curse at Iphigenia's request Pylades is now wearing his words.

The curse as a garment in Ancient Greek thought

Is it possible that the pervasive analogy of the curse as a garment in ancient Semitic oath contracts might be the predecessor of the deadly robes of Greek tragedy? There is no doubt that the language of Near Eastern oath-curses influenced the form of self-imprecations that guaranteed ancient Greek oaths. For example the use of wax images (*kolosoi*) as surrogates for perjurers in the Foundation Oath of Cyrene, as Faraone explains, has its antecedent in Near Eastern practices,

[5] See Wright's discussion (1994, 385–404) of the use of analogy and symbolism in Psalm 109 in the context of ritual.
[6] Hittite text KUB 26.25 dated c. 1200 to 1180 BCE (discussed in Kitz 2004, 316).
[7] The inscription is edited and translated by Grayson 1987, 183.
[8] Bachvarova 2007, 183.
[9] See, for example, φᾶρος περιβάλλεσθαι, Hdt. 1.152.

and so do the conditional curses on the descendants of an oath taker, as Priest and Katz have recognized.[10] Scholars think in terms of a Near Eastern/Aegean *koinē*, a cultural community that encompassed the Eastern Mediterranean Basin and that shared a common vocabulary of myth and ritual. Cultural exchange occurring from the Bronze Age to the Hellenistic period accounts for the many similarities between Mesopotamian (*i.e.* the Sumerian, Babylonian and Assyrian societies of southwest Asia), Anatolian (*e.g.* Hittite), Biblical and what we call 'Greek' religion and literature.[11] West's magisterial study of this east/west diffusion acknowledges the influences of Near Eastern ideas about the daemonic power of oath-curses, which includes the persecution of perjurers, ideas which would have been diffused by treaty language between different cultures.[12]

Examples of garment similes in formal Greek oaths and curses are hard to find, but they do exist. The abundant lead *defixiones* and curse tablets from ancient Greece offer other formulae for binding their victims, but their impetus is more an effort to put one individual under the control of another for judicial, commercial or amatory purposes.[13] The Near Eastern garment curses are generally publically oriented, as part of treaty oaths, and their language does not seem to have been transferred to the more personal curse tablets of the Greeks. There is possibly a hint of the curse/garment analogy in Alcaeus (129.17–18 Voigt), which has the format of a curse, and, according to Bachvarova, features other elements of Semitic curses.[14]

Alcaeus remembers 'stern oaths' of alliance (sworn by himself and Pittacus) that included a self-imprecation to 'lie in cloaks of earth' (γᾶν ἐπιέμμενοι κείσεσθαι). There are more explicit examples of garments being used in public oath and curse rituals, however. Pseudo-Lysias describes how priests and priestesses cursed Andocides for defiling the mysteries by shaking purple cloaks, *phoinikida*, towards the west (*Andoc.* 51):

> καὶ ἐπὶ τούτοις ἱέρειαι καὶ ἱερεῖς στάντες κατηράσαντο πρὸς ἑσπέραν
> καὶ φοινικίδας ἀνέσεισαν,κατὰ τὸ νόμιμον τὸ παλαιὸν καὶ ἀρχαῖον.

> And for these transgressions, priestesses and priests stood up and cursed him, facing the west, and shook their purple garments according to the ancient and time-honored custom.

The phrase "according to the ancient and time-honored custom" suggests that this gesture was a relatively familiar one to the Athenians.

As we know from other sources, these purple robes, *phoinikida*, named after the Phoenicians who produced the purple dye, were part of the priestly garments of some priestesses and priests of Demeter.[15] Having a cloak shaken to curse a man for his sins is not the same thing as wearing a curse cloak, but it does seem to be conceptually related, and I shall return to it. We find a significant use of purple robes in Plutarch's description of a Syracusan oath ritual taken in the sanctuary of Demeter. Recounting the perjury of Callippus, the murderer of Dion, tyrant of Syracuse, in 354

[10] Faraone 1993, 60–80; Priest 1964, 48–56; Katz 1998, 183–217. See also Karavites' discussion (1992, 109–16) of the Near Eastern elements in Homeric oath curses.

[11] See Noegel 2007, 25 for bibliography and theories of how this cultural exchange occurred.

[12] West 1997, 19–23.

[13] Scholars working in this area now categorize the curse tablets under categories based on Audollent's 1904 collection (*Defixionum tabellae*). These include judicial, commercial, theatrical, love and 'prayers for justice.' Eidinow 2007, 154 explains the motives for cursing among the ancient Greeks as "a way for people to gain the initiative in a situation of essential powerlessness."

[14] See Bachvarova 2007, 179–88 for a discussion of the Near Eastern elements in the fragment. For a general discussion of the fragment as a curse see Watson 1991, 63–65.

[15] See Connelly 2007, 92 on the purple cloaks worn by religious official at Eleusis.

BCE, Plutarch says that he put on the purple robe of the goddess and held a blazing torch to swear his oath (περιβάλλεται τὴν πορφυρίδα τῆς θεοῦ καὶ λαβὼν δᾷδα καιομένην ἀπόμνυσι, *Dio.* 56.5).[16] Notice that Plutarch uses the verb *periballetai* ("he put on") here, a word that we have seen used of the binding power of the oath as if the oath itself were a garment.

Curses and garments in the *Oresteia*

Surviving Greek historical texts hint at the association between garments and curses but the metaphor is, I contend, quite pronounced in Greek tragedy. The purple fabric that Clytemnestra lays at Agamemnon's feet for his homecoming will be our first example. These textiles are part of a rich system of cloth imagery sustained throughout the trilogy, and they have multiple significance: they are like a river of blood from all Agamemnon's many victims, including his own daughter, Iphigenia; they symbolize his hubris, his excesses, and his all too willing acceptance that he deserves the kind of triumphal homecoming that Priam would expect. According to Goheen, their colour may have been similar to dried blood, and they contribute to the recurring motif of blood on the ground, a symbol of the irreversible nature of death.[17] The fabrics are polysemous and one of their meanings, I would suggest, is that of a curse.

At least one curse is among the interlocking forces that contribute to the death of Agamemnon. He is the victim of his own folly and he makes his own choices, yet the play exemplifies the tragic concept of multiple determination.[18] He has inherited the curse on his household from his father Atreus, and has also incurred the "civic curse" of the Argive citizens (δημοκράντου δ' ἀρᾶς, *Ag.* 459) who, according to the Chorus, blame him for the loss of their husbands and sons at Troy. Would the audience read the purple cloth as a symbol of a curse or of a magical spell? After she has succeeded in getting her husband to walk on the cloth Clytemnestra uses language with elements of magical incantations (*Ag.* 958–73), as McClure has shown.[19] So there is some sense of the supernatural in the air as the king steps onto the purple cloth that leads to the house of death. The exact nature of these textiles is a matter of contention, but Agamemnon refers to them as garments (εἵμασι 921), and Morrell has revived the idea that the king is walking on clothing. [20] If shaking purple robes was an ancient curse ritual, as pseudo-Lysias says it was, then the very act of putting the garments on the ground would have some significance to the audience. In any case, for the ancient Greeks

[16] "Parallels to the use of a burning torch in oath ceremonies in the Near East suggest that this Sicilian oath might have involved a similar self-curse, in which those swearing the oath likened themselves to the burning torch in a conditional execration." Faraone 1993, 64.

[17] The textile imagery begins with the saffron robes of Iphigenia pouring onto the ground as if they were blood (we can imagine the festive colour drenched in sacrificial gore) (*Ag.* 328 ff.). See Goheen 1955, 115–126 for the relationships between the recurring motif of blood on the ground (*Ag.* 1018–21; *Cho.* 48; *Cho.* 66–7; *Eum.* 261–3; *Eum.* 652–4) and the fabrics, and Morrell 1997, 141–42, n.4 for the bibliography on the meaning and nature of the textiles.

[18] Scholars debate how much influence the curse exerts. Lloyd-Jones 1962, 199 suggests that it is because of the curse that Agamemnon finds himself in the dilemma at Aulis. West 1999, 37, on the other hand, objects to putting excessive importance on the curse, and Sewell-Rutter 2007, 75–76 explains the curse as one aspect of the multiple determination that befalls the descendants of Atreus. Mardikes' 1994 dissertation on curses in the *Oresteia* provides a cogent overview of the topic. See Bakola, this volume, for an interpretation of the 'tapestry scene'.

[19] McClure 1999, 81–90 notes that the use of assonance, repetition and metaphor in the context of weaving and cloth production (*cf.* the Homeric Circe), and the appeal to Zeus Teleos at the end of the speech have a particularly incantatory quality. See also Neustadt 1929, 243–65 who discusses language as a form of magic throughout the play.

[20] Morrell 1997, 147 suggests that the garments function as a public display of Clytemnestra's "authority and position within the traditionally intimate confines of the *oikos*".

clothing was just textile shaped, draped and wrapped. Since garments were not tailored, clothes and textiles looked very much alike.

A more obvious example of the curse as a robe, as I have already suggested, is the garment in which Agamemnon is captured. He dies wrapped in what Clytemnestra describes as "an inescapable net (…) an evil wealth of clothing" (ἄπειρον ἀμφίβληστρον (…) πλοῦτον εἵματος κακόν 1382–83). Net imagery runs throughout the *Oresteia* as a signifier of the ineluctable doom that ensnares the family of Agamemnon, but it is most concretely manifested as the robe that captures Agamemnon. Cassandra envisions the robe as a "net of Hades" (ἦ δίκτυόν τί γ' Ἅιδου, 1115), Aegisthus refers to the "net of Justice" (τῆς Δίκης ἐν ἕρκεσιν, 1611) in which Agamemnon now lies, and Orestes, searching for the right word, will describe the robe as a "hunting net" (δίκτυον, *Ch.* 999–1000).[21] Aeschylus uses the image of the net throughout the trilogy with his characteristic allusiveness, but also with particular reference to the robe in which Agamemnon dies. The net is an image, among other things, of the curse which grips the son of Atreus, and it is significant that the net motif is one of the oldest types of maledictions in Sumerian oath curses from the beginning of the 3rd millennium and showing up in the Old Testament.[22]

The binding robe becomes more explicitly associated with curses when Aegisthus refers to it as "the woven robes of the Erinyes" (ὑφαντοῖς ἐν πέπλοις Ἐρινύων 1580) just before he narrates his father Thyestes' curse against Atreus. The Erinyes, especially in this trilogy, are more than just curse goddesses,[23] but this function is well supported both by pre-Aeschylean literature,[24] and by the Erinyes themselves in the *Eumenides*: "we are called curses in our home beneath the earth," (*Eum.* 417) they tell Athena. The menacing quality of their own clothing is obvious in the *Eumenides* when they sing their binding song as they encircle Orestes and threaten him with "black-robed assaults" (ἐφόδοις μελανείμοσιν, *Eum.* 370). Thus when Aegisthus calls the deadly garment a "robe of the Erinyes" especially in the context of his father's curse against Agamemnon's family, there is a good reason to think of the robe as the material manifestation of the curse.

This curse is not the conditional malediction of an oath, but the Near Eastern prototypes are not necessarily always oath curses (*e.g.* Psalm 109). At any rate, Aeschylus seems to have inherited ancient curse language and has adapted it in a spectacularly brilliant way for his dramatic purposes. It is worth noting, however, that the curse of Thyestes is linked to an oath sworn by Clytemnestra to kill Agamemnon for sacrificing Iphigenia. As she stands over the corpses of her husband and Cassandra, she announces to the Chorus:

καὶ τήνδ' ἀκούεις ὁρκίων ἐμῶν θέμιν:

[21] The image of the net is also used of the conquest of Troy (*Ag.* 357–61) in the first stasimon, but in the meditations of the Chorus it becomes the net in which Agamemnon is caught.

[22] The earliest example of the net as an oath curse dates to the Sumerian king E-an-na-tum of Lagash (c. 2900 BCE) who invokes the 'net' of a host of gods, to punish the perjurer (Mercer 1913, 48). At *Exekiel* 17.19–20 God says that because Zedikiah broke his oath, "I will spread My net over him, and he shall be taken in My snare."

[23] On the nature of the Erinyes see Parker (1983, 107) who notes that the distinction between the Erinyes as "spontaneous products of transgression" or of a spoken curse is unnecessary. As Faraone (1983) shows, their binding song or *hymnos desmios* (i.e. the first stasimon of *Eumenides*) bears many similarities to *defixiones* employed to tongue tie litigants and prosecutors in court. They are also among the powers invoked on curse tablets of the fifth and fourth century. For further discussion of their role as curses in the *Oresteia* see Fletcher 2012, 62–66.

[24] In pre-Aeschylean literature the Erinyes are associated with curses: Althea curses her son Meleager by striking the earth so that the Erinys below can hear (*Il.* 9.453–56). In two virtually identical passages (*Il.* 3.276–80 and 19.258–60), Agamemnon invokes them as "those who avenge the dead who have been forsworn" to guarantee oaths in their role as curse goddesses.

μὰ τὴν τέλειον τῆς ἐμῆς παιδὸς Δίκην,
Ἄτην Ἐρινύν θ', αἷσι τόνδ' ἔσφαξ' ἐγώ (...)

(*Ag*. 1431–1433)[25]

And you hear the rightness of my oath-sacrifices: by Justice accomplished for my daughter, and Ate and the Erinys, to whom I sacrificed this man (...)

The lines not only suggest that Clytemnestra swore an oath to avenge her daughter's death – this is made more explicit by Orestes in the following play (*Ch*. 977–80) – but also that Agamemnon is somehow the sacrifice that guarantees this oath. Clytemnestra has perverted the ritual of sacrifice and distorted the formulae of oath swearing.[26] If indeed the tableau of the dead king wrapped in a garment refers to the curse on the house of Atreus, her reference to an oath has a particularly unsettling implication. Agamemnon does not suffer the consequences of perjury, but rather Clytemnestra's oath of vengeance helps to accomplish Thyestes' curse that is now physically present as the deadly garment.

The Poisoned Robe of Sophocles' *Trachiniae*

Let us turn now to the fatal robe that Deianeira sends to her husband Heracles in *Trachiniae*. Sophocles did not invent this part of the story, which dates back to Hesiod (fr. 25.17–25 MW from the *Catalogue of Women*).

(...). καὶ δείν' ẹ̓ρ̣ξ̣[', ἐπεὶ ἀάσατ]ọ μέγα θυμῶι
ὀππότε φάρμακον .[ἐπιχρί]ṣασα χιτῶνα
δῶκε Λίχηι κήρυ[κι] φ[έρειν· ὃ δὲ δῶ]κεν ἄνακτι
Ἀμφιτρυωνιά[δ]ηι Ἡ[ρακλῆϊ πτολιπό]ρθωι.
δ[εξ]αμένωι δέ ο[ἱ αἶψα τέλος θανάτοι]ο παρέστη·
καὶ] θάνε καί ῥ' Ἀΐδ[αο πολύστονον ἵκε]το δῶμα.

(fr. 25.19–25)

(...) and she did terrible deeds, greatly deluded in mind, when she sprinkled the destructive poison on the robe and gave it to the herald Lichas to convey. He took it to his master Heracles, sacker of cities, son of Amphitryon. And when Heracles received it, death's end came quickly to him, and he died and went to the grievous house of Hades. (tr. March)

Sophocles had several versions of the death of Heracles upon which to draw, and it is clear that he adapted a pre-existing story of a poisoned robe dating back to Hesiod or perhaps even earlier.[27] March argues that Sophocles transformed the malevolent, man-killing Deianeira into the innocent dupe of the centaur, a revision that would come as a surprise to the audience.[28] Whatever traditions Sophocles might have been referencing and manipulating, it is also clear that he wants his audience to remember the robe that Clytemnestra used to kill Agamemnon.[29] Heracles describes that robe

[25] This text is Page's 1972 OCT edition.
[26] Both Zeitlin (1965, 474) and Lebeck (1971, 83) note how Clytemnestra perverts the ritual of sacrifice in this scene. See Fletcher 2012, 47–51 for a discussion of her distorted oath ritual.
[27] Davies (1991, xxii–xxxiv) discusses other sources from which Sophocles might have drawn including the lost epic *The Sack of Oechalia*.
[28] March 1987.
[29] As Lee 2004, 264 points out, both garments are called *peploi*, a term that in the context of tragedy signifies a perversion

as a "woven net of the Erinyes" (Ἐρινύων ὑφαντὸν ἀμφίβληστρον 1051–1052). The language is pointedly similar to that of the *Agamemnon*, and conflates both Clytemnestra's description of the deadly robe (ἄπειρον ἀμφίβληστρον (…) πλοῦτον εἵματος κακόν, "an endless net (...) an evil wealth of garment," 1382–1383) and that of Aegisthus (ὑφαντοῖς ἐν πέπλοις Ἐρινύων, "in the woven robes of the Erinyes," 1580). Sophocles has animated the robe, so that it consumes Heracles "like a deadly viper" (ἐχθρᾶς ἐχίδνης), another allusion to the *Oresteia* (used by Orestes of his mother at *Ch.* 929). Garner appropriately describes the garment as a "vampire-robe, which has taken on a parasitic life of its own, drinking his blood and robbing his lungs of breath (1054–56)."[30] Essentially the robe has taken on the properties of the Erinyes, and its action evokes the ancient curse that would seep into the skin of its victim. The "garment clung to his sides close-glued at every joint, as if by a sculptor" (προσπτύσσετο πλευραῖσιν ἀρτίκολλος, ὥστε τέκτονος, χιτὼν ἅπαν κατ' ἄρθρον, 768–769) according to Hyllus. Heracles groans that it is "plastered to my sides, and has devoured my inmost flesh" (πλευραῖσι γὰρ προσμαχθὲν ἐκ μὲν ἐσχάτας βέβρωκε σάρκας, 1052–53). The latter description calls to mind the Psalmist's curse-cloak that "entered his inward parts like water."

While Sophocles has joined together the Hesiodic tale of the poisoned robe and the vocabulary and dramatic situation of Aeschylus, he also seems to be drawing on and exploiting motifs that predate both these Greek works. The robe as a fatal gift that inflicts skin disease has a long history extending from the Old Testament, for example the leprous garment in the Book of Kings, and showing up in the early modern folklore of India as deadly khilats (ceremonial robes).[31] Sophocles also seems to have been aware of the anointing curses that are found in the Near Eastern contracts (e.g. Assyrian treaties), or at least of some descendant of them.

Certainly the corrosive effects of the robe bring to mind Near Eastern garment curses, which are also constructed as "anointing rituals." Deianeira has rubbed the fatal garment with a potion that she thinks is a love charm; it is really a deadly poison given to her by the centaur Nessus. Ancient love magic bears a significant similarity to binding spells or curses – as Faraone puts it, "If eros is a disease, then erotic magic is a curse – thus the blurring of categories here is comprehensible.[32] When Heracles puts it on to make a sacrifice the fire activates the poison and eats into his flesh: "my entire body is wasted" (διέφθαρμαι δέμας τὸ πᾶν, 1055–1056). The robe effectively melts away the flesh of Heracles' body, an outcome that sounds strikingly similar to the melted wax images that represent the workings of the oath curse, a form of sympathetic magic used in some Near Eastern cultures, which I discuss below.

Unlike the *Agamemnon*, however, we are not dealing with an explicit curse here. While we might be inclined to think that Nessus cursed Heracles as he was dying, there is nothing in the text to substantiate this. As we have seen, the robe can be used in different kinds of spells, not just curses, in Hittite magical rituals; and Nessus had persuaded Deianeira to use the garment to cast a spell on her husband. The robe acts like a curse garment, and the language of physical disintegration

of marriage, although other vestimentary terms are used as well.

[30] Garner 1990, 108.

[31] Maskiell and Mayor, 2001a and 2001b catalogue and discuss these legends which were influenced by ancient Iranian customs of presenting robes of honour (khilats) to friends and enemies. As they demonstrate, the Indian legends (dating from around 1600 CE) have similarities with the poisoned robes of Deianeira and Medea.

[32] Faraone 1999, 43. As he further states "On a strictly forma basis, the, the techniques of many forms of erotic magic are quite indistinguishable from those of hostile curses used against enemies or of self-curses used in especially fearful oaths." (55)

suggests that Sophocles has been influenced by the ancient vocabulary of curses transmitted over many centuries. But why are the Erinyes mentioned here? The best answer to this question comes from Garner who writes:

"Heracles' Clytemnestra is not Deianeira, after all; if anything, it is himself. He has produced his own Furies. He has captured women and beasts, tamed and killed beasts and men, and the poison of one of his monstrous conquests [Nessus] has blended with the blood of another [the Hydra] so that one captured woman could use it to try and regain him from the latest female prize."[33]

Medea's fatal robe

Since the dating of the *Trachiniae* is unknown – although the style suggests an early production – it is impossible tell if it came before or after Euripides' *Medea* (produced in 431 BCE). Like *Agamemnon* and *Trachiniae*, *Medea* features a woman who must confront the fact that her husband has selected a younger female for his bed. The disturbed households that result from the husbands' alienation are symbolized by a perversion of the normative use of cloth, a domestic product, and therefore a fitting symbol of the *oikos*. Sophocles clearly wants his audience to think of Clytemnestra and her robe, but he makes no textual reference to *Medea*, nor does Euripides make reference to *Trachiniae*. Certainly one poet could have borrowed the idea of the flesh eating robe from the other, and the situational similarity with the jilted wife and the poisoned gift makes this a strong possibility. Lee is undoubtedly justified in suggesting that Aeschylus was the first to make the deadly peplos a signifier of the dysfunctional household, but there is no way to determine whether Sophocles or Euripides was the next poet to pick up the motif.

My interests lie more in how each tragedian used the enduring motif of the toxic garment that was part of an ongoing process of cultural transmission in the Eastern Mediterranean Basin. Aeschylus exploited the connotation of the curse as a garment, and worked in an even more ancient idea of the curse as a net. Sophocles developed the robe's related association with magic spells, and went further than Aeschylus by expropriating and dramatizing the corrosive qualities of the robe. Euripides combines both these attributes of the poisoned robe; of the three toxic garments, his is the most closely aligned with the Near Eastern prototype of the garment curse appended to oaths.

It is significant that Euripides' version of the poisoned robe actually does occur in the context of a perjured oath. The fate of Jason – the loss of his sons and the woman who would bear him more sons – is most certainly a penalty for breaking an oath to Medea, sworn before the play begins.[34] She accuses him of precisely this at several points (161–163, 492, 1392), and the Chorus concurs (409–445). The destruction of a man's family is a typical malediction in Greek oaths, and Jason's failure to take his oath (and Medea's power) seriously leads to the catastrophe. But why must the Corinthian princess pay the penalty as well? Burnett suggests that both she and her father, Creon, are guilty by association; they are accessories to perjury, as it were.[35]

Perhaps, however, we can go further than this. In a recent study Holland suggests that both Jason and Creon were targeted by a curse on the house of Aeolus, passed down through Sisyphus

[33] Garner 1990, 109.
[34] The importance of the perjury of Jason was first noted by Burnett 1973, 1–24. Kovacs (1994, 45–70) notes that the oath of Aegeus allows Medea to act as the retributive agent for this perjury.
[35] Burnett 1973, 14–15.

whose family was "linked in tragedy to an inherited curse" involving maternal infanticide, *e.g.* Melanippe.[36] Medea makes reference to Sisyphus at strategic points in the drama (404–407 and 1382), and thus, argues Holland, signals Jason's implication in this inherited curse. The curse is therefore also the unfortunate legacy of the princess. Innocent herself of any wrongdoing, she nonetheless exemplifies how a curse can visit the descendants of its original target. On the other hand, we need to heed West's caution about attaching too much significance to ancestral curses in the interpretations of tragedy. We cannot assume that the audience would automatically expect Euripides to be dramatizing this ancient curse. As West notes, Greek myth is a very flexible and dynamic tradition that provides basic outlines of a plot, but any myth can be manipulated and changed according to the agenda of the individual poets.[37] Holland's thesis is intriguing, and it adds nuance to the death of Creon's daughter, but we do not need to construct a hypothetical curse, since the curse of Jason's perjured oath is operating quite manifestly.

The curse garment that Medea anoints with poison (789, along with the crown), recalls the anointing ritual associated with Near Eastern curses (for example the Hittite oath curse mentioned above which required the oath taker to "rub yourself down with oil" and put on the oath curses as a garment.) The effects of Medea's poisoned gifts upon the princess bear some resemblances to the curses of Near Eastern oaths. The crown bursts into flames on the young woman's head, and the poisoned cloak causes the skin of the princess to melt from her body like "like tears of sap oozing from pinewood" (1186–1200); the flesh of her father is pulled from his bones when he touches his daughter's corpse. The fiery quality of the gifts is akin to the Near Eastern oaths and treaties that threaten the oath breaker with fire and disintegration. The image of the disintegration or melting of the target of curse occurs in a Hittite oath (1450–1350), "Just as this wax melts, and just as this mutton fat dissolves, whoever breaks these oaths (…) let him melt lik[e wax], let him dissolve like [mutton fat]." An Aramaic text known as the "Sefire Inscription" sworn in the 8th century BCE is typical of a group of curses that probably used wax images as surrogates for the perjurers. "As this wax is consumed by fire, thus Ma[tti'el] shall be consumed b[y fi]re."[38] The action of the cloak on Creon and his daughter is remarkably like these vivid curses.

This physical manifestation of the curse is a variant of the skin aliment that would be imposed on perjurers and targets of curses in Near Eastern treaties. The destruction of the family of Creon set in motion by the ancient curse, or by their association with Jason's perjury, is now complete, and it sets the stage for the contingent punishment of Jason, violator of oaths and possible co-inheritor of a curse. While Medea kills her sons the Chorus name her as an Erinys (1260); she is the embodiment of a curse that has visited Creon, his daughter, and Jason, and perhaps it is significant that this Asian woman has activated a curse formula that has its origins in the East.[39] We do not have the terms of the oath that Jason swore to Medea, but the oath that Medea extracts from Aegeus (746–53) suggests what its format might have been. "Swear by the plain of Earth and the Sun," she demands. These guarantor gods, Earth and Sun, are only used in loyalty or treaty

[36] Holland 2003, 263.

[37] West 1999.

[38] See Faraone 1993, 62 who explores the similarities between these types of oath curses and the oath of the Theran colonists in Cyrene in late 7th century BCE (SEG ix 4).

[39] Holland (2003, 269) notes, however that "in one sense Medea does embody the Erinyes as its agent of destruction, but the demonic spirit should also be seen as its own entity." Luschnig makes similar observations (2006, 2), but also notes (156) that Medea functions as an Erinys who punishes Jason for transgressions against "the rules of family life."

oaths, whose terminology can be traced back to Near Eastern treaty oaths.[40]

In conclusion: the tragic poets seem to have been aware of the ceremonial language that represented spells and curses as skin diseases coating the body like a cloak, or with anointing rituals that were metaphoric cloaks. The curse as a garment simile was widely used in oaths and treaties from Hittite and Kassite contracts from c.1400 BCE to Akkadian and Neo-Assyrian documents 900 years later. Although this form of malediction did not seem to be used in the treaty oaths of the Greeks, who apparently preferred the standard imprecation on a man's progeny, it does seem to have seeped into the cultural imagination of 5th century Athens. The tragic poets are not interested in replicating an ancient treaty, but they have mined the language of curses for the cause of poetry. Note that it is female characters who wield the toxic textiles: the poets have exploited a long standing association of women and cloth production and have married that convention to an ancient curse or spell formula that conceives of the garment as an instrument of pain and death.

[40] The earliest Greek example would be the temporary truce between the Greeks and Trojans at Iliad 3.85–301. Bachvarova 2007, 180 usefully summarizes the scholarship on the similarities between the Iliadic treaty oath and its Near Eastern precedents which include invocations of Earth and Sun.

Bibliography

Audollent, A. (ed.) 1904 *Defixionum tabellae quotquot innotuerunt: tam in Graecis Orientis quam in totius Occidentis partibus praeter Atticas in Corpore Inscriptionum Atticorum.* Paris.

Bachvarova, M. 2007 Oath and Allusion in Alcaeus fr. 129. In A. H. Sommerstein & J. Fletcher (eds), *Horkos: The Oath in Greek Society*. Exeter, 179–188, 258–264.

Burnett, A. P. 1973 *Medea* and the Tragedy of Revenge. *CP* 68, 1–24.

Connelly, J. B. 2007 *Portrait of a Priestess: Women and Ritual in Ancient Greece*. Princeton NJ.

Davies, M. 1991 *Sophocles* Trachiniae *with Introduction and Commentary.* Oxford.

Eidinow, E. 2007 *Oracles, Curses, and Risk among the Ancient Greeks*. Oxford.

Faraone, C. A. 1985 Aeschylus' *hymnos desmios* (*Eum.* 306) and Attic Judicial Curse Tablets. *JHS* 105, 150–154.

Faraone, C. A. 1993 Molten Wax, Spilt Wine and Mutilated Animals: Sympathetic Magic in Near Eastern and Early Greek Oath Ceremonies. *JHS* 113, 60–80.

Faraone, C. A. 1999 *Ancient Greek Love Magic*. Cambridge MA.

Fletcher, J. 2012 *Performing Oaths in Ancient Greek Drama*. Cambridge.

Garner, R. 1990. *From Homer to Tragedy: the Art of Allusion in Greek Poetry.* London–New York.

Goheen, R. F. 1955 Aspects of Dramatic Symbolism: Three Studies in the *Oresteia. AJPh* 76, 113–137.

Grayson, A. K. 1987 Akkadian Treaties of the Seventh Century B.C. *Journal of Cuneiform Studies* 39, 127–160.

Holland, L. 2003. *Pas Domos Erroi*: Myth and Plot in Euripides' *Medea. TAPhA* 133, 255–279.

Karavites, P. (with the collaboration of T. Wren) 1992 *Promise-Giving and Treaty-Making Homer and the Near East.* Leiden.

Katz, J. R. 1998 *Testimonia Ritus Italici*: Male Genitalia, Solemn Declarations, and a New Latin Sound Law. *HSCP* 98, 183–217.

Kitz, A. M. 2004 An Oath, Its Curse and Anointing Ritual. *JOAS* 124, 315–321.

Kovacs, D. 1993 Zeus in Euripides' *Medea. AHPh* 114, 45–70.

Lebeck, A. 1971 *The* Oresteia*: A Study in Language and Structure*. Cambridge.

Lee, M. 2004 Evil Wealth of Raiment: Deadly *Peploi* in Greek Tragedy. *CJ* 99, 253–279.

Lloyd-Jones, H. 1962 The Guilt of Agamemnon. *CQ* 12, 187–199.

Luschnig, C. A. E. 2007 *Granddaughter of the Sun: A Study of Euripides' Medea*. Leiden.

McClure, L. 1999 *Spoken Like a Woman: Speech and Gender in Athenian Drama*. Princeton NJ.

March, J. 1987 *The Creative Poet: Studies on the Treatment of Myths in Greek Poetry.* London.

Maskiell, M. & Mayor, A. 2001a Killer Khilats, Part 1: Imperial Collecting of Poison Dress Legends in India. *Folklore* 112, 23–45.

Maskiell, M. & Mayor, A. 2001b Killer Khilats, Part 2: Imperial Collecting of Poison Dress Legends in India. *Folklore* 112, 163–182.

Mardikes, C. M. 1994 Curses and Conspiratorial Oaths in the *Oresteia* of Aeschylus. Diss. Chicago IL.

McClure, L. 1999 *Spoken Like a Woman: Speech and Gender in Athenian Drama.* Princeton NJ.

Mercer, S. A. B. 1913 The Oath in Cuneiform Inscriptions. *JOAS* 33, 33–55.

Meir, G. (ed.) 1937 *Die assyrische Beschwörungssammlung Maqlu.* Berlin.

Morrell, K. S. 1997 The Fabric of Persuasion: Clytaemnestra, Agamemnon, and the Sea of Garments. *CJ* 92, 141–165.

Neustadt, E. 1929 Wort und Geschehen in Aischylos' *Agamemnon. Hermes* 64, 243–265.

Noegel, S. B. 2007 Greek Religion and the Ancient Near East. In D. Ogden (ed.) *A Companion to Greek Religion.* Oxford, 21–37.

Parker, R. 1983 *Miasma: Pollution and Purification in Early Greek Religion*. Oxford.

Priest, J. F. 1964 *Horkia* in the *Iliad* and Consideration of a Recent Theory. *JNES* 23, 48–56.

Sewell-Rutter, N. J. 2007 *Guilt by Descent: Moral Inheritance and Decision Making in Tragedy*. Oxford.

Watson, L. 1991 *Arae: The Curse Poetry of Antiquity.* Leeds.
West, M. L. 1999 Ancestral Curses. in J. Griffin (ed.) *Sophocles Revisited.* Oxford, 31–45.
West, M. L. 1997. *The East Face of Helicon.* Cambridge.
Wright, D. 1994 Ritual Analogy in Psalm 109. *JBL* 113, 385–404.
Zeitlin, F. I. 1965 The Motif of the Corrupted Sacrifice in Asechylus' *Oresteia. TAPhA* 96, 463–508.

7. Textile Symbolism and the 'Wealth of the Earth': Creation, Production and Destruction in the 'Tapestry Scene' of Aeschylus' *Oresteia* (*Ag.* 905–978)

*Emmanuela Bakola**

Spinning, weaving and the symbolism of production, generation and creation

A recurring theme in Greek thought is the reflection on the way the production and consumption of wealth structures the relationship between humanity and nature. There is an ongoing dialogue in the poetry and philosophy of archaic and classical Greece that explores the ethical connotations of agricultural labour, manufacturing, technology, warfare, commerce and other types of economic activity in relation to how they affect the balance between humans and environment.[1] Different forms of human economic activity thus become invested with profound – if constantly re-negotiated – symbolic values that relate not just to the community, but to the natural environment more widely.

Textiles constitute one of the most prominent forms of wealth in the ancient world. In Greece, spinning and weaving were regarded as the primary female economic activity within the *oikos*, alongside child-bearing, child-rearing and overall household management, with which they were closely associated. Female slave labour generated the largest amount of textile-wealth in the domestic space, but even outside it, commercial textile-production was almost exclusively connected with female (free or slave) labour.[2] The dominant role of spinning, weaving and textile-making in the

* This paper is part of a wider research project, entitled *Aeschylean tragedy and early environmental discourse*, which employs spatial criticism and cultural anthropology to demonstrate that the relationship of humans, earth and natural resources is a central theme in Aeschylean tragedy. I would like to thank Marie-Louise Nosch, Mary Harlow, Giovanni Fanfani, Judith Fletcher and Chris Carey for their generous comments on this paper and the Leverhulme Trust for funding my research.

[1] See especially Vernant 1983b, 1983c and Nelson 1998, 162–170, with analyses of Homer, Hesiod, Aeschylus, Xenophon, the Hippocratic and the Aristotelian corpuses. For further insights into the relationship of humans and nature in Greek literature and the issues of productivity, labour, use and control of resources, see Segal 1963.

[2] See Xen. *Oec.* 7.36; Xen. *Mem.* 2.7.2–12; Ar. *Lys.* 571–582, 728–750; *Inscr. Cret.* 4.72 col. 2 48–52, col. 3 17–24; Aeschin. 1.97. The bibliography on this topic is enormous. For iconographical material and analyses, see Bundrick 2008 and Lewis 2002, 59–65. For general discussions, see Neils 2011, 92–97; Håland 2004; Cottica 2004; Carr 2000; and Pomeroy 1994, 61–65. For the connection of female domestic labour with the necessities of childcare, see Neils 2011, 94 and Brown 1970. For Attic iconography connecting the production of children with the production of textiles, see Bundrick 2008, 304–306, 316–317, 320–2, and Lewis 2002, 81–82. For women's woolwork outside the house, see Brock 1994, 338 and Rosivach 1989, 366 with nn. 6–9. For the involvement of males in commercial textile production, see Thomson 1982 and Pomeroy 1994, 61–62. However, as Lyons 2003, 106 rightly notes, "the centrality of the male/female divide to the binarism so prevalent in Greek mythic thinking leads to a mapping of gender onto the division of labor, even in contexts where this may not reflect social realities."

lives of Greek women endowed these activities with rich symbolic associations which are echoed extensively in both literature and material culture. Scholars have so far explored the symbolic use of textiles mainly by looking at the ideological association of the female with guile and deception, as well as in relation to a woman's perceived role in marriage and domestic and social harmony.[3] However, only minimal attention has been paid to the textiles' symbolism in relation to concepts of wealth-creation, generation of life and growth.

As I will show, this is a very promising line of enquiry, since both the domestic space and the female gender, categories with which textile-making is closely associated, possess immense symbolic capital in relation to the concepts of creation, production and procreation. This suggestion is supported by cross-cultural comparisons. In traditional cultures of all periods, textiles are symbolically associated with female spheres of activity, especially childbearing and childrearing. This is similarly the case in their connection with natural materials and the way these are processed to generate commodities: non-invasive procedures, such as animal and plant tending, are used to obtain wool and flax, the primary materials of textiles. Thus, in many traditional cultures, textiles and textile-making are understood as 'creative' wealth and are fundamentally connected with growth and life.[4] Like agricultural activity, textile-making is conceptualised as one of the most 'natural' of human interventions in the natural environment since it adds value to materials without destroying them. The fact that this process is often connected to female labour enhances this symbolism. Furthermore, the activity of weaving frequently has positive associations with nature, earth and life, and even takes on religious symbolism when it is used metaphorically to conceptualise the generation of life and even the creation of the cosmos.[5] Textile-making is thus connected with ecological forms of thinking in a nexus that includes the female, nature, harmony and balance with the external world.

Several mythical paradigms from ancient Greece suggest that this line of enquiry is an appropriate one to pursue in relation to the Greek context. The most suggestive parallel is the tradition of the Fates, in which the creation of life is likened both to the spinning of a thread and to the weaving of a garment.[6] Furthermore, life and textile-making processes seem to have constituted an important association of Orphic thought, which compared the structure of a living being to a woven net and the male's copulation with the female to the process of weaving.[7] The association that underpins the link between 'human tissue', 'fabric', 'structure' and 'net' is attested in poetry and prose of archaic and classical times, including Aeschylean tragedy.[8] Related religious and mythical traditions are equally suggestive of the generative symbolism of wool-working and textile-making. Pherecydes, for

[3] See, for example, Bundrick 2008; Lyons 2004; Lee 2004; Scheid & Svenbro 1996 *passim*; McClure 1996/7, 127–130; Jenkins 1985. Textiles are also associated with material status and power (see *e.g.* Thalmann 1980) and textile-making with poetic craft (see Guilleux Coward, Fanfani & Harlizius-Klück in this volume, with bibliography).

[4] See especially Weigle 1989, ch. 3 on 'Spinners and Spinsters', where she discusses cosmogony, parturition and nature in North and Middle American societies, as well as societies in India and the South Pacific. Weaving as a creative process in relation to the earth has a central symbolism in the cultures of the Navajo people and the Kogi Indians. See M'Closkey 2002, 205–255; Reichel-Dolmatoff 1990, 12–17, 30–31.

[5] See the 2006 volume *Weaving and Cosmology*, ed. by E. Lyle. See also Eliade 1961, 113–121; Eliade 1985, 3–4. For cosmic weaving in Talmud, Estonia and India in relation to early Greek philosophy, see West 1971, 54–55, with references.

[6] See *SEG* 15.670 and 796; 27.615; 39.1132; fr. 1018 *PMG*; *cf.* Orph. *H.* 59.16. See Onians 1951, 416–419.

[7] Orph. fr. 407 *PEG* 2.1; *cf.* Vernant 2001, 85.

[8] See Arist. *Gen. Anim.* 734a20–21 (= Orph. fr. 404 PEG); E. *Bacch.* 746–747. Rosenmeyer 1982, 120 rightly argues that this link underlies *Ag.* 866–8. For the idea see also Emped. fr. 50 Wright, Pind. *O.* 2.55–77, Hdt. 2.123, Arist. *de An.* 407b21–23, and *cf.* Scheid & Svenbro 1996, 157–163.

example (mid-6th century), imagines the creation of the cosmos as a process of weaving a garment. It envisages the creator-god Zas weaving a richly designed robe for his bride Chthoniē, with which she becomes Gē, the earth of our familiar cosmos (frr. 14 and 68 Schibli). It is remarkable that Pherecydes' poem uses the metaphor of weaving in order to depict a harmonious cosmogony, thus departing from the violence so often present in the cosmogonies of other traditions.[9] Both this myth and other mythical traditions highlight the connection of spinning and weaving with the institution of marriage, which uses the bonding of male and female to ensure the creation of new life.[10]

Further evidence for the connection of textiles with the processes of production and growth is found in the semantic connotations of ἄνθος and related terms, like ἄωτος. As Borthwick 1976 and Raman 1975 have shown, these terms, which denote 'flower' and 'blossom', are used in some poetic contexts (including Aeschylean tragedy) to denote the soft nap of woven wool, the part of the fabric that gives it its sheen and bloom. At the heart of the association between the imagery of 'flowers' and 'softly woven fabric' is the idea of 'freshness' and of 'young age'. This is evident, for example, in the allusion to the wearing away (lit. 'scouring') of the Greek youth in the harsh conditions at Aulis, as described by the Herald in the *Agamemnon* (τρίβῳ κατέξαινον ἄν/θος Ἀργείων, "they wore away the flower of the Argives / they scoured the soft nap of the Argives" *Ag.* 192–198).[11] The connection between young blooming life and freshly woven fabric seems to be based, among other things, on their shared natural origins and the idea of creation and growth. As we will see, the same connection is exploited at length in Aeschylus' *Persians*, which uses the imagery of fabrics to explore in detail the idea of natural wealth and life.

This chapter argues that the symbolic capital of textiles and textile-making as source of generation, creation and production is key to understanding one of the most famous and heavily discussed scenes of ancient Greek theatre, the 'tapestry scene' of the *Oresteia.* It argues that Aeschylus capitalises and expands on this capital – including on his own deployments of the imagery in earlier plays – to portray wealth and life as natural wealth. He does so, in particular, by evoking two processes of textile production in this scene, namely weaving and dyeing with murex purple. The chapter demonstrates that through the gendered, spatial and symbolic associations of these processes, as well as their power to evoke concepts of production, destruction, value-addition and waste, the 'tapestry scene' explores natural productivity in relation to human economic activity and in contrast to human greed, destructiveness and wastefulness of resources. Textiles, in other words, play a crucial role in Aeschylus' symbolic reflection on humanity's place within the natural order.

Wealth, natural processes and textile symbolism before the *Oresteia*

Like much archaic and early classical poetry, Aeschylean tragedy dwells extensively on questions of acquisition, production, use and abuse of wealth. In relation to other treatments of the theme, these plays in particular are distinctive in that they consistently treat the idea of natural growth

[9] The portrayal of the arch-god Zas using feminine and creative crafts is consistent with the fact that he abstains from bloodshed in his acquisition of power (the battle against Orphioneus is left to Chronos-Kronos). See Granger 2007, 152–153. For a discussion of the fragment, see Schibli 1990, 50–77 and Purves 2010, 101–108. For other uses of weaving in relation to the earth and the cosmos in the Greek context, see Orph. fr. 288 *PEG*; Plato *Rep.* 616c-d.

[10] For analyses, see, for example, Scheid & Svenbro 1996, 53–82; McNeil 2005.

[11] See also Fraenkel 1950 *ad loc:* "It is just this destructive attrition of the natural faculties as an effect of idleness (...) that is the main point of τρίβῳ".

within the overarching idea of wealth. Wealth is conceptualised as anything generated by either human labour or natural productive powers or – most often – both. Thus, material resources, natural growth and human life are amalgamated under a composite understanding of 'wealth'. Within this constellation of ideas, human life, whose creation is connected to both the fertility of humanity and that of the earth, is presented as the most valuable manifestation. These concerns are particularly prominent in the *Persians*, the *Seven against Thebes* and the *Oresteia.*[12]

The *Persians* is particularly significant for my reading of the 'tapestry scene' because it connects all three concepts – wealth, natural growth and life – with the imagery and symbolism of textiles. Textiles are among of the most prominent symbols through which the play reflects upon the issue of use and abuse of material and human resources. The overarching theme of the play, the downfall of the Persians from prosperity to destruction is captured by the devaluation of their textiles, which are prominent both as images and as visible objects. As the play unfolds, the image of the luxurious clothes worn by the Persians is gradually replaced by images of black funerary garments (*Pers.* 114–115; *cf.* 605–606), ripped cloaks in the hands of mourners (*Pers.* 122–125, 537–539), torn fabrics on floating youthful corpses (*Pers.* 274–277), and rags on the body of Xerxes (*Pers.* 199, 468, 834–836, 1030) and the chorus (*Pers.* 1060). Crucially, the catastrophe of wealth in this play does not include solely material wealth. One of the terms used to refer to textiles in the *Persians*, *kosmos*, is key to understanding the semantic and symbolic spectrum of 'wealth' in the play.[13] Through the manipulation of this multifaceted term in the *Persians*'s imagery and stage action, Aeschylus shows that it is not just Persia's material *kosmos* that is lost and mourned, but above all the *kosmos* of its youth, the wealth that blossomed from its earth:

τοιόνδ' ἄνθος Περσίδος αἴας
οἴχεται ἀνδρῶν,
οὓς πέρι πᾶσα χθὼν Ἀσιῆτις
θρέψασα πόθῳ στένεται μαλερῷ

Pers. 59–62

Such is the flower of manhood, such the flower of the Persian land which has gone. The entire land of Asia which nurtured them grieves with violent yearning

Indeed, as the play draws towards its climax, the chorus mourns the irreplaceable loss of young men as both *kosmos* that has been scythed away / sheared, and *anthos* that has withered and gone:

ὀτοτοῖ, βασιλεῦ, στρατιᾶς ἀγαθῆς
καὶ περσονόμου τιμῆς μεγάλης
κόσμου τ' ἀνδρῶν,
οὓς νῦν δαίμων ἐπέκειρεν.
γᾶ δ' αἰάζει τὰν ἐγγαίαν
ἥβαν Ξέρξᾳ κταμέναν, Ἅιδου

[12] See *Pers.* 59–60, 165–168, 249–255, 512, 821–822, 842, 919–927, 978; *Ag.* 198, 525–528, 545, 872, 954–955, 1008–1021; *Eum.* 781–787 = 811–817, 909–912; *Sept.* 16–20, 412–416, 477, 584–588, 769–771, 815–819. The connection between natural growth and the growth of life in relation to wealth has been discussed by scholars of Greek religion, *e.g.* Parker 2005, ch. 18 'The Growth of Plants and Men'. This association is common in societies that are heavily dependent on the earth, like ancient Greece. I have explored these associations in the *Oresteia* in Bakola 2013 and in the *Persians* in Bakola 2014.

[13] As an abstraction, *kosmos* means beautiful arrangement and order, but as a physical item it refers to ornaments of the body, especially clothing, and has connotations of luxury and wealth. *Cf.* Nagy 1990, 145.

> σάκτορι Περσᾶν· ἀγδαβάται γὰρ
> πολλοὶ φῶτες, χώρας ἄνθος
> (...) ἐξέφθινται
>
> *Pers.* 918–927

Ototoi, my king, for that fine army, and for the great honour of Persian empire and the *kosmos* of its men, whom now the god has scythed away / sheared![14] The land laments its native youth (or: the youth in its depths), killed by Xerxes, who crammed Hades with Persians: many men who were marched away, the blossom of the land (...) have perished and gone. (tr. Sommerstein 2008, adapted)

At the very end of the play, Xerxes' entry on stage dressed in rags is also crucial for understanding the connection between textiles, human life and wealth (see fig. 7.1). The king's *kosmos*, as Dareios and the Queen had feared earlier, is gone forever (*cf. Pers.* 833, 849). The Queen's intention to bring Xerxes a new outfit (*Pers.* 849–850) is never fulfilled. As the ragged king points to the pitiful remnant of his dress and refers to it as *stole* (*Pers.* 1017), there is a poignant wordplay with *stolos*, the army that he lost.[15] Instead of following him, the youth of Persia is now crammed in the depths of the earth like a wasted harvest (*Pers.* 922–924). The chorus' final decision to rend their garments in mourning (*peplon, Pers.* 1060) as they painfully tread on the earth (*Pers.* 1069, 1073) that generated this 'harvest of tears' (*Pers.* 819–826) provides a climax to the play's concern with

Fig. 7.1 Image from the 2006 production of the Persians, *dir. L. Koniordou, National Theatre of Greece. Photo: Takis Diamantopoulos. This particular manipulation of Xerxes' costume follows the Aeschylean text in that it has clear connotations of lifeblood that is shed on the earth, devalued and ruined.*

[14] Ἐπικείρω evokes scything crops (natural wealth, since the youth is imagined as born from the earth, *cf. Pers.* 59–62) as much as it evokes shearing wool (see *LSJ* s.v.), including shearing to remove the nap of the fabric, making it lose its fine downy pile and hence its beauty: see Borthwick 1976, 3.

[15] Rightly so Saïd 2007, 91–92.

loss.[16] Persia has irredeemably lost its prosperity because it has lost its most valuable asset, the lives of the men born from its earth (*Pers.* 59–62).

Textiles as 'wealth of the oikos' in the 'tapestry scene' of the *Oresteia*

Aeschylean tragedy frequently returns to the powerful metaphor of textiles, reworking it and enhancing it with extra layers of symbolic value. The most striking instance is to be found in Aeschylus' last trilogy, the *Oresteia*, where alongside regular invocations of weaving and spinning, textiles take on central importance in four key scenes: the famous 'tapestry scene' of the *Agamemnon* (*Ag.* 905–974), the two mirror scenes which close the first and the second play (*Ag.* 1372–end; *Cho.* 973–end) and the finale of the trilogy, where the Erinyes are admitted into the *polis* of Athens (*Eum.* 996–end). The present chapter will explore how the symbolism of textiles as a form of wealth and their symbolic associations with production and creation encourage a deeper understanding of the 'tapestry scene', as well as a more profound understanding of the *oikos* interior, whose multiple semantic functions dominate the entire trilogy. The scene (see fig. 7.2) is located halfway through the first play, the *Agamemnon*. It follows a dense reflection on the Trojan war as an economic enterprise, which, like all military actions in Aeschylean tragedy, is portrayed as wasteful of life and disruptive of nature.[17]

Agamemnon returns to his house victorious from the invasion of Troy, standing on his chariot, accompanied by the human symbol of Troy's destruction and looting, Cassandra, "the flower of the booty" (πολλῶν χρημάτων ἐξαίρετον / ἄνθος, στρατοῦ δώρημ', "a gift from the army, the choice flower of its rich booty" *Ag.* 954–955). Clytemnestra insists that Agamemnon does not touch his foot on the earth (*Ag.* 906–907),[18] but that he enters the house by trampling, and hence symbolically destroying, the household's intricately woven, purple-dyed and precious textiles spread in front of him, by way of sacrifice to the divine (*Ag.* 908–930). Agamemnon hesitates to commit an act which he perceives as hubristic and barbaric (*Ag.* 914–943); but after a tense exchange, Clytemnestra persuades him. As Agamemnon is about to set foot on the textiles, he expresses hesitation once more and describes his act as "destruction of the wealth of the *oikos*", a concept encapsulated in the striking term δωματοφθορεῖν:

> καὶ τοῖσδέ μ' ἐμβαίνονθ' ἁλουργέσιν θεῶν
> μή τις πρόσωθεν ὄμματος βάλοι φθόνος.

[16] The 'harvest of tears' image has a pivotal role in the *Persians*. Most recently, Rosenbloom 2006 has titled the chapter which closes his analysis of the play 'The Harvest of Tears'. He argues (at 2006, 122–127) that this striking metaphor constitutes the epilogue of a dense nexus of ominous agricultural and natural imagery which pervades the characterisation of the Persian expedition.

[17] The parodos (esp. *Ag.* 126–166, 206–249), as well as the first (*Ag.* 369–384, 433–455, 461–474) and the second choral odes (*Ag.* 688–736), invoke the human cost of the war. For the ideas of youth, natural growth and their abuse, see also *Ag.* 197–198, 659–660 (*cf. Pers.* 59–60, 252, 511–512, 821–826, 922–927, 978 and the analysis above, as well as *Sept.* 16–20, *Supp.* 659–666). For the commodification of life and the violation of the natural processes as a result of the war, see *Ag.* 207, 359, 438, 525–528, 709–711. For the angry reaction of nature to the expedition, see *Ag.* 187–201, 555–574, 648–673.

[18] *Ag.* 904–906 show that the chariot must have stood very close to the edge of the fabrics, so that Agamemnon would not have to step on the earth at all. The NT production (shown in figs. 7.2 and 7.3) was otherwise very faithful to the stage directions implied by the Aeschylean text, largely because it was influenced by Taplin's *The Stagecraft of Aeschylus*; *cf.* Parker 1986 *passim*, esp. 343–346; Taplin 2001.

πολλὴ γὰρ αἰδὼς δωματοφθορεῖν ποσὶν
φθείροντα πλοῦτον ἀργυρωνήτους θ' ὑφάς.

Ag. 946–949

and as I walk on these purples of the gods, may no envious eye strike me from afar! For I feel a great sense of impropriety about despoiling this house under my feet, ruining its wealth and the woven work bought with its silver. (tr. Sommerstein 2008, adapted)

Less than ten verses later, Agamemnon starts trampling, and symbolically 'destroying', the textiles. As he tramples, verse-by-verse, beat-by-beat, his steps are directed by the highly cryptic and disquieting words of Clytemnestra[19]

ἔστιν θάλασσα–τίς δέ νιν κατασβέσει; –
τρέφουσα πολλῆς πορφύρας ἰσάργυρον
κηκῖδα παγκαίνιστον, εἱμάτων βαφάς·
οἶκος δ' ὑπάρχει τῶνδε σὺν θεοῖς, ἄναξ,
ἔχειν· πένεσθαι δ' οὐκ ἐπίσταται δόμος.
πολλῶν πατησμὸν δ' εἱμάτων ἂν ηὐξάμην,
δόμοισι προυνεχθέντος ἐν χρηστηρίοις,
ψυχῆς κόμιστρα τῆσδε μηχανωμένη.

Ag. 958–965

The sea is there – and who shall quench it? – nurturing the juices which yield much purple worth its weight in silver, wholly renewable, the dye of vestments. The *oikos* has an abundance of these with the gods' help, my lord. This house does not know how to be poor. To contrive a means of bringing this life back, I would have vowed to trample many garments, if that had been prescribed by an oracle.[20]

An enormous, and ever-increasing, amount of scholarly ink has been expended in analyses of this scene. The various readings attest to the scene's manifold layers of meaning. Most studies have privileged the textiles' symbolic evocation of blood spilt on the ground, and consequently, of the loss of life caused by and within the Atreid *oikos*.[21] Others have emphasised the connection between these woven objects and the imagery of nets, snares and hunting that populate the trilogy, as well as with being caught by the Eriny(e)s, and by extension, with the curse of the *oikos* as embodied by the Erinys.[22] Other studies have highlighted the themes of marriage, female authority, oriental luxury and pomp, the motif of perverted sacrifice and the custom of apotropaic destruction of wealth.[23]

Although all interpretations of the textiles are valuable, some are more successful than others at dealing with a central element of the scene, namely the representation of the destruction of the textiles as destruction of the 'wealth of the *oikos*'. That the element of wealth is crucial for our understanding of the profound symbolism of the textiles and of the scene as a whole is suggested not only by the words of Agamemnon at *Ag.* 946–949 and Clytemnestra at *Ag.* 958–965 (both

[19] For this stage direction, see below, n. 27. For Clytemnestra's multiple characterisation, see below, pp. 129–131.
[20] The edition I follow is Page, *OCT*. The translations have been adapted from Collard 2003 and Sommerstein 2008.
[21] See Goheen 1955, 115–126, with analysis of the colour symbolism. This thesis has been accepted (and further elaborated) by nearly all subsequent scholars.
[22] This is also a universally prevalent symbolism. See, most recently, Lee 2004, 262–267; McClure 1996/7, 128 and n. 23. For the connection between the fabrics and the curse, see Fletcher in this volume.
[23] For nuptial imagery see McNeil 2005 *passim*; Rabinowitz 1981, 174; for the various associations of the apotropaic destruction of wealth, see Crane 1993; for cloth and female authority, Morrell 1996/7. For the theme of sacrifice, see Zeitlin 1965, 478.

Fig. 7.2 Drawing of the 'tapestry scene' of the National Theatre Oresteia *(1981–3), directed by P. Hall (filmed by Channel 4). See n. 18 for the physical distance between chariot and fabric as suggested by the Aeschylean text. Image credit: Rosa Wicks*

quoted above), but also by the ode which closes, and climaxes, the 'tapestry scene' and which focuses heavily on the idea of destruction and redemption of wealth. *Ag.* 1008–21 are particularly telling of this, and alert us to the importance of the motif 'abuse of wealth' in earlier choral odes and across the *Oresteia*:[24]

καὶ τὸ μὲν πρὸ χρημάτων
κτησίων ὄκνος βαλὼν
σφενδόνας ἀπ' εὐμέτρου,
οὐκ ἔδυ πρόπας δόμος
πλησμονᾶς γέμων ἄγαν,
οὐδ' ἐπόντισε σκάφος.
πολλά τοι δόσις ἐκ Διὸς ἀμφιλα-
φής τε καὶ ἐξ ἀλόκων ἐπετειᾶν
νῆστιν ὤλεσεν νόσον.
τὸ δ' ἐπὶ γᾶν πεσὸν ἅπαξ θανάσιμον

[24] For an analysis of the motif in the *Oresteia*, see Bakola 2013, 236–248.

πρόπαρ ἀνδρὸς μέλαν αἷμα τίς ἂν
πάλιν ἀγκαλέσαιτ' ἐπαείδων;

Still, if caution casts forth part of the wealth in his possession from a sling of generous dimensions, the whole house does not founder when crammed too full in surfeit, nor does he wreck the ship: the gifts of Zeus are surely great, coming abundantly from furrows teeming year after year to destroy the plague of hunger. But once the black blood of death has fallen on the earth in front of a man, who by any incantation can summon it back again?

Among the most incisive approaches to the 'tapestry scene' in this regard are those by Jones, Taplin and Goldhill.[25] By highlighting that the economic value of a Greek *oikos* includes not only its material possessions, but equally the life of its members, these scholars have argued that the image of the destruction of the dark red textiles captures the depletion of the *oikos* through the shedding of its lifeblood, the most valuable asset for its survival. For these scholars, the 'wealth of the *oikos*' is a highly complex and multifaceted category that subsumes both material wealth and human life.

Although this line of argumentation is widely recognised, I believe that its potential has not been fully realised. We are able to gain a deeper understanding of the complex connections between 'wealth' and *oikos* if we consider the textiles not just as valuable (and heavily symbolic) objects, but also, as the text suggests, as evocative of processes of production. Crucially, these processes involve not only human labour, but also natural labour, and not only human resources, but also natural ones. Thinking closely about the processes of production that are evoked in this scene will give us a better understanding of the blood- and wealth-symbolism instantiated by the textiles. Furthermore, an examination of the imagery of natural productive powers in the words of Clytemnestra (*Ag.* 958–963) as well as of the semantics of the interior space, which is presented as containing and generating wealth, will provide insight into the role of textiles, which dominate this scene.

Textiles, productive and reproductive labour

Our modern experience is one of alienation from the processes of production. Living in industrialised and capitalist societies where goods are readily and economically available, we have become oblivious to the laborious manufacturing processes that were the norm only a century ago. In particular, easy access to cheap, ready-made clothing and linen, has greatly removed us from the processes of textile-production that were so central to ancient societies. We especially disregard the fact that the production of textiles was one of the most time-consuming economic activities in the ancient world. Research into ancient techniques and tools, supported by experiments with methods and processes used by traditional and indigenous societies shows that spinning and weaving alone constituted most of an ordinary woman's everyday activity. Carr 2000, for example, has estimated that the production of 40 sq m (around 430 sq ft) of plain wool fabric, an amount corresponding to the average needs of a Greek household per year, required around 2000 hours of spinning and 160 hours of weaving. Even before spinning and weaving could begin, raw sheep's wool needed gathering, washing, combing and roving into bundles.[26]

[25] Jones 1962, 82–93. Taplin 1977, 313–314 and Goldhill 1986, 11, 69, 171 have expanded on Jones. For other discussions of wealth, see Sailor-Stroup 1999; Seaford 1998; Morrell 1996/7, 149–150, 158–161; Crane 1993; and Flintoff 1987.
[26] See Neils 2011, 94. See also the timings provided by Andersson et al. 2008, 172–174; Costin 2013, 182. Flax required even more time and labour: see Carr 2000, 164–165. It had to be plucked by hand, dried, soaked in water, retted,

The fabric that is unfolded and 'destroyed' in the 'tapestry scene' is likely to have been of considerable size (between 90 and 260 sq ft, possibly closer to the latter).[27] Ancient audiences – for whom the connection between labour and value would have been much more obvious than it is for modern ones – would not have been oblivious to the enormous number of labour-hours required for the production of such an object. This impression would have been reinforced by the fact that the fabric is presented as anything but plain. Two processes especially, which place it at the highest end of textile-making, are emphasized by the words of the characters. One of them is the technique of weaving variegated and intricate patterns. This is suggested by πετάσματα at *Ag.* 909 (used by Clytemnestra), which means fabrics woven in intricate patterns, as are made to be spread and appreciated in their entirety. It is also suggested by ποικίλα, a term repeatedly used by Agamemnon (*Ag.* 923, 926, 936) which means tapestry weave, pattern weave or in-woven patterns, different techniques which all produce a variegated result.[28] Although, as some scholars have noted, the nature of the textiles referred to in the 'tapestry scene' (and thereafter) is vague, what remains constant are the references to their intricacy.[29]

Furthermore, both Clytemnestra and Agamemnon allude frequently to the purple dye that has been used on the fabrics. Clytemnestra refers to them as πορφυρόστρωτος πόρος "path spread with purple dye" (*Ag.* 910), while Agamemnon calls them ἁλουργή "purple-dyed" (*Ag.* 946) and πορφύρας (*Ag.* 957).[30] The image of the purple dye becomes central to the scene when Agamemnon begins to trample the fabric and Clytemnestra simultaneously evokes the image of the many sea-molluscs from which the purple dye is extracted: πολλῆς πορφύρας ἰσάργυρον / κηκῖδα παγκαίνιστον, εἱμάτων βαφάς, "the juices which yield much purple worth its weight in silver, wholly renewable, the dye of vestments" (*Ag.* 959–960).

In order to gain a better understanding of the symbolic and ideological value of these references, it is important to consider the two techniques in terms of time and labour requirements. The laboriousness of the art of making ποικίλα can be illustrated comparatively by considering labour- and time-investment in equivalent techniques used today, for example tapestry-weaving, one of the techniques meant by the Greek term ποικίλλειν. From approximate figures provided by the American Tapestry Alliance, we may estimate that the production of 260 sq ft (24 sq m) of tapestry required up to 3840 hours of weaving.[31] Although equating modern with ancient timings bears risks, especially

scutched, and heckled.

[27] It is of course impossible to be certain about the fabric's dimensions. However, we know that it would have stretched from the threshold of the *skene* to Agamemnon's chariot (see above, n. 18). The traditional direction is that Agamemnon would have travelled the length of the cloth while Clytemnestra recited *Ag.* 958–972 (see Fraenkel 1950, iii 813, n. 1; Taplin 1977, 309). If Agamemnon took one step for every foot of the iambic trimeter, and considering that, on average, a human step measures 25–30 inches, he would have travelled between 87 and 105 feet by the time he entered the *skene*, all the time moving on the fabric. Even the less likely scenario of Agamemnon taking one step for every trimeter would mean that the fabric was no shorter than 30 feet. The fabric's width is impossible to estimate, but given the distance of the action from the audience and the nature of the textiles, it seems safe to suppose that it was at least three feet wide.

[28] See Fraenkel 1950, ii 412–413; for the technique, see Vickers 1999 *passim; cf.* Villacèque 2008.

[29] Agamemnon speaks about the fabrics as εἵματα 'garments' (*Ag.* 921) and, more vaguely, ὑφάς (*Ag.* 949). Clytemnestra also uses εἵματα (*Ag.* 960, 963, 1383). Other characters refer to them as πέπλοι (*Ag.* 1126, 1580, *Cho.* 1000, *Eum.* 635) and φᾶρος (*Cho.* 1011, *Eum.* 634). For the fabrics' vague nature, see Taplin 1977, 314–315; Flintoff 1987, 120–126. However, Vickers 1999, 28 has argued that variegated fabrics could be either worn or hanged for display, so the references to the nature of the fabrics may not be as conflicting as we think. The fabrics' intricate nature is evoked even when they are referred to as nets: *Ag.* 1127, 1381–1382, 1492, *Cho.* 997–1004, *Eum.* 633–635.

[30] On the ambiguous meaning of ἁλουργή and πορφύρας, see below, p. 127 n. 37 and pp. 132–133.

[31] 'A skilled, professional tapestry weaver who works 35–40 hours a week at the loom can weave about one square meter a month. The rate of weaving varies considerably and depends upon the level of detail and the number of warps and wefts per inch.' Data provided by http://americantapestryalliance.org/resources/faq/ (last accessed 10.12.2014). On the same measurements, even the conservative estimation of 90 ft^2 of tapestry would require 1330 hours of work. Naturally,

since we are not sure that it is the same technique we are comparing, there is no reason to suppose that the production of ποικίλα would be significantly less laboursome than a modern equivalent. The impression of laboriousness conveyed by ποικίλα is accentuated by the reference to the use of royal purple, the production and application of which were also notoriously labour-intensive processes. Based on a recent project of experimental archaeology by Ruscillo 2005, we may estimate that to colour wool for the quantity of fabric presented in the 'tapestry scene' (possibly up to 260 sq ft), nearly 20000 murex molluscs would be needed. Even a conservative estimate of 90 sq ft of cloth would still require a staggering amount of murex (around 6700), which becomes even more striking if we consider that these would have to be fished from depths up to 150 m. These thousands of shells would subsequently have to be crushed in order to extract the hypobranchial glands from the molluscs' meat. Once extracted, the glands would have to be processed, first by mixing with salt in open-air vats and then through a long heating process, lasting up to ten days. Moreover, the fabric would have to be dipped more than once into the dye in order to achieve optimum results.[32]

Given the total amount of labour involved in these processes, it is not difficult to see why a 5th-century audience would view the gratuitous destruction of such textiles as hybristic and shocking. There is, however, something more sinister about the destruction of the wealth-generating labour of this scene. Productive labour (and the resulting wealth) is not the only issue at stake. For this, one should take into account not only the words of the characters, but also, since this is a theatrical representation, the scene's spatial configuration (see figs. 7.2 and 7.3). The tableau of the 'tapestry scene' unmistakably evokes the female womb and menstrual blood, suggesting the female generative capability. Alongside productive labour, therefore, the evocation of reproductive labour – that is, the labour generated in the interior of the female body – is equally powerful in this scene. As we saw earlier, both types of labour are conceptualised in Greek culture through the imagery of weaving.[33] In the 'tapestry scene', the precious generative powers of the female are destroyed in the same way as the precious textiles. This understanding finds support in the words of Clytemnestra about the destruction of the fruit of her own womb, Iphigeneia, at the hands of Agamemnon (*Ag.* 1415–1418).[34] It also finds support in the earlier associations between the lives of young soldiers and the imagery of fabrics, both in the *Agamemnon* (*Ag.* 192–198) and in the *Persians*.[35] Through striking usages of

the quality of the actual theatrical prop need not have been of the quality it is presented to be.

32 For the production of purple in antiquity, see the excellent chapter by Marzano 2013, 143–160, with recent bibliography, and Cardon 2007, 566–575. The principal references to ancient texts include Plin. *HN* 9.125–42, Arist. *HA* 5.547, Ael. *NA* 7.34 and Oppian. *Hal.* 5.598–9. For the rationale behind the estimates provided here, see Ruscillo 2005, 104 whose conclusions are based on actual experiments (For the present purposes, I adopted Ruscillo's lowest estimate of 100 murex per 48 square inches of fabric. The fabric may be presented as a ποικίλον, but the dyes could have been different shades of murex purple). For further estimations, see Marzano 2013, 145, 156–160, and below, p. 132. For the difficulties and laboriousness of fishing murex, see Ruscillo 2005, 101–103.

33 See pp. 117–119. For the sexual symbolism, see Goheen 1955, 121. In the present volume, MacMahon further explores the symbolism of weaving in relation to the space of the womb.

34 Iphigeneia's sacrifice is certainly recalled in the 'tapestry scene': *cf.* Lebeck 1971, 85; Scodel 1996, 120. J. Fletcher reminds me that Agamemnon calls Iphigeneia δόμων ἄγαλμα 'treasure of my house' (*Ag.* 207) before he sacrifices her. This is symbolically evoked by Agamemnon's destructive action against the 'wealth of the house' in the 'tapestry scene'. Furthermore, the image of saffron dye / saffron-dyed textiles (*Ag.* 239 κρόκου βαφάς) that are shed on the ground during the sacrifice is recalled by the location of the tapestries on the ground. Κρόκου βαφάς must be taken to evoke not only Iphigeneia's clothing (*cf. Ag.* 233; Mitchell-Boyask 2006, 279) but especially her blood (see Lebeck 1977, 84–85; Garvie 1986, 332). The meaning of κρόκου βαφάς is debated, but it often escapes scholarly notice that it evokes the *liquid* saffron dye, which is not necessarily yellow, but ranges from shades of deep red to shades of orange and yellow.

35 See above, p. 117 for the *Agamemnon* passage, and pp. 117–119 for the *Persians.* Clytemnestra is paradoxically portrayed as encouraging the destruction of labour although her own (reproductive) labour has been destroyed. That her characterisation here evokes the Erinys, see below, pp. 129–131.

word and action, space and colour, imagery and movement, and by reworking earlier associations both within the play and in the Aeschylean oeuvre, the dramaturgy of the 'tapestry scene' offers a poignant comment on the destruction of both material and human labour. The central concern of Aeschylean tragedy with wealth, life and their destruction by the elite becomes all the more sharply delineated by this representation of life and wealth-giving labour as extremely valuable but subject to waste by the rich and powerful.

Thus far I have argued that extending the acclaimed line of argument centering on the wealth of the *oikos* and its destruction can lead to a more profound understanding of the symbolism of textiles and their relationship to the concepts of generation and waste. However, I have analysed the wealth of the *oikos* that is embodied by the textiles merely as the wealth of a household, and humanly-produced.[36] As I will now argue, this understanding of the *oikos* does not do full justice to the scene. As an analysis of the words of Clytemnestra will reveal, the main issue of the scene is not the waste and abuse of human labour – whether productive or reproductive – nor of human wealth. Running through the scene, and directly related to the trilogy's central concern with the violation of the cosmic order, is the larger issue of the waste and destruction of natural labour and natural wealth.

The natural generative powers in the words of Clytemnestra and the role of the Erinys. Interior spaces (womb, house, earth), and the concepts of the natural and the generative

Clytemnestra's words occupy the central focus of this scene and direct Agamemnon's steps; simultaneously, they prove a turning point in the Aeschylean discourse on wealth and labour. Rather than stressing the human labour expended in making textiles, they focus on the toil of nature, which is portrayed as inexhaustible and, as such, easily susceptible to catastrophic waste:

> ἔστιν θάλασσα–τίς δέ νιν κατασβέσει; –
> τρέφουσα πολλῆς πορφύρας ἰσάργυρον
> κηκῖδα παγκαίνιστον, εἱμάτων βαφάς·
> οἶκος δ᾽ ὑπάρχει τῶνδε σὺν θεοῖς, ἄναξ,
> ἔχειν· πένεσθαι δ᾽ οὐκ ἐπίσταται δόμος.

Ag. 958–962

> The sea is there – and who shall quench it? – nurturing the juices which yield much purple worth its weight in silver, wholly renewable, the dye of vestments. The *oikos* has an abundance of these with the gods' help, my lord; this house does not know how to be poor.

These words trope the vast sea as inexhaustible nurturer of life, and more specifically, of the thousands of murex molluscs which generate the purple in natural conditions. They thus draw attention to the natural productive powers which create this wealth *prior* to man's intervention. In contrast to the earlier perspective, therefore, these words elevate the focal point from the human agent, his labour and his purchasing potential to the larger principle of the natural world and the ultimate female principle of generation. The concept of τρέφω and the emphasis on inexhaustibility (παγκαίνιστον) encapsulate the universal notion of natural growth, which implicates humans and natural resources together as products of natural processes. Water, liquid, and juice (esp. κηκῖδα) are universal images of nature's self-generation. The expensive liquid dye used on the fabrics is ultimately traced back to the sea.[37] Even ἄργυρος, given the evocation

[36] Note also Agamemnon's reference to monetary acquisition ἀργυρωνήτους θ᾽ ὑφάς (*Ag.* 949).
[37] At *Ag.* 946 ἁλουργέσιν could have been understood merely as 'worked with purple, purple-dyed'. At this point, however, one may retrospectively appreciate that, literally, the word evokes the labour of the sea. *Cf. LSJ* s.v. ἁλουργής.

Fig. 7.3 Drawing of the 'tapestry scene' from the National Theatre's Oresteia *(1981–3), directed by P. Hall (filmed by Channel 4). Image credit: Rosa Wicks.*

of weighing (ἰσ-άργυρον), refers more to silver as precious metal in its pure form (as extracted from the earth) than to its face value as minted coinage (as Agamemnon had used it earlier).[38] What makes human wealth possible and what makes the human *oikos* rich are the productive powers of nature, or 'earth'.

I use the terms 'nature' and 'earth' (as natural environment) and not 'sea', because the spatial semantics of the scene (and other considerations which we will analyse below) suggest that the focus here is much broader than the sea, and that it concerns more generally *the generative powers of nature*. There are several reasons for this. In Aeschylean theatre, the sea is not conceptualised only in opposition to land. In its generative capability, it is also understood and portrayed as part of a larger, all-encompassing whole, the 'earth' or environment, which includes the land and all other natural spaces and elements: the winds, the weather, celestial bodies like the sun and the stars. The key connection between all these elements is their generative powers and their ability to facilitate

[38] πολλῆς πορφύρας ἰσάργυρον evokes weighing, and hence the purity of the precious metal; *cf.* Theop. *FGrH* 115 F 117 'for the purple was being valued as equal in weight against silver'. Athenian coins were struck on a specific standard of purity and weight, with a drachm weighing 4.3 grams of silver. πολλῆς πορφύρας ἰσάργυρον, may refer to the fact that what belongs to the earth (purple) is acquired again with materials extracted from the earth (silver). For the earth as 'environment', see below.

the growth of life and natural wealth.[39] Naturally – and usually in response to human action – this ability can also turn destructive. Sea and land constitute particularly prominent aspects of this all-encompassing whole because they are associated with 'deep' spaces, which readily incorporate both generative and deadly connotations.[40] As we will explore below, this all-encompassing, generative space is evoked by Clytemnestra's reference to the *oikos*: this interior space turns out to have an even broader spectrum of reference than the domestic space alone, and comes much closer to what we would understand as the 'earth'.

The symbolism of the deep interior where the textiles originate is crucial for understanding Clytemnestra's words about natural generative powers. Anthropological research shows that in spatial terms, the natural and the generative are often conceptualised in relation to interior 'deep' spaces, including the female body (on a microcosmic level) and the *oikos* (as unit of production). The association between interiority, the natural and the generative (as well as its twin concept of the deadly) is amply demonstrated in the spatial associations found in the myths and rituals of many pre-industrialised societies, including ancient Greece. One of the most influential studies concerning the Greek context has been Vernant's article 'Hestia-Hermes: The religious expression of space and movement in ancient Greece', where the author has examined the anthropological symbolism inherent in the polarity of the 'inside' and the 'outside'. Through a study of the hearth's symbolic location in the earth, Vernant has observed that female and natural productivity are strongly linked to interior spaces, especially deep ones. More recently, Ruth Padel has provided a thorough treatment of the connection between interiority and the earth in an acclaimed study of Greek tragedy's anthropological symbolism, *In and Out of the Mind*. Throughout the book, but particularly in the section entitled 'Mind, Earth, Womb, Hades', Padel has argued forcefully for the essential association between these interior spaces. She has linked the house, the underworld and the innards of the human body to both generative and deadly powers, natural and human alike. Padel does not distinguish between the underworld and the earth, but uses the latter term to capture the significances of both. According to Padel, 'earth' is a mythical, religious and especially spatial concept. As an interior, it captures the essence of the 'earth' by emphasizing the dark space of death and the productive space of generation and growth.

It has long been argued using a cross-cultural methodology that interiors have strong symbolic and gendered associations with the earth. Bourdieu's study of the cosmic and gendered symbolism of the Berber house and its spatial organisation (1970) has probably been the most influential approach, and has been followed by studies of the built environment in other pre-industrial contexts.[41] There are, of course, vast variations between social practice, rituals and beliefs in

[39] For the passage's evocation of the generative powers of not only sea but also land, see Goheen 1955, 121 and n. 17, Segal 1963, 34. This holistic understanding of 'earth', and the attribution of generative powers to these elements (which also appear as ominous and destructive) is attested in the closing scene of the *Oresteia, Eum.* 903–913. I have explored this holistic meaning of 'earth' in Bakola 2014, 25–29. Purves 2010, 101–106 and Schibli 1990, 53–56 have also argued that the idea of 'earth' in Pherecydes includes land, sea and heavens. For insights into the modern use of the term 'earth' and its connotations of fertile ground/soil as well as environment, see Cosgrove 2001, 5–8, esp. 7.

[40] See Padel 1992, 99–101. For a cross-cultural theological approach, see especially Keller 2003 and Patton 2006, 48–49.

[41] See also Bourdieu 1990, ch. 3, with revised structuralist principles. Other studies include Errington 1979 on the Buginese house; Ellen 1986 on the Nualu house in Indonesia; Lok 1987 on the Mexican Indian house; Gillespie 2000 on the Maya house; for a wide cross-cultural survey of the symbolism of vernacular architecture from America, Asia and Africa, see Oliver 1987, 153–170.

these different contexts. However, it emerges that deep and dark spaces tend to be invested with a symbolic connection to fertility, preservation and increase of resources in relation to the ideas of the female and the earth, as well the more general concept of the natural environment. 'Natural' activities, including sleep, sexual intercourse, childbirth, as well as activities such as cooking and calendar rites, are consistently associated across cultural boundaries with the darkest, dampest and least accessible spaces.

What emerges from such studies is that interior, enclosed and deep spaces constitute primary symbolic media through which societies that have a direct relationship with the natural environment reflect on the idea of the natural and the generative. The female womb and body as a whole is envisaged as one such space, the house as another. Both are connected to nature. Indeed, anthropological research has shown that in many such cultures the natural environment itself is conceptualised as a house, and therefore, that 'earth as house' and 'house as earth' constitute different perspectives of the same cultural construct.[42] Notably, the correlation concerns the two 'enclosures', namely household and natural environment, in terms of the economic relationship that man has to both. Production, preservation, consumption, enhancement and the balance which should characterise these processes are central to the association.

Interiority and its associations with the generative are entirely coherent with the function of Greek theatre space, and the symbolic configuration of space in the 'tapestry scene', in particular. As has been demonstrated by leading Greek theatre critics, theatre space does not have only a representational function. It also functions on a symbolic level, and interior spaces are especially susceptible to multivalent symbolic associations. In Greek tragedy, interiors are generally imagined as dark and deep, and are used to evoke categories of thought such as female, private, past, mind, psyche, especially in opposition to exterior spaces, which in turn are associated with the male, public, present, known, *etc.*[43] In the 'tapestry scene', as we observed, the tableau of the red stream flowing out of the opening of the interior evokes the female generative capability. However, the allusion to the sea's generative powers adds a powerful additional dimension. Through these words, female and natural are merged in this stage tableau. This is further enhanced by the image of the tree which Clytemnestra describes rising from the depths of the *oikos* and shading the entire space (*Ag.* 966–972). Trees, with their changing foliage in the course of the year are symbols of vigour, fertility and the regenerative capacity of nature. Schibli, who explores this image in the context of Pherecydes' work, echoes Eliade in suggesting that 'there is a natural link between earth and tree, as both express an inexhaustible source of cosmic fertility'.[44] The image of the tree rooted in the *oikos* suggests that the generative space Clytemnestra refers to as inexhaustibly rich is thus not just the domestic *oikos*, nor even the human body, but something much larger. It becomes at once a female interior and a cosmic interior, and it represents the womb of the natural space which generates natural wealth, animal life and human life: namely, the earth.

Considering the different roles and identities of Clytemnestra at this point sheds further light on the evocation of the natural productive powers. Scholars have often noted that the play forges

[42] See the studies cited in n. 41.

[43] These are only some of the categories that are symbolically associated with the interior. The bibliography on this topic is extensive, but among the most influential works have been Segal 1999, esp. chs. 4–8 (1st ed. 1981); Zeitlin 1985a and 1985b; Padel 1990; Wiles 1997, ch. 7.

[44] Schibli 1990, 76; Eliade 1949, 233–246.

a link between Clytemnestra and the Eriny(e)s.[45] The textiles play a key part in this identification: the link between Clytemnestra and Erinys is made explicit at *Ag.* 1580, where the textiles that Clytemnestra had used to lure Agamemnon into the interior, and subsequently to murder him, are called "woven robes of the Erinyes" (ὑφαντοῖς ἐν πέπλοις Ἐρινύων, *Ag.* 1580). Even earlier, in the scene of the revelation of the corpses (*Ag.* 1372–1576), Clytemnestra's image and voice have steadily merged with those of the Erinys. For example, as she stands over the corpses, the chorus sees her as the "*daimōn* of the house" (*Ag.* 1468–1474), an image which heavily evokes that of the "Erinys of the house" since the second choral ode.[46] As this scene reaches its climax, the image of the Erinys conflates completely with that of Clytemnestra when the latter identifies herself explicitly with the (*daimōn*) *alastōr* of the house (*Ag.* 1497–1504).[47]

However striking this conflation is, it is only the development of an identification which the play has constructed from its very beginning. The images of Clytemnestra and the Erinys are first linked as early as *Ag.* 155:

μίμνει γὰρ φοβερὰ παλίνορτος
οἰκονόμος δολία, μνάμων Μῆνις τεκνόποινος

Ag. 154–155

for there awaits, to arise hereafter, a fearsome, guileful keeper of the house, a Wrath that remembers and will avenge a child.

In this passage, both females are imagined as embedded in the *oikos*, guarding its interior and ready to act as its avengers. The term *oikonomos,* in other words, constructs a double image of Clytemnestra and the Erinys as the female presence in the interior who is responsible for the use and abuse of its wealth.[48] And while for Clytemnestra the characterisation *oikonomos* may be understood on the basis of the destruction of Iphigeneia's life (*cf.* τεκνόποινος *Ag.* 155 and δόμων ἄγαλμα *Ag.* 207), it has a much wider application for the chthonic Eriny(e)s. As I have shown elsewhere, in the *Oresteia* the Erinyes, agents of the earth, punish the destruction of life as *destruction of natural wealth.* Throughout the trilogy, the crimes the Erinyes react to and the punishments they inflict are in every case connected to a violation of the 'earth's wealth', which is almost always conceptualised as violation of life, and especially murder.[49] The Erinyes of the *Oresteia* are, as it is often said, agents of the cosmic order; but this order is profoundly connected with the broad

[45] See McClure 1996/7, 138–140; Fowler 1991, 91–95, 99. For the Erinyes as both singular and plural entities, see Henrichs 1994, 52; Padel 1992, 165.

[46] See *Ag.* 750–776, esp. 769–771, with Collard 2002, 138. For the Erinyes' connection with the house, see *Ag.* 155, 717–772, 1186–1193, 1500–1503; *Cho.* 566, 698–699, 800–802.

[47] For the chorus' identification of Clytemnestra and the *daimōn* at *Ag.* 1468–1474, consider the occupying sense of ἐμπίτνεις at *Ag.* 1468. For the Erinyes as subsuming the inner *daimones* of tragedy, see generally Padel 1992, 164–192. The most relevant passages are *Ag.* 768 (*cf.* Collard 2002, 138; Fraenkel 1950, ii.753); *Ag.* 1475–1479 (*cf. Cho.* 577–578 and the image of the Erinyes licking blood at *Eum.* 106; *cf.* Fowler 1991, 95, 98); *Ag.* 1482 (*cf.* Dodds (1960) 26; *contra* Sewell-Rutter (2007) 84). From *Ag.* 1567, Clytemnestra's attitude changes and she treats the *daimōn* as a force external to herself. For the identification of the Erinys with a male subject, see Finglass 2005, esp. 41.

[48] For the Erinyes as essentially interior powers, see Padel 1992, 171–172, 189–192, and for the *kosmos*' relation to the (human) interior, *passim*. The Erinyes are primarily powers of the earth (*Eum.* 417), which is also conceptualised as interior space.

[49] However, an essential feature of the concept of the Erinyes in Greek culture is that they avenge crime and destruction by causing more crime and destruction. See Padel 1992, 167, 177. See Bakola 2013, 231–233, 236–248, on the Hesiodic *daimones plutodotai* (Hes. *Op.* 122–126), the Erinyes of the *Oresteia* and other chthonic deities as watchers of the 'wealth of the earth'. For the concept of the chthonic powers as guardians of the earth and its resources, see also Burkert 1985, 200.

concept of wealth. By alluding to the Erinyes' relationship to wealth, the term *oikonomos* at *Ag.* 155 thus sets out the basis on which they should be understood from early on in the drama.

The conflation of Clytemnestra and the Erinys under the characterisation *oikonomos* provides a crucial key for understanding the meaning of vv. 958–965, the multiple identities of the female voice in these lines and the symbolism of the textiles in relation to different treatments of the earth's generative powers. As we saw, the textiles, symbols of wealth and life which are generated by the earth and by female labour, are spread at the destructive feet of Agamemnon (fig. 7.3). By invoking the allegedly inexhaustible generative powers of nature, the powerful female *oikonomos* of the scene invites Agamemnon to commit an act of hubris against them. In a profoundly symbolic gesture, she calls on him to be destructive towards life and nature in the same way as he and his *oikos* had acted during the Trojan expedition, and earlier (as narrated in the parodos and the first stasimon).[50] This is where the image of the Erinys becomes clearly discernible behind the figure of Clytemnestra: Agamemnon is encouraged to abuse wealth, again. Though his decision to tread on the textiles is ostensibly excusable in view of the sea's inexhaustible powers, Aeschylus represents it as an unmistakable act of destruction against the cosmic scheme of wealth, the 'wealth of the earth'. Lured to walk on the path that brings him home (*cf.* the sinister *Ag.* 963–965) he finds himself in the snares of the Erinyes (*Ag.* 1580; *cf.* 1382–1383), powers internal to the earth and the *oikos*, and will be punished accordingly. A space normally generative of wealth and life (see fig. 7.2), the *oikos* appears as a terrifying space of death, a space wherefrom he (and other characters) will not emerge alive again. Later on, Cassandra will articulate the nature of the interior not as generative earth, but as Hades, the deadly earth:

> Ἅιδου πύλας δὲ τάσδ' ἐγὼ προσεννέπω.
>
> *Ag.* 1291
>
> I address these gates as the gates of Hades.[51] (tr. Sommerstein 2008)

Textiles and the production, consumption and destruction of natural wealth

Once we acknowledge the association of the interior with the earth, the Erinys and the naturally generative capabilities of the cosmos, we gain a better understanding of the textiles in their role as 'wealth'. Furthermore, we gain a deeper understanding of why this scene focuses especially on the material wealth of textiles and purple dye. We saw earlier that textiles evoke wealth by alluding to the processes of creation and natural production. This is largely because of their close connection to the female and female productive labour, including reproductive labour. It is also because the human activity of textile-making seems to add value to the materials taken from the earth, thus providing an extension to the natural processes of production and growth. Not only does the gathering of materials such as wool and flax avoid invasive or destructive procedures, but the processing of wool into thread and subsequently the joining of the threads into useable objects of exchangeable value and beauty readily suggests human activity that is in harmony with nature.[52]

[50] See above, n. 17.

[51] *Cf. Ag.* 1090–1092, 1096–1097, 1122–1124, 1160–1161, 1217–1222, 1314. I explore the notion of earth as *oikos* more extensively in my monograph *The Erinyes and the wealth of the earth* (in preparation).

[52] Interestingly, Xenophon compares textile-making with the activity of the bees in the bee-hive: Xen. *Oec.* 7.32–6.

In the 'tapestry scene', these symbols of natural and generative labour are symbolically destroyed, evoking the destruction caused by Agamemnon's imperialist expedition, which resulted in a massive loss of young human life, and in violations of nature.[53] The play portrays the abuser of these different sorts of wealth throughout as representative of the greedy and destructive aristocratic *oikoi*; it thus presents war as a wealth-making enterprise that destroys natural, sacred wealth for the profit of a few.[54] However, this symbolic destruction of natural wealth is further accentuated by the emphasis on another process of wealth production, namely purple dye. Making the purple dye, as we said earlier, required enormous quantities of living molluscs to be crushed and their natural purpose ended. What is particularly striking about this process is that the proportionate rate of living specimens which have to be destroyed is enormous in comparison with the resulting amount of the luxury product. According to one study, the production of just one pound of dye required the extraction of glands from up to 60000 murex molluscs.[55] Ruscillo's recent experiments with dye application on wool attest that to produce dye enough for 1.25 ft^2 of woollen cloth, 100–160 *murex brandaris* needed to be used; based on her experiment, we estimated that to make 260 ft^2 of the extremely expensive purple cloth, up to 20000 murex would need to be crushed.[56] Above and beyond destruction of human toil, this constitutes enormous wastage of natural life. These reflections square with the impression conveyed by waste sites connected with the purple industries, which have been excavated by archaeologists. At Sidon, for example, a heap of crushed murex discovered in the 19th century measured 110 m long and up to 7.5 meters deep. The size of crushed shells heaps in such sites must have been as evocative of destruction in ancient times as it is now.

Careful consideration of the ambiguous wording used by the characters in the scene shows contemporary awareness of the destructive dimension of the dye production. This is brought into sharp focus in Agamemnon's final words before he puts his foot on the textiles:

> εἶμ' ἐς δόμων μέλαθρα πορφύρας πατῶν
>
> *Ag.* 957
>
> I will go to the halls of my house treading on *porphyras*.

Throughout the trilogy, 'treading' terms have connotations of crushing and destroying.[57] In this context, since the word πορφύρα does not connote only purple dye but also the murex mollusc, the juxtaposition of πορφύρας and πατῶν and the mental image which it creates is striking.[58] The possibility that πορφύρας πατῶν may evoke the destructive aspect of the production of purple has never been considered, and this is because it has generally been thought that in this line, πορφύρα refers only to the dyed fabrics. Nevertheless, only two lines later, at *Ag.* 959, πορφύρα evokes both 'murex molluscs' (because of κηκῖδα) and 'purple dye'. Since in that context the focus is the molluscs and the nurture of their purple-producing glands, it seems plausible to read Agamemnon's reference πορφύρας πατῶν as having various connotations, and to connect it also with the destruction of animal life. Thus, his subsequent act of treading evokes not only the destruction of textiles, the

53 See above, n. 17.
54 For the class discourse of the trilogy, see Rose 1992, 204–210, 220–221.
55 See Reese 1980, 84–85.
56 See Ruscillo 2005, 103–4, and above, n. 32.
57 Lebeck 1971, 75–79.
58 See *LSJ* s.v. πορφύρα, I.

labour that has gone into them and, symbolically, the natural wealth they represent, but also the crushing and destruction of life on a vast scale for the production of luxury.

Such considerations also suggest that in this scene there is marked contrast between the human and the natural contribution to the production of wealth. As opposed to the natural principle that generates the precious natural wealth in the first place, namely the precious gland secretions of the murex molluscs, the human labour required to extract it – evoked here by Agamemnon's words πορφύρας πατῶν and his trampling act (fig. 7.3) – constitutes destruction and wastefulness. In this light, Agamemnon's act acquires extra layers of symbolic value. As natural life (like the life of the murex molluscs) is treated like a disposable commodity when it comes to acquiring wealth, so human life (especially young life, like that of Iphigeneia and the soldiers sacrificed in the Trojan war) is treated in the same way if it is to serve the acquisitiveness and greed of its elite leaders. This is an idea which Aeschylus obsessively revisits in his corpus, from the *Persians*, to the *Seven against Thebes*, the *Oresteia* and many of the fragmentary Aeschylean plays: namely, that the products of the earth, human and natural wealth, are abused and wasted by the powerful in their pursuit of wealth.

From a modern perspective, what is perhaps most striking in this scene, is the realisation of how far this stance towards nature is from the anthropocentric view that takes unhindered use of natural resources as normal, a view which one might expect in a society which was not faced with profound ecological challenges. The belief in the inexhaustibility of nature that Clytemnestra articulates is embedded in a context of invitation to transgression; it evokes the Erinys' lure of human agents into destructive crime and their subsequent punishment. As the destructive and wasteful attitude towards the earth is condemned forcefully, Aeschylean tragedy contemplates the relationship to nature in a way that is remarkably similar to contemporary, 21st century concerns. Having presented the elite households of power and wealth, especially the Atreid *oikos*, as agents pursuing their own narrower interests at the expense of community and nature alike, Aeschylean tragedy refocuses the concept altogether and reflects the view that the earth is the ultimate household, the ultimate reserve and source of wealth and life. Envisaging the earth as a larger *oikos* which human *oikoi* can destroy at will comes strikingly close to modern considerations of power, wealth, and the use and abuse of resources.

Bibliography

Andersson, E., Mårtensson, L., Nosch, M.-L., & Rahmstorf, L. 2008 New research on Bronze Age textile production. *Bulletin of the Institute of Classical Studies* 51, 171–174.

Bakola, E. 2013 Crime and punishment: Cratinus, Aeschylus' *Oresteia*, and the metaphysics and the politics of wealth. In E. Bakola, L. Prauscello & M. Telò (eds) *Greek comedy and the discourse of genres*. Cambridge, 226–255.

Bakola, E. 2014 Interiority, the 'deep earth' and the spatial symbolism of Darius' apparition in the *Persians* of Aeschylus. *Cambridge Classical Journal* 60, 1–36.

Borthwick, E. K. 1976 The 'flower of the Argives' and a neglected meaning of ἄνθος. *Journal of Hellenic Studies* 96, 1–7.

Bourdieu, P. 1970 The Berber house or the world reversed. In J. Pouillon & P. Maranda (eds) *Échanges et communications: Mélange offerts à Claude Lévi-Strauss à l'occasion de son 60e anniversaire*, vol. 2. The Hague 151–170.

Brock, R. 1994 The labour of women in classical Athens. *Classical Quarterly* 44, 336–346.

Brown, J. K. 1970 A note on the division of labor by sex. *American Anthropologist* 72, 1073–1078.

Bundrick, S. D. 2008 The fabric of the city: Imaging textile production in classical Athens. *Hesperia* 77, 283–334.

Burkert, W. 1985 *Greek Religion: archaic and classical*, tr. J. Raffan. Oxford.

Cardon, D. 2007 *Natural Dyes: Sources, Tradition, Technology and Science*. London.

Carr, K. 2000 Women's work: spinning and weaving in the Greek home. In D. Cardon, & M. Feugère (eds) *Archéologie des Textiles des origines au Ve siècle*. Montagnac, 163–166.

Collard, C. 2003 *Aeschylus Oresteia, translated with an introduction and notes*. Oxford.

Cosgrove, D. E. *Apollo's eye: a cartographic genealogy of the earth in the western imagination*. Baltimore MD.

Costin, C. L. 2013 Gender and textile production in prehistory. In D. Bolger (ed.) *A Companion to Gender Prehistory*. Chichester, 180–203.

Cottica, D. 2004 The symbolism of spinning in Classical art and society. In E. Lyle (ed.) *Weaving and Cosmology. Cosmos* 20. Edinburgh, 185–210.

Crane, G. 1993 Politics of consumption and generosity in the carpet scene of the *Agamemnon*. *Classical Philology* 88, 117–136.

Dodds, E. R. 1960 Politics and the Oresteia. *Proceedings of the Cambridge Philological Society* 6, 19–31

Eliade, M. 1949 *Traité d'histoire des religions*. Paris.

Eliade, M. 1961 *Images and symbols: studies in religious symbolism*, trans. P. Mairet. London.

Eliade, M. 1985 *Symbolism, the sacred, and the arts* (ed. D. Apostolos-Cappadona). New York.

Ellen, R. 1986 Microcosm, macrocosm and the Nuaulu house: concerning the reductionist fallacy as applied to metaphorical levels. *Bijdragen tot de Taal-, Land- en Volkenkunde* 142, 1–30.

Errington, Sh. 1979 The cosmic house of the Buginese. *Asia* 1, 8–14.

Finglass, P. J. 2005 Erinys or Hundred-Hander? Pindar, fr. 52i(a). 19–21 Snell-Maehler = B3.25–7 Rutherford (Paean 8a). *Zeitschrift für Papyrologie und Epigraphik* 154, 40–42.

Flintoff, E. 1987 The Treading of the Cloth. *Quaderni Urbinati di Cultura Classica* 25, 119–130.

Fowler, B. H. 1991 The Creatures and the Blood. *Illinois Classical Studies* 16, 85–100.

Fraenkel, E. 1950 *Aeschylus: Agamemnon,* edited with a commentary. Oxford.

Garvie, A. F. 1986 *Aeschylus: Choephori*, with introduction and commentary. Oxford.

Gillespie, S. D. 2000 Maya 'nested houses': The ritual construction of place. In R. A. Joyce & S. D. Gillespie (eds) *Beyond kinship: social and material reproduction in house societies*. Philadelphia PA, 135–160.

Goheen, R. F. 1955 Aspects of dramatic symbolism: Three studies in the *Oresteia*. *American Journal of Philology* 76, 113–137.

Goldhill, S. 1986 *Reading Greek Tragedy*. Cambridge.

Granger, H. 2007 The theologian Pherecydes of Syros and the early days of natural philosophy. *Harvard Studies in Classical Philology* 103, 135–163.

Håland, E. J. 2005 Athena's peplos: Weaving as a Core Female activity in ancient and modern Greece. In E. Lyle (ed.) *Weaving and Cosmology. Cosmos* 20. Edinburgh, 155–182.

Henrichs, A. 1994 Anonymity and polarity: unknown gods and nameless altars at the Areopagos. *Illinois Classical Studies* 19, 27–58.

Jenkins, I. D. 1985 The ambiguity of Greek textiles. *Arethusa* 18, 109–132.

Jensen, L. B. 1963 Royal purple of Tyre. *Journal of Near Eastern Studies* 22, 104–118.

Jones, J. 1962 *On Aristotle and Greek tragedy.* London.

Keller, C. 2003 *The face of the deep: a theology of becoming.* London.

Lebeck, A. 1971 *The Oresteia.* Washington.

Lee, M. M. 2004 Evil wealth of raiment: Deadly πέπλοι in Greek tragedy. *Classical Journal* 99, 253–279.

Lewis, S. 2002 *The Athenian woman: an iconographic handbook.* London.

Lok, R. 1987 The house as microcosm: some cosmic representations in a Mexican Indian village. In de R. Ridder & J. A. Karremans (eds) *The Leiden Tradition in Structural Anthropology: Essays in Honour of P.E. de Josselin de Jong.* Leiden, 211–223.

Lyle, E. (ed.) 2006 *Weaving and Cosmology. Cosmos* 20. Edinburgh.

Lyons, D. 2003 Dangerous gifts: Ideologies of marriage and exchange in ancient Greece. *Classical Antiquity* 22, 93–134.

M'Closkey, K. 2002 *Swept under the rug: a hidden history of Navajo weaving.* Albuquerque NM.

Marzano, A. 1969 *Harvesting the sea: the exploitation of marine resources in the Roman Mediterranean*. Oxford.

McClure, L. 1996/7 Clytemnestra's binding spell (*Ag.* 958–974). *The Classical Journal* 92, 123–140.

McNeil, L. 2005 Bridal Cloths, cover-ups, and *kharis*: The "carpet scene" in Aeschylus' *Agamemnon, Greece & Rome,* 52, 1–17.

Mitchell-Boyask, R. 2006 The marriage of Cassandra and the *Oresteia*: Text, image, performance. *Transactions of the American Philological Association* 136, 269–297.

Morrell, K. S. 1996/7 The fabric of persuasion: Clytemnestra, Agamemnon, and the sea of garments. *Classical Journal* 92, 141–165.

Nagy, G. 1990 *Pindar's Homer: the lyric possession of an epic past.* Baltimore MD.

Neils, J. 2011 *Women in the ancient world.* London.

Nelson, S. A. 1998 *God and the land: the metaphysics of farming in Hesiod and Vergil.* New York.

Oliver, P. 1987 *Dwellings: the house across the world.* Austin TX.

Onians, R. B. 1951 *The origins of European thought*. Cambridge.

Padel, R. 1990 Making space speak. In J. Winkler and F. Zeitlin (eds) *Nothing to do with Dionysus? Athenian drama in its social context.* Princeton NJ, 336–65.

Padel, R. 1992 *In and out of the mind: Greek images of the tragic self.* Princeton.

Parker R. 2005 *Polytheism and Society at Athens.* Oxford.

Parker, R. B. 1986 The National Theatre's *Oresteia*, 1981–82. In M. Cropp, E. Fantham & S. Scully (eds) *Greek tragedy and its legacy. Essays presented to D. J.Conacher*. Calgary, 337–357.

Patton, K. 2006 *The sea can wash away all evils: modern marine pollution and the ancient cathartic ocean.* New York.

Pomeroy, S. B. 1994 *Xenophon, Oeconomicus: a social and historical commentary*. Oxford.

Purves, A. C. 2010 *Space and time in ancient Greek narrative.* Cambridge.

Rabinowitz, N. S. 1981 From force to persuasion: Aeschulus' *Oresteia* as cosmogonic myth. *Ramus* 10, 159–191.

Raman, R. A. 1975 Homeric ἄωτος and Pindaric ἄωτος: a semantic problem. *Glotta* 53, 195–207.

Reese, D. S. 1980 Industrial Exploitation of Murex Shells: Purple dye and Lime Production at Sidi Khrebish, Benghazi (Berenice), *Libyan Studies: The Society for Libyan Sudies 11th Annual Report* (1979–1980), 79–93.

Reichel-Dolmatoff, G. 1990 *The sacred mountain of Colombia's Kogi Indians*. Leiden–New York.

Rose P. W. 1992 *Sons of the gods, children of earth: ideology and literary form in ancient Greece*, New York–London.

Rosenbloom, D. 2006 *Aeschylus: Persians.* London.

Rosenmeyer, T. G. 1982 *The art of Aeschylus*. Berkeley CA.

Rosivach, V. J. 1989 *Talasiourgoi* and *paidia* in IG22 1553–78: a note on Athenian social history. *Historia* 38, 365–370.

Ruscillo, D. 2005 Reconstructing murex Royal Purple and Biblical Blue in the Aegean. In D. E. Bar-Yosef Mayer (ed.) *Archaeomalacology: Molluscs in former environments of human behaviour.* Oxford 99–106.

Saïd, S. 2007 Tragedy and Reversal. In M. Lloyd (ed.) *Oxford Readings in Aeschylus.* Oxford, 71–92.

Sailor, D. & Stroup, S. C. 1999 ΦΘΟΝΟΣ Δ' ἈΠΕΣΤΩ: The translation of transgression in Aiskhylos' *Agamemnon*. *Classical Antiquity* 18, 153–182.
Scheid, J. & Svenbro, J. *The craft of Zeus: myths of weaving and fabric,* tr. by C. Volk. Cambridge MA.
Schibli, H. S. 1990 *Pherekydes of Syros*. Oxford–New York.
Scodel, R. 1996 *Domon agalma*: Virgin sacrifice and aesthetic object, *Transactions of the American Philological Association* 126, 111–128.
Seaford, R. 1998 Tragic Money. *Journal of Hellenic Studies* 118, 119–139.
Segal, C. P. 1963 *Nature and the World of Man in Greek Literature. Arion* 1, 19–53.
Segal, C. P. 1999. *Tragedy and civilization: an interpretation of Sophocles*. Norman OK.
Sewell-Rutter, N. J. 2007 *Guilt by descent: moral inheritance and decision making in Greek tragedy.* Oxford.
Sommerstein A. H. 2008 *Aeschylus,* edited and translated in 3 vols. Cambridge MA–London.
Taplin, O. 1977 *The stagecraft of Aeschylus: the dramatic use of exits and entrances in Greek tragedy*. Oxford.
Taplin, O. 1989 *Greek tragedy in action*. London.
Taplin, O. 2001 The experience of an academic in the rehearsal room, *Didaskalia* 5. http://www.didaskalia.net/issues/vol5no1/taplin.html
Thalmann, W.G. 1980 Xerxes' rags: some problems in Aeschylus' *Persians*. *American Journal of Philology* 101, 260–282.
Thomson, W. 1982 Weaving: a man's work. *Classical World* 75, 217–222.
Vernant, J. P. 1983a Hestia-Hermes: The religious expression of space and movement in ancient Greece. In *Myth and thought among the Greeks*. London, 127–175 (originally published in French: *L'Homme* 3.3 (1963) 12–50).
Vernant, J. P. 1983b Work and nature in ancient Greece. In *Myth and thought among the Greeks*. London, 248–270.
Vernant, J. P. 1983c Some remarks on the forms and limitations of technological thought among the Greeks. In *Myth and thought among the Greeks*. London, 279–301.
Vernant, J. P. & Curtis, D. A. 2001 Weaving Friendship. *Salmagundi* 130/131, 75–87.
Vickers, M. J. 1999 *Images on textiles: the weave of fifth-century Athenian art and society*. Konstanz.
Villacèque, N. 2008 Histoire de la ποικιλία, un mode de reconnaissance sociale dans la démocratie athénienne. *Revue des Études Anciennes* 110, 443–459.
Walton, J. M. 2005 Translation or Transubstantiation. In F. Macintosh, P. Michelakis, E. Hall & O. Taplin (eds) *Agamemnon in Performance, 458 BC to AD 2004,* Oxford, 189–206.
Weigle, M. 1989 *Creation and procreation: feminist reflections on mythologies of cosmogony and parturition*. Philadelphia.
West, M. L. 1971 *Early Greek Philosophy and the Orient*. Oxford.
Wiles, D. 1997 *Tragedy in Athens: performance space and theatrical meaning*. Cambridge.
Zeitlin, F. 1965 The motif of the corrupt sacrifice in Aeschylus' *Oresteia*. *Transactions of the American Philological Association* 96, 463–508.
Zeitlin, F. I. 1985a The power of Aphrodite: eros and the boundaries of the self in Euripides' *Hippolytus*. In P. Burian (ed.), *Directions in Euripidean criticism*. Durham, 52–111, 189–208.
Zeitlin, F. I. 1985b Playing the other: theater, theatricality, and the feminine in Greek drama. *Representations* 11, 63–94.

8. The Cloak of Deianeira or the Shirt of Nessus?

Andrea Doyle

Who then devised the torment? Love.
Love is the unfamiliar Name
Behind the hands that wove
The intolerable shirt of flame
Which human power cannot remove.
We only live, only suspire
Consumed by either fire or fire.[1]

Cloth in various forms is an important motif in the story of Hercules and his relationship with women. One thinks first of the distinctly unheroic method of his death in the poisoned cloak or shirt of Nessus sent him by Deianeira. At the time of Sophocles there were at least two contradicting versions concerning the culpability of Deianeira, wife of the great hero and thus two contrasting characterisations of this figure, or heroine, if we may. According to some traditions, it is Deianeira herself who weaves the cloak which she later anoints with the love charm in an attempt to bind him to her fearing she is about to be replaced in her own house by a younger woman.[2] In other traditions Nessus persuades her to take his own shirt, already soaked in his poisonous blood from the fatal wound, which he tells her will act as a love charm on Hercules. The love charm turns out to be a deadly poison that will cause him an agonizing death. And so, the cloak woven by Deianeira and impregnated with a deadly poison also by her own hands, is not interchangeable with the shirt of Nessus pre-poisoned and simply sent by Deianeira to Hercules under the misapprehension that it would secure her husband's love and fidelity. Deianeira's weaving of the shirt and poisoning it (wittingly or unwittingly) before sending it to Hercules, emphasises and indeed allows for a higher degree of agency on her part. Whereas if she simply sends her husband a pre-poisoned garment which she did not make or tamper with by adding the poison, then the agency of Deianeira is minimised as is her culpability and she becomes merely the means of transfer from one male adversary to another.[3]

[1] T. S. Eliot, *Four Quartets Little Gidding IV,* 207–213.

[2] Sophocles' *Women of Trachis* is the most well-known example of this.

[3] The relationship between Hercules and cloth is not confined to the method of his death but figures in a curious episode during his period of atonement, often referred to as 'The Twelve Tasks'. I refer to his bondage to the Libyan Queen Omphale a year-long enslavement to the queen during which time he was made to wear women's garments and learn the craft of weaving. This is period of Hercules' life is regarded as a shameful time, the ultimate reversal of his heroic *modus vivendi,* wherein the demi-god is emasculated at the beck and call of the Libyan Queen. Not all sources mention this curious 'labour' of the great hero, indeed many writers in late antiquity and the Middle Ages confused or conflated

The Greek Classical Tradition

In the Sophoclean version of Hercules' death, Deianeira's story is poignantly told and treated with great sensitivity and complexity.[4] The figure of Deianeira in Classical literature came to be most closely associated with Sophocles' *Trachiniae* but she is mentioned in other genres of Greek literature. In the fragments of Hesiod and the *Odes* of Bacchylides (to name both an earlier and a contemporary source), we find evidence of a different version of Sophocles' innocent heroine.[5] From these sources and other later ones which we shall discuss presently, we find ourselves staring at Deianeira the manslayer who slew her husband in the throes of jealousy, plotting his downfall deliberately so as to punish him for taking as his mistress, the younger and more beautiful Iole.

Hesiod

Hesiod, in his *Catalogue of Women* composed in the 6th century anywhere between 580 and 520 BC, gives us a jealous Deianeira who actively smeared the robe with poison, and gave it to Lichas the herald to take to Hercules:

> (...) κ[αὶ ὑπέρ]φ[ρ]ονα Δηϊάνειραν,
> ἣ τέχ᾽ ὑπομηθεῖ[σα βίηι Ἡρ]ακλη[ε]ίηι
> Ὕλλον καὶ Γλῆνον καὶ [Κτή]σιππον καὶ Ὀνείτην·
> τοὺς τέκε καὶ δείν᾽ἔρξ[᾽, ἐπεὶ ἀάσατ]ο μέγα θυμῶι,
> ὁππότε φάφμακον ο[ὗλον ἐπιστά]ξασα κιτῶνι
> δῶκε Λίχηι κήρυ[κι] φ[έρειν· ὃ δὲ ἔνει]κεν ἄνακτι
> Ἀμφιτρυνωιά[δ]ηι Ἡ[ρακλῆϊ πτολιπό]ρθωι.
> δ[εξ]αμένωι δέ ο[ὶ ἆψα τέλος θανάτοι]ο παρέστη·
> καὶ] θάνε καί ῥ᾽ Ἀΐδ[αο πολύστονον ἵκε]το δῶμα.[6]

fr. 25.17–25 M-W

> (…) and shrewd Deianeira, who, overpowered by the force of Herakles,
> bore Hyllus and Glenus and Ktessipus and Onites;
> these she bore and committed terrible deeds, in reckless passion
> when she smeared the robe with poison
> and gave it to Lichas to take. And he gave it to his lord and master
> Herakles, sacker of cities, son of Amphitryon,
> and, upon receiving it, the finality of death came swiftly upon him,
> and he died and came to the house of Hades full of laments.[7]

Omphale and Iole treating them as one and the same. Hercules' enslavement to Omphale falls beyond the scope of this study and so I do not cover the incident in this paper. Suffice it to say that in the Middle Ages in particular, Hercules' submission to the queen was regarded variously as a sign of his moral degeneracy or the power of lust symbolized by Woman. See for example, Lactantius, *Divinarum Institutionum: De Falsa Religione Deorum* (9) and Fulgentius, *Mythologiae* (2.2).

[4] The Sophoclean Deianera has been much discussed and analysed. See particularly discussions by Scott (1995; 1997), Wohl (1998), Ormand (1999), and Foley (2003).

[5] Diodorus (60–30 BC) tells of a robe and shirt, which were Hercules' customary sacrificial attire; Deianeira anoints the shirt with the poison believing it to be a love charm to secure Hercules' dwindling affections (*The Library of History.* 4.38.1). The tale is also mentioned in Antoninus Liberalis, *Collection of Metamorphoses* 2.

[6] I have consulted Most's edition of Hesiod's text from the Loeb as well as Merkelbach's and West's from the *TLG*, but have decided to reproduce the emended version of March 1987. See pp. 50–51 for March's full comments on her textual emendations.

[7] Unless specified, all translations are my own.

There is no question here that the deed is Deianeira's (fr. 25.20).[8] She has committed these terrible (δείν') deeds in great passion (μέγα θυμῶι). And, while there is no mention of Nessus in this fragment, the episode that follows fr. 25 contains an account of Hercules' sack of Oechalia. There is enough here to discern the cause of the destruction: Iole,

> (…) ξανθὴν Ἰόλειαν,
> τ[ῆς ἕ]νεκ' Οἰχ[αλ]ίη[ν
> Ἀμφι]τρυωνιάδης [
> τ]ὴν [δ'] αὐτέων παρὰ πα[τρ
> Θέσ[τ]ιος ἱππόδ[α]μος δ[
> ἠγάθ' ἵππ[ο]ισίν τε [καὶ ἅρμασι κολλητοῖσι
> μυρία ἕ[δ]να [π]ορώ[ν [9]

fr. 26.31–37 M-W

> (…) he begot golden haired Iole
> for whose sake Oechalia[was sacked]
> the son of Amphitryon,
> she, beside their father, Thestius tamer of horses,
> he led off (in marriage) with his horses and closely joined chariots
> presenting thousands of gifts

Edwin Carawan's study on the guilt of Deianeira examines the evidence on Greek vases in order to offer a more martial pre-Sophoclean figure. On the question of Nessus' involvement in the death of Hercules before the 5th century, Carawan reveals that the Deianeira of epic conceived of the fatal robe on her own, motivated either by delusion or desire for vengeance.[10]

Indeed, the connection between Nessus, the 'love charm' and Hercules' death was not a fixed one in pre-classical times. March, after reviewing the extant depictions of Nessus' death in pre-classical art and literature, concludes that there is no connection between the death of the centaur and Hercules' later death.[11] We are reminded, instead, of the etymology of Deianeira's name, 'man-slayer' (δηϊ-άνειρα), and violent parental tendencies being the offspring (according to some traditions) of none other than the god Dionysos and, of course, Althea who murdered her own son, Meleager, in revenge for killing her brothers during the Calydonian Boar hunt.[12]

Sophocles

Sophocles forged a subtler version of the 'Amazonian' Deianeira than her counterpart in Hesiod. The Deianeira he creates in his *Trachiniae* is fearful, subservient, and seemingly utterly conventional in her approach to her marriage. So great was the impact of the Sophoclean Deianeira on audiences and readers alike that she has all but banished traces of the earlier, far less fearful and far less innocent version.[13]

---

[8] Haubold 2005, 92, 95.

[9] Hesiod, *Catalogue of Women*, fr. MW 26.31–37.

[10] Carawan 2000, 195. See also Stafford 2012, 81.

[11] "Therefore we cannot assume that behind Hesiod's reference in 25 M-W to Deianeira's φάφμακον and κιτών there was any knowledge of Nessos and his love charm." 1987, 56.

[12] Dio. 4. 34; Apd. i.8.1 Δηιάνειραν, ἥν Ἀλθαίαν λέγουσιν ἐκ Διονύσου γεννῆσαι. αὕτη δ'ἡνιόχει καὶ τὰ κατὰ πόλεμον ἤσκει,/καὶ περὶ τῶν γάμων αὐτῆς Ἡρακλῆς πρὸς Ἀχελῷον / ἐπάλαισεν. "Deianeria, whom they say Althaia had with Dionysos. This daughter drove chariots and trained for war; Heracles wrestled with Acheloos to see who would marry her" and ii.7.5; Lib. *Met*, *2*.

[13] Davies 1989, 469–72.

In Sophocles' Deianeira we first find introduced the obsession with her status as wife of πάντων ἀρίστου φωτὸς ("the best of men" 177). In Sophocles, Deianeira's feelings for Hercules are complex. She is a fearful woman whose life is characterized by passive acceptance of her role as mother and wife to a man who is, for the most part, absent and unfaithful (460; 542–43). However, when she learns the true identity of Iole and the truth behind her arrival, her initial empathy with the captive princess is eroded by jealousy. Her fear of being replaced by Iole is described in a striking image of she and Iole lying under a shared blanket awaiting the embrace of Hercules, καὶ νῦν δύ' οὖσαι μίμνομεν μιᾶς ὑπὸ / χλαίνης ὑπαγκάλισμα (539–40). This extraordinary description comes when we learn of the jealousy that Deianeira is trying unsuccessfully to suppress for the young spear-won bride. Ten lines later, Deianeira will admit, ταῦτ' οὖν φοβοῦμαι μὴ πόσις μὲν Ἡρακλῆς / ἐμὸς καλῆται, τῆς νεωτέρας δ' ἀνήρ "And this is why I am afraid that he may / be called my husband but be the younger woman's man." (550–551)

Iole threatens to supplant Deianeira. This is the first time Hercules has sent one of his concubines into the home, the domestic space governed by Deianeira. Iole's invasion of her home and her bed prompts the good wife to act. She has already promised Lichas that she has messages and gifts for him to take back to her husband:

> ὡς λόγων τ' ἐπιστολὰς φέρῃς,
> ἅ τ' ἀντὶ δώρων δῶρα χρὴ προσαρμόσαι,
> καὶ ταῦτ' ἄγῃς.
>
> 493–495
>
> I have messages for you to take and
> gifts in exchange for the gifts your brought –
> these you must also take with you.[14]

This apparently innocent promise takes on layers of new meaning only later, once she re-emerges from the house to tell of "the work my hands have done" (τὰ μὲν φράσουσα χερσὶν ἀτεχνησάμην 534). While Hercules has been absent, Deianeira has been weaving a special robe for him. Carawan invites us to imagine that in *Women of Trachis* the robe that Deianeira has been working on for so long, the emblem of wifely devotion and chastity occupies the centre stage and is only removed when Deianeira takes it indoors (492) to put the final touches on it as a fitting gift for her husband.[15] This is compelling – the cloth as visual symbol and sign: as Deianeira waits out a 'Penelope-like' loneliness she weaves an elaborate robe for her husband, a ceremonial garment to celebrate the success of all his labours. She tells Lichas of its conception,

> οὕτω γὰρ ηὔγμην, εἴ ποτ' αὐτὸν ἐς δόμους
> ἴδοιμι σωθέντ' ἢ κλύοιμι πανδίκως
> στελεῖν χιτῶνι τῷδε καὶ φανεῖν θεοῖς
> θυτῆρα καινῷ καινὸν ἐν πεπλώματι.
>
> 610–613
>
> For this I swore, if I should ever see him returning
> safely home, as is my duty, I would clothe him in this robe
> to appear before the Gods
> so that he may perform sacrifice in new robes.

[14] Translations of *Trachiniae* are my own. The text is taken from Lloyd Jones & Wilson 1990 (TLG online).
[15] Carawan 2000, 205.

Sometime between her earlier promise of gifts and messages of 493–95 and her emergence from the house, Deianeira's plans have changed. Using the robe she has so long been weaving to honour his homecoming, she conceives of a daring plot, one that she reveals to her companions unobserved by anyone else (ἦλθον ὡς ὑμᾶς λάθρᾳ "I have come out secretly" 533). She has resorted to magic, infusing her gift with a love charm. She justifies her act by describing her inability to cope with Iole's arrival in her house, τὸ δ' αὖ ξυνοικεῖν τῇδ' ὁμοῦ τίς ἂν γυνὴ / δύναιτο, κοινωνοῦσα τῶν αὐτῶν γάμων; "for to live in the same house, to share the same marriage? What woman could endure that?" (545–546).

In fact, if we look back to the seemingly anodyne promise of gifts made by Deianeira to Lichas (494), we find more undercurrents revealed: προσαρμόσαι, the verb used of Deianeira's preparation of the gift means 'to fit' or 'attach closely to' in the literal sense as well as 'to suit', 'to match' or 'agree with'. This anticipates the fastening of the deadly robe on Heracles at 1051.[16]

The complexity of Deianeira's feelings and the undercurrents of destructive impulses towards her husband are mirrored in the terminology used to describe the robe and its instability as unwavering sign of female virtue and skill. The robe and the poison are described with the intermingled vocabulary of technical skill and sorcery: it is the finished product of Deianeira's "cunning of hand" or "craft" (ἀτεχνησάμην 534),[17] it is also a device or contrivance (μεμηχάνηται 586; τόνδε μηχαναῖς πέπλον 774). These are terms that tap into a deep-lying suspicion, manifest throughout Greek literature, toward the products of female labour.[18]

The use of the charm is described by Deianeira herself as a "remedy" (ᾗ δ' ἔχω, φίλαι,/ λυτήριον λώφημα, τῇδ' ὑμῖν φράσω 553–54); an "enticement" (κηλητήριον 575); a "love charm" or "potion" (φίλτροις 584); a *spell* (θέλκτροισι 585). The role of the cloak, which is to convey stature upon its male wearer, while at the same time bearing evidence of a virtuous, invisible, and domesticated woman in the *oikos* has been perverted and becomes a weapon of murder, covert and deceitful. Deianeira's cloak, like Clytemnestra's carpets and bathrobe are benign domestic items turned deadly.

Discussing Deianeira's silent withdrawal when confronted by Hyllus' accusation, Mary Scott asks,

> What is it that is hindering her from giving overt expression to her grief and pain? We must remember that the message delivered by her son is not that Heracles is dead through her action but that he is dying in agony. If she was motivated by nothing more than love in her actions, why is her first impulse not to rush to Heracles' aid, to strive with all her power to ease his pain? Surely hers is a reaction appropriate rather to one overwhelmed with shame or guilt?[19]

---

16 I am indebted to Wohl 1989, 24 for this observation.

17 *tekhnē* encompasses a multitude of meanings – art, skill, cunning, wiles.

18 Wohl 1989, 24. In addition she remarks, "This play shows a profound ambivalence toward *tekhnē* in general (…) Deianeira's death, too, is a *tekhnē* (928). The products of this *oikos' tekhnai* are fatal: its one literal product is the poisoned robe. The poison is preserved carefully in the home (556, 578–79, 685–86, 692), guarded as if it were a household treasure; the wool it is rubbed on is home-produced (689–90). Both the disintegration and the melting of the robe are described with craft metaphors: the wool disintegrates like the sawdust of cut wood (699–700); the robe sticks close to Hercules' limbs (ὥστε τέκτονος, 768), an obscure phrase (…) that clearly connects the disease to artisanship. (…) This is expressed most vividly by the imagery of consumption associated with Hercules' disease, which is repeatedly described as devouring him (1054–56; cf. 676, 700, 771, 987, 1084, 1088). See also Bergren 2008, 13–43.

19 Scott 1992, 62.

Hercules' own description of the robe clinging to his flesh and causing his death is revealing in itself,

> ὦ πολλὰ δὴ καὶ θερμά, καὶ λόγῳ κακά,
> καὶ χερσὶ καὶ νώτοισι μοχθήσας ἐγώ·
> κοὔπω τοιοῦτον οὔτ'ἄκοιτις ἡ Διὸς
> προὔθηκεν οὔθ' ὁ στθυγνὸς Εὐρυσθεὺς ἐμοὶ
> οἷον τόδ' ἡ δολῶπις Οἰνέως κόρη
> καθῆψεν ὤμοις τοῖς ἐμοῖς Ἐρινύων
> ὑφαντὸν ἀμφίβληστρον, ᾧ δόλλυμαι.

1046–1052

> Many and savage even terrible to speak of
> have been the labours of my arms and back!
> And yet never has the wife of Zeus or loathsome Eurystheus
> set such a thing on me as this woven encircling net of the Furies
> which the false-faced daughter of Oineus has put upon my shoulders, which is now killing me.

Heracles' describes the robe as a 'trap' or 'encircling net' of the Furies as he curses his wife. This web and net imagery here and in Greek tragedy as a whole functions as a metaphor for feminine guile and thus contributes to reading a more subversive Deianeira than traditional scholars have put forward. For indeed Sophocles' 5th century reconstruction of the epic Deianeira as 'the good wife' who waits at home and weaves for her husband has much in common with Homer's Penelope yet her experiences are not unlike those of Aeschylus' Clytemnestra. The parallels between this play and *Agamemnon* are striking. In both instances the concubine or 'spear-won' bride of the husband intrudes upon the household, the wife kills the husband by means of cloth. Deianeira's reception of the silent Iole is clearly modeled on Aeschylus' similar scene in Agamemnon when Cassandra the prophetess arrives at the king's side. In 542 Deianeira speaks of Iole's presence as a poor recompense for her fidelity and duty towards Heracles. The word she uses suggests recompense for household duties (οἰκούρια "rewards for keeping the house"). Her image of herself and Iole under one 'cloak' (literally χλαίνης 539–40) most commonly interpreted as sheet or blanket, waiting for the shared embrace of Heracles is ultimately transmuted into reality when she proposes to send him a gift in the form of a cloak (494).

Bacchylides

Bacchylides, who may well have been influenced by Sophocles' Deianeira, does not forget her darker pre-classical existence.[20] In *Ode* 16, he describes Deianeira as a victim of a plot, woven by a nameless, immortal god or powerful force of fate (ἄμαχος δαίμων (...) ὕφανε 23–24) and then mentions her own plan devised out of jealousy for Iole (24–29). Deianeira is 'ill-fated'

[20] Stafford (2012, 77) makes the point that since neither Bacchylides' *Odes* nor Sophocles' play can be precisely dated, it is not sure which source influenced the other if at all. Maehler, however (2004, 167) argues in his introduction to Ode 16, that Bacchylides was influenced by Sophocles' treatment of the myth. He says, "as most scholars now agree, Bacchylides' version would simply not have been intelligible to an audience who were unfamiliar with the version which connected the death of Nessos by Heracles' arrow with Heracles' own death by the poisoned robe." Suffice it to say that both poets had in their minds a Deianeira who fell prey to jealousy. See also March 1987, 62.

(δύσμορος 30), the power of jealousy (φθόνος 31) works as an external force which clouds her judgement and so causes her state of nescience or ignorance which Platter calls, "mortal blindness",[21]

> τότ᾽ἄμαχος δαίμων
> Δαϊανείρα πολύδακρυν ὕφανε
> μῆτιν ἐπίφρον᾽ἐπεὶ
> πύθετ᾽ἀγγελίανω ταλαπενθέα,
> Ἰολαν ὅτι λευκώλενον
> Διὸς υἱὸς ἀταρβομάχας
> ἄλοχον λιπαρό [ν π]οτὶ δόμον πέμποι.
> ἆ δύσμορος, ἆ τάλαιν᾽, οἷον ἐμήσατο:
> φθόνος εὐρυβίας νιν ἀπώλεσεν,
> δνόφεόν τε κάλυμμα τῶν
> ὕστερον ἐρχομένων,
> ὅτ᾽ἐπὶ ῥοδόεντι Λυκόρμᾳ
> δέξατο Νέσσου πάρα δαιμόνιον τέρ [ας.]

Ode 16 = *Dithyramb* 2.23–35[22]

> Then a powerful force
> wove for Deianeira a plot deceitful and sorrowful,
> when she heard the hurtful news
> that the son of Zeus, fearless in battle,
> was sending white-armed Iole to his house to be his wife.
> O ill-fated, o wretched woman! What a deed she planned!
> The force of jealousy destroyed her,
> and a dark shroud of things to come,
> when on the rose-clad banks of the Lycormus river
> she took from Nessus the fateful monstrosity

Deianeira's jealousy is explicit here as the cause of Hercules' death as it clouds her judgement sufficiently into receiving the "fateful monstrosity" (35) from the centaur.

What is striking about this passage is the vocabulary specific and evocative of weaving: the "unnamed god" or "force of fate" (δαίμων, 23) who "weaves" (ὕφανε, 24) the plot in which Hercules and Deianeira are caught and the "dark shroud" or "veil" δνόφεόν τε κάλυμμα (32) of future events are bold, verbal portents of the robe of death Deianeira will weave for her husband. The verb μήδομαι ('to plot', 'contrive' 29) itself evokes the weaving of plots, and with its etymological connections to μῆτις ('skill', 'plan' or 'craft') together with its personification in the form of the primordial goddess, Metis, the poet has managed in all brevity to question the innocence of Deianeira. Furthermore, he highlights the role of Athena in the story of Hercules earlier in the ode.[23]

---

[21] 1994, 343. Platter argues that the φθόνος (31) refers to the resentment of Nessus towards Hercules and not to Deianeira's jealousy of Iole (346–348), he further remarks: "Even if *phthonos* could refer to Deianeira's jealousy, the connotations of the word would remain clearly hostile. She would have to be perceived as responding to Heracles' infidelity as an avenging Medea figure. Yet there is no trace of this representation in the Deianeira of *Ode* 16. The poet stresses her ignorance concerning the effects of her actions. This ignorance is the second cause of her destruction, the "dark veil of future things" (33–44).

[22] Irigoin, J. 1993 ed online TLG, henceforth referred to as *Ode* 16.

[23] "As daughter of Metis and goddess of women's skills, Athena is the patron of weaving, and so when Bacchylides specifies the action of the daimon as that of weaving (24), he [Bacchylides] reminds us not only of the robe that Daianeira has made (though not yet poisoned) but also of the divinity who would normally supervise all such work." Burnett 1985, 125.

This subtle emphasis on sphere of Athena leads the reader all to quickly to an awareness of the corruption that has occurred in the sacred space of women's work: the potentially dangerous combination of the powers of Metis and Athena has resulted in a prosaic household item becoming a deadly weapon. The robe is a manifestation of the literal and figurative union of the skills over which Metis and Athena preside.

In Bacchylides *Ode* 5 Deianeira is described by her brother Meleager as beautiful and innocent. Stafford states that the description of Deianeira as φυὰν ἀλιγκία "resembling Meleager in nature" may be taken as an indication that Deianeira is as strong as her brother thereby alluding to a family trait that reveals itself as "dangerous ruthlessness in the family's women-folk."[24]

> Ἦρά τις ἐν μεγάροις Οἰ-
> νῆος ἀρηϊφίλου
> ἔστιν ἀδμήτα θυγάτρων,
> σοὶ φυὰν ἀλιγκία;
> Τάν κεν λιπαρὰν ἐ θέλων θείμαν ἄκοι-
> τιν. Τὸν δὲ μενεπτολέμου
> ψυχὰ προσέφα Μελεά-
> γρου· Λίπον χλωραύχενα
> ἐν δώμασι Δαϊάνει-
> ραν, νῆϊν ἔτι χρυσέας
> Κύπριδος θελξιμβρότου.
> Λευκώλενε Καλλιόπα,
> στᾶσον εὐποίητον ἅρμα
> αὐτοῦ·
>
> Tell me, is there within the halls
> of Oineus, dear to Ares,
> one of his daughters still unwedded,
> bearing your likeness in her form?
> Her I would gladly make my lustrous wife."
> To him the shade of Meleagros staunch in war
> replied: "I left one with youth's bloom about her neck
> there in the house, Deianeira,
> as yet without experience of golden
> Cypris who casts her spells on mortal men."[25]

Here the conversation comes to an end, leaving the readers all too aware of how Hercules' solicitation of Deianeira's hand in marriage from the shade of her dead brother will end. Readers are reminded of Meleager's grim death at the hands of his own mother and will thus draw parallels between the women looking ahead to Hercules' equally macabre death by burning at the hands of Deianeira.[26]

Much is made of Deianeira's lineage by the Latin authors Ovid and Seneca as a justification for latent vengeful and murderous tendencies.[27] Sophocles, however, chooses to avoid this emphasis and creates instead the ultimate passive aggressive wife of Hercules.

These subtleties in the characterisation of Deianeira and in the myth itself are echoed in Sophocles' portrayal as well. Much of which is unfortunately left out of later writers in their blunt attempts to squeeze a moral or an allegory out of each pagan story.

[24] Bacchyl. *Od.* 5. 168. Stafford 2012, 81.
[25] Bacchyl. *Od.* 5 165–175 (trans. Trzaskoma, S. M., Scott Smith, R., & Brunet, S.).
[26] See Platter 1994, 343, and March 1987, 52.
[27] This is discussed in the section on the Latin literary tradition.

The Roman Literary Tradition

Ovid

Ovid's *Heroides IX, Deianeira to Hercules*, is written at the moment when Deianeira receives Iole who has been sent on ahead by Hercules. This epistle has a unique feature in the collection in that news reaches the writer from the outside world while she is writing and this causes her task to be interrupted and altered. Ovid has used Sophocles' *Trachiniae* as his source text here and so, just as the messenger arrives with the news that Hercules is dying from the poisoned robe in the play, here in *Heroides IX*, a messenger arrives with the news that changes what Deianeira will write in her letter. The Ovidian heroine writes of how the great conqueror has himself been conquered by his love for Iole and how this affair is damaging his great name. Much is made of his ignominious enslavement to Omphale and that Queen's insistence on Heracles' donning of feminine garments, taking up the pursuits of wool-working while regaling her court with tales of his manly adventures. The letter has a poignancy and sadness of tone as Deianeira recounts not once but twice the glorious achievements of her hero husband, a desperate reminder of the crumbling source of her self-worth and status in society. When she receives news that Heracles is dying, *scribenti nuntia venit / fama virum tunicae tabe perire meae* "Rumour whispers that you are dying of the poison in that cloak" (143–4), she resolves to die in a manner that befits the wife of Hercules and the sister of Meleager, *tu quoque cognosces in me, Meleagre, sororem!* "And you also, Meleager shall recognise me as a true sister" (151).

What Ovid achieves *Heroides IX* is the doubling impact of the motif of weaving and all its connotations-Deianeira has literally already woven the cloak and now she is weaving words, words which only serve to remind Hercules of his great deeds and of the danger he is in of losing all the glory he has gained by becoming ensnared and enslaved to another woman.[28] Like all the letters in this collection, the letter receives no response – however, many centuries later, in le Fèvre's epic, his much more gallant Hercules does respond to Deianeira's missives, albeit with no better results for the tragic couple.[29]

In his *Metamorphoses* (AD 8), Ovid concerns himself more closely with the story of Hercules' death and resultant apotheosis (9.101–173). An important deviation in the narrative of the gathering of the poison from Nessus reveals a departure from his Sophoclean source at this point. Ovid uses the version where Nessus gives Deianeira his own blood-soaked garment ready-poisoned thus removing the motif of the good wife weaving with her own hands a prosaic garment for her husband. Nessus tears out the barbed arrowhead protruding from his breast,

> *exstabat ferrum de pectore aduncum.*
> *quod simul evulsum est, sanguis per utrumque foramen*
> *emicuit mixtus Lernaci tabe veneni.*
> *Excipit hunc Nessus "ne" que enim "moriemur*
> *Inulti" secum ait, et calido velamina tincta cruora*
> *Dat munus raptae velut inritamen amoris.*
>
> 128–133[30]

> The barbed point protruded from his breast. This he tore out, and spurting forth from both wounds came the blood mixed with the deadly poison of the Lernean hydra. Nessus caught this, and muttering, "I shall not die unavenged," he gave his tunic, soaked with his blood, to Deianeira as a gift, potent to revive waning love.

[28] Lindheim 2003, 68.
[29] Le Fèvre will be the last work discussed in this paper.
[30] Trans. Miller (3rd edition.)

A few lines later when she hears of her new rival, she first contemplates killing her, calling up her 'man-slaying' lineage,

> (…) *Quid si me, Meleagre, tuam memor esse sororem*
> *forte paro facinus, quantumque iniuria possit*
> *femineusque dolor, iugulata paelice testor?*
>
> 149–151

What if, O Meleager, remembering that I am your sister, I make bold to plan some dreadful deed, and by killing my rival prove how much a woman's humiliation and grief can do?

> *in cursus animus varios abit. omnibus illis*
> *praetulit inbutam Nesseo sanguine vestem*
> *mittere, quae vires defecto reddat amori*
>
> 152–154[31]

her mind is divided between various courses; but to all other plans she prefers to send to her husband the tunic soaked in Nessus' blood, in the hope that this may revive her husband's failing love

We cannot help but recall the Deianeira of Bacchylides 5 as opposed to Sophocles' Deianeira who abjures all knowledge of anger, ἐγὼ δὲ θυμοῦσθαι μὲν οὐκ ἐπίσταμαι ("I for my part, do not know how to get angry" 543)

Seneca

Seneca's *Hercules Oetaeus*, written somewhere between AD 50 and 62, is an adaptation of Sophocles' play. It is not clear whether any of Seneca's plays were written for dramatic performance they were primarily literary platforms for the expression and development of his Stoic and political philosophies.

For the stoics and for Seneca himself, Hercules' embodied the ideal of Stoic *virtus* and as such, it is not surprising that the Deianeira of Seneca's play was written to provide the 'counterbalance' to this ideal.[32] As such, Seneca's Deianeira is ruled by passions while Hercules is ruled by reason. The bipartite structure of the play reflects the struggle between the two principles. Thus the first half concerns itself with Deianeira consumed by jealousy and passionate rage and ends with her suicide. Seneca's Deianeira enters with curses and thoughts of terrible vengeance on her philandering husband (256–258), while the nurse vainly tries to calm her down in her frenzy; she rants on about Iole and the indignities heaped upon her (278–290):

> *Iole meis captiva germanos dabit* 278
> *natis Iovisque fiet ex famula nurus?*
> (…)
> *no ibo inulta. gesseris caelum licet*
> *totusque pacem debeat mundus tibi,*
> *est aliquid hydra peius: iratae dolor*
> *nuptae. quis ignis tantus in caelum furit*
> *ardentis Aetnae?quidquid est victum tibi*
> *hic vincet animus. capta praeripiet toros?*
> *adhuc timebam monstra; iam nullum est malum,*

[31] Trans. Miller (3rd edition.)
[32] See King 1971, 217.

cessere pestes, in locum venit ferae
invisa paelex.

Shall Iole the captive produce brothers for my sons? Shall a slave become daughter-in-law to Jove? (…) I shall not go unavenged. Though you have borne the heavens and the whole world owes its peace to you, there is something worse than a hydra: the pain of an angry wife. Does any fire as great as that rage toward heaven from burning Etna? Anything you have conquered, this spirit will conquer. Is a mistress to steal my marriage bed? I was still fearing monsters, but now such evils are no more, scourges have vanished, and in place of a beast has come this hateful mistress.[33]

In order to prevent her mistress pursue her threats of revenge, ironically it is the character of the Nurse who suggests the idea of a love potion. While Deianeira takes this suggestion up, she never articulates a rejection of her revenge plans. This leaves ambiguity about her intentions and her understanding of the so-called love potion especially when we examine the vocabulary she uses to describe the circumstances of acquisition,

tunc verba moriens addit: "Hoc" inquit "magae
dixere amorem posse defigi malo;(…)
inlitas vestes dabis
hac" inquit "ipsa tabe, si paelex tuos
invisa thalamos tulerit (…)

523–524

Then while he was dying he added these words,
"With this vile stuff, sorceresses say that love can be fixed;(…)
you shall give him clothing smeared with this very gore,
if a hateful mistress should take your marriage bed" (…)

Nessus' description of the love charm as a *malus* (524) is sinister in the extreme.

More sinister is Deianeira's own description of how the potion will act,

Tu, quam meis admittit arcanis fides,
Perge ut nitentem virus in vestem datum
Mentem per artus adeat et tacitum intimas
Intret medullas

535–538

[to the Nurse]
You have been admitted to my secrets through your faithfulness,
proceed, so that the venom as it is placed on the shining garments
may work its way to his mind through his limbs, and spread in stealth into his deepest marrow.

This is a most accurate description of what will happen to Hercules when he puts the charmed robe on his body! Despite the chillingly prophetic words of Deianeira in passages such as this, critics such as King still argue that Deianeira has no prior knowledge of what the effects of the potion on Hercules will be, arguing that Seneca is simply emphasizing the destructive nature of passion.[34]

King's argument notwithstanding, there is more than enough linguistic evidence to suggest that Seneca's passionate and frenzied heroine wished destruction upon her husband. I do not venture to

[33] Translation Fitch.
[34] King 1971, 218.

suggest that this was the playwright's main concern, however, since he was more concerned with portraying the Stoic man's ascension to a level of divinity and thus for him, Deianeira had to be the foil for Hercules. This is the reason for the frenzied suicide of the heroine (934–1024) – such a departure from the powerful and silent exit Sophocles' heroine makes from the stage once Hyllus accuses her of murdering his father (Soph. *Trach.* 807–811).

It was as a Stoic hero that Hercules became so popular with readers of late antiquity and beyond. Christian writers found parallels in this hero whose immoral deeds are vindicated through tremendous suffering. In his new life as a demi-god Heracles' new role is in aid of mankind – a theme which was most attractive for writers aiming to mould the classics to a Christian frame and cleanse it of aspects which were, for the new religion, immoral and, more importantly, against the law.

Deianeira in Late Antiquity and the Middle Ages

By the time of Theodosius I in 391 AD, all forms of paganism had been prohibited and in 435 during the reign of Theodosius II, pagans were subject to the death penalty. Despite these legal rulings, paganism did not simply disappear but with Christianity becoming the dominant religion in areas previously presided over during Roman imperial rule in the west, it's voice was heard less through the channels of office and more through those of the literary and philosophical arts. The two renaissances which occurred during the Middle Ages witnessed a resurgence of interest and preservation of classical texts and artefacts; the first being the Carolingian Renaissance which occurred from the late 8th to 9th Centuries at the courts of Charlemagne and Louis the Pious, and the second which occurred during the High Middle Ages known as the 12th Century Renaissance. By this latter renaissance, classical myth had become the favourite medium for Christian allegorists.[35]

Due to the hero's ubiquity in the Roman Empire, the figure of Hercules and his cult initially vied with the figure of Christ in terms of popularity. Analogies were drawn between the two figures based on the overt similarities between them: both were the products of a god and a mortal woman, both overcame death after banishing evil from the earth. Many writers of Late Antiquity and the Middle Ages do not dwell on Deianeira, concentrating only on revising the great pagan hero under the aspect of supreme exemplar of virtue and justice.[36]

The search for literary treatments of Heracles' story which include Deianeira and the tale of the poisoned robe yields a relative paucity of texts.[37] Those under discussion here reveal a striking

[35] See Seznec 1971, 84–122.

[36] An excellent study of Hercules through the ages is Galinsky 1972. In Augustine's *De Civitate Dei contra Paganos*, (5th century), Hercules is an example of pagan virtue in contrast with Christian virtue (Bk. 14). Fulgentius in *Mythologiae* (Bk.2, 2) identifies Hercules as *exemplum virtutis*, the symbol of manly virtue overcome by the evil of lust wielded by the wicked Omphale (6th century). His contemporary Boethius focuses on the glorious deeds of the hero ending in a triumphant apotheosis in *Consolations of Philosophy* (Bk 4, *Song 7, The Hero's Path*). Lactantius, however, (circa 240–circa 320), in Book 1 of *Divinarum Institutionum* entitled *De Falsa Religione Deorum* (ch. 9), discusses the life and death of Hercules in a negative manner. He rejects the pagan hero's virtue and apotheosis, questioning the divinity of a figure who became enslaved to his own vices of lust, indulgence, gluttony and violence. He also uses the incident of Hercules' servile entanglement with Omphale as a sign of the worst moral degeneracy. In the scriptural allegory of the *Ovide Moralisé,* the author likens Hercules' conquests with the likes of Busiris, Diomedes, the Hydra, Geryon, the Amazons and Achelous to the victorious battles of Christ, the Saviour (9.989–95). See chapter 7 "Post-Classical Variations" in Stafford 2012, 201–244 for a more detailed discussion of the 'afterlife' of Hercules as well as Heinrichs 1990.

[37] *Contra Iudices* (195–200) *by Theodulf of Orleans* composed between 799 and 800 at the court of Charlemagne mentions the death of Hercules but not much is made of Deianeira and the robe. The author is more concerned to portray

ambiguity on the part of the authors in their approach to this character. Her reputation amongst Medieval and early Renaissance authors contains the same ambiguities we find in the ancient sources albeit shaped by the conventions of genre and Christian morality. It is not surprising therefore, to find that in accordance with the practice of applying allegorical modes of interpretation to pagan stories she is found to be a Delilah equivalent, the exemplum of destructive female jealousy, an allegory of the flesh that seeks to entrap the spirit and much is made of her employment of sorcery. Conversely, other writers find in her that gullible innocence which caused her to destroy her husband in error and she is praised for her love and commitment to Heracles.

Theodulus

Nothing beyond the name of this author is known (and this too is not beyond doubt) although his *Eclogue* came to be used as scholastic text for centuries. Written somewhere between the 9th and 10th centuries, the *Ecloga Theoduli*, inspired doubtlessly by Virgil's *Eclogues*, is a debate cast in a pastoral setting between classical pagan myth and Christian myths from the Bible. It is presented as a verse contest between *Pseustis* (Falsehood) and *Alitheia* (Truth), which is presided over by *Phronesis* (Wisdom). It goes without saying that Christianity represented by *Alitheia* wins the debate. Once the sets of myths are countered against one another, *Alithea* announces that she will rely on the Gospels which causes *Pseustis* to submit to defeat.

Theodulus pairs the pagans Hercules and Deianira against the biblical Samson and Delilah from the book of *Judges* (16.4–18),

> ***Pseustis***
> *Alcidae vigilem spoliavit clava draconem;*
> *Gerionis pompam rapit et consumpserat ydram;*
> *Cacus cessit ei, succumbit ianitor Orci:*
> *Incendit demum paelex Deianira superbum.*
>
> ***Alithe***
> *Samson exuviis indutus membra leonis*
> *Sternit mille viros, devastat vulpibus agros,*
> *Urbis claustra tulit, nervorum vincula rupit:*
> *Fraude sua tandem praecidit Dalila crinem.* [38]
>
> **Falsehood**
> The club of Hercules despoiled the vigil of the dragon;
> He stole the pride of Geryon pride and killed the Hydra;
> Cacus fell to him as did the gatekeeper of Hell.
> In the end his mistress Deianeira burned him in his arrogance.
>
> **Truth**
> Samson, his body clothed in lions skins,
> Slew a thousand men, he laid waste to the fields with foxes,

Hercules as a divided and flawed hero contrary to writers such as Boethius and Fulgentius for example. See Nees 1991, for further details.

[38] *Theoduli Eclogam* digital collection on Archive.org. Harvard University 1902, 41–42.

> He took the lock of the city, and broke the restraining chains:
> In the end Delilah cut off his hair in deceit.

The famous *Roman de la Rose* by *Guillaume de Lorris* and completed by *Jean de Meun* in the early 14th century portrays Deianeira in the same manner as Theodulus,

> Hercules had many struggles: he conquered twelve horrible monsters, and when he had overcome the twelfth he could never finish with the thirteenth his sweetheart Deïaneira, who, with her poisonous shirt, lacerated his flesh, all enflamed with the poison. His heart had already been made mad with love for Iole. Thus Hercules, who had so many virtues, was subdued by a woman.
>
> 9191–9202[39]

This is a brief cameo appearance and so just a few remarks will be made: the shirt is Deianeira's and with it she is responsible for the lacerations of his flesh. It is thus she who prevented Heracles from completing his labours which are presented here as noble and he, full of virtue. No condemnation is found for him as the faithless husband only the ignominy of being destroyed by a woman. There are echoes of the Sophoclean Hercules in the last line of this passage, not to mention Fulgentius.

Berchorius/Bersuire

Pierre Bersuire, the French Benedictine, also known as Petrus Berchorius wrote his *Ovidius Moralizatus* between 1340 and 1342. His work is not to be confused with the *Ovide Moralisé* written anywhere between 1305 and 1329 the author of which is anonymous.[40] Berchorius sought to impose allegorical meaning upon a series of classical myths structured on Ovid's *Metamorphoses* thereby drawing out Christian interpretations of these myths. His work is possibly the most influential of the moralising commentaries on Ovid's work albeit not the best known.

After describing the battle between Hercules and Achelous for the hand of Deianeira, Bersuire interprets the victory of Hercules thus,

> *Sic cum filia regis calidonie id est anima humana, filia dicit patris. Acheloo cornuto id est diabolo fuisset per ei peccatum coniuncta et ipsi per complacentias viciolas matrimonialiter copulate. Herculem filius dei dicitur cuius fortitude est quasi Rinocerontis illum qui nunc erat anguis per fallaciam nunc Taurus per proterviam vicit & expugnavit. Cui cornu sue potentiae fregit et sic animan humanam ab eo liberavit & eam sibi per gratiam copulavit. [.ps.]Cornua peccatorum confringam.*
>
> 9.67 A[41]

> So the daughter of the king of Calydonia – that is the human soul, the daughter of the Father – had been linked to the horned Achelous – that is the devil – through sin and joined to him in marriage through sinful pleasures. The Son of God is called Hercules whose fortitude is like a rhinoceros'. Hercules conquered and subdued the one who at one time was a snake through deceit and at another a bull through wantonness. He broke his horn of power and thus freed the human soul from him and joined her to himself through grace. Psalm 74.11: 'I will break the horns of sinners.'

In this first allegory of the union between Hercules and Deianeira, we have something along the lines of a *hieros gamos*, Deianeira representing the human soul is freed from bondage with the devil,

---

[39] Brumble (trans.) 1998, 163–4.

[40] Berchorius' work is sometimes also entitled *Ovide Moralisé* but to avoid confusion its Latin title will be used exclusively in this paper.

[41] Trans. Reynolds 1971, 323–24.

represented by the shape-shifting Achelous, and is joined to Hercules, the "Son of God" (*filius dei*) "through grace" (*per gratiam*). Yet Bersuire does not pursue this allegory as he continues with the story of Heracles' demise. His version of the poisonous shirt which follows has the dying centaur dip a shirt in his own poisonous blood and give it to Deianeira under the pretext of it being an inducement of love. Deianeira sends it off to her husband, through Lichas after she hears of Hercules' marriage with Iole and a curious incident of dressing in each others' clothes is reported to her.[42]

Speaking of infidelity amongst spouses Bersuire first offers some mundane facts, saying,

> *Ex quo etiam saepe contigit quod Hercules id est maritus ipsos occidit. Ipso vero maritum camisia venenata id est uxor zelotipa comburit et dolor & invidia ipsum angit.*
>
> 9.67 B
>
> It also happens often that because of this [sc. unfaithful wives] Hercules, that is a husband, kills them [wives]. But a poisoned shirt, that is a jealous wife, burns up this husband and grief and envy torture him.

Hercules can be read as all jealous husbands while the poisoned shirt now stands for the jealous wife whose burning jealousy literally tortures her spouse. Bersuire then moves on to a more Christian allegory in which he sees Hercules as the soul, Deianeira as the flesh and the poisoned shirt the symbol of carnal pleasures,

> *Vel dic moraliter quod Hercules est animus uxor eius caro. Ista igitur est quem viro suo id est animo dat camisiam venenatam id est objectamenta carnalia quibus per temptationes urit ita quod finaliter institutus igne luxuriem concremant Nessus id est diabolo non potuit aliter Hercules id est spiritus per peccatum occidere: quam uxori suem id est carni camisiam venenatam daret ut sic mediante sensualitate vel carne possit animum per concupiscentiam urere et tandem per vicia cremare. Quod bene fuit figuratum Gen.3. In serpent quod primo mulieri obtulit pomum vetitum qua mediante decepit virum. Quia vero serpens id est diabolus primo offert poma vetita id est mundi & carnis delectabilia mulieri id est sensualitati & ea mediante spiritui vel menti.*
>
> 9.67 B–C
>
> Or say morally that Hercules is the soul, his wife the flesh. It is she who gives to her husband, that is the soul, a poisoned shirt, that is carnal pleasures – because of which he so burns through temptations that he is finally consumed in the fire of luxury. Nessus, that is the devil, is able to kill Hercules, that is the spirit, through sin in no other way than by giving a poisoned shirt to his wife, that is the flesh, so that by means of sensuality or the flesh he may be able to set the soul on fire through concupiscence and at last consume it through sins. This was well figured in Genesis 3 in the serpent which offered the first woman the forbidden apple by means of which she deceived her husband. The serpent, that is the devil, first offers the forbidden apple, that is the delights of the world and of the flesh, to a woman, that is to sensuality and by this means to the spirit or the mind.[43]

Bersuire has mastered the art of allegory here, he goes on to offer some more convoluted explanations providing parallels to the death of Hercules but they depart from Deianeira. This is the first instance where the shirt itself is given a meaning all of its own. One could offer that since this allegory was not taken up by many other writers in quite the same depth and detail that it was not popular either as it was too complex or the story itself had too many other more simple delights to be taken from it.

---

[42] This is yet another example of the erroneous or inventive transmission of the Omphale affair whereby Omphale and Iole have been conflated (9.67 A).

[43] Trans. Reynolds 1971, 325.

Boccaccio

Although Giovanni Boccaccio wrote at a similar time and with a similar moral purpose, his work *De mulieribus Claris / On Famous Women* does not offer such a clear moral parallel when it comes to his narration of the story. Boccaccio's purpose in writing his compilation in 1361–62 was to record for posterity the stories of women who were renowned for great deeds both good and bad. In his dedication to *mulieri clarissime Andree de Acciarolis de Florentia Alteville comitisse* the 'Illustrious Lady Andrea Acciaiuoili of Florence, Countess of Altavilla', he expresses his wish that his work will cause her spirit to emulate the deeds of women in the past (8), exhorting her not to be put off by his inclusion of wicked deeds with pure ones (*lasciva(...)immixta sacris*) implying that her virtuous nature is discerning enough,

> *Et esto nonnunquam lasciva comperias immixta sacris – quod ut facerem recitandorum coegit oportunitas – ne omiseras vel horrescas; quin imo perseverans, uti viridarium intrans, eburneas manus, semotis spinarum aculeis, extendis in florem, sic, obscenis sepositis, collige laudanda;*
>
> You will find, at times, that an appropriate recital of the facts has compelled me to mix the impure with the pure. Do not skip over these parts and do not shy away from them, but persevere in your reading. As on entering a garden you extend your ivory hands towards the flowers, leaving aside the thorns, so in this case relegate to one side offensive matters and gather what is praiseworthy;[44]

Many of Boccaccio's lives conclude with a moral lesson or a moral exhortation or a passage of philosophical reflection. Deianeira's life is lacking any of these conclusions,

> *Quod sentiens Nessus seque mortuum abritratus, vestem sanguine suo infectam confestim Deyanire tradidit, asserens, sice cruentam si induat, posse Herculem ab omni extero in suum amorem retrahere. Quam Deyanira credula, loco pregrandis muneris, summens, clam aliquandiu servatam, Herculi Omphalem, seu Yolem, amanti, per Lycam servulum caute transmisit.*
>
> *ipse autem cum sudore cruorem, veneno infectum, resolvisset porisque bibisset, versum in rabiem se igni comburendum ultro concessit. Et sic Deyanira, tanto viduata viro, dum retrahere speraret, perdidit et Nessi cedem etiam expiavit.*
>
> 25.4–5[45]
>
> When he felt the wound, Nessus knew that we would die and at once gave Deianira the cloak coloured with his blood, telling her that if she put it on Hercules, bloody as it was, she would win him back from any other love. Deianira was all too willing to believe this story; she accepted the cloak as she would a great gift and kept it hidden for some time. When Hercules fell in love with Omphale (also called Iole) she took care to send the cloak to him by his servant Lichas.
>
> Once the poisoned blood on the cloak mixed with his own sweat and was absorbed through is pores, Hercules went mad, and of his own will he threw himself into the fire. Thus was Deianira widowed of her mighty husband: hoping to recapture his love, she destroyed him and also avenged Nessus' death.

Boccaccio does not give us much with which to condemn Deianeira, if anything she is guilty of naivety. Nessus gives the *credula* "gullible" Deianeira his cloak already soaked with his poisoned blood, (*vestem sanguine (...) infectam*). She accepts the cloak as if it were a very powerful gift (*pregrandis muneris*) and uses it in the hopes that Hercules will be restored to her. As the robe is

[44] Trans. Brown 2001, 5.
[45] Trans. Brown 2001, 99.

neither woven by Deianeira nor is she guilty of gathering the poison and storing it or applying it years later, according to instructions, her guilt is minimised. There is no mention of jealousy or anger on her part. The deed this noble woman commits is that of murder yet it is presented as a mistake, and she was the victim of such a crime in the sense that Boccaccio takes care to describe her as deprived of such a man *tanto viduata viro*. Curiously though, Deianeira's deed is also described as the inadvertent atonement for Nessus' death *Nessi cedem etiam expiavit*

Chaucer

Heracles appears for the first time in English literature in Chaucer's *Canterbury Tales* (circa 1390s). In *The Monk's Tale*, the narrator entertains the audience with a series of tragic tales taken from classical and biblical sources. In the story of Heracles, the monk demonstrates how Fortune in all her caprice and guile brings the great hero Heracles to his end. After praising the great and worthy tasks of Hercules, the monk tells of the love Heracles had for Deianeira, calling her his *lemman* or "beloved",

A lemman hadde this noble championn,
That highte Dianira, fressh as May;
And as thise clerkes maken mencioun,
She hath hym sent a sherte, fresh and gay.
Allas, this sherte—allas and weylaway!—
Envenymed was so subtilly withalle
That er that he had wered it half a day
It made his flessh al from his bones falle.

But nathelees somme clerkes hire excusen
By oon that highte Nessus, that it maked.
Be as be may, I wol hire noght accusen;
But on his bak this sherte he wered al naked
Til that his flessh was for the venym blaked.
And whan he saugh noon oother remedye,
In hoote coles he hath hymselven raked,
For with no venym deigned hym to dye.

Thus starf this worthy, mighty Hercules.
Lo, who may truste on Fortune any throwe?
For hym that folweth al this world of prees
Er he be war is ofte yleyd ful lowe.
Ful wys is he that kan hymselven knowe!
Beth war, for whan that Fortune list to glose,
Thane wayteth she her man to overthrowe
By swich a wey as he wolde leest suppose.

This noble champion had a beloved,
Named Deianeira, fresh as Spring, and
As the scholars say,
She sent him a shirt, all fresh and fine.
Alas, this shirt, alas and alack!
It was poisoned so subtly
That before he had worn it half a day
It made his flesh fall away from his bones.

But nonetheless some scholars excuse her
And blame Nessus who made it.
Be that as it may, I will not accuse her;
But he wore the shirt upon his naked back
Until his flesh was blackened with the venom.
And when he saw there was no other remedy,
He raked a bed of hot coals for himself,
For he chose not to dye by means of poison.

Thus died, this worthy and mighty Hercules.
Lo, who may trust Fortune for any length of time?
For he who follows this dangerous world
Is often laid low before he knows.
Wise is he who knows himself!
Beware, for when Fortune chooses to beguile,
She waits to overthrow her man
In such a way as he would least expect.

2119–2142

Chaucer does not blame the 'sweetheart' Deianeira, he blames the Centaur. The shirt, he says, was made by Nessus (2130), already subtly poisoned (2124) and thus he is unwilling to accuse her (2129). It is pertinent to mention here that in Chaucer's *Hous of Fame*, the narrator meditates upon the propensity of men to betray women, listing many heroines from Ovid's *Heroides* including Deianeira.[46]

Raoul le Fèvre

Raoul le Fèvre's epic romance the Trojan cycle, the *Recueil des histoires de Troyes* written *circa* 1460 was the first book printed into English. It was printed by William Caxton who was also responsible for the translation circa 1473–74.[47] It proved to be immensely popular and thus was reprinted numerous times as *The Ancient History of the Destruction of Troy* or simply, *The Destruction of Troy*. The first two of the three books contain accounts of the deeds of Hercules and his death by means of the poison shirt. Le Fèvre's Hercules is the virtuous hero of allegorical literature so popular in Medieval times – a combination of the intellectual and physical hero championed by authors before him, "being a keen student of an academic version of Atlas, and later being sought after as teacher. He is, however, a much more flesh-and-blood character, owing much to the medieval tradition of the chivalrous knight."[48]

Le Fèvre relates the tale in considerable detail and embellishes it with some innovations of his own. It is these that are of interest for our purposes here;

We are privy to Nessus' dying thoughts which are intent on avenging his death and using Deianeira in malice for this purpose. le Fèvre thus sets Deianeira up as the unwitting instrument of Nessus' revenge,

[46] Heinrichs 1990, 60.
[47] Le Fèvre's Recueil was not only the first book printed in the English language it was also the first book from the press of William Caxton, the father of printing in England. Interestingly enough it wasn't printed in England but in Bruges or Ghent.
[48] Stafford 2012, 208.

> *Nessus by the Wound that Hercules gave him, began to feel the Approaching of Death, and to suffer sharp Anguish: He ran a great while unto a Valley where he fell down; and considering that his Life had no Recovery, he employed the End of it to imagine how he might do displeasure unto Hercules calling to remembrance, that he had a most terrible and mortal Poyson about him, he said to Dejanira in great Malice; Fair Lady, the Love of you hath caused me to receive Death, which displeaseth me not so much as that Hercules shall enjoy you, who deserve a far worthier Man. Hercules is no true Husband, but the unfaithfullest to his Wife that ever was. Forasmuch as I have singular Pity of you, as your Beauty constraineth me to do you Pleasure, I will give you here a precious thing, I having such Vertue, that if you boyl it with one of the Shirts of Hercules, with the Blood that runneth out of my Wound, and give the Shirt to Hercules that he wear it, he shall never after love any other Woman but you.* (2.17.42) [49]

Nessus gives Deianeira strict instructions with regards to its application but he does not supply her with the shirt. When hearing about Heracles' infidelity, Nessus' potion is not Deianeira's first recourse.

Deianeira, devastated with the news of Heracles' love affair with Iole writes Heracles not one but two letters. Heracles, in his Medieval guise as the chivalrous knight, has the grace to blush when he reads Deianeira's first epistle in which she berates him over his behaviour with Iole, and is suitably shamed into writing back and assuring her of his affections.[50] One would be hard pressed to imagine a Sophoclean or Ovidian Hercules doing the same. le Fèvre has undoubtedly read his *Heroides*. As access to the classical world was gained largely through Latin literary sources rather than Greek this is hardly surprising. Indeed, glosses on the *Heroides* for example, began to proliferate around the late 12th century, at the time of their renewed popularity. As Katherine Heinrichs relates, "(…) the epistles were read as a guide to the proper conduct of love, a series of exempla offered for the instruction of Christian readers, who were to abhor the evil and imitate the good."[51] Deianeira's letters to Hercules in le Fèvre have much in common with her Ovidian predecessor's in content and structure.[52] For the purposes of this paper, our focus must remain on the details surrounding the gift of the robe and the poison.

Chapter 31 contains the most dramatic innovation of any author dealing with the Heracles and Deianeira story. In response to Deianeira's second letter, which is much longer and follows *Heroides IX* very closely, le Fèvre narrates how Heracles is "smitten with Remorse of Conscience" (80), so much so that he ends his affair with Iole and retreats to Mount Oeta to sacrifice to Apollo with none but the loyal Philotes for company. On his way, so the story goes,

> *he met Lycas, who made him Reverence, and demanded him if it pleased him any thing to send unto Dejanira. Hercules answered Lycas, that he would go and make his Sacrifice to Apollo, and at his Return again, he would go, or else would send unto her.*
>
> 2.31.80

When Deianeira receives her husband's response, she is overjoyed,

[49] All excerpts from le Fèvre are transliterated from the 18th edition of Caxton's translation. At times I have resorted to the 13th edition when legibility was a problem as they are digitized photocopies in the Bodleian accessible online.
[50] 2.30.77–78.
[51] Heinrichs 1990, 63.
[52] Both letters dwell on the imagery of Hercules' metaphorical enslavement to Iole and what an outrageous role reversal this is in the light of his heroic and valorous deeds of the past. Both Deianeira's are concerned with their own fall from grace and status as the wives of such a great man. For the most part the letters emphasise the indignity that Hercules himself suffers by allowing a woman to gain such power over him and remind him of his former glory in the hopes that these memories will galvanise him into abandoning his position of weakness and return to his 'legitimate' wife.

Dejanira comforted with these good Tydings, went unto her Chamber, and thanked God and Fortune. After she began to think on their Estate; and thus thinking, she remembered her of the Poyson that Nessus had given her, when at the Point of Death; she had kept it in one of her Coffers, and forthwith incontinently she opened the Coffer, and took the cursed Poyson and one of the Shirts of Hercules: as she had imagined by the Vertue of the Poyson to draw again to her the Love of Hercules, like as Nessus had said unto her, she made the Shirt to be boyled with the Poyson, and gave the Charge thereof to one of her Women, When the Shirt was boyled enough, the Woman took the Vessel and set it to cool: after she took out the Shirt openly and wrung it, but so soon as she wrung it, the Fire sprang in her Hands so vehemently, that she cast it upon a Perch to dry, and fell down dead.

In Process of time, Dejanira desired to have the Shirt, and seeing the Woman that had Charge thereof brought it not, went into the Chamber where the Shirt had been boyled, and found the Woman dead, whereof she had great Marvel. Nevertheless, she passed the Death lightly, and one of her Damsels she made take the shirt that hanged on the Perch and was dry, commanding her that she should fold it, and wind it in a Handkerchief.

At the Commandment of Dejanira the Damsel folded the Shirt; but so doing she was served with the Poyson in such wise that she lost her speech and died soon after. Nothwithstanding, Dejanira that thought nothing but to come to her Intention, took the Shirt and delivered it to Lycas, and charged him to carry it to Hercules.

2.31.81–82

This chapter is the most fascinating given the detail it reveals of the boiling process of the poison and the two casualties of the 'charm's' completion. Is le Fèvre intending for us to glimpse a ruthless side to his desperate damsel? She disregards the sudden deaths of both her handmaidens who were charged with the preparation of the shirt and sends it off to Hercules. The author himself does not comment on this but it is startling. In the passage quoted above, more is made of the shirt than in any literary source since Sophocles. Indeed, Le Fèvre's narrative detailing surpasses that of the tragedian – the instructions for the preparation of the poison and its application are more intricate as Le Fèvre's Deianeira remembers that the shirt must be boiled in the noxious mixture. We then watch how the shirt is transformed into a deadly weapon once it is boiled: the servant who is put in charge of boiling it, sets it to cool, and then takes it out and wrings it dry whereupon, she catches fire and dies from contact with the garment but not before fortuitously throwing it upon a perch where it dries. So deadly has this shirt become that another of Deianeira's handmaidens, charged with folding and packaging the charmed gift is struck dumb after handling it for so short a time, and dies shortly afterwards.

Following closely with the ancient sources, once this chivalric Hercules has donned his gift and begins to burn, he resorts to the faithless husband that Nessus had predicted to Deianeira earlier in chapter 17. He curses his wife, accusing her of jealousy and sorcery,

Oh! Dejanira, Thy Malice as unhappy and cursed Serpent, hath wrought this malicious and reproachful Murder: Thy false Jealousy hath more Power to terminate my Life than all the Monsters of the World. By thine Offence, and by thy mischievous Slight, where from I could not keep me, I must die, and pass out of this World. Since it is so, I thank Fortune, and ask of God no Vengeance against thee; but surely to the end it be not said, that the Vanquisher of Men be vanquished by a Woman; I will not pas the bitter Passage of Death by thy mortal Sorceres full of Abomination, but by the Fire that is neat and clear, and the most excellent of Element.

2.31.82

Wretched Deianeira, learning of this, ends her life with the point of a knife, declaring her innocence and her love for Hercules, blaming herself for her '*Simpleness*' (2.31.83) presumably in believing the evil centaur.

Conclusion

Heracles has suffered some rather dramatic metamorphoses under the pens of moralisers, allegorists and imitators of the golden age. He is compared with no less than Christ, imbued with virtues and learning, stripped of a darker streak of violence and hung about with the cloak of chivalry. Deianeira remains curiously that innocent figure about whom there is just enough contrary evidence to build a case against but much that speaks for her innocence. The two classical traditions portraying a woman who slays her husband in guile out of rage and jealousy and one who makes a fatal error in trying to win back her husband's love are still very much in evidence in Medieval and indeed early Renaissance times. Various writers have manipulated mythical material in accordance with their own agendas in portraying the wife of arguably the most famous hero in antiquity and beyond. There are few in the western world who do not know to some degree who Hercules was, but there are comparatively few who know how he died and why.

The most subtle characterisation of Deianeira and the most controversial, remains, in the opinion of this paper, that of Sophocles. Indeed, the ancient sources have made her motives less accessible than have later writers who sought maybe to simplify pagan stories according to their own mechanisms of analysis. While it is less rewarding to read Deianeira as one of many Delilah figures it is perhaps most rewarding to find a composite of classical and chivalric culture of the high middle ages in the Deianeira of Raoul le Fèvre who finally gives us a response to Ovid's unanswered 'one-way' letters of Heroides although with similar results.

To turn to the question posed in the title of this paper, the garment itself and its provenance: is it the poison cloak of Deianeira or the shirt of Nessus? With each narration of the myth over the centuries the answer has changed according to the guilt of Deianeira versus the role of the Centaur in the death of Hercules. This paper has traced the literary beginnings of the tale from Hesiod, who writes robe into the nimble fingers of a darkly jealous Deianeira and so begins the journey of this seemingly prosaic piece of cloth. Over the centuries the role of centaur becomes more prominent and Deianeira's guilt concomitantly more questionable as her role becomes less active and more passive as she changes into the conduit between her husband and his mortal enemy, merely facilitating the deathblow. But this is by no means a linear progression: in Sophocles Deianeira's culpability is presented in the most complex and arguable of ways, by the time of Ovid she has very little to do with the weapon, receiving it pre-envenomed from Nessus all those years ago, she sends it on in ignorant hope. By the time of the *Roman de la Rose* in the early 14th century, the guilt is back on Deianeira's shoulders and it is once again "her poisonous shirt"[53] that brings the great hero to an ignominious end. It is Bersuire who endows the shirt with allegorical meaning autonomous from its creator when he likens the garment first to a jealous wife and then refines the allegory further whereby he claims that the poisoned shirt is the very symbol of carnal pleasures and thus causes its victims to burn from its temptations and ultimately perish in the "fire of luxury".[54] Boccaccio and

[53] 9191–9202.

[54] 9.67 B-C.

Chaucer, in their respective works, exculpate Deianeira by inserting a Nessus shirt pre-poisoned, *prêt-à-porter*, so to speak. This brings us to le Fèvre whose *Recueil* reveals new detail on the shirt and its transformation into the deadly weapon as discussed in the previous section. Le Fèvre's manuscripts contain a wealth of visual images, typical of texts produced in late Medieval France. The experience of the reader and the narrative itself was heavily influenced by the images produced in such manuscripts[55] and thus I conclude with an illustration (fig. 8.1 below) of a woodcut from an *incunabula* of le Fèvre's *Recueil* printed in 1494–5, which shows Deianeira's robe in two narrative frames. In the first section of the woodcut, Deianeira hands the garment, seemingly innocuous, to Lichas. The second section shows Hercules being consumed by the sacrificial fire, he is clutching at the robe as if to tear it off while he looks directly at the reader with an expression of pain mingled with resignation and dignity in his suffering as his fellow companions look on. The robe has become the deadliest of weapons while his iconic club rests inactive in the hands of Philotes.

Fig. 8.1 Deianeira, Hercules and the Poisoned Robe. Woodcut, ex incunable le Fèvre Recueil 104v (1494–5 BnF).

55 See Desmond & Sheingorn 2006.

Bibliography

Anyonymous 1915 Ovide Moralisé. In C. De Boer (ed.). Amsterdam.

Bacchylides 1993 *Ode 16 / Dithyramb 2* online *Thesaurus Linguae Graecae.* In J. Irigoin (ed.). Paris.

Bacchylides 1993 *Ode 5* online *Thesaurus Linguae Graecae.* In J. Irigoin (ed.). Paris.

Bergren, A. 2008 *Weaving Truth – Essays on Language and the Female in Greek Thought*. Cambridge MA.

Bersuire, P. 1509 in S. Orgel, (intro.), 1979 *Metamorphosis Ovidiana Moraliter (...) Explanata* & Albricus *Libellus.* New York–London.

Boccaccio, G. 2001 *De mulieribus Claris.* In V. Brown (ed. & trans.), *Giovanni Boccaccio Famous Women,* The I Tatti Renaissance Library. Cambridge MA.

Boethius 1897 *The Consolation of Philosophy of Boethius.* In H. R. James (trans.). London.

Burnett, A. P. 1985 *The Art of Bacchylides.* Harvard MA.

Carawan, E. 2000 Deianeira's Guilt. *Transactions of the American Philological Association*, vol. 130, 189–237.

Chaucer, G. 1987 The Monk's Tale. In L. D. Benson (ed.), *The Riverside Chaucer*. Oxford.

Davies, M. 1989 Deianeira and Medea: A Footnote to the Pre-History of Two Myths. *Mnemosyne* 42, 469–472.

De Lorris, G. & de Meun, J. 1994 *The Romance of the Rose* (F. Horgan trans. & intro.). Oxford.

Desmond, M. & Sheingorn, P. 2006 *Myth, Montage, and Visuality in Late Medieval Manuscript Culture – Christine de Pizan's Epistre Othea.* Michigan.

de Pizan, C. 1990 *L'Epistre d'Othèa.* 1399–1400 in J. Chance (ed. & trans.), *Christine de Pizan's Letter of Othea to Hector*. Cambridge.

Eliot, T. S. 1959 *Four Quartets.* London.

Foley, H. 2003 *Female Acts in Greek Tragedy*. Princeton NJ.

Fulgentius 1971 Mythologiae. In L. G. Whitbread (trans.), *Fulgentius The Mythographer*. Ohio.

Galinsky, G. K. 1972 *The Herakles Theme – The Adaptations of the Hero in Literature from Homer to the Twentieth Century*. Oxford.

Haubold, J. 2005 Heracles in the Hesiodic Catalogue of Women. In R. Hunter (ed.), *The Hesiodic Catalogue of Women – Constructions and Reconstructions.* Cambridge.

Heinrichs, K. 1990 *The Myths of Love: Classical Lovers in Medieval Literature*. Pennsylvania.

Hesiod 2007 Catalogue of Women. In G. W. Most (ed.), *The Shield. Catalogue of Women. Other Fragments Most.* Cambridge MA.

King, C. M. 1971 Seneca's 'Hercules Oetaeus': A Stoic Interpretation of the Greek Myth, *Greece & Rome, Second Series*, Vol. 18. No. 2, 215–222.

Lactantius, L. C. F. circa 300 *De Falsa Religione,* in *Divinarum Institutionum* 1524 Basel. Digital Library – Munich Digitization Centre.

le Fèvre, R. 1464–65 *Le Recueil des hystoires troyennes, contenant trois livres.* Lyon. Public domain digitised manuscript: https://gallica.bnf.fr

le Fèvre, R. 1460 *Recueil des histoires de Troye / The destruction of Troy.* Caxton (trans.) 13th Ed. 1708 London.

le Fèvre, R. 1460 *Recueil des histoires de Troye / The destruction of Troy.* Caxton (trans.) 18th Ed. 1738 Dublin.

Lindheim, S. H. 2003 *Mail and Female – Epistolarity, Narrative and Desire in Ovid's Heroides*. Wisconsin.

Maehler, H. 2004 *Bacchylides – A Selection.* Cambridge.

March, J. 1987 Deianeira and Heracles in *The Creative Poet – Studies on the Treatment of Myths in Greek Poetry. Bulletin of the Institute of Classical Studies* Suppl. no. 49. Vol 34. Issue S.49, v–vi, 1–141.

Nees, L. 1991 *A Tainted Mantle Hercules and the Classical Tradition at the Carolingian Court*. Philadelphia PA.

Ormand, K. 1999 *Exchange and the Maiden—Marriage in Sophoclean Tragedy*. Texas.

Ovid 1915 *Heroides.* In G. Showerman (trans.), *Heroides and Amores*. London.

Ovid 1984 *Metamorphoses.* In F. J. Miller (trans.), *Ovid IV Metamorphoses Books IX–XV.* Cambridge MA.

Platter, C. 1994 Heracles, Deianeira, and Nessus: Reverse Chronology and Human Knowledge in Bacchylides. *American Journal of Philology* 115.3, 337–349.

Reynolds, W. D. 1971 The Ovidius Moralizatus of Petrus Berchorius an Introduction and Translation. PhD thesis, Illinois.

Scott, M. 1992 Psychoanalysis and Sophoclean Tragedy. *Acta Academica* 24.2, 55–66.

Scott, M. 1995 The Character of Deianeira in Sophocles' *Trachiniae*: 1. Her attitude to her Marriage. *Acta Classica* 38, 17–27.

Scott, M. 1997 The Character of Deianeira in Sophocles' *Trachiniae*: 2. Where is the Anger? *Acta Classica* 40, 33–47.
Seneca 2004 Hercules Oetaeus. In J. Fitch (ed. & trans.), *Seneca IX Tragedies II*. Cambridge MA.
Seznec, J. 1971 *The Survival of the Pagan Gods – the Mythological Traditions and Its Place in Renaissance Humanism and Art*. New Jersey.
Sophocles 1990 *Trachiniae* online *Thesaurus Linguae Graecae*. In H. Lloyd-Jones & N. G. Wilson. Oxford.
Stafford, E. 2012 *Herakles*. London–New York.
Theodulus 1902 (ed.) *Theoduli Eclogam*. Library of Harvard University on the Archive.org: https://archive.org/details/theodulieclogam.
Trzaskoma, S. M., Scott Smith, R. & Brunet, S. (eds & trans.), 2004 *Anthology of Classical Myth – Primary Sources in Translation*. Indianapolis IN.
Wohl, V. 1998 *Intimate Commerce – Exchange, Gender and Subjectivity in Greek Tragedy*. Austin, TX.

9. Weaving for the People Not a *Peplos*, but a *Chlaina*: Wool-Working, Peace, and Nuptial Sex in Aristophanes' *Lysistrata*

*Jennifer C. Swalec**

As Hector prepares to return to the battlefield in Book 6 of Homer's *Iliad*, he bids his wife Andromache a long, tender farewell. He urges her to busy herself at home with weaving and spinning, decreeing, "war shall be the care of men" (πόλεμος δ' ἄνδρεσσι μελήσει, *Il.* 6.492).[1] Andromache obeys Hector reluctantly, weeping great tears while repeatedly turning back to glimpse him leaving.[2] The Homeric antithesis of spinning and fighting that Hector establishes in the *Iliad* endured for many centuries in the Greek literary tradition, reappearing, for example, within the central *agon* ('debate') of Aristophanes' wartime comedy *Lysistrata*.[3] That play, performed for the first time in 411 BC,[4] depicts its eponymous female protagonist's efforts to unite Greek women from Athens, Sparta, and elsewhere in pursuing the establishment of a peaceful resolution to the ongoing Peloponnesian War.[5] The comedy stands out among extant Greek dramatic works not only for its women, with their anti-war agenda, but also for its references to textiles and their production.

* I owe special thanks to Johanna Hanink for first encouraging me to pursue this project, as well as to the fellow participants at the conference sponsored by the Danish National Research Foundation's Centre for Textile Research where I first presented this paper. I am grateful also for suggestions on subsequent drafts from Marie-Louise Nosch, Mary Harlow, Giovanni Fanfani, Adele Scafuro, Steve Kidd, Mitchell Parks, Joe McDonald, Matt Wellenbach, Ross Brendle, and the participants of the 2014–2015 Brown University Mellon Workshops facilitated by Katherine Harrington, Linda Gosner, and Tara Mulder. Finally, I am much indebted to the American School of Classical Studies at Athens for the opportunity, made possible by the Virginia Grace Fellowship, to spend April 2014 working on this chapter at the Blegen Library.

[1] I print Monro and Allen's text. Translations here and throughout the chapter are my own. Hector's full set of instructions for Andromache occupies Hom. *Il.* 6.490–493.

[2] Hom. *Il.* 6.495–496.

[3] On the Homeric antithesis of spinning and fighting, see Henderson 1987, 134. The *agon* occupies *Lys.* 476–613.

[4] For the debate on whether the play was performed at the Lenaia or the City Dionysia of 411, see Sommerstein 1977; Hubbard 1991, 243–245. I follow the consensus that the *Lysistrata* was performed at the Lenaia. Although *Lysistrata* is the comedy's accepted title, see Henderson 1987, 119 on 389; 197 on 1114, for discussion of alternative titles.

[5] On Lysistrata's status as the first known female protagonist in Greek comedy, see Henderson 1980, 169; Taaffe 1993, 48–49; MacDowell 1995, 225; Lowe 2008, 45–46; Kanavou 2010, 129–132; Hall 2010, 29. Hall highlights in particular the contrast between the sex strike in Aristophanes' *Lysistrata* and that of Alcmena in the short epic *The Shield of Heracles*, initiated in order to inspire Alcmena's husband to avenge violently the deaths of her brothers.

Lysistrata leads her female colleagues in pursuing two complementary peace strategies: younger wives from multiple Greek *poleis* ('cities') initiate a conjugal sex strike, while older Athenian women seize the Athenian Acropolis in order to control access to war funds located there.[6] John Vaio links the two strategies by interpreting the occupation of the sacred Acropolis as that site's symbolic transformation into a large *oikos* ('household').[7] In doing so, he highlights the series of speeches that Lysistrata delivers in response to the question of why Greek women are so concerned with war and peace.[8] She recounts numerous occasions on which her husband has thwarted her attempts to challenge the management of political business by Athenian men, culminating with his efforts to silence her by quoting Hector's famous decree to Andromache: "war shall be the care of men."[9] Lysistrata has not forgotten how her husband altered Hector's affectionate tone, threatening to beat her if she should not cease to question his war-related decisions. Ultimately, she articulates in metaphorical terms her strategy for untangling the Athenians' knotted wartime relations with the Spartans, describing the complete production – from raw fibres to finished garment – of a *chlaina* for the Athenian people. Lysistrata's wool-working metaphor occupies two main passages (*Lys*. 567–570; 574–586) and will be discussed in depth later in the chapter.[10]

I extend Vaio's emphasis upon domestic themes to the symbolic *chlaina* whose production Lysistrata depicts. The wool-working metaphor has received two main interpretations: first, as offering a specific political solution for the Peloponnesian War's resolution, and second, as evoking the sacred *peplos* that Athena receives annually during her Panathenaic festival.[11] With respect to the second interpretation, it is essential to distinguish the *chlaina* from the *peplos*. The *chlaina* is "a large piece of cloth for a blanket or a man's coat,"[12] while the *peplos* is a heavy woollen garment typically worn by women.[13] Moreover, as Mireille Lee has demonstrated, within Athenian tragedy the *peplos* acquires particularly strong "destructive capacities" for dissolving

[6] Age is not the only variable quality among the women who are seeking peace: *cf.* Konstan 1995, 47. For the scarcity of opportunities for real Greek married women to meet, see Cartledge 1990, 38.

[7] Vaio 1973, 372–374; *cf.* Hulton 1972; Foley 1982; Henderson 1987, xxxii; Konstan 1995, 50. See especially Foley 1982, 12, on how the collapse of boundaries between domestic and political spheres in the *Lysistrata* serves to reinforce limitations on female intrusions into male politics.

[8] The question is posed at *Lys*. 502.

[9] See *Lys*. 506–520 for the full exchange that takes place between Lysistrata and the Proboulos, an Athenian magistrate. On the appointment of a board of *probouloi* at Athens after the disastrous Sicilian expedition of 413 BC, see Whitman 1964, 201; Henderson 1987, 117. For the debate as to whether Lysistrata refers to her own husband or speaks on behalf of women generally, see Henderson 1987, xxxvii–xxxviii. For the view that Lysistrata lacks a husband, see Wilamowitz-Moellendorff 1927, 54. Hall 2010, 33, wonders whether Lysistrata lacks a husband because her name alludes to a priestess who could not marry. Lewis 1955 first recognised the relationship of Aristophanes' protagonist Lysistrata to Lysimache, the contemporary priestess of Athena Polias. Foley 1982, 9, suggests that beyond evoking the priestess Lysimache, "Lysistrata becomes almost an incarnation of her patron deity." See Sidwell 2009, 256, for the view that Lysistrata was meant to serve as a satirical representation of the priestess Lysimache; *cf.* Schaps 1977 on the etiquette of naming Greek women publicly.

[10] Scholars have long noted the length and novelty of Lysistrata's wool–working metaphor: *e.g.*, Taillardat 1962, 393–395.

[11] For the first interpretation see Hugill 1936, 39–102; Henderson 1987, 141–142 on 567–586. For the latter interpretation, see Henderson 1987, 141 on 567–586; 145 on 582–586; Scheid & Svenbro 1996, 18–21; *cf.* Foley 1982, 9. Neils 2009 offers an introduction to the ancient Greek practice of offering gifts of votive textiles to gods. For select information on the Panathenaic *peplos*, see Mansfield 1985; Brulé 1987, 99–116; Barber 1992; Lefkowitz 1996; Shear 2001; Aleshire & Lambert 2003; Robertson 2004; Reuthner 2006, 295–320.

[12] Henderson 1987, 145 on 582–586; *cf.* Stone 1980, 160–163; Cleland *et al.* 2007, 33 (*chlaina*).

[13] Cleland *et al.* 2007, 143 (*peplos*). Persian men are also depicted in Greek literature as wearing the *peplos*.

the stability of the *oikos*.[14] Because the *peplos* possesses religious and dramatic associations of such a public nature at Athens, Aristophanes' choice of a *chlaina* for the garment that Lysistrata symbolically weaves appears to be deliberate. Consequently, I propose a third interpretation for the wool-working metaphor: that the *chlaina* whose creation Lysistrata uses it to describe functions as a nuptial coverlet intended for a marital *klinē* ('bed').[15] The *peplos* either sustains the Athenian *polis* ('city') or destroys the Greek *oikos*, whereas the *chlaina* revitalises the *oikos* through its complementary abilities either to clothe a man or to cover a much-desired nuptial bed and the couple that enjoys it.[16]

In order to construe Lysistrata's wool-working metaphor and its *chlaina* within a nuptial framework, I address three themes: the comedy's initial framing of sexual conflict in elemental terms (*i.e.*, as the opposition of fire to water, culminating in a 'nuptial' bath); the drama's frequent imaginings and eventual staging of marital bed(room)s; and the literary tradition of depicting the *chlaina* as a coverlet that married couples share. All women who engage in either of Lysistrata's peace strategies revert to a maiden-like (*i.e.* unmarried) status by refusing sex, as the younger women do, or by recalling the ritual activities that they performed as unwedded maidens, as the older women do.[17] Thus, (re)marriage, evoked by rites of nuptial bathing, as well as attempted marital consummation,[18] will reconcile all women with the *oikos* – and by extension, with the *polis*.

Fire vs. water: an elemental battle of (the) sex(es)

Throughout the first half of the *Lysistrata*, the natural opposition of fire to water gradually comes to represent the sexual conflict that war creates between men and women. The symbolic opposition develops from the play's first scene, in which Lysistrata proposes that the younger Athenian wives follow her in establishing a conjugal sex strike.[19] Lysistrata's plan initially meets with disbelief and desperation, prompting one wife – an Athenian woman named Calonice – to beg for any other option, declaring, "If I must, I'm willing to walk through fire!" (κἄν με χρῇ, διὰ τοῦ πυρὸς / ἐθέλω

[14] Lee 2004, 253; *cf.* Jenkins 1995, 118. The πέπλος serves as "an instrument of feminine destruction" in Aeschylus' *Oresteia*, Sophocles' *Trachiniae*, and Euripides' *Medea*: *cf.* Lee 2004, 262–276. Lee observes that "garments other than πέπλοι are rare in tragedy but common in comedies, histories and the Hippocratic corpus" (2004, 261) and presents evidence that "the πέπλος was not a garment in everyday use in the fifth century" (2004, 275). See also E. Bakola's chapter in this volume.

[15] For discussion of the bride's customary gift of a woollen garment to her bridegroom, see Poll. 3.39–40; *cf.* Redfield 1982, 194; Oakley & Sinos 1993, 39. On the value of cloth exchange or production during weddings and marriages as representations of female wealth, specifically from the perspective of Solonian restrictions on φερναί 'bridal gifts'), see Wagner–Hasel 2010; 2013.

[16] See Henderson 1987, 145, for the *chlaina*'s connotations of a 'new suit'.

[17] Lysistrata's older female colleagues review their maidenly ritual credentials at *Lys.* 641–647. They include serving as an *arrēphoros* at age seven (641–642); serving an *aletris* at age ten (643–644) and shedding saffron as a 'bear' at Brauron (645); serving as a *kanēphoros* "once a beautiful girl" (ποτ' οὖσα παῖς καλὴ, 646); and wearing a necklace of figs, presumably to symbolise fertility (647). On the duties that service as an *arrēphoros* entails, see Robertson 1983; Brulé 1987, 79–98; Donnay 1997; Dillon 2002, 57–60. For discussion of the *arkteia* ('playing the bear') at Brauron, see Kahil 1983; Brulé 1987, 179–261; Sourvinou-Inwood 1988, esp. 136–148; Dillon 2002, 220–222; Marinatos 2002; Perusino 2002; Cole 2004, 198–230. Regarding the shift from play to work that marriage causes, see Carson 1990, esp. 150–151.

[18] For discussion of ancient Greek wedding rituals (including nuptial bathing and consummation), begin with Redfield 1982 and Oakley & Sinos 1993; *cf.* Avagianou 1991, 1–18; Smith 2005; Sabetai 2008, 291–297.

[19] On the metatheatricality that is inherent in the conjugal sex strike, see Taaffe 1993, 51; 60. At the beginning of the comedy, women fulfil men's stereotypes of them: *cf.* Perusino 2001, 37–40.

βαδίζειν, 133–134).[20] Another wife immediately echoes the preference for walking through fire.[21] Nevertheless, the wives do yield and swear a lengthy oath of commitment to Lysistrata's conjugal sex strike.[22] One element of the oath makes explicit the wives' desired outcome from participating in the strike: "so that my man is inflamed especially for me" (ὅπως ἂν ἀνὴρ ἐπιτυφῇ μάλιστά μου, 221; 222). The wives prime their spectators to associate the destructive power of fire with the violent men whom their older female colleagues will face at the Acropolis. The older men will employ physical fire in their siege attempts, as well as burn with anger, while their younger male counterparts burn with lust.[23] Thus, the motif of fire eases the transition from one peace strategy to the other.

The older men attack the occupied Acropolis with fire

The younger wives conclude their oath-taking just before an offstage cry signals that the older women have successfully seized the Acropolis. Lysistrata, in the course of taking the younger wives with her to assist in barring the Acropolis entrance, identifies their male attackers' two chief weapons: threats and fire.[24] The men march onstage chanting a *parodos* (a choral entry song) in which they menace the women who are occupying the Acropolis with heavy logs meant for firewood, ludicrously oversized symbolic versions of the phalluses that they wear as a part of their comic costumes.[25] The *double entendre* that the men create with their firewood recalls the recent establishment of the women's conjugal sex strike.[26] The men carry not only logs, but also torches and a bucket of charcoal, from which to ignite the torches.[27] While approaching the Acropolis, they lament their logs' weight and how smoke stings their eyes.[28] The latter complaint, as Henderson persuasively observes, "symbolizes the old men's angry, warlike spirit."[29] While lighting their torches, the men pray to Lady Victory (δέσποινα Νίκη, 317) for assistance in overpowering their female adversaries.

The women fight fire with water

The members of the women's semi-chorus first pour onstage with the intention of smothering their male rivals' "thick smoky fire" (λιγνὺν, 319). During their opening song, the women refer explicitly to the water-pitchers (τὴν ὑδρίαν, 327) that they carry, lamenting with what great difficulties they struggled to fill them that day.[30] The water-pitchers evoke the contributions that women make to

[20] I print Wilson's 2007 text for the *Lysistrata* here and throughout this chapter. For further discussion of the women's initial resistance see Vaio 1973, 369–70.

[21] *Lys*. 136.

[22] *Lys*. 181–239.

[23] *Cf.* Dorati 1999, 83–85, for emphasis upon the men's fire as it relates to sexual potency, and for discussion of ancient biological and medical theories of male dryness.

[24] *Lys*. 240–251, with threats and fire occurring at 249.

[25] The *parodos* occupies *Lys*. 254–386, with references to the men's firewood occurring at *Lys*. 254–255; 266–270; *cf.* Robson 2010, 50–51; Beale 2010, 66. The men claim that their firewood is meant for building a pyre. It is rather the women, however, who will later achieve the symbolic burial of the Proboulos who comes to bind them. The wood that the men use is Athena's olive wood: *cf. Lys*. 255.

[26] For the *double entendre* of the words for 'gates' (*e.g.* τὰς πύλας, 250) and 'door' (*e.g.* τὴν θύραν, 309) that occur at *Lys*. 246–310, see Loraux 1993, 138–139.

[27] *Lys*. 286–298; 306–309.

[28] See *Lys*. 286–295 for the complaint about weight and *Lys*. 296–301 for the smoke's sting. The men characterise the smoke particularly as being Lemnian: *cf.* S. *Ph*. 800; Taillardat 1962, 154; Martin 1987.

[29] Henderson 1987, 105 on 296–298.

[30] Henderson 1987, 109–110, discusses the κρήνης ("fountain") of *Lys*. 328–329, suggesting that the women attempted to fill their water-pitchers at the Enneakrounos or Kallirhoe fountain, "whose water was used for various religious purposes

the *polis*, in the form of public ritual service, and to the *oikos*, in the form of private domestic duties. The women pray to their patron goddess Athena for her aid, highlighting that goddess's cultic connection to water by addressing her with the epithet "Tritogeneia."[31] The prayer concludes with a wish for water, to protect Athena's sacred precinct from any man's fire.[32]

Shortly after the women conclude their prayer to Athena, the men issue threats that prompt the women to set aside their pitchers (τὰς κάλπιδας, 358). Nevertheless, the Coryphaea (the leader of the female semi-chorus) soon steers the conflict between the men and women back to the opposition of fire and water, instructing her colleagues to take their pitchers in hand once more.[33] The Coryphaeus (the leader of the male semi-chorus) asks the Coryphaea why she has brought water, while she in turn asks why he has brought fire.[34] When the Coryphaea promises to drown the men's fire with her water, her threats acquire an explicitly nuptial nature (*Lys*. 375–378):

Χο. Γε. τοὐμὸν σὺ πῦρ κατασβέσεις;
Χο. Γυ. τοὔργον τάχ' αὐτὸ δείξει.
Χο. Γε. οὐκ οἶδά σ' εἰ τῇδ' ὡς ἔχω τῇ λαμπάδι σταθεύσω.
Χο. Γυ. εἰ ῥύμμα τυγχάνεις ἔχων, λουτρόν γέ σοι παρέξω.
Χο. Γε. ἐμοὶ σὺ λουτρόν, ὦ σαπρά;
Χο. Γυ. καὶ ταῦτα νυμφικόν γε.

Men's Leader: You'll put out my fire?
Women's Leader: The deed soon will show itself.
ML: I don't know why I shouldn't roast you with this torch that I have!
WL: If you've got any soap with you, I'll give you a *bath*.
ML: You give me a bath, you rotten crone?
WL: Indeed – a *nuptial* one!

The quick pace of the chorus leaders' threats, traded in *stichomythia* (the exchange of single lines of verse by characters in a dramatic dialogue) and – most strikingly – *hemistichomythia* (the exchange of half-lines of dramatic verse), underscores the significance of the bath's nuptial quality. The Coryphaea draws additional attention to the adjective "nuptial" (νυμφικόν, 378) by following it with the particle γε, which emphasises the word that precedes it.[35] After the agitated Coryphaeus screams out an order to torch the Coryphaea's hair, she douses him with her pitcher.[36] This onstage bathing constitutes the climax of the *parodos*. The promise of a nuptial bath has not only been made, but realised.[37]

---

and for weddings," as opposed to the Klepsydra; *cf.* Owen 1982. The women claim that they had to compete with slaves for space while attempting to fill their pitchers: *cf. Lys*. 327–332.

[31] The prayer occupies *Lys*. 341–349, while the epithet "Tritogeneia" occurs at *Lys*. 347; *cf.* Henderson 1987, 111–112. Regarding prayers generally in the *Lysistrata*, see Horn 1970, with discussion of prayers to Athena at 47; 52; 60.

[32] *Lys*. 347–349.

[33] The Coryphaea issues the instruction to pick up the pitchers at *Lys*. 370, only 12 lines after giving the command to set them down.

[34] *Lys*. 371–372.

[35] I have italicised the referents of "bath" (λουτρόν, 377) and "nuptial" (νυμφικόν, 378) in order translate the particle γε in those verses; for this translation strategy see LSJ *s.v.* γε I. ("with single words(…)often only to be rendered by italics in writing, or emphasis in pronunciation(…)."). While the γε that follows the adjective "nuptial" is secure, the γέ σοι that follows "bath" is Resig's suggestion, reprinted by Wilson 2007 and Henderson 1987. The adjective "nuptial" of Lys. 378 is doubly emphasised by the preceding phrase καὶ ταῦτα within that verse; *cf.* LSJ *s.v.* οὗτος VIII.2.

[36] *Lys*. 381–384.

[37] Faraone 1997, 48, observes a shift from the "Alcmene model" (the rescue of females taking refuge in a sanctuary) to the "later type of Heracles story, in which a formerly heroic but not dying male is saved and then reborn or reinvigorated in some fashion."

The afterlife of the 'nuptial' bath

The *parodos* ends when an Athenian *proboulos* (a magistrate) enters with slaves and Scythian archers, intending to resolve the ongoing dispute between the men and women. The Proboulos accuses all women present of acting with "licentiousness" (ἀκολαστάσματα, 398).[38] The outraged Coryphaeus echoes the charge, declaring, "these women have committed every outrage" (αἳ τἄλλα θ' ὑβρίκασι, 400). He stresses in particular how the women have "even bathed us with pitchers" (κἀκ τῶν καλπίδων / ἔλουσαν ἡμᾶς, 400–401). The responsibility for the single 'nuptial' bath that the Coryphaea forced upon him now extends to all of her female colleagues. The complaint of the Coryphaeus immediately and substantially reinforces his onstage bath's significance.

Subsequent attempts by the Proboulos to bind Lysistrata meet with successful resistance from the armed older women who rush to her aid. The Coryphaeus continues to dwell upon the insult of his 'nuptial' bath, asking, "don't you know what kind of bath these women gave us just now in our *himatia*, and even without soap?" (οὐκ οἶσθα λουτρὸν οἷον αἵδ' ἡμᾶς ἔλουσαν ἄρτι / ἐν τοῖσιν ἱματιδίοις, καὶ ταῦτ' ἄνευ κονίας, 469–470). Using the emphatic adverb "just now" (ἄρτι, 469) to refer to an event that has occurred 85 lines earlier reveals the enduring fixation of the Coryphaeus upon his 'nuptial' bath. He explicitly evokes the earlier exchange during which the Coryphaea offered to give him a bath if he had any soap,[39] as well as repeats the emphatic phrase καὶ ταῦτα ("Indeed!") with which the Coryphaea began to characterise her promised bath as 'nuptial.'[40]

With the memory of bathing the Coryphaeus renewed, Lysistrata and the Proboulos launch into a vigorous *agon* (*Lys.* 476–613). As a consequence of the complaints made by the Coryphaeus about his bath, as well as Lysistrata's order for her companions to move away from their water-pitchers,[41] the 'nuptial' bath of the Coryphaeus will remain on spectators' minds as Lysistrata and the Proboulos begin to argue. Thus, it is notable that Lysistrata begins her metaphorical discussion of wool-working by imagining the *polis* as a woollen fleece freshly shorn,[42] "having been washed in a bath" (ἐν βαλανείῳ / ἐκπλύναντας, 574–575).

Imagined spaces, desired actions: the marital bed(room), wool-working, and nuptial sex

In the course of articulating her metaphorical description of wool-working, Lysistrata speaks not only of bathing her woollen fleece, but also of placing it "upon a bed" (ἐπὶ κλίνης, 575). The *klinē* evokes the comedy's opening scene, in which young wives make numerous references to their husbands, their marital bed(room)s, and the sex that happens there. Lysistrata initiates the motif of marital sex when she locates absent wives at home in bed, ambiguously "sleeping" (εὕδουσι, 15). Lysistrata herself even contributes to a long series of *double entendres* when she describes having spent many nights sleeplessly, tossing restlessly in bed.[43] After Lysistrata has revealed her plan for a conjugal sex strike, a Spartan wife named Lampito asserts, "It's hard(…)for women to sleep alone without a penis" (χαλεπὰ μὲν(…) / γυναῖκάς ἐσθ' ὑπνῶν ἄνευ ψωλᾶς μόνας, *Lys.*

[38] Compare the Proboulos's accusations with Lysistrata's initial harsh criticisms of her absent co-conspirators at *Lys.* 1–5.
[39] *Lys.* 377.
[40] *Lys.* 378.
[41] *Lys.* 539.
[42] *Lys.* 574.
[43] *Lys.* 26–27. For discussion of Aristophanes' employment of *double entendre* here, see McLeish 1980, 100–103.

142–143).[44] Hesitant Athenian wife Calonice inquires, "And what if they take us and drag us into the bedroom by force?" (ἐὰν λαβόντες δ' εἰς τὸ δωμάτιον βίᾳ / ἕλκωσιν ἡμᾶς;, 160–161). The marital bed(room) furnishes the mutual imaginary setting for sleeping, sleeplessness, and unwanted sexual force.

After Lysistrata addresses Calonice's concerns, the wives vow their commitment to the conjugal strike by promising that "at home, an unwedded maiden, I shall pass my life" (οἴκοι δ' ἀταυρώτη διάξω τὸν βίον, 217; 218).[45] Later in their oath the women recall Calonice's initial concerns, providing for the possibility that their husbands may act forcefully toward them.[46] The oath remains relevant even once the play's action has shifted from the conjugal strike's inception to the Acropolis occupation. When the Coryphaeus complains for a second time about his nuptial bath, the Coryphaea cautions him against engaging his neighbours in violent conflict, "since I would rather be sitting chastely like a maiden" (ἐπεὶ 'θέλω 'γὼ σωφρόνως ὥσπερ κόρη καθῆσθαι, 473). Although the Coryphaea is not participating in the sex strike proper, her warning succinctly elides the women's two complementary peace strategies.

Wool, yarn, and spindles: Lysistrata weaves peace for the Proboulos in the agon

At the outset of the *agon*, a debate concerning women's interests in times of war and peace, the Proboulos opposes Lysistrata's right to speak on political matters by citing the veil (κάλυμμα, 530) that she wears as a married woman. The objection prompts Lysistrata, in an unsuccessful attempt to silence the Proboulos, to remove her veil and give it to him.[47] In doing so, she evokes the nuptial ritual of *anakalypterion* ('unveiling'), albeit in reverse, since Lysistrata transitions from being married to unmarried.[48] One of Lysistrata's companions hands the Proboulos a wool-basket (τὸν καλαθίσκον, 535) to complete his wifely makeover.[49] Lysistrata orders the Proboulos to begin combing wool, declaring, "War shall be the care of women!" (πόλεμος δὲ γυναιξὶ μελήσει, 538).[50] After modifying Hector's famous decree to Andromache, Lysistrata begins to discuss her plans for concluding the Peloponnesian War by articulating a metaphorical description of wool-working, set forth in two parts. She first envisions the wartime relations between Greeks cities as a tangled "ball of yarn" (κλωστῆρ', 567), echoing multiple words previously uttered by her opponent, the

---

[44] *Cf.* MacDowell 1995, 230.

[45] The adjective "unwedded" (ἀταυρώτη) is first attested in a choral ode of Aeschylus' *Agamemnon*, describing the chaste voice of unwedded Iphigenia singing (ἁγνᾶι δ' ἀταύρωτος αὐδᾶι, 254).

[46] *Lys*. 225; 226.

[47] *Lys*. 530–534. Foley 1982, 10, highlights a link between Aristophanes' *Lysistrata* and Euripides' earlier *Wise Melanippe*: *cf.* Hall 2010, 30–31. Much later in the play (*Lys*. 1126–1127), Lysistrata elaborates upon her own education in the art of persuasive speaking: *cf.* McClure 1999, 26; Silk 2000, 229.

[48] On the nuptial ritual of *anakalypterion* ('unveiling'), see Toutain 1940; Oakley 1982; Oakley & Sinos 1993, 25–26; Gherchanoc 2006. For the veiling of Greek women generally, see Kardulias 2001; Cairns 2002; Llewellyn-Jones 2003; 2012.

[49] Dorati 1998, 51, compares the Proboulos's makeover to Heracles donning women's clothing.

[50] *Lys*. 536–538. Barber 1991, 260–282, offers a thorough explanation of the steps and terminology involved in ancient wool-working, treating the *Lysistrata* scene in particular on p. 262; *cf.* Forbes 1956. For analysis of iconographical and ideological trends in vase-painting depictions of wool-work, particularly concerning women's roles in domestic textile production, see Lewis 2002, 62–65; Bundrick 2008; Larsson Lovén 2013. For discussion of later depictions of weaving and politics in the works of Plato, see Blondell 2005; *cf.* North 1977; Thompson 1982, 217; Harlizius-Klück 2004.

Proboulos.[51] He challenges the plan to effect peace "with wool and balls of yarn and spindles" (ἐξ ἐρίων δὴ καὶ κλωστήρων καὶ ἀτράκτων, 571). Lysistrata herself has cited only yarn and spindles, but now begins to speak of wool.

In the lengthier second portion of her metaphorical description of wool-working, Lysistrata depicts select actions that contribute to the production of a symbolic *chlaina*, omitting the spinning that she has already demonstrated.[52] After highlighting the acts of bathing a woollen fleece and placing it upon a bed,[53] Lysistrata uses four distinct but related verbs that signify acts of mixing. Considered collectively, the four 'mixing' verbs evoke sex that is properly associated with and situated at a marital *klinē*. First, Lysistrata describes "mixing together" (καταμειγνύντας, 580) the clean, combed wool in a wool-basket.[54] She suggests "mixing in" (ἐγκαταμεῖξαι, 581) a variety of people with connections to Athens: "metics, anyone who is foreign or a friend, and anyone who owes money to the state" (τούς τε μετοίκους κεἴ τις ξένος ᾖ φίλος ὑμῖν, / κεἴ τις ὀφείλῃ τῷ δημοσίῳ, 580–581).[55] As she concludes her metaphorical description of wool-working, Lysistrata pictures people in colonies as "balls of wool" (τὰ κατάγματα, 583), advocating for "bringing together and gathering into one mass" (ξυνάγειν καὶ ξυναθροίζειν, 585) a large bobbin from which to "weave a *chlaina* for the people" (τῷ δήμῳ χλαῖναν ὑφῆναι, 586). The verb σύναγω, in particular, can be used with the plural object γάμους ('weddings') to speak about contracting a marriage.[56] In combination, Lysistrata's descriptions of bathing a woollen fleece and placing it upon a *klinē*, as well as of engaging in several modes of 'mixing', suggest traditional nuptial rites of bathing and consummation.[57]

In spite of Lysistrata's rhetorical efforts, the newly domesticated Proboulos remains unconvinced that women share in war's burdens. Lysistrata objects, emphasising first how women bear children whom they lose in wars.[58] Furthermore, Lysistrata identifies two distinct hardships that women bear in the absence of men who have undertaken military campaigns. First, women who are already married "sleep alone" (μονοκοιτοῦμεν, 592).[59] Second, husbandless maidens who wish to marry instead grow old "in their inner chambers" (ἐν τοῖς θαλάμοις, 593).[60] For all women, both married and desirous of being so, the bedroom serves as the site of their lonely suffering.[61] While men are at war, women can work wool, but they cannot contribute to their households by bearing children. Classical Attic vase-painting iconography that predates the performance of Aristophanes' *Lysistrata* attests a similar emphasis upon wool-work and children as primary markers of wives' productivity. For example, a hydria from the collection of the Arthur M. Sackler Museum at Harvard University depicts a domestic

[51] Lysistrata picks up on the Proboulos's mention of "agitation" (her τεταραγμένος of 567 echoes his τεταραγμένα of 565) and "untangling" (her διαλύσομεν of 569 recalls his διαλῦσαι of 566).

[52] *Cf. Lys.* 574 (ὥσπερ πόκον, "like a fleece") and 586 (χλαῖναν, "*chlaina*"). See Pantelia 1993 for a comparison of spinning with weaving, particularly as depicted in the Homeric epics.

[53] *Lys.* 574–575, as discussed earlier this chapter.

[54] *Lys.* 579–580. The Proboulos would, of course, still have the wool-basket that was handed to him at *Lys.* 535. For the verb καταμειγνύντας (*Lys.* 580), the codices preserve καταμίγ- *vel* κᾆτα μίγ-.

[55] *Lys.* 581–582.

[56] *Cf. LSJ s.v.* συνάγω A.I.4.

[57] *Cf.* n. 18 above.

[58] *Lys.* 587–590: *cf.* E. *Med.* 250–251; Shaw 1975, 264–265; Morwood 2010, 22.

[59] As Henderson 1987, 146, notes, the verb μονοκοιτοῦμεν is a *hapax legomenon.*

[60] Henderson 1987, 146, characterises the "omen-interpreting" (ὀττευομένη, *Lys.* 597) that women perform "in their inner chambers" particularly as the divining of marriage omens.

[61] *Cf.* Konstan 1995, 48, for emphasis upon what war puts at stake: not just sex, but marriage.

Fig. 9.1 Red-figure Hydria: Family Scene in a Domestic Setting, 440–430 BC, Vari. Courtesy of Harvard Art Museums/Arthur M. Sackler Museum, Bequest of David M. Robinson.

scene of a seated mother handing her baby to a woman who is most likely his nurse (fig. 9.1).[62] Behind the group of mother, child, and nurse stands an upright loom, as well as the child's father.

The afterlife of the wool-working metaphor within the Lysistrata

A nuptial interpretation of Lysistrata's metaphorical wool-working derives additional support from subsequent episodes: the numerous attempted desertions of wives who have committed to the conjugal sex strike, on the one hand, and on the other, one wife's successful rejection of her husband's desperate sexual propositioning. The first wife who wavers directly evokes Lysistrata's metaphorical description of wool-working in her excuse for returning home, thereby reinforcing its sexual connotations. She claims in her misery that she has Milesian wools (ἔριά (...) Μιλήσια, 729) at home that moths are devouring.[63] Lysistrata scoffs. The wife insists, "But I'll come right back, by the Two Goddesses, just let me do some spreading on the bed!" (ἀλλ' ἥξω ταχέως, νὴ

[62] Attic red-figure hydria, Harvard University, Arthur M. Sackler Museum 1960.342, c. 440–430 BC, from Vari; *cf.* Robinson 1937, 31–32, Plate XLIII. The vase is attributed to the circle of Polygnotos.

[63] *Lys*. 728–730. Milesian textiles were famous throughout the Greek world in antiquity. See *SEG* 38–1210 for a text and commentary for an inscribed sacred inventory from Miletus that records the possession of several textiles and dates to c. 100 BC; *cf.* Günther's *editio princeps* of 1988. On the spread of Milesian textiles as luxury items, see Van Wees 2005, 49.

τὼ θεώ, / ὅσον διαπετάσασ' ἐπὶ τῆς κλίνης μόνον, 731–732). Her omission of a direct object for the participle "spreading" (διαπετάσασ', 732) produces a *double entendre*.[64] Furthermore, her phrasing of "upon the bed" (ἐπὶ τῆς κλίνης, 732) recalls the setting for Lysistrata's metaphorical wool-working: a *klinē*.[65]

The second wife who wavers alleges that she has left her flax (τῆς αμόργιδος, 735) at home "unscutched" (ἄλοπον, 736). She promises that she too will return quickly, insisting that she just needs to do a little "scutching" (ἀποδείρασ', 739), which consists of preparing flax fibres by beating them. Once more, the omission of a direct object produces a *double entendre*. The proposed beating of flax fibres recalls not only the instruction for any wife who is ignored by her husband during the conjugal sex strike to masturbate,[66] but also Lysistrata's earlier metaphorical description of beating wool fibres upon a *klinē*.[67] Lysistrata denies the second wife's request, explaining that the other women will also want to scutch.[68] Were any wives granted their requests to 'spread' or 'scutch', they might realise the third wavering wife's excuse: pregnancy.[69]

After a short choral interlude, desperate Athenian husband Cinesias comes to beg his wife Myrrhine to relieve him of his agony.[70] As Cinesias approaches, Lysistrata instructs Myrrhine to "roast" (ὀπτᾶν, 839) him, reprising with a culinary twist the fiery promises of her prologue oath.[71] Myrrhine relents in her refusal to see Cinesias only after he thrusts forward their neglected baby.[72] As Myrrhine approaches him, Cinesias remarks that she looks "even younger" (καὶ νεωτέρα, 885) than he had remembered. She has reverted to a maiden-like status while honouring her commitment to the conjugal sex strike. Myrrhine ignores Cinesias, focusing instead upon their baby. When Cinesias tries to catch Myrrhine's attention by asking whether it bothers her that all of their belongings are in disarray, he highlights in particular the destruction of her woollens (τῆς κρόκης, 896) by hens, but she feigns nonchalance.

Only after Cinesias promises to consider pursuing peace does Myrrhine agree to return home – but not instantly.[73] Frustration consumes Cinesias, who begs Myrrhine to lay down (κατακλίνηθι, 904) with him right where they stand. Myrrhine interrupts Cinesias's subsequent advances by insisting that she must fetch certain items of bedding that she would have at home. The items include a little bed (κλινίδιον, 916), a mattress (ψίαθός, 921), a pillow (προσκεφάλαιον, 926), and a coverlet (σισύραν, 933).[74] At the height of his sexual frustration Cinesias exclaims, "That woman will ruin

---

[64] See Henderson 1987, 165, for the omission of a direct object as producing a *double entendre*; *cf.* Whitman 1964, 207–208; Taaffe 1993, 66–67.

[65] *Cf. Lys.* 575.

[66] Lysistrata repeats Pherecrates' advice to "skin the skinned dog" (κύνα δέρειν δεδαρμένην, *Lys.* 158); *cf.* Henderson 1987, 166.

[67] *Lys.* 575–576. Barber 1991, 274, offers commentary on the verb ἐκραβδίζειν ("to beat out") of *Lys.* 576.

[68] *Lys.* 740–741.

[69] *Lys.* 742–747.

[70] For the erotic connotations of the names Myrrhine and Cinesias, see Kanavou 2010, 133; 139, respectively. See Plin. 4.12.23 and Elderkin 1940, 394, for the name of Myrrhine as evoking the Amazon queen by the same name (*cf. Lys.* 72, when Aristophanes' Myrrhine claims to be late for the women's meeting because she cannot find her "girdle," τὸ ζώνιον), and for the name of Myrrhine as evoking a Lemnian town (*cf. Lys.* 299). Consider also the fact that Myrrhine was a contemporary priestess of Athena Nikē. For the view that the Cinesias of Aristophanes' *Lysistrata* was known to Athenians as the poet previously mocked in the *Birds* of 414, see MacDowell 1995, 243.

[71] *Cf. Lys.* 221, 222; Taillardat 1962, 160.

[72] *Lys.* 877–884; *cf.* Whitman 1964, 206.

[73] *Lys.* 898–903.

[74] The type of coverlet that Myrrhine retrieves is usually made of goat hair; *cf.* Cleland *et al.* 2007 (*sisyra*). See Ar. *V.* 1138 for a depiction of confusion between a *chlaina* and a *sisyra*; *cf.* Stone 1980, 166.

me with bedding!" (ἄνθρωπος ἐπιτρίψει με διὰ τὰ στρώματα, 936).[75] When Myrrhine insists that they need to use something scented, Cinesias grumbles about her selection "not smelling of marriage" (κοὐκ ὄζον γάμων, 943).[76] After Myrrhine finally sneaks away, Cinesias declares, "she has killed me" (ἀπολώλεκέν με, 952). She has abandoned him after she has "flayed" – or "scutched" – him (κἀποδείρασ', 953).[77] Cinesias's complete defeat within Myrrhine's improvised bedroom underscores the only option for the play's resolution. Within each household, husbands and wives must reconcile by having nuptial sex in real bedrooms.

The *chlaina* as a 'nuptial coverlet' in the ancient Greek literary tradition

The Greek tradition of depicting – and often, of worrying about – lovers coupled together beneath a single shared coverlet spans works diverse in both genre and date.[78] The chief aim of this section is to situate the nuptial symbolism of Lysistrata's metaphorical *chlaina* within a broader literary tradition, particularly as it relates to depictions of a nuptially symbolic *chlaina* in 5th century tragedy. As a result, the *chlaina* of the *Lysistrata* can be seen to contribute to general associations between sex and coverlets, as well as specific connections between marriage and coverlets. Lysistrata's imaginary *chlaina* stands out for its use in a social context that is larger than a pair of lovers or a single household: the *polis*.

The nuptial chlaina *as a symbol of marriage in 5th century tragedy*

The *chlaina* appears as a coverlet for couples engaging in marital sex once each in the extant works of Euripides and Sophocles.[79] Both depictions predate the 411 BC debut of Aristophanes' *Lysistrata*. The work of earliest certain date is a fragment of Euripides' *Peliades* that Stobaeus preserves.[80] The play is known from the ancient *Life of Euripides* to have been part of Euripides' first production at the City Dionysia in 455 BC.[81] Collard and Crop suggest that the fragment possibly constitutes part of a dialogue between Pelias and his eldest daughter Alcestis.[82] The fragment consists of five lines, with at least one additional line missing between the fourth and fifth preserved lines. The fragment's speaker advises a girl (ὦ τέκνον, 1) on how to think and act according to her age at different stages in her life. When she is a child (παῖς, 2), her thoughts should not exceed a child's thoughts. Among maidens (ἐν παρθένοις, 3), she is to exhibit the characteristics of a maiden. The next line, although incomplete, begins to provide advice for the addressee as a married woman:

[75] Contrast with Cinesias and Myrrhine's makeshift bedding the immovable bed of Penelope and Odysseus, discussed at the end of the *Odyssey*; *cf.* Bergren 1995, 213–214; 2008, 229–232.

[76] The complaint prompts Myrrhine to offer a second unguent that is stored in an *alabastron* (*cf. Lys.* 947). Henderson 1987, 182, associates the *alabastron* with marriage; *cf.* Keuls 1983, 120, on Ar. *Ach.* 1063.

[77] *Cf. Lys.* 158 and 737–741.

[78] *Cf.* Scheid & Svenbro 1996, 62–63; 70–73. Scheid and Svenbro do not situate the *chlaina* of Aristophanes' *Lysistrata* within their discussion of erotic cloaks (1996, 53–82); rather, they discuss it during one of two chapters organised under the rubric of "Peplos" (*cf.* 1996, 18–21).

[79] I discuss in this section only the *chlaina*; for the *peplos* and *peploi* within tragedy generally see Lee 1999, 2004, and 2005. For a comparison of the *peplos* of Sophocles' *Trachiniae* with the central cloth of Aeschylus' *Agamemnon* see Lyons 2003; 2012, 77–90; and Fletcher's chapter in this volume; *cf.* Morrell 1997. McNeil 2005 persuasively characterises the cloth of the *Agamemnon* as a bridal cloth; *cf.* Taplin 1977; Seaford 1984; Jenkins 1985; Rehm 1994; McClure 1997. For a discussion of nuptial textiles in Euripides' *Medea*, see Mueller 2001.

[80] Euripides fr. 603 Nauck = 603 Collard and Cropp = Stob. 4.23.25.

[81] *TrGF* 5.1, T 1.1A.9; *cf.* Lefkowitz 2012, 87–103, for discussion of the *Vita* tradition for Euripides.

[82] Collard & Cropp 2008, 69.

"when you have slipped beneath the *chlaina* of a well-bred husband(...)" (ὅταν δ' ὑπ' ἀνδρὸς χλαῖναν εὐγενοῦς πέσῃς, 4). The final line preserved by Stobaeus advocates leaving to men "other contrivances" (τὰ δ' ἄλλ'(...) μηχανήματ', 5) of some kind.[83] Although the preserved passage is fragmentary, its speaker unequivocally uses a husband's *chlaina* to represent marital sex, which is equated with marriage generally.

Among extant tragedies the *chlaina* next appears as a nuptial symbol in Sophocles' *Trachiniae*. The tragedy's original performance date remains unknown, but suggestions have ranged from c. 457 to c. 425/424 BC.[84] The *chlaina* of the *Trachiniae* is envisioned by Deianeira, the lovelorn wife of Heracles. After hearing that her husband will return home with his mistress Iole in tow, Deianeira's imagination begins to race. She questions Iole's maidenly status, suggesting that Iole has already been "yoked" (ἐζευγμένην, 536), *i.e.* in marriage.[85] Deianeira envisions herself, her husband, and his mistress nightmarishly entangled in a lurid *ménage à trois*, with just one nuptial cloth to cover both women yoked in marriage to Heracles. As Deianeira exclaims, "And now, although we are two, beneath a single *chlaina* do we await our beloved" (καὶ νῦν δύ' οὖσαι μίμνομεν μιᾶς ὑπὸ / χλαίνης ὑπαγκάλισμα, 539–540). Deianeira then emphasises how she has faithfully kept house for Heracles for quite a long time in his absence.[86] She asks in desperation what woman could endure living with Iole, "sharing the same marriage" (κοινωνοῦσα τῶν αὐτῶν γάμων, 546). For Deianeira – as for the speaker in the aforementioned fragment of Euripides' *Peliades* – lying with one's husband beneath a nuptial coverlet represents marriage generally. Thus, Deianeira becomes undone at the thought of two wives waiting together beneath a single nuptial coverlet.

The **chlaina** *as a coverlet for lovers in genres beyond drama*

While 5th century tragedy offers a fundamental dramatic precedent for the *chlaina*'s nuptial symbolism in Aristophanes' *Lysistrata*, the earliest extant literary attestation for the *chlaina*'s erotic symbolism may have appeared in the 7th century "first Cologne epode" of Archilochus.[87] The poem's speaker recounts how he criticised a girl named Neobule while attempting to seduce a different maiden (παρθένον, 42).[88] He reclined with the unnamed maiden among blooming flowers and began to caress her neck, "having covered her with a soft *chlaina*" (μαλθακῆι δ[έ μιν / χλαί]νηι καλύψας, 44–45). I follow Merkelbach and West's restoration for the incomplete transmission of "*chlaina*" (χλαί]νηι), inspired by the later account of Hieronymus of Rhodes that once while seducing a boy, the tragic playwright Sophocles used both the boy's *himation* (as a blanket for the couple to recline upon) and his own *chlanis* (as a single coverlet for the pair

[83] Collard & Cropp 2008, 69, supports Headlam's supplement for the missing fifth line of "sit at home and take care of domestic matters" (τἄνδον φυλάσσειν ἐν δόμοις καθημένην).

[84] Easterling 1982, 19–23, summarises arguments surrounding the dating of the *Trachiniae*, ultimately highlighting the period between 457 and c. 430 BC; *cf.* Webster 1936, 4; Hoey 1979; Davies 1991, xviii. Vickers 1995 argues for the low date of c. 425/424 BC due to perceived reflections of Peloponnesian War events.

[85] *Cf.* Seaford 1986; 1987, 119, 128–129; Ormand 1993. For all quotations from Sophocles' *Trachiniae* I reprint the Greek text from Lloyd-Jones & Wilson 1990.

[86] S. *Tr.* 542.

[87] I reprint West's 1989 text. Archilochus fr. 196a W^2 = S 478 Page 1974; *cf.* Merkelbach & West 1974; Van Sickle 1975; Henderson 1976.

[88] See Eckerman 2011, 11, for the proposal that the young girl whom the poem's speaker attempts to seduce is not Neobule or Neobule's sister, but a slave.

of lovers).[89] Moreover, I follow Scheid and Svenbro's suggestion that the seducer in Archilochus's epode covers his quarry with a *chlaina* in an effort "to suggest that the union of the two is conjugal in nature."[90] Although the poem's speaker leaves much to the imagination, he does clearly divulge his goal of engaging in sexual acts with a young, unmarried maiden.

Three texts that postdate the *Lysistrata* of 411 BC attest the endurance of the *chlaina*'s erotic associations.[91] Theocritus composed a 3rd century BC *Epithalamium of Helen* that hails Menelaus as fortunate because "Zeus's daughter came under your single *chlaina*" (Ζανός τοι θυγάτηρ ὑπὸ τὰν μίαν ἵκετο χλαῖναν, 18.19).[92] That such praise follows the performance of a wedding song by a chorus of Spartan maidens confirms the *chlaina*'s nuptial nature. In the *Palatine Anthology*, the speaker of an epigram that is attributed to Meleager uses a *chlaina* to identify the rival lover of Heliodora against whom he beseeches Night to bear out ill wishes: "If someone having been placed beneath the *chlaina* of Heliodora grows hot there, warmed by sleep-cheating flesh(…)" (εἴ τις ὑπὸ χλαίνῃ βεβλημένος Ἡλιοδώρας / θάλπεται, ὑπναπάτῃ χρωτὶ χλιαινόμενος, 5.165.3-4).[93] The adjective "warmed" (χλιαινόμενος, 5.165.4) puns upon the noun "*chlaina*" (χλαίνῃ) of the preceding line.[94] The speaker's closing wish for any overheated lover of Heliodora to fall into the eternal sleep of Endymion underscores the poem's vestimentary emphasis.[95] A separate epigram that is attributed to Asclepiades discards various poetic juxtapositions in favour of the view that it is "sweeter, when one *chlaina* covers two lovers and Cypris is praised by both" (ἥδιον δ', ὁπόταν κρύψῃ μία τοὺς φιλέοντας / χλαῖνα καὶ αἰνῆται Κύπρις ὑπ' ἀμφοτέρων, 5.169.3–4).[96]

Two peace strategies united under one chlaina: a return to Aristophanes' *Lysistrata*

By the time that Myrrhine scutches sex-starved Cinesias, Lysistrata has succeeded in her two-pronged scheme for achieving peace.[97] Men of all ages have burned fiercely with desire or with anger, a nuptial bath has saved Athena's Acropolis from a male threat of fire, and Myrrhine has demonstrated what bedding is necessary for a successful reconciliation of all Greek women to the *oikos*, and thus to the *polis*. In the course of Cinesias's subsequent exchange with an equally

---

[89] Merkelbach & West 1974, 109, citing Ath. 13.604d-f; *cf.* Gentili 1976, which confirms the restoration of Merkelbach & West and rejects that of Ebert & Luppe 1975.

[90] Scheid & Svenbro 1996, 70.a

[91] I consider here only explicitly heterosexual couples, or couples whose lovers' identities are not known. Therefore, I exclude, *e.g.*, various depictions of Alcibiades and Socrates sleeping beneath the same coverlet (*cf.* Pl. *Smp.* 219b-c; Luc. *Am.* 49; Ath. *Epit.* 5.219b).a

[92] Scheid & Svenbro 1996, 62, invoke the explicit reference by Theocritus to a nuptial *chlaina* as confirmation for the hypothesis that the Homeric epics twice depict married couples sleeping together beneath one shared garment (Hom. *Il.* 1.611; *Od.* 4.304–305). I quote the Greek text from Gow 1952.

[93] *AP* 5.165; *cf.* Scheid & Svenbro 1996, 70–71, for the argument that Heliodora is a courtesan who enters myriad 'marriages' with her various lovers. For quotations from the *Palatine Anthology*, I print Beckby's text from his second edition of 1965–1967.

[94] I owe thanks first to Matthew Johncock for suggesting this pun to me on 6 June 2013, and second, to Mitchell Parks, for independently making the same suggestion to me a few weeks later.

[95] *AP* 5.165.5–6; *cf.* Scheid & Svenbro 1996, 14–15, on the etymology of 'Endymion' as a name of vestimentary significance. *Cf. AP* 5.173, a second epigram that is attributed to Meleager. Its speaker addresses Night's counterpart of Dawn, asking why she delays "when another grows warm beneath Demo's *chlanis*" (ἄλλος ἐπεὶ Δημοῦς θάλπεθ' ὑπὸ χλανίδι, 5.173.2). The tone and topic closely echo those of the Heliodora epigram.

[96] The marital status of this unnamed pair of lovers remains unknown, but the poem's placement within the anthology suggests a non-conjugal context: *cf.* Scheid & Svenbro, 1996, 71.

[97] *Cf. Lys.* 952–953.

miserable, hyper-aroused Spartan herald, it emerges that the Spartans have also capitulated to the strains of their wives' peace schemes.[98] The remainder of the comedy depicts Lysistrata's supervision of peace talks between the Athenians and the Spartans. The negotiations concentrate upon the onstage appearance of a personified Reconciliation (ἡ Διαλλαγή), who is female and naked.[99] At Lysistrata's behest, the drunken, lustful men assert claims to various parts of Reconciliation's body, equating her physical features with actual geographic locations.[100]

After the men have settled the allocation of the most alluring parts of Reconciliation's body, the Athenians and the Spartans demonstrate their readiness to (re)marry by offering a series of celebratory songs. The Athenian chorus consists of pairs of reunited husbands and wives standing side by side to hymn several gods.[101] The Spartans perform an ode that abounds with details of their local religious customs.[102] After honouring a catalogue of typically Spartan deities, they describe Leda's daughter Helen leading a chorus of beautiful, lively maidens (κόραι, 1307) in performing a joyous dance.[103] As Henderson explains, Helen appears here "as the ideal Spartan maiden and bride."[104] The only previous mention of Helen in the *Lysistrata* occurs during the prologue, when the Spartan wife Lampito recalls Menelaus throwing down his sword at the sight of his wife Helen, naked.[105] The comedy's second reference to Helen thus reinforces a connection between peace and marital sex. Through the combination of the nuptial bathing of the fire-wielding Coryphaeus with Lysistrata's weaving of a nuptially evocative *chlaina*, the Athenians restore to her proper *oikos* each of their "maidens," young and old. Once the Spartans demonstrate a similar readiness for achieving stability within their households, a peace can be forged between the two cities. Interpreting the wool-working metaphor within a nuptial framework imbues the *Lysistrata* with a stronger dramatic and thematic unity.

---

[98] *Lys*. 980–1013.

[99] *Lys*. 1106–1188, with Lysistrata's presentation of a personified Reconciliation commencing at *Lys*. 1114; *cf.* Newiger 1957, 106–108; 1981; Dover 1972, 47; Moulton 1981, 75; Dillon 1987; Konstan 1995, 60; Hall 2010, 35. The ending of the *Lysistrata* shares many features with the endings of Aristophanes' two other complete, extant peace plays: the *Acharnians* of 425 and the *Peace* of 421.

[100] The Spartans request, *e.g.*, that the Athenians return the "abutment" (τὤγκυκλον, 1162) of Pylos. The Spartans, the Athenians, and Lysistrata discuss this request at *Lys*. 1162–1167.

[101] The Athenians' call for song occupies *Lys*. 1273–1278, while the song itself comprises *Lys*. 1279–1290. For discussion of the assignment of lines in the *exodus*, see Moulton 1981, 70; Russo 1994, 169–173. For a comedy ending with such a pronounced focus on song and dance, *cf.* Ar. *V*. 1484ff.; Vaio 1971.

[102] The Spartans' call for song comes at *Lys*. 1295; the song then occupies *Lys*. 1296–1321. The pro-Spartan nature of this final ode has sparked vigorous debate: *cf.* Dover 1972, 154; MacDowell 1995, 246; Revermann 2006, 254–260; Morwood 2010, 26; Robson 2010, 53–54.

[103] On the custom of marriageable *parthenoi* ('maidens') dancing and singing as a chorus, see Dillon 2002, 211–215. For the relationship of this choral ode to Alcman's 7th century BC *Partheneion*, see Bierl 2007.

[104] Henderson 2000, 439, fn. 119. Compare the Helen of Aristophanes' *Lysistrata* with the Helen portrayed by Euripides' *Trojan Women* of 415 and *Helen* of 412; *cf.* Moulton 1981, 79. Battezzato 2013, 102–110, offers a new interpretation of the 'Demeter ode' of E. *Hel*. 1301–1367 as a Spartan cultic dithyramb, which Helen is supposed to join as she returns home. For parthenaic imagery in the same choral ode of Euripides' *Helen*, see Swift 2009. Consider also Helen's gift of a nuptial textile to Telemachus at Hom. *Od*. 15.104–106; 123–130; *cf.* Sutton 1997–1998, 27–28; Bergren 1983, 79–86; Mueller 2010. Oakley 1995, 66, observes the iconographical popularity of Helen's depiction as Menelaus's bride upon vases; *cf.* Austin 1994; Sutton 1997–1998, 28.

[105] *Lys*. 155–156.

Bibliography

Aleshire, S. B. & Lambert. S. D. 2003 Making the *peplos* for Athena: a new edition of *IG* II^2 1060 + *IG* II^2 1036. *ZPE* 142, 65–86.

Austin, N. 1994 *Helen of Troy and Her Shameless Phantom*. Ithaca NY.

Avagianou, A. 1991 *Sacred Marriage in the Rituals of Greek Religion*. Bern.

Barber, E. J. W. 1991 *Prehistoric Textiles: The Development of Cloth in the Neolithic and Bronze Ages with Special Reference to the Aegean*. Princeton.

Barber, E. J. W. 1992 The Peplos of Athena. In J. Neils (ed.), *Goddess and Polis. The Panathenaic Festival in Ancient Athens*. Hanover, 103–117.

Battezzato, L. 2013 Dithyramb and Greek Tragedy. In B. Kowalzig & P. J. W. Wilson (eds), *Dithyramb in Context.* Oxford, 93–110.

Beale, A. 2010 Fantasy and Plot in *Lysistrata*. In D. Stuttard (ed.), *Looking at Lysistrata: Eight Essays and a New Version of Aristophanes' Provocative Comedy*. London, 61–69.

Beckby, H. (ed.) 1965–1967 *Anthologia Graeca, I: Buch I–VI*. 2nd ed. Munich.

Bergren, A. L. T. 1983 Language and the Female in Early Greek Thought. *Arethusa* 16, 69–95.

Bergren, A. L. T. 1995 The (Re)marriage of Penelope and Odysseus: Architecture Gender Philosophy. A Homeric Dialogue. In J. P. Carter & S. P. Morris (eds), *The ages of Homer: a tribute to Emily Townsend Vermeule*. Austin TX, 205–220.

Bergren, A. L. T. 2008 *Weaving Truth. Essays on Language and the Female in Greek Thought*. Cambridge.

Bierl, A. F. 2007 L'uso intertestuale di Alcmane nel finale della *Lisistrata* di Aristofane. In F. Perusino & M. Colantonio (eds), *Dalla lirica corale alla poesia drammatica: forme e funzioni del canto corale nella tragedia e nella commedia greca*. Pisa, 259–290.

Blondell, R. 2005 From Fleece to Fabric: Weaving Culture in Plato's *Statesman*. *OSAPh* 28, 23–75.

Brulé, P. 1987 *La fille d'Athènes. La religion des filles à Athènes à l'époque classique. Mythes, cultes et société.* Paris.

Bundrick, S. 2008 The Fabric of the City: Imaging Textile Production in Classical Athens. *Hesperia* 77.2, 283–334.

Cairns, D. L. 2002 The meaning of the veil in ancient Greek culture. In L. Llewellyn-Jones (ed.), *Women's Dress in the Ancient Greek World.* London–Swansea, 73–94.

Carson, A. 1990 Putting Her in Her Place: Woman, Dirt, and Desire. In D. Halperin, J. Winkler, & F. Zeitlin (eds), *Before Sexuality: The Construction of Erotic Experience in the Ancient Greek World.* Princeton, 135–169.

Cartledge, P. 1990 *Aristophanes and his Theatre of the Absurd.* Bristol.

Cleland, L., Davies G. & Llewellyn-Jones L. 2007 *Greek and Roman Dress from A to Z*. London–New York.

Cole, S. G. 2004 *Landscapes, Gender and Ritual Space.* Berkeley CA.

Collard, C. & Cropp M. 2008 *Euripides. Fragments. Oedipus-Chrysippus. Other Fragments*. Cambridge.

Davies, M. (ed.) 1991 *Sophocles. Trachiniae*. Oxford.

Dillon, M. 1987 The *Lysistrata* as a Post-Deceleian Peace Play. *TAPhA* 117, 97–104.

Dillon, M. 2002 *Girls and Women in Classical Greek Religion*. London–New York.

Donnay, G. 1997 L'arrhéphorie: initiation ou rite civique? Un cas d'école. *Kernos* 10, 177–205.

Dorati, M. 1998 Lisistrata e la tessitura. *QUCC* 58, 41–56.

Dorati, M. 1999 Acqua e fuoco nella *Lisistrata. QUCC* 63, 79–86.

Dover, K. J. 1972 *Aristophanic Comedy*. Berkeley CA.

Easterling, P. E. (ed.) 1982 *Sophocles. Trachiniae*. Cambridge.

Ebert, J. & Luppe W. 1975 Zum neuen Archilochus-Papyrus: Pap. Colon. inv. 7511. *ZPE* 16, 223–233.

Eckerman, C. C. 2011 Teasing and Pleasing in Archilochus' 'First Cologne Epode'. *ZPE* 179, 11–19.

Elderkin, G. W. 1940 Aphrodite and Athena in the *Lysistrata* of Aristophanes. *CP* 387–396.

Faraone, C. 1997 Salvation and Female Heroics in the Parodos of Aristophanes' *Lysistrata. JHS* 117, 38–59.

Foley, H. 1982 The Female Intruder Reconsidered: Women in Aristophanes' *Lysistrata* and *Ecclesiazusae. CP* 77.1, 1–21.

Forbes, R. J. 1956 *Studies in ancient technology, IV: Fibres and fabrics in antiquity. Washing, bleaching, fulling and felting. Dyes and dyeing. Spinning. Weaving*. Leiden.

Gentili, B. 1976 Nota ad Archiloco, P. Col. 7511; Fr. 2 Tard., 2 West. *QUCC* 21, 17–21.

Gherchanoc, F. 2006 Le(s) voile(s) de mariage dans le monde grec: se voiler, se dévoiler. La question particulière des *anakaluptêria*. *Mètis* 4, 239–267.

Gow, A. S. F. 1952 *Bucolici Graeci*. Oxford.

Günther, W. 1988 'Vieux et inutilisable' dans un inventaire inédit de Milet. In K. Denis (ed.), *Comptes et inventaires dans la cité grecque. Actes du colloque international d'épigraphie tenu à Neuchâtel du 23 au 26 septembre 1986 en l'honneur de Jacques Tréheux*. Geneva, 215–237.

Hall, E. 2010 The Many Faces of *Lysistrata*. In D. Stuttard (ed.), *Looking at Lysistrata: Eight Essays and a New Version of Aristophanes' Provocative Comedy*. London, 29–36.

Harlizius-Klück, E. 2004 *Weberei als episteme und die Genese der deduktiven Mathematik in vier Umschweifen entwickelt aus Platons Dialog*. Politikos, Berlin.

Henderson, J. 1976 The Cologne Epode and the Conventions of Early Greek Erotic Poetry. *Arethusa* 9, 151–179.

Henderson, J. 1980 *Lysistrate*: The Play and Its Themes. *YClS* 26, 153–218.

Henderson, J. (ed.) 1987 *Lysistrata*. Oxford.

Henderson, J. (ed.) 2000 *Birds. Lysistrata. Women at the Thesmophoria. Aristophanes*. Cambridge.

Hoey, T. F. 1979 The Date of the *Trachiniae*. *Phoenix* 33.3, 210–232.

Horn, W. 1970 *Gebet und Gebetsparodie in den Komödien des Aristophanes*. Nürnberg.

Hubbard, T. K. 1991 *The Mask of Comedy: Aristophanes and the Intertextual Parabasis*. Ithaca NY.

Hugill, W. 1936 *Panhellenism in Aristophanes*. Chicago IL.

Hulton, A. O. 1972 The Women on the Acropolis: A Note on the Structure of the *Lysistrata*. *G&R* 19, 32–36.

Jenkins, I. D. 1985 The Ambiguity of Greek Textiles. *Arethusa* 18, 109–132.

Kahil, L. 1983 Mythological Repertoire of Brauron. In W. G. Moon (ed.), *Ancient Greek Art and Iconography*. Madison WI, 209–230.

Kanavou, N. 2010 *Aristophanes' Comedy of Names: A Study of Speaking Names in Aristophanes*. Berlin.

Kardulias, D. R. 2001 Odysseus in Ino's Veil: Feminine Headdress and the Hero in *Odyssey* 5. *TAPhA* 131, 23–51.

Keuls, E. C. 1983 "Attic Vase-Painting and the Home Textile Industry". In W.G. Moon (ed.), *Ancient Greek Art and Iconography*. Madison WI, 209–230.

Konstan, D. 1995 *Greek Comedy and Ideology*. Oxford.

Larsson Lovén, L. L. 2013 Textile production, female work and social values in Athenian vase painting. In A.-L. Schallin (ed.), *Perspectives on ancient Greece. Papers in celebration of the 60th anniversary of the Swedish Institute at Athens*. Stockholm, 135–151.

Lee, M. M. 1999 The Myth of the Classical Peplos. PhD diss. Bryn Mawr.

Lee, M. M. 2004 'Evil Wealth of Raiment': Deadly Πέπλοι in Greek Tragedy. *CJ* 99.3, 253–279.

Lee, M. M. 2005 Constru(ct)ing Gender in the Feminine Greek *Peplos*. In L. Cleland, M. Harlow & L. Llewellyn-Jones (eds), *The Clothed Body in the Ancient World*. Oxford, 55–64.

Lefkowitz, M. R. 1996 Women in the Panathenaic and Other Festivals. In J. Neils (ed.), *Worshipping Athena: Panathenaia and Parthenon*. Madison WI, 78–91.

Lefkowitz, M. R. 2012 *The Lives of the Greek Poets*. 2nd ed. London.

Llewellyn-Jones, L. 2003 *Aphrodite's Tortoise. The Veiled Woman of Ancient Greece*, Swansea.

Llewellyn-Jones, L. 2012 Veiling the Spartan Woman. In M. Harlow (ed.), *Dress and Identity*. Oxford, 17–35.

Lewis, D. M 1955 Notes on Attic Inscriptions (II), XXIII: Who Was Lysistrata? *ABSA* 1, 1–12.

Lewis, S. 2002 *The Athenian woman. An iconographic handbook*. London–New York.

Lloyd-Jones, H. & Wilson N. G. (eds) 1990 *Sophoclis Fabulae*. Oxford.

Loraux, N. 1993 *The Children of Athena: Athenian Ideas about Citizenship and the Division between the Sexes*. Transl. C. Levine. Princeton NJ.

Lowe, N. 2008 *Comedy*. Cambridge.

Lyons, D. J. 2003 Dangerous Gifts: Ideologies of Marriage and Exchange in Ancient Greece. *ClAnt* 22.1, 93–134.

Lyons, D. J. 2012 *Dangerous Gifts: Gender and Exchange in Ancient Greece*. Austin TX.

MacDowell, D. M. 1995 *Aristophanes and Athens*. Oxford.

Mansfield, J. M. 1985 The Robe of Athena and the Panathenaic *Peplos*. PhD diss. Berkeley CA.

Marinatos, N. 2002 The *Arkteia* and the Gradual Transformation of the Maiden into a Woman. In B. Gentili & F. Perusino (eds), *Le orse di Brauron. Un rituale di iniziazione femminile nel santuario di Artemide*. Pisa, 29–42.

Martin, R. P. 1987 Fire on the Mountain: *Lysistrata* and the Lemnian Women. *ClAnt* 6.1, 77–105.
McClure, L. K. 1997 Clytemnestra's Binding Spell (*Ag*. 958–974). *CJ* 92, 123–140.
McClure, L. K. 1999 *Spoken Like a Woman: Speech and Gender in Athenian Drama*. Princeton NJ.
McLeish, K. 1980 *The Theatre of Aristophanes*. London.
McNeil, L. 2005 Bridal Cloths, Cover-ups, and *Kharis*: The 'Carpet scene' in Aeschylus' *Agamemnon*. *G&R* 52.1, 1–17.
Merkelbach, R. & West M. L. 1974 Ein Archilochus-Papyrus. *ZPE* 14, 97–113.
Monro, D. B. & Allen, T. W. (eds) 1931 *Homeri Opera. Tomus I.* Oxford.
Morrell, K. 1997 The Fabric of Persuasion: Clytemnestra, Agamemnon, and the Sea of Garments. *CJ* 92, 141–165.
Morwood, J. 2010 The Upside-Down World of Aristophanes' *Lysistrata*. In D. Stuttard (ed.), *Looking at Lysistrata: Eight Essays and a New Version of Aristophanes' Provocative Comedy.* London, 20–28.
Moulton, C. 1981 *Aristophanic Poetry*. Göttingen.
Mueller, M. 2001 The Language of Reciprocity in Euripides' *Medea*. *AJP* 122.4, 471–504.
Mueller, M. 2010 Helen's Hands: Weaving for *Kleos* in the *Odyssey*. *Helios* 37.1, 1–21.
Nauck, A. (ed.) 1964 *Tracigorum Graecorum Fragmenta: Supplementum continens nova fragmenta Euripidea et adespota apud scriptores veteres reperta adiecit Bruno Snell*. Hildesheim
Neils, J. 2009 Textile Dedications to Female Deities: The Case of the Peplos. In C. Prêtre (ed.), *Le donateur, l'offrande et la déesse. Systèmes votifs dans les sanctuaires de déesses du monde grec. Actes du 31e colloque international organisé par l'UMR HALMA-IPEL.* Liège, 135–147.
Newiger, H.-J. 1957 *Metapher und Allegorie. Studien zu Aristophanes*. Munich.
Newiger, H.-J. 1981 War and Peace in the Comedy of Aristophanes. *YClS* 26, 219–237.
North, H. F. 1977 The Mare, the Vixen, and the Bee. Sophrosyne as the Virtue of Women in Antiquity. *ICS* 2, 35–48.
Oakley, J. H 1982 The Anakalypteria. *AA*, 113–118.
Oakley, J. H. 1995 Nuptial Nuances: Wedding Images in Non-Wedding Scenes of Myth. In E. D. Reeder (ed.), *Pandora: Women in Classical Greece.* Baltimore MD, 63–73.
Oakley, J. H. & Sinos R. H. 1993 *The Wedding in Ancient Athens.* Madison WI.
Ormand, K. 1993 More wedding imagery: *Trachiniae* 1053 ff. *Mnemosyne* 26, 224–227.
Owen, E. J. 1982 The Enneakrounos Fountain House. *JHS* 102, 222–225.
Page, D. L. (ed.) 1974 *Supplementum lyricis Graecis: Poetarum lyricorum Graecorum fragmenta quae recens innotuerunt.* Oxford.
Pantelia, M. 1993 Spinning and Weaving: Ideas of Domestic Order in Homer. *AJPh* 114.4, 493–501.
Perusino, F. 2001 Leggendo la *Lisistrata* di Aristofane. *QUCC* 69, 37–45.
Perusino, F. 2002 Le orse di Brauron nella *Lisistrata* di Aristofane. In B. Gentili & F. Perusino (eds), *Le orse di Brauron. Un rituale di iniziazione femminile nel santuario di Artemide.* Pisa, 167–174.
Redfield, J. 1982 Notes on the Greek Wedding. *Arethusa* 15, 181–201.
Rehm, R. 1994 *Marriage to Death: The Conflation of Wedding and Funeral Rituals in Greek Tragedy.* Princeton.
Reuthner, R. 2006 *Wer webte Athenes Gewänder? Die Arbeit von Frauen im antiken Griechenland.* Frankfurt–New York.
Revermann, M. 2006 *Comic Business. Theatricality, Dramatic Technique, and Performance Contexts of Aristophanic Comedy*. Oxford.
Robertson, N. 2004 The Praxiergidae Degree (*IG* 1^3 7) and the Dressing of Athena's Statue with the *Peplos. GRBS* 44.2, 111–161.
Robertson, R. 1983 The Riddle of the Arrephoria at Athens. *HSPh* 87, 241–288.
Robinson, D. M. 1937 *Corpus Vasorum Antiquorum. United States of America, Fascicule 6. The Robinson Collection, Baltimore, MD. Fascicule 2*. Cambridge.
Robson, J. 2010 Friends and Foes: The People of *Lysistrata*. In D. Stuttard (ed.), *Looking at Lysistrata: Eight Essays and a New Version of Aristophanes' Provocative Comedy*. London, 49–60.
Russo, C. F. 1994 *Aristophanes: An Author for the Stage*, trans. K. Wren. London.
Sabetai, V. 2008 Women and the Cycle of Life. In N. Kaltsas & A. Shapiro (eds), *Worshiping Women: Ritual and Reality in Classical Athens.* New York, 286–297.

Schaps, D. 1977 The Woman Least Mentioned: Etiquette and Women's Names. *CQ* 71, 323–330.
Scheid, J. & Svenbro J. 1996 *The Craft of Zeus. Myths of Weaving and Fabric*, trans. C. Volk. Cambridge.
Seaford, R. 1984 The Last Bath of Agamemnon. *CQ* 34, 247–254.
Seaford, R. 1986 Wedding Ritual and Textual Criticism in Sophocles' *Women of Trachis. Hermes* 114, 50–59.
Seaford, R. 1987 The Tragic Wedding. *JHS* 107, 106–130.
Shaw, M. 1975 The Female Intruder: Women in Fifth-Century Drama. *CP* 70, 255–266.
Shear, J. 2001 Polis and Panathenaia: The History and Development of Athena's Festival. Ph.D. diss. University of Pennsylvania.
Sidwell, K. 2009 *Aristophanes the Democrat. The Politics of Satirical Comedy during the Peloponnesian War*. Cambridge.
Silk, M. S. 2000 *Aristophanes and the Definition of Comedy*. Oxford.
Smith, A. C. 2005 The Politics of Weddings at Athens: An Iconographic Assessment. *LICS* 4.1, 1–32.
Sommerstein, A. H. 1977 Aristophanes and the Events of 411. *JHS* 97, 112–126.
Sourvinou-Inwood, C. 1988 *Studies in Girls' Transitions: Aspects of the Arkteia and Age Representation in Attic Iconography*. Athens.
Stone, L. M. 1980 Costume in Aristophanic Comedy. PhD. diss. University of North Carolina.
Sutton, R. F. 1997–1998 Nuptial Eros: The Visual Discourse of Marriage in Classical Athens. *JWAG* 55–56, 27–48.
Swift, L. A. 2009 How to Make a Goddess Angry: Making Sense of the Demeter Ode in Euripides' *Helen*. *CP* 104.4, 418–438.
Taaffe, L. K. 1993 *Aristophanes and women.* London.
Taillardat, J. 1962 *Les Images d'Aristophane: études de langue et de style*. Paris.
Taplin, O. 1977 *The Stagecraft of Aeschylus. The Dramatic Use of Exits and Entrances in Greek Tragedy*. Oxford.
Thompson, W. 1982 Weaving: A Man's Work. *CW* 25, 217–222.
Toutain, J. 1940 Le rite nuptial de l'anakalypterion. *REA* 345–353.
Vaio, J. 1971 Aristophanes' *Wasps*. The Relevance of the Final Scenes. *GRBS* 12, 335–351.
Vaio, J. 1973 The Manipulation of Theme and Action in Aristophanes' *Lysistrata. GRBS* 14, 369–380.
Van Sickle, J. 1975 Archilochus: A New Fragment of an Epode. *CJ* 71.1, 1–15.
Vickers, M. J. 1995 Heracles Lacedaemonius. The Political Dimensions of Sophocles *Trachiniae* and Euripides *Heracles*. *DHA* 21.2, 41–69.
Wagner-Hasel, B. 2010 The Veil and Other Textiles at Weddings in Ancient Greece. In L. Larsson Lovén & A. Strömberg (eds), *Ancient marriage in myth and reality*. Newcastle-upon-Tyne, 102–121.
Wagner-Hasel, B. 2013 Marriage Gifts in Ancient Greece. In M. L. Satlow (ed.), *The Gift in Antiquity. The ancient world: comparative histories*. Malden MA–Oxford, 158–172.
Webster, T. B. L. 1936 *An Introduction to Sophocles.* Oxford.
Wees, H. van 2005 Trailing Tunics and Sheepskin Coats: Dress and Status in Early Greece. In L. Cleland, M. Harlow, & L. Llewellyn-Jones (eds), *The Clothed Body in the Ancient World.* Oxford, 44–51.
West, M. L. (ed.) 1989 *Iambi et elegi Graeci ante Alexandrum cantati. Volumen I: Archilochus. Hipponax. Theognidea.* 2nd ed. Oxford.
Whitman, C. H. 1964 *Aristophanes and the comic hero.* Cambridge.
Wilamowitz-Moellendorff, U. von 1927 *Aristophanes. Lysistrate*. Berlin.
Wilson, N. G. (ed.) 2007 *Aristophanis Fabulae. Tomus II.* Oxford.

10. Clothing the Self in Metaphor: The Spartan *Phoinikis* as Sign of Double Identity

Luca Sansone di Campobianco

In this chapter I wish to propose an alternative interpretation of the *phoinikis*, the crimson/purple military outfit associated with the Spartan hoplite as early as 421 BC (unless otherwise stated, all dates hereafter are BC).[1] To this end I will begin presenting the existing descriptions of the item and the interpretations proposed by ancient authors to explain its adoption. My aim is to establish whether the item possessed a feature that all authors, regardless of how they explained it, have come to identify as its most qualifying and distinctive trait. I will then set this feature at the ground of my interpretative attempt, and propose to see the *phoinikis* as part of a more articulated metaphor meant to work in synergy with the equally traditional Spartan *aspis* (shield). Bound to the *syntagma* formed by garment and shield together, the citizen of Sparta would see his own self defined by two interrelated identities forced onto him at the same time: that of a *soldier*, fated to become a *sacrifice* on the battlefield.

The *phoinikis*: sources and interpretations

I will begin my discussion on the *phoinikis* by presenting in chronological order the extant accounts that offer a description as well as an interpretation of the item.

1. For battle, he [Lycurgus] established the following: they should have a *stole phoinikis* [crimson cloak], because he thought that this colour would have least in common with any feminine apparel, but would be *polemikotate* [the most warlike]. (Xen. *Lac.* 11.3, early 4th century).[2]
2. Aristotle in the Spartan constitution says that the Spartans made use of *phoinikides* [crimson garments] against their enemies. Partly because of the *andrikon* [male character] that is inherent to the colour, partly because its *aimatodes* [chromatic resemblance to blood] accustoms [the Spartan soldiers] to disregard its flowing. (Arist. fr. 8.44.542 late 4th century).[3]

[1] The garment might be referenced for the first time by Aristophanes in his *Pax* (*cf.* Ar. *Lys. Pax* 302) performed in 421, and is clearly mentioned in his *Lysistrata* in 414 (*cf.* Ar. *Lys.* 1139; Harvey 1994, 44). It is not possible to determine with certainty when Sparta would have adopted the garment as its signature military outfit, whether in the 5th century (Sekunda 1998, 20–22), or even earlier (*cf.* Anderson 1970, 39; Cartledge 1977, 15; Hodkinson 2000, 243, arguing for its adoption in funerary practices as early as the 550). However, its possible use in *Pax* as a metonymy for the on-going conflict with the Spartans suggests that at that time the garment might have already become a sign of identity for the Spartan army, thus hinting at its adoption at an earlier time.

[2] Lipka 2002. If not otherwise stated, all subsequent translations of this text are taken from Lipka's edition. For the chronology of the text *cf.* Lipka 2002, 9–13, who identifies in 371 a safe *terminus ante quem* for its composition, although proposing to consider a much earlier date based on the style and on the appreciation of other chronological references.

[3] The translation is mine. The Aristotelian Spartan constitution, of which only fragments have survived, belongs to a collection of constitutions probably composed by Aristotle's students under his supervision (*cf.* Mathieu & Haussoullier 1967, II–IV).

3. In wars they [the Spartans] made use of *phoinikides* [crimson garments]: partly because the colour seemed to them to be *andrike* [virile], partly because its *aimatodes* [chromatic resemblance to blood] causes great terror in the mind of the inexperienced [enemy]. Also, if anyone of them [Spartan soldiers] receives a wound, it is advantageous that the enemy does not easily discover it, but that it is unperceived because of the *omochroun* [chromatic identity to blood] [of the item]. (*Inst. Lac.* 24 = *Mor.* 238f1-5, written sometime after 146 but drawing from earlier material).[4]
4. [the Spartans] also wore *punicae tunicae* [scarlet tunics] in battle to disguise and conceal the blood from their wounds, not lest the sight of them bring terror to themselves but lest it bring some measure of confidence to their foes. (Val. Max. 2.6.2, early 1st century AD).[5]
5. They [the Spartans] had to clad themselves in *phoinikides* [crimson garments] for beside the colour had in itself a *semnotetos ti* [a certain *quid* of solemnity], after the blood had erupted from the wounds its flowing on the garment would have exceedingly struck the enemy with fear, appearing more intense to the eye and more dreadful. (Ael. *VH* 6.6.16, end 2nd/beginning 3rd century AD).[6]

At a first reading we may notice that the sources show no particular interest in describing the shape and make of the garment. Aside from the very general term *stole* used by Xenophon and a potentially unreliable *tunica* from Valerius, the sources give no valuable information on what type of clothing the *phoinikis* actually was.[7] Instead, they consistently highlight the colour of the garment as its main distinctive trait, the feature by which one can successfully identify a *phoinikis* and account for its adoption in the Spartan army. Focusing their attention on the colour, sources would then explain it mainly in two ways. Sparta would have 'gone crimson' because of the existence of a strong cultural association between the dye *phoinikoun* [crimson] and the *andrikon* [male principle]; and/or because the chromatic identity perceived betw een blood and the dye, consistently described in our sources as *aimatodes* [similar to blood], would have warranted its use as a ruse in a rudimentary psychological warfare. For the sake of brevity, I will henceforth refer to the first as *gender* interpretation, to the second as *psychological warfare* interpretation.

At first, all sources would seem to generally agree on the *gender* interpretation. At a second reading, however, this initial impression begins to falter.

Table 10.1 suggests the possibility that the endorsement of the *gender* interpretation may be related to the cultural background of the source. Authors that share an Athenian background tend

[4] The translation is mine. The *Instituta Laconica* is a standalone section preserved in the *Moralia* alongside the equally independent sections of the *Apophthegmata Laconica* (henceforth simply *Sayings*) and the *Apophthegmata Lacaenarum* (henceforth *Sayings of Spartan women*). Scholarship speculated at first that all sections had been edited by the same author (*cf.* Tigerstedt 1974, 88–89; Santaniello 1995, 19 n. 20 with a brief sum of the earlier positions). Subsequently scholars tentatively attributed its composition to Plutarch, and considered the material collected within as the notes he would have personally taken while preparing his *Lycurgus* and *Agesilaus* (*cf.* Fuhrmann 1988, 134–135 *et al.*). The latest scholarship rejected this hypothesis and began to consider each section as the result of an independent work (*cf.* Hodkinson 2000, 40), sometimes even proposing to invert the chronological relationship between the *Sayings* and the *Life of Lycurgus* (*cf.* Ducat 2006, 31, but not followed). Now scholars are inclined to see the *Instituta Laconica* as a heterogeneous collection of material derived from different works on the Spartan constitution from the late 4th century, including Aristotle's, and edited by an anonymous author sometime after the establishment of Roman hegemony in Greece in 146. (*cf.* Kennel 1995, 107; Santaniello 1995, 20; Hodkinson 2000, 48).

[5] Bailey 2000. If not otherwise stated, all subsequent translations of this text are taken from Bailey's edition. According to Bailey, the text was still being revised in 30 AD, but was likely finished within the first half of the century (*cf.* Bailey 2000, 2–4).

[6] The translation is mine. For a short biography and chronology of the author *cf.* Wilson 1997, 2–6.

[7] So far, scholarship has proposed two reconstructions of the garment: something meant to go under the armour, thus serving mainly as padding (*cf.* Anderson 1970, 39; Cartledge 1977, 15) or under and over the armour, thus serving also as a cloak (Lazenby 1985, 32; Sekunda 1998, 20–22; Hodkinson 2000, 224, 233 n.30).

Table 10.1: Overview of the *gender* interpretation

Athenian cultural background		Non-Athenian cultural background	
Xenophon	Aristotle / *Instituta*	Valerius Maximus	Aelian
to andrikon [male principle]		n/a	*semnotes* [solemnity]

to move towards it,[8] whereas the non-Athenians either ignore this aspect or dismiss it in favour of solemnity. This Athenian preference raises a certain degree of concern, as it may suggest an interpretative framework influenced by an Athenian binary conception of gender of which the *gender* interpretation could be a by-product. This influence emerges more clearly in Xenophon's description of the *phoinikis* as a clothing that has "the least in common with the feminine apparel". Through his words the difference between genders is rendered in terms of a gap, a distance that may reflect the difference perceived in their nature and the ensuing function and value they had in society.[9]

If an Athenian cultural filter is indeed at work, then both Xenophon and Aristotle's opinions on the *phoinikis* should be received with caution. They might be trying to make sense of an item through conceptual categories that simply would not match those used by the culture that created and adopted it. As extensively researched in recent times by Ducat, Spartan females were subjected to a regime that had almost no parallels in other Greek cities.[10] Furthermore, this system was fiercely criticised by the Athenians, who had set different standards of behaviour for their own female counterparts.[11] Spartans girls used to run naked, throw the discus and the javelin, wrestle in gyms,[12] compete in strength with each other,[13] exercise aerobically by performing a series of rhythmic leaps,[14] while none of these practices could be seen in Athens. According to tradition, this training was designed to increase what authors refer to by nouns that share the root *hisch-, seemingly perceived as

[8] I group the passage of the *Instituta Laconica* together with the fragment from the Aristotelian *Spartan constitution* as both the similarity between the passages quoted and the nature of the *Instituta* as a collection of items from Spartan constitutions of different authors point towards a relationship between the two texts (as generally accepted by scholarship, *cf.* Santaniello 1995, 21; Lipka 2002, 191; Ducat 2006, 31).

[9] More precisely, a male nature culturally perceived as better and opposed to an inferior female nature, where 'nature' stands here for "a set of characteristics, real or imaginary, which in the writing of 5th and 4th-century Athens men commonly attributed to women as natural of their sex" (Just 1989, 153; *cf.* also Henry & James 2012, 84). This cultural scheme organises the relationship between genders in a binary structure male *vs.* female, where the female nature is defined by the lack of the positive qualities that define men, as in the pair 'brave man' *vs.* 'cowardly woman' (see Arist. *Pol.* 1277b21–23, for other examples of similar oppositions *cf.* Pritchard 2004, 183–184). Overall, the existence of differences in how Sparta and Athens dealt with women seems a fairly acknowledged fact in gender studies (*cf.* Blundell 1995, 150 *et al.*). It is worth noting that the Athenian writers of the 5th century appeared to be proud of their different frame of mind, and took it as an evidence of the Athenian cultural superiority over Sparta (*cf.* Pritchard 2004, 187; Neils 2012, 154).

[10] *Cf.* Ducat 2006, 223–247.

[11] *Cf.* for instance Eur *Andr.* 595–609, where the Spartan habit of allowing the women outside their household for physical practice is chastised due to the promiscuity between genders it would favour.

[12] *Cf.* Theoc. *Idyll.* 18.22–25. Euripides and Plutarch mention female races and *palaistras* [gyms] (*cf.* Plut. *Lyc.* 14.2–3; Eur. *Andr.* 599).

[13] *Cf.* Xen. *Lac.* 1.4.

[14] *Cf.* Poll. 4.102.5 and the physician Antyllos quoted in Oribasius *Coll. Med.* 6.31, both describing the exercise of the *bibasis* as a "Spartan type of dance, whose movements were imposed not only to boys but also to girls". The custom was "sufficiently well-known during the classical era to constitute one of the features typical of the Spartan's woman demeanour" (Ducat 2006, 230; *cf.* 227–231; Arrigoni 1985, 88–92).

gendered and habitually used to express the notion of dynamic male strength.[15] The aim of this type of training-education was to ensure that women could give their gendered contribute to the common cause of war through the *teknopoiia,* i.e. the production of fit and healthy offspring.[16] Although very brief and general,[17] this sortie into Spartan female education should be enough to give a rough perception of how differently Athens and Sparta related to the biological diversity between the sexes and associated gendered behaviours.[18] Spartan women may have been as helpless in battle as their Athenian counterparts,[19] yet Sparta never seemed to evaluate their usefulness on this ground. There is then a plausible chance that the gendered interpretation featured in the Athenian sources may actually derive from the Athenian cultural bias against this particular gendered behaviour. This bias may contribute to partially explain the admittedly somewhat unclear remark of Xenophon on the item as having the "least in common with any feminine apparel". Since the author considers the *phoinikis* to be *polemikotate* [most warlike], and war itself is the activity that better defines men in opposition to women, then the garment could have been easily interpreted as a sign of manliness because of its privileged cultural connection to war. In short, as war has nothing to do with women, neither does a type of clothing the identifying feature of which is its exceptionally war-suited dye.

We may now proceed to consider the *psychological warfare* interpretation somehow foreshadowed in Xenophon's adjective but developed by the following authors. Again, our sources would seem to agree on it, but on closer examination this agreement is again less real than apparent.

As Table 10.2 should show, each author seems to have a different understanding of the psychological effects of the *phoinikis*. Xenophon, theoretically the best bet owing to his first-hand experience of Spartan military matters, states that the item is *polemikotate* but does not explain what properties would make it so and why. The other sources pick up where he left off and further develop the adjective into an explanation of the psychological effects that the colour would have had on Spartans and enemies alike. However, while they all agree on highlighting the item's chromatic similarity to blood through the use of the adjective *aimatodes*, they seem to disagree on almost everything related to its use. There is dispute on: 1) the purpose of the dye, either to

15 As noted by Chantraine, the root denotes the "force du corps, force physique" while its formations are used "notament dans le vocabulaire militaire" in reference to the dynamic strength used in assaulting the enemy (*cf.* Chantraine 1977, 472). The process of identification between the concepts of strength and army conjured up by the two terms will become so tight that Xenophon could use the term *hischus* as a metonymy for a fully fielded army (*cf.* Xen. *cyr.* 1.4.19).

16 For a possible origin of the term *cf.* Xen. *Lac.* 1.4, for its analysis *cf.* Napolitano 1985, 34. According to Napolitano, later on Philostratus would even take the concept one step further, redefining *teknopoiia* into *eu-paidia* [good-rearing], hinting more directly at procreation as an operation of eugenics. The Spartan legislation traditionally presents procreation accordingly as a duty, mentioning several laws meant to create the best conditions for its fulfilment (*cf.* Crit. *Vorsokr.* 2.81; Xen. *Lac.* 1.4; Plut. *Lyc.* 14.2; Philostr. *Gym.* 27) and going as far as prescribing the best time for sexual intercourse (*cf.* Xen. *Lac.* 1.5). Likewise, polygamy would also have been allowed to ensure a better outcome in procreation, *cf.* MacDowell 1986, 82–88.

17 It is not possible to deal with Spartan female condition or education in a footnote, and a detailed discussion on the subject does not seem to be necessary to prove the point being made, namely the different cultural perception of the women in Athens and Sparta. Beside this point, the rest is a matter of quantity of information. Therefore, I would rather refer to Pomeroy 2002, 19–26 and especially Neils 2012, 153–166 for other concrete examples of the difference between Athens and Sparta in dealing with genders.

18 Although Sparta would have also acknowledged the biological differences between genders; *cf.* Ducat 1999, 169; Loraux 1981, 36.

19 *Cf.* the hysterical reaction of the Spartan women during the Theban invasion in Xen. *Hell.* 6.5.28; Plut. *Ages.* 31.5 (on which *cf.* Powell 2004, 116–120). Napolitano proposes to see a direct connection between Plato's negative judgement on the usefulness of women in war expressed in *Leg.* 806a2b10 and Xenophon's account (*cf.* Napolitano 1987, 132).

Table 10.2: Overview of the *psychological warfare* interpretation

	+Hide	+Enhance	+Confidence	+Fear	+Self	+Enemy
Xenophon	n/a	n/a	n/a	n/a	n/a	n/a
Aristotle	Y	N	Y	N	Y	N
Inst. Lac.	Y	Y	Y	Y	Y	Y
Val. Max.	Y	N	Y	N	N	Y
Aelian	N	Y	N	Y	N	N

Table 10.3: Overview of the metaphor 'a cup is an altar'

Domain of symposium	Metaphorical Shift (→)	Domain of Ritual
Jug	→	Offerings
To open	→	To tear apart
Wine	→	Blood
Cup	→	Altar of sacrifice

[+hide] or [+enhance] the visibility of the blood gushing from the wounds; 2) the psychological effect pursued, either to strike [+fear] in the enemy or to build up [+confidence] in the army, or both; and 3) the target itself of this psychological ruse, either the [+self] or, as polemically argued by Valerius, the [+enemy], or again both. These inconsistencies may perhaps confirm the use of the *phoinikis* as a rather original Spartan custom, yet they also suggest caution in embracing *tout-court* the interpretation of the item as a tool of psychological warfare. In fact, the authors following Xenophon's *polemikotate* seem to be working their personal way through the understanding of an adjective that has been left without comment by their main source. The prevailing impression is that the original meaning was lost to 4th century and later authors who, challenged by its survival and renown, are somehow trying to rationalise and make sense of it.[20]

The *phoinikis* and the Spartan *aspis* as a composite metaphor for double identity

Regardless of their internal inconsistencies or the perplexities they may raise, both the *gender* and the *psychological warfare* interpretations contribute to the discussion on the *phoinikis* by identifying and exploring what can arguably be considered its distinctive quality. There is unanimous appreciation of the *aimatodes* [bloodlike] dye as what mainly identifies the item, regardless of the different opinions entertained on its *raison d'etre*. This cultural association between the dye *phoinikoun* and blood is strongly entrenched within the connection that was perceived existing between the two words. As Eustathius will remark, "the well-known origin

[20] As Lipka remarks in relation to Aristotle's passage "it cannot be decided with certainty how far Arist. 542[R] is based on autopsy or a rationalistic explanation of our Xenophontic passage" (Lipka 2002, 191).

of the word *phoinikoun* [the crimson dye of the *phoinikis*] is *phonos*, that is to say blood",[21] especially the blood dripping from a fresh wound.[22] Although the linguistic connection supporting the mental association between crimson dye and blood might not be entirely correct,[23] it seems to have been unanimously received from Homer onward.[24] Finally, owing to this cultural association the dye could be perceived accordingly in rituals as a double for blood.[25] In conclusion, I receive this association as a *datum* that is crucial for the analysis of the *phoinikis*, and set it as the foundation of my own interpretative attempt. However, I also wish to propose an additional and alternative approach. Rather than considering the *phoinikis* as a 'stand-alone' object, I wish to enquire after its meaning by taking into account the cultural interactions it entertains with other related objects that have equal standing in Spartan culture. With this in mind, the second part of Xenophon's account on the *phonikis* may yield a useful hint:

> For battle, he [Lycurgus] established the following: they should have a *stole phoinikis* [crimson cloak], because he thought that this colour would have least in common with any feminine apparel, but would be most *polemikotate* [warlike]. And they should have a *chalke aspis* [bronze shield], for bronze is polished to shine very quickly and tarnishes very slowly.
>
> Xen. *Lac.* 11.3

Xenophon's words seem to suggest that at Sparta the functional identity of the soldier was formulated onto the citizen using not just the *phoinikis* but also the *aspis*.[26] Both objects could then be considered as related to each other, belonging to the conceptual domain of war and forming a single conceptual unit. Using a metaphor from the field of linguistics, if Spartan culture were to be regarded as a language, the 'word' soldier might then be understood as the result of the interactions set between the concepts conveyed by the 'words' *phoinikis* and *aspis*. In other terms, in Spartan culture the concept of 'soldier' could be spelled out as a *syntagma* formed by the synergy resulting from the interaction between '*aspis*' and '*phoinikis*'. But why is the identity of a soldier signified by these two items and not, say, by a spear and a helmet, a helmet and a shield and so on? Alternatives were certainly limited, yet abundant enough to suggest seeing this *syntagma* less as a chance than a choice. Xenophon may in fact be recording here the existence of a mental association rooted into the Spartan culture and formed over a relationship

---

[21] Eust. 1.720.24, the translation is mine.

[22] In opposition to the *melan* [black] blood mingled with the sand/dust on the ground, *cf.* Capponi 2009, 22–29.

[23] *Cf.* Chantraine 1974, 1220, expressing doubts on the linguistic link between the two words. However, he would still link *phoinikis* to the Indo-European root *bher-, which conveys the idea of 'beating to death'.

[24] *Cf.* Hom. *Il.* 16.158–159; 23.717. In the 5th century the dye can be defined as *phoinou* [suited to murder] (*cf.* Aesch. *Ch.* 1013), and Aristophanes can forge the expression 'to dye someone crimson' as a periphrasis for stoning (*cf.* Ar. *Ach.* 320). This mental association will later find its way into the *Interpretation of Dreams*, where dreaming of a cloth of colour *phoinikoun* would preannounce wounds (*cf.* Artem. 2.33.44).

[25] *Cf.* Glotz 1979, 112–115 with other sources and their discussion.

[26] The origin of this mental association may be traced back to the Spartan concern for standard military equipment (*cf.* Pritchett 1974, 43; Cartledge 1977, 27; Finley 1986, 166–167, *contra* Chrimes 1952, 27; Hodkinson 2000, 221–222). According to Cartledge, Sparta would have started a process of militarisation in response to a situation of severe social unrest (*cf.* Arist. *Pol.* 1306a-b). A key element of this ideological process would have been the adoption of standard military equipment, featuring the bronze shield and the crimson garment. Chrimes and Hodkinson would rather place Sparta's militarization at the beginning of the 5th century, rejecting Cartledge's position. In recent times Hodkinson has newly criticised the traditional view of Sparta as a military society, setting the creation of this image in the 4th century (*cf.* Hodkinson 2006, 120).

binding hence defining these two items. If this premise is plausible, then an explanation of the *phoinikis* could be found in the *raison d'être* of the relationship that the item entertains with the Spartan hoplite shield. A short detour, however, might be necessary to better pinpoint meaning and purpose of this association. I will now consider the dynamics of another mental association, whose existence and characteristics should prove less difficult to explore. I will then use any insight gained from this analysis as a conceptual map, useful to better understand the association *phoinikis-aspis*.

While explaining in his *Poetics* dynamics and functions of metaphor, Aristotle classifies its instances in four types. One of them is the metaphor by analogy, where four terms are placed into a relationship of identity according to the scheme $b : a = d : c$, allowing the poet to change b with c and *vice versa*.[27] Perhaps aware that further clarifications were needed he offers the well-known example:

> (…) for instance, the *phiale* [wine-cup, *b*] is to Dionysius [*a*] what the *aspis* [shield, *d*] is to Ares [*c*]; so he will call the wine-cup [*b*] 'the shield of Dionysus' [*d* : *a*] and the shield 'the wine-cup of Ares' [*b* : *c*].
>
> Arist. *Po.* 1457b21–22[28]

Despite the fame it earned owing to Aristotle, the mental association between cup and shield through a metaphorical device does not originally belong to philosophy. This association can also be found in the expression "the cups of Ares" used by Anaxandrides,[29] a poet contemporary to Aristotle who himself is quoting Antiphanes, one generation older.[30] In turn, Antiphanes seems to have found inspiration in a longer passage by Timotheus of Miletus: "then give here the *phiale* [cup] of Ares".[31] Finally, Timotheus' metaphorical translation of the mental association cup-shield finds a far more detailed example in the verses of his contemporary Aristophanes:

Ly.	Put that shield down in front of us – no, *yptia* [hollow up, thus with the concave side uppermost], and someone give me the severing.
Ka.	Lysistrata, what kind of oath are you planning to make us swear?!
Ly.	What kind? The kind they say Aeschylus once had people swear, slaughtering an animal over an *aspis* [shield],
Ka.	Lysistrata, please don't take any oath about peace over a shield!
Ly.	Then what kind of oath will it be?
	(…)
Ka.	Well, how *are* we to swear?
Ly.	(…) we put a *melaina kylix* [big black wine-cup][32] *yptia* [hollow up], right there; We slaughter a magnum of Thasian wine into it; and we swear not to pour any water into the cup!

[27] *Cf.* Arist. *Po.* 1457b16–19: "Metaphor 'by analogy' is a case where the relation of *b* to *a* is the same as that of *d* to *c*: the poet will use *d* instead of *b*, or the reverse". (Halliwell 1997).
[28] Halliwell 1997. If not otherwise stated, all subsequent translations of this text are taken from Halliwell's edition.
[29] *Cf.* Anaxandr. *CAF* 2.80: *phialas Areos* According to the tradition (*cf. Suid.* α. 1982, *Marm. Par. Ep.* 34) and the remaining fragments of his plays (*cf.* Nesselrath 1993, 190), Anaxandrides would have been active from the 374 to the 349.
[30] *Cf.* Antiph. *CAF* 2.112.1. According to the tradition (*cf. Suid.* α. 2734, 1), Antiphanes would have lived between the 93rd and 112th Olympiad, hence between 404 and 330, probably staging his first plays between the 388/7 and 385/4 (*cf.* Konstankatos 2000, 180–185).
[31] Tim. fr. 797.1 = Ath. 10.42.15. Timotheus of Miletus is the most important exponent of the New Dithyramb at the end of 5th century, and lived between the 446 and 357 (*cf.* Hordern 2002, 3–8; Berlinzani 2008, 140–1).
[32] One may object that Aristophanes presents here a *kylix* and not a *phiale*. However, the two objects are so similar in their identifying physical features that the former term may be used as a synonym of the latter, *cf. Suid.* κ. 2665.1.

La. Oh da, I can't find words to praise that oath!
(...)
Ly. Hands off! Now all join me in laying hands upon this boar. Mistress persuasion and Cup of Fellowship, graciously receive *ta sphagia* [the entrails, this sacrifice] from the women.
Ka. The *aima* [blood] is a good colour and spurts out nicely!

Ar. *Lys.* 185–204[33]

In this scene[34] Aristophanes sets forth a *melaina kylix*, a wide black wine-cup as the unusual receiver of a sacrifice and the seat of the women's oath. As it has been remarked, a "sacrifice (...) is amongst the most potent acts of ritual metaphor" (Fletcher 2012, 229), where the term 'metaphor' refers not just to the use of a rhetorical device but to a conscious operation of category shifting. The aim of this metaphorical process is the recreation of reality by presenting a domain of experience in terms of another.[35] Through her choice of words Lysistrata/Aristophanes is wittingly redefining the objects involved by projecting over them a coherent set of different functional identities drawn from the domain of ritual. Thus, by the on-going metaphorical process the "jug" (v. 196) becomes the animal ("boar", v. 202) to be killed and its meat ready to be offered ("entrails", v. 204)[36]; the action of "opening" is redefined as a "slaughtering" (v. 196); the wine gushing from the jug is redefined as the inside of the animal, therefore "blood" (v. 204). At the end of this metaphorical process, the wine-cup *kylix* is redefined as its functional identity is forced to shift between domains. Thus, the 'wine-cup' is transferred into the domain of ritual, becoming the altar on which the sacrifice is performed. The conceptual mapping[37] underlying this metaphor is rather straightforward (see Table 10.3).

However, Aristophanes is playing here an even more complicated metaphorical game. At vv. 185–186 we may read that the 'hollow up' wine-cup has been chosen to act as a substitute for a *yptia aspis* ('hollow up' shield), which the women avoid using due to the item's cultural link to the domain of war.[38]

These passages show the existence, at least within the Athenian culture of the 5th and 4th centuries, of a fairly established mental association between the objects 'shield' and 'wine-cup'. In addition, by following its chain of transmission backward from Aristotle to Aristophanes we may also notice that while later authors seem to regard this association mainly as metaphorical, Aristophanes seems to understand it in rather concrete terms: once we flip a shield and lay it down *yptia, i.e.* on its convex side, with the concave side facing upwards, its identity shifts along with its purpose and the item becomes a recipient, looking and acting precisely as a *phiale/kylix*. Finally, as stated by Lysistrata herself the 'black wine-cup' references an even earlier instance of this mental

[33] Henderson 2000. If not otherwise stated, all following translations of this text are taken from Henderson's edition.

[34] The novelty and role of this scene in Aristophanes' *Lysistrata* have been meticulously studied elsewhere (*cf.* Fletcher 2012, 228–234 with relevant bibliography; also Culpepper 2004, 46–49; 47 f. 22 with bibliography), and will not be considered in this paper as not relevant to the enquiry pursued here.

[35] As clarified by Lakoff, a metaphorical projection from one domain to another allows the understanding of one kind of thing in terms of another kind of thing (*cf.* Lakoff 2003, 172). However, the purpose of a ritual is not the understanding of reality, rather its alteration or even creation.

[36] Coherent with the domain of experience evoked by its use, the term refers specifically to the part of the slain victim offered in a sacrifice *cf.* Chantraine 1977, 1073.

[37] For the definition of metaphorical mapping *cf.* Lakoff 2003, 256–258.

[38] Hence Kalonike's remark at vv. 189–90, *cf.* also Ar. *Pax* 1018–9. Sommerstein believes the 'shield' carried by the Scythian slave-girl to be an actual cup (*cf.* Sommerstein 1990, 165). However, his view would not only contrast with the traditional understanding of the passage (*cf. Schol. in Lys.* 195) but it would also render Kalonike's remark meaningless.

association. In Aeschylus' *Seven against Thebes* (467) the herald recounts that once arrived at the Theban gates Polynices and his companions:

> (…) slaughtered a bull, let its blood run into a *melandeton sakos* [black-rimmed shield]
> and touching the bull's blood with their hands
> swore and oath to Ares, Enyo, and blood-loving terror
>
> Aesch. *Th.* 42–44[39]

The derivation of Aristophanes' scene as a parody of the ritual described by Aeschylus has been discussed elsewhere.[40] This ritual lives only in the description of the herald, yet we may easily imagine a shield set on the ground over its convex side, 'belly-up', and darkened by the blood poured over it.[41] Laid down in an unusual horizontal position the shield receives the sacrificial blood of the victim, the hands of the Seven and finally their oath, becoming the physical seat of the new bond of alliance binding them.[42] Not simply a grim poetic invention, Aeschylus could have found his inspiration in a type of ritual described by Herodotus,[43] where Scythians lay on the ground a *megale kylix* [wide wine-cup] and use it to drink a mixture of wine and blood drawn from the oath takers.[44] We find the same ritual later on described by Xenophon, but this time the wine-cup disappears, substituted by a shield whose echoes Aeschylus:

> (…) and the two parties – the Greek officers, and Ariaeus together with the highest in rank of his followers – made oath that they would not betray each other and that they would be allies (…) These oaths they sealed by sacrificing a bull, a boar and a ram over a shield [*aspis*], the Greek dipping a sword into the blood and the barbarians a lance.
>
> Xen. *An.* 2.2.8–9[45]

As in the rituals described by the other sources, the sacrifice revolves around a shield that once again is laid down on the ground 'belly-up'. However, the rituals so far considered suggest that we re-evaluate whether we are still dealing with a shield. In language we may see that the ordinary[46] meaning of the term 'shield' derives from the emphasis given to a selected set of features, the importance of which is reflected in its definition. In linguistics terms, the lexeme 'shield' is thus defined by emphasising features such as [+sturdy], [+round], [+convex], [+vertical], whereas other features like [+concave], [+horizontal] or [+deep] are played in a minor key. Although they still belong to the object, these features are irrelevant to what is perceived as the item's main function: that of [+protecting] its bearer by [+deflecting] the incoming blows. They are somehow felt as 'secondary' and muted in its definition.[47] We may observe similar dynamics at

[39] Sommerstein, 2008.

[40] On the similarities between the ritual described in *Lysistrata* and Aeschylus' *Seven cf. Schol. ad Lys.* 189; Fletcher 2012, 233; Sommerstein 1990, 165; Henderson 1987, 91; Foley 1982, 18; Dover 1972, 18.

[41] On the interpretation of the adjective 'black' as 'filled with blood' rather than 'of black colour' *cf.* Centanni 1995, 128.

[42] For the interpretation of the ritual slaughtering of an animal as an act aimed at establishing new bonds *cf.* von Staden 1991, 230 quoted in Fletcher 2012, 233.

[43] *Cf.* Hdt. 4.70. Scholarship has suggested the existence of a connection between the two passages (*cf.* Centanni 1995, 128) which the chronology of the authors would somehow make possible.

[44] For a comment on the ritual see Corcella 1997, 288–89; Asheri & Lloyd & Corcella 2007, 632.

[45] Brownson 1968.

[46] By 'ordinary' I mean what Aristotle refers to as *kyrios*, *i.e.* the "name used by everybody" when referring to an object (Arist. *Po.* 1457B3–4).

[47] *Cf.* the standard definition of *shield* as "a broad piece of metal or another suitable material, held by straps or a handle attached on one side, used as a protection against blows or missiles" (Stevenson 2010, s.v. *shield*). In this definition there is no mention of the other physical qualities because they have no use in view of the ordinary function of the item.

work for the three main terms used in ancient Greek to refer to a shield: *aspis*, *sakos* and *pelton*. There might be differences between them in shape, material or size, yet they can all be grouped together under the definition of 'shield' due to the identifying feature of [+deflecting] blows which they all share and that justifies their use as a protection.[48] But the action of flipping and laying down a shield 'hollow up' creates a new object by reversing the relationship between main and secondary features. The physical similarities between shield and cup allow to force onto the shield an identity constructed over its secondary features of [+concavity] and [+horizontality] and leading to its new function of [+receiving].[49] Then, a shield may change into a container,[50] and if the shifting happens in a ritual context this new shield / container may easily become the *kylix*/*phiale* of the comic poets. It turns into a sacrificial cup, redefined by its use as an altar while still retaining the shield's original connection to the domain of war. The merging of these two functional identities within the same object, along with the ability of shifting from one to another, are hallmarks of an on-going metaphorical process. Both objects are together there, affecting each other by means of what has been defined as a "stereoscope of ideas",[51] *i.e.* the ability to entertain two different points of view at the same time. One sees a shield, yet at the same time he may also see the ritual wine-cup, foreshadowed by secondary features.[52] The outcome of this metaphorical process is a dynamic, conceptual shifting between the two functional identities that cause their cultural connotations to collapse and merge together.[53] The shield becomes a cup, acquiring its key function of recipient and its ritual connotations, yet still retaining its identity as a shield hence its cultural associations to war. Aeschylus' passage showcases a possible outcome of this metaphorical process: the shield becomes a "sacrificial mess of the Chthonian sphere",[54] an item perceived as the true 'cup of Ares' because it is filled with blood that ritually foreshadows the killing to come.

[48] The etymological dictionaries show that the items *aspis*, *sakos* and *pelton*, regardless of their physical differences, are culturally perceived as synonymous and interchangeable precisely because of their common function (*cf.* for instance the equivalence between *sakos* and *aspis* in *Etymologicum Gudianum* σ. 495.62, or the definition of *pelton* in *Suid.* π. 951.1 as an *aspis* without the curved rim). The term *aspis* seems to refer to the shield's round shape, allowing the etymologists to extend its signification to denote the snake when it gathers itself in spirals, yet again so as to better defend itself (*cf.* *Etymologicum Magnum* 157.7).

[49] A *phiale* could be defined as "un bassin peu profond, à très large overture (…) une coupe évasée" (Chantraine 1977, 123). As such, its physical similarities with an *aspis* laid on its convex side have always been so strikingly plain that in the antiquity the two items were often thought as interchangeable. For instance, an epigram quoted by Pausanias and celebrating the offering at Olympia of a golden shield made by the Spartans begins with the words "the temple now possesses a golden *phiale* [cup]" (Paus. 5.10.4), where the term '*phiale*' is now clearly used as a mental substitute for '*aspis*'.

[50] Another interesting example of this mental process, although not within a ritual context, could be spotted in Thuc. 7.82.3, where Syracusan soldiers use four *yptias aspides* [hollow up shields] to carry the money of the surrendering Athenians.

[51] Stanford 1936, 105.

[52] A sign of the true genius, "to 'metaphorise well' implies an intuitive perception of the similarity within the dissimilarity" (Arist. *Po.* 1459a; *cf.* *Rh.* 1412a7–8). A metaphor is a heuristic device that violates an order of reality in order to create another, forcing two concepts together in a way that qualities, features and functions of the first are transferred to the other, operating an "apprehension of reality within the differences between two terms" (Ricoeur 2008, 26).

[53] *Cf.* Berggen 1962, 242: "That is to say, the perspectives prior to and subsequent to the transformation of the metaphor's principal and subsidiary's subjects must be conjointly maintained".

[54] Guidorizzi 2001, 67, the translation from the original Italian text is mine. A further evidence supporting the cultural link between the *phiale* and funerary rituals, Homer reports the habit of using the item to collect the bones of the dead (*cf.* Hom. *Il.* 23.243).

A third interpretation: the ritual unit of the *aspis* + *phoinikis*

Back to the Spartan shield, the analysis of the metaphorical process underlying the association cup/shield offers further insight for understanding its cultural meaning. The *Sayings* present the custom of handing down the shield as a sign of obligation, from one generation to another, to defend the city with it.[55] Refraining from a discussion on its historical authenticity, this apophthegm showcases the usual understanding of the object 'shield' in its main and ordinary meaning. In Sparta, 'shield' may be defined as usual as a [+convex] surface, which the soldier holds [+verticality] to [+deflect] any incoming blow. Ideally, this is the type of shield that Xenophon reports given to the Spartan citizen by Lycurgus to 'spell' him as a soldier, alongside his crimson uniform. However, the Spartan tradition also records the existence of the functional shift leading to the other type of shield. Along with others,[56] the author of the *Sayings of Spartan women* reports the well-known Spartan apophthegm:

> Another, as she handed the shield to her son, exhorted him saying "either (…) with this or over this."
>
> *Sayings of Spartan women* 6.16[57]

Here, the shield is forced to change into something else by shifting the emphasis from its main features of [+convexity] and [+verticality] to their opposites.[58] However, the rituals discussed above should now help recognising in this shield set 'hollow up' another instance of the mental connection shield/cup operating in Aeschylus and Xenophon's texts. Furthermore, as seen in Aristophanes' *Lysistrata* this first functional shift is pivotal to another metaphorical change of identity that affects and redefines the identity of what is laid inside the 'cup of Ares'. While the Spartan citizen is alive on the battlefield, the shield he carries in its vertical and ordinary position works as a badge of functional identity. It defines him as a soldier, translating into a concrete *sema* the obligation of defending the city that he inherited with it. But as death on the battlefield makes him a corpse, the

[55] *Cf.* Plut. = *Mor.* 241f. It is worth noting that no shield has so far been recovered from a Spartan burial. The only exceptions would be two epigrams dedicated to Leonidas, where the shield is presented either as a decoration [*kosmos*] of the tomb (*cf. Ant. Pal.* 9.293) or as part of the equipment buried with him (*Ant. Pal.* 9.294). However, the circumstances of Leonidas's death suggest considering these two passages to be poetic inventions, proving at best the existence of a tight relationship between the shield and its bearer (*cf.* Missoni 1986, 76).

[56] *Cf.* Stob. 3.7.30; Val. Max. 2.7; Sen. *Suas.* 2.8. The first certain attestation could be dated back to the 3rd century (*cf.* Hammond 1980, 104, n. 29), however both Stobaeus's source and other authors (Val. Max. 2.7; Sen. *Suas.* 2.8 allegedly reporting a statement from the orator Blandus) agree on considering the saying fairly earlier. Valerius Maximus attributes it to the Spartan general Clearchus, whereas Aristotle/Ariston and Blandus would push it even earlier, at the beginning of the 5th century. For a more comprehensive discussion on the genesis and transmission of this saying, see Hammond 1980, 99–101.

[57] Babbitt 1968.

[58] A justified objection is that the saying features an *epi* [over], whereas an *en* [inside] would have made more sense. This inconsistency was probably acknowledged as early as the first Latin translations of the saying. Valerius translates it with *aut vivi cum armis in conspectum earum venirent, aut mortui in armis referrunt* (Val. Max. 2.7), amending the *epi* into an *in* and adding a *referrunt* as to suggest an actual application of the saying. The reception of his interpretation is mixed: Silius seems to echo the original *epi* by choosing a *supra arma* (*cf. Sil. Pun.* 5.584–5), but later on Ausonius follows Valerius and translates *aut cum hoc aut in hoc redi* (*Ep.* 43, 44). A clever observation from Missoni (*cf.* Missoni 1986, 73–7; Nafissi 1991, 296–300; *contra* Hodkinson 2000, 253–4) offers a reasonable way out: the saying would have originated in the archaic age, when the double eight shield was still in use. In a time when Spartans used to fight not far from their land this type of shield would have still allowed the carrying of the fallen soldiers over a short distance (*cf.* Missoni 1986, 81). If we accept Missoni's interpretation, then we may imagine that the *epi* featured by the saying and suited to a double eight shield has remained even after the switch to the round and convex hoplite shield, because already established as tradition.

ritual action of laying his body within the horizontal cavity of the 'other' shield forces his identity to follow the item's functional shift. As for Aristophanes' jug of wine metaphorically slain over the shield-cup, once he is laid down into the 'cup of Ares', the citizen undergoes a similar change in his identity. His body is ferried into the domain of ritual, reconfigured as a sacrificial victim slain on the battlefield on behalf of Sparta.

Reverting to the Spartan uniform, an alternative interpretation for its adoption can now be formulated on the ground of the relationship it entertains with the shield. As with the shield, the *phoinikis* also features in two cultural domains, shifting between them in the same way the shield does. We read from Plutarch that Lycurgus:

> (...) permitted nothing to be buried with the dead. They simply wrapped around [*peristello*] the body a *phoinikis* and olive leaves, when they laid it away.
>
> Plut. *Lyc.* 27.1[59]

Plutarch's account of the Spartan funerary ritual has found support in recent archaeological findings.[60] However, the item he describes cannot be conceptually identified with the item discussed by Xenophon. The verb "to wrap around" [*peristello*][61] chosen by Plutarch implies the occurrence of a shift between domains, which causes the functional identity of the uniform to change into something different. As noted at the beginning, the military *phoinikis* is described by Xenophon as a *stole*, essentially a [+dress] culturally associated with war and used accordingly to define the [+living] Spartan citizen as a soldier alonsgise the shield. On the other hand, the *phoinikis* discussed by Plutarch is [+wrapped] around the body of the Spartan citizen during a rite of burial, ritually denoting his status as [+dead]. In this transition from one context to another, the identity of the *phoinikis* also shifts from that of a military uniform to that of a perhaps uncommon funerary *pharos*.[62] Finally, as for the shield that turns into a cup, the trigger for this functional shift is again the death of the carrier. An overview may help showing how the dynamics governing this identity shift are in fact the same for both items (see Table 10.4).

The appreciation of this feature and its dynamics strengthen the impression of a conceptual cooperation between *aspis* and *phoinikis*. Together, the two items bind the Spartan citizen to a double identity of soldier and sacrificial victim, managing his transition from one to another by means of a metaphorical shift. The 'stereoscopic vision' embedded within the metaphorical process forces these two identities to merge together, their temporal dimensions to collapse onto one another. Bound to the shield and uniform he is given, the Spartan citizen seems himself as a soldier. Yet at the same time he can already foresee his corpse, wrapped within the same *phoinikis* and ideally carried onto his shield as a slain sacrifice. Once the correct interpretative frame is set, the identifying blood-

[59] Perrin 1959. If not otherwise stated, the subsequent translations of this texts are taken from Perrin's edition.

[60] Corpses of Spartan soldiers sewn into a shroud conserving traces of crimson dye would have been found and dated by the end of the 5th century (*cf.* Kennel 2010, 157). These striking findings support Plutarch's account, and would allow dating the customs he describes back to the Peloponnesian wars. We have no definitive evidence supporting an even earlier date, but Hodkinson (*cf.* Hodkinson 2000, 211) notes that Spartan burials, conspicuous throughout the 8th and first mid-7th centuries, would strangely disappear once excavations hit the second mid-7th century layer. This seeming absence of burials may in fact be explained as a consequence of a change in the Spartan burial practice, which would be consistent with the funerary custom described by Plutarch.

[61] *Cf.* Garland 2001, 24 noting that the verb should be regarded not as a general but rather as a technical term that belongs to the lexical field of funerary ritual.

[62] Garland notes that the usual, although not only, colour of choice for the shroud in antiquity was white (*cf.* Garland 2001, 24–27). In support of his remark, the use of *phoinikoun* shrouds is attested anywhere but Sparta.

Table 10.4: Identity shifts for shield and uniform

Original functional identity	Domain of war	Trigger (→)	Domain of burial	New functional identity
	[+vertical]	Death	[+horizontal]	Cup (of Ares)
Shield	[+convex]	(→)	[+concave]	
	[+reflecting]		[+receiving]	
Uniform	[+dress]	Death	[+wrapped around]	Shroud
	[+living body]	(→)	[+corpse]	

like dye of the *phoinikis* can be interpreted by appreciating the contribution given to the process of identity shifting. Several sources shows that aside from its connection to [+blood],[63] the colour *phoinikoun*[64] is culturally perceived throughout antiquity as a sign of [+eminence]. When [+worn] in a non-religious context by a [+living] man, the dye always marks a person's high standing in military hierarchy. Thus, Roman[65] and Athenian[66] generals, members of the Athenian cavalry,[67] even Persians commanders wear crimson as a visual sign of their rank.[68] However, when featured in a religious context the original notion of [+distinction] gains an additional range of nuances, depending on the circumstances and contexts of its use. Worn as a [+dress] by a [+living] man, this eminence is further defined as a privileged relationship between the wearer of the dye and the divine. The colour becomes the hallmark for the existence of this communion, also standing as a visual sanction for the right of the wearer to perform ritual actions. For instance, in the classical age the Athenian priestesses wear *phoinikides* while bestowing their most solemn curse,[69] the *metoikoi* while attending the religious procession of the Panathenaea,[70] the city magistrate of Plataea while performing a heroic ritual on the graves of the fallen.[71] However, when the dye is used in a burial context as the colour of choice for the cloth [+wrapped] around the body of a [+dead] man, the connection with the divine seems to become stronger. In a funerary context the dye becomes the visual sign of a [+consecration], either in progress or completed, that affects the identity and status

[63] *Cf.* Grand-Clément 2011, 350–352 with sources.

[64] For a discussion on the process of production of the colour and its ingredients *cf.* Grand-Clément 2011, 168–9.

[65] *Cf.* Plut. *Cras.* 23.1. According to tradition, Romulus would have been the first to wear crimson dye as a sign of his standing (*cf.* Plin. *HN* 9.136). Numa would have extended the custom to Roman senators, magistrates and high officers (Malalas, 266–167 ed. Reifferscheid; *cf.* Serv. *in Aen.* 7.612). Coriolanus would have chosen the crimson as well to mark his new status of *dictator* (Val. Max. 5.2.1).

[66] *Cf.* Ar. *Pax* 303.

[67] *Cf.* Xen. *Eq. mag.* 1.22–3.

[68] *Cf.* Hdt. 9.22.3, where Maxentius wears *phoinikis* that makes him stand out amongst the other soldiers, dressed in multi-coloured tunics.

[69] *Cf.* Lys. *in And.* 6.51.6, registering the interesting idiomatic expression "to shake the *phoinikides*" as in 'to curse' someone.

[70] See Phot. s.v. *skaphas*.

[71] Plut. *Arist.* 21.2–5. Other ritual wearers of the dye in later times and different cultures include: the petitioners approaching the oracle of Trophonius (*cf.* Athen. 5.27); a seer in first century Rome while giving her divination (*cf.* Plut. *Mar.* 17.2); Pagan satyrs celebrating their master Dionysus (*cf.* Val. Max. Tyr. *Dial.* 8.2a.4).

of what is wrapped within it. Sources show that the crimson dye could be used to mark the passage of the dead to a semi-divine condition.[72] For instance, in the Egyptian city of Sais Herodotus claims to see a theriomorphic sarcophagus covered by a *phoinikis*, allegedly containing the remains of Menkaure's daughter who now receives divine 'honours'.[73] Plutarch shows an Antonius ordering to cover the dead body of Brutus with a *phoinikis*, and its theft is later considered equal to a blasphemous act punishable by death. Finally, in some cases this ritual overlapping between planes develops into full identity. A grandiose display of crimson covers Alexander's body and his entire grave goods[74] to convey his true identity as a god,[75] and in his final hours Christ will also wear a scarlet cloth that foreshadows his divinity.[76] The heterogeneity of these sources illustrates that the meanings [+eminence] and [+consecration] to the divine are so deeply rooted into the crimson dye that they remain understood and used by very different cultures throughout a remarkable span of time. These meanings seem consistent with the shifting of the *phoinikis* between domains and identities in Spartan culture. While the Spartan citizen is alive, the crimson operates as a double for blood, a sign for the wounds to come[77] that defines the wearer in his military identity. Once the citizen is dead, the blood-like dye becomes a sign of consecration, contributing to the identity shift by which the citizen is redefined as a sacrifice.

Deeply embedded in Spartan culture, these two interconnected items would seem to play a role in how a citizen perceives himself. Their synergy creates a world-view meant to define him: bound to the two identities they carry, he is led to see himself as a soldier, yet at the same time he can already see his fate as a sacrifice of blood for his city. It very much feels like a tool of social control, devised to force an expected behaviour on the receiver by creating a perception of reality that somehow justifies and warrants it. Under this aspect the *phoinikis* would then be part of a more complex 'political metaphor', as political sciences have recently named it: an argumentative device,[78] used to manipulate the perception of the surrounding reality in a way that a certain *status quo* could find its legitimisation or justification, and become what "it [the reality] ought to be".[79]

Bibliography

Akrivoulis, D. E. 2008 The ways of stargazing: Newtonian metaphoricity in American foreign policy. In T. Carver & J. Pikalo (eds), *Political language and metaphor: interpreting and changing the world.* London–New York 12–67.

Arrigoni, G. 1985 Donne e sport nel mondo Greco. In G. Arrigoni (ed.), *Le donne in Grecia*. Rome–Bari, 55–201.

Asheri, D. L & Corcella, A. 2007 *A commentary on Herodotus: books I-IV*. Oxford.

Bailey, D. R. S. 2000 *Valerius Maximus. Memorable doings and sayings, vol. I*. London.

[72] On this point, *cf.* Glotz's interesting remarks on the use of crimson in ancient Greece as a symbol of the divine that has "le don d'élever les hommes à l'immortalité ou, pour le moins, de les sauver d'une mort menaçante" (Glotz 1979, 117).

[73] See Hdt. 2.132.1

[74] Diod. 18.26. It is worth noting that the display of divine honours was organised by Eumenes of Cardia, satrap of Cappadocia, hence tied to a more Persian cultural background, in contrast with the wishes of the Macedonians who vehemently protested against his decision (*cf. Curt.* 8.6.2). From Alexander onward, the dye *phoinikoun* will become characteristic of the Hellenistic kingship (*cf.* Plut. *Pyr.* 8.1; Bessone 1998, 161).

[75] On the literary *topos* of Alexander's *deus atque rex Graecorum cf.* Vigilio 1999, 40–3.

[76] *Cf.* Mt. 27.27–8, where the colour *kokkinon* is a synonym for *phoinikoun*, *cf.* Hesyc. φ. 706.1.

[77] *Cf.* Artem. 2.3.44: "a crimson cloth [*kokkine*] and completely dyed in blood-like crimson [*phoinikobaphes*] brings either wounds or high fever".

[78] *Cf.* Akrivoulis 2008, 19.

[79] *Cf.* Akrivoulis 2008, 24.

Berggren, D. 1962–63 The use and abuse of metaphor I. *The Review of Metaphysics* 16, 237–258.
Berlinzani, F. 2008 Timoteo di Mileto. In G. Zanetto (ed.), *Nova vestigia antiquitatis: seminari 2006–2007*. Milan, 115–142.
Babbitt, F. C. 1968 *Plutarch's Moralia in fifteen volumes, vol. III*. Cambridge.
Bessone, L. 1998 La porpora a Roma. In O. Longo (ed.), *La porpora: realtà e immaginario di un colore simblo ico*. Venice 149–202.
Blundell, S. 1995 *Women in Ancient Greece*. London.
Brinton, J. 2000 *The structure of modern English: a linguistic introduction*. Amsterdam.
Brownson, C. L. 1961 *Xenophon in seven volumes, vol. III Anabasis, books I-VII*. Cambridge.
Cartledge, P. 1977 Hoplites and heroes: Sparta's contribution to the technique of ancient warfare. *JHS* 97, 11–27.
Centanni, M. 1995 *Eschilo: i Sette contro Tebe*. Venice.
Chantraine, P. 1977 *Dictionnaire étymologique de la langue Grecque: historie des mots*. Paris.
Chrimes, K. M. T. 1952 *Ancient Sparta*. Manchester.
Corcella, A. 1993 *Erodoto: le Storie, libro IV*. Milan.
Culpepper, S. S. 2004 Designing women: Aristophanes *Lysistrata* and the "hetairization" of the Greek wife. *Arethusa* 34.1, 37–73.
Ducat, J. 1999 La femme de Sparte et la guerre. *Pallas* 51, 159–171.
Ducat, J. 2006 *Spartan education*. Wales.
Figueura, T. J. 2010 Gynecocracy: how women policed masculine behaviour in Archaic and Classical Sparta. In A. Powell & S. Hodkinson (eds), *Sparta: the body politic*. Swansea, 265–96.
Finley, M. I. 1986 *Ancient history: evidence and models*. New York.
Fletcher, J. 2012 *Performing oaths in Classical Greek drama*. Cambridge.
Foley, H. P. 1982 The 'female intruder' reconsidered: women in Aristophanes' *Lysistrata* and *Ecclesiazusai*. *CP* 77, 1–22.
Garland, R. 2001 *The Greek way of death*. London.
Glotz, G. 1979 *L'ordalie dans la Grèce primitive*. New York.
Grand-Clément, A. 2011 *La fabrique des couleurs: histoire du paysage sensible des Grecs anciens (VIII – début du v s.av.n.è)*. Paris.
Guidorizzi, G. 2002 Uno scudo pieno di sangue. In A. Aloni (ed.) *I Sette a Tebe: dal mito alla letteratura*. Bologna, 63–72.
Halliwell, S. 1997 *The Poetics of Aristotle. Translation and commentary*. London.
Hammond, M. 1980. A famous *exemplum* of Spartan toughness. *CJ* 75.2, 97–109.
Henderson, J. 1987 *Aristophanes' Lysistrata*. Oxford.
Henderson J. 2000 *Aristophanes: Birds, Lysistrata, Women at the Thesmophoria*. Cambridge.
Henry, M. M. & James, S. L. 2012 Woman, city, state: theories, ideologies, and concepts in the Archaic and Classical periods. In S. L. James & S. D. Malden (eds) *A companion to women in the ancient world*. Malden WA.
Hodkinson, S. 2000 *Property and wealth in classical Sparta*. London.
Hodkinson, S. 2006 Was Classical Sparta a military society? In S. Hodkinson & A. Powell (eds) *Sparta & war*. Swansea 111–162.
Hordern, J. H. 2002 *The fragments of Timotheus of Miletus*. Oxford.
Just, R. 1989 *Women in Athenian law and life*. London.
Kennel, N. M. 1995 *The gymnasium of virtue. Education and culture in ancient Sparta*. Chapel Hill NC.
Kennel, N. M. 2010 *Spartans: a new history*. New York.
Konstankatos, I. M. 2000 Notes on the chronology and career of Antiphanes. *Eikasmos* 11, 173–196.
Lakoff G. & Johnsen M. 2003 *Metaphors we live by*. London.
Lipka, M. 2002 *Xenophon's* Spartan Constitution. *Introduction, text, commentary*. Berlin–New York.
Loraux, N. 1981 Le lit, la guerre. *Homme* 21, 37–67.
Mathieu G. & Haussoullier B. 1967 *Aristotle: Constitution d'Athenes*. Paris.
Missoni, R. 1986 Idealità e prassi degli Spartani circa i caduti in guerra. *MGR* 10, 61–81.
Nafissi, M. 1991 *La nascita del kosmos*. Naples.
Napolitano, M. L. 1985 Donne Spartane e τεκνοποιία. *Aion* 7. 19–50.

Napolitano, M. L. 1978 Le donne Spartane e la guerra: problemi di tradizione. *Aion* 9, 127–144.
Nesselrath, H.-G. 1993 Parody and later Greek comedy. *HSCP* 95, 181–95.
Perrin, B. 1959[4] *Plutarch's lives in eleven volumes, vol. I.* Oxford.
Powell, A. 2004 The women of Sparta – and of other Greek cities – at war. In A. Figueira (ed.) *Spartan society*. Swansea, 103–36.
Pritchard, D. 2004 A woman's place in Classical Athens. *Ancient History* 34, 170–191.
Pritchett, W. K. 1974 *The Greek states at war II*. Berkeley CA.
Ricoeur, P. 2008 *The rule of the metaphor*. London.
Sommerstein, A. H. 1990 *Lysistrata*. Warminster.
Sommerstein, A. H. 2008 *Aeschylus. Persians, seven against Thebes, suppliants, Prometheus bound.* Cambridge.
Staden, H. von 1991 The discovery of the body: human dissection and its cultural contexts in ancient Greece. *The Yale journal of biology and medicine* 65, 223–241.
Stanford, W. B. 1936 *Greek metaphor: studies in theory and practice.* Oxford.
Stevenson, A. (ed.) 2010 *Oxford dictionary 3rd ed.* New York.
Tigerstedt, E. N. 1974 *The legend of Sparta in Classical antiquity vol. 2.* Göteborg.
Virgilio, B. 1999 *Lancia, diadema e porpora: il re e la regalità Ellenistica*. Pisa.
Wilson, N. G. 1997 *Aelian: historical miscellany.* Cambridge.

11. Erinna's Loom

*Camillo Neri**

A meaningful image

A childhood spent at the loom, overseen by a stern mother. A bosom friend, Baucis, fellow in (playing) joys and (weaving) sorrows. A thread broken soon, because of death at nineteen, as if Hades' bride, just as many mythical heroines.[1] A poetic talent, crest-fallen by both her persistent weaving and her bitter end, and yet capable of a fine-grained and thorough lyrical weft, cleverly gone by the name of *Distaff*.[2]

All the essential ingredients for a promising biographic novel were to appear soon in the mysterious *dossier* on Erinna's life.[3] Were we to flip through the collection – including Hellenistic epigrams, late antique and Byzantine lives – we would suddenly realise how tricky it proves to be setting real data apart from autoschediams and speculation: in other words, to separate reality from fiction.

Spinning Erinna

The act of weaving played a major role in Erinna's life: at least three epigrammatic *testimonia*, which praise her, make this clear. Answering to Asclepiades (*AP* 7.11 [*HE* 28] = test. 4),[4] who sang Erinna's

* I should like to thank Barbara Fero and Valentina Garulli for their help with the English translation of this contribution.

[1] For an overall outline of Erinna, poetess from Telos (near Rhodes) or from Tenos (in Laconia), living in the first half of 4th century BC and primarily authoress of a poem entitled *Distaff* (on the discussed meaning if this title, see below pp. 206–207) now preserved by a 2nd century AD papyrus (*PSI* IX 1090) and dedicated to her friend and peer Baucis, died young at the time of marriage, see Neri 2003. In the following all the references to Erinna's *testimonia* ('test(t).') and fragments ('fr(r).') – the latter definition is conventional, since epigrams are obviously not fragments) follow this edition. I will also provide some bibliographical updating. Although the original form of the poet's name seems to be Ἤϱιννα (hence 'Herinna'), I will use the form 'Erinna', ratified by *AP* 9.190 = test. 7 (on this problem, *cf.* Neri 2003, 35–37). The full text and an English translation of the main passages (from fragments and *testimonia*) quoted below are given in the Appendix.

[2] Elizabeth Barber (1991, 263–264) and most recently Nosch (2010, 341) argued that ἠλακάτη indicates 'spindle', not 'distaff': as I shall explain below, their arguments are strong although not decisive in my opinion.

[3] They all are summarised in a short biographical entry of *Suda* (η 521 A. = test. 16a) that the compilers drew from the epitome of Hesychius of Miletus' *Onomatologos* (6th century AD): this entry attests the title *Distaff* for the first time (hence Eust. *Il.* 326.43–327.9 = test. 16b, who adds *AP* 9.190). *Cf.* Neri 2003, 212–213. Recurrent attempts to deny the existence of such epitome, or even of the *Onomatologos*, or else its connection with *Suda* (*cf. e.g.* Schamp 1987), 57–67; Kaldellis 2005, 386–388; Costa 2010, 43–55; more reasonable appears Chávez Reino 2010, 239–243) did not affect the *communis opinio* seriously.

[4] The epigram's large reception includes also *App. Anth.* II 400 C. (on Oppian of Apamea, implicitly compared with Erinna): see Magnelli 2004, 269–274.

γλυκὺς (...) πόνος, "the sweet labour" (v. 1), an anonymous epigrammatist (*AP* 7.12 [*FGE* 39] = test. 5) laments her premature death – even though compensated by the ἐπέων (...) καλὸς πόνος, "the beautiful labour in verses", which made her immortal (vv. 5f.) – when Μοῖρα, λινοκλώστου δεσπότις ἠλακάτης, "mistress of flax-spinning distaff" (v. 4),[5] forced her to "Acheron's wide wave" (v. 3).

This expression clearly refers to the untimely breaking of the thread of life; the image of the "spinning wheels of Fates" bears a high degree of iconicity,[6] yet the reference could here hint at the distaff (be it already perceived as a 'title' or not) which, accordingly, would have been associated to this minute masterpiece since the early Alexandrian age. An allusion to the feminine act of spinning is anyway apparent in the *hapax* λινόκλωστος, built on akin compounds, such as λινόκροκος, λινορραφής, λινουλκός, etc. (*cf.* LSJ[9] 1051–1052) on the one side, εὔκλωστος (*H. Hom. Ap.* 203), δικτυόκλωστος (Soph. *Ant.* 346) on the other, as well as *iuncturae* like the ones in Aesch. *Ch.* 507 κλωστῆρα (...) λίνου, Eur. *Tr.* 538 κλωστοῦ (...) λίνοιο, fr. 1001.1 K. λίνου κλωστῆρα. Further echoes in the *Anthologia*, as in Antip. Sid. *AP* 7.174.4 (*HE* 5.193 ἐργάτιν εὐκλώστου νήματος ἠλακάταν, "the distaff that works the well-spun thread"), anon. *AP* 6.284.3f. (*HE* 40.3824f. εὔκλωστον δὲ γυναικῶν / νῆμα καὶ ἠλακάτην κτλ, "women's well-spun thread and distaff"), Macc. *AP* 7.33.3 (*GP* 6.2502 εὐκλώστοιο λίνου), Antip. (Thess.) *AP* 7.109.7 (*GP* 54.369 τρίκλωστον).

Erinna's work is flattered even more by another anonymous epigrammatist (*AP* 9.190 [*FGE* 38] = test. 7), whose elegiacs certainly inspired the anonymous compiler of Erinna's biography in the *Suda*. Writing perhaps the prologue of a circulating edition of the poem (as also revealed by τόδε at v. 1), he celebrates the Λέσβιον (...) κηρίον (v. 1), "Lesbian honeycomb", "all coated in the Muses' honey" (v. 2) of the nineteen-year-old poetess (v. 4); her three hundred verses are equal to Homer (v. 3) and her hexameters could compete with Sappho's μέλη (vv. 7f.).[7] He shapes a quasi-biographical portrait of the poetess (vv. 5f.): ἢ καὶ ἐπ' ἠλακάτῃ μητρὸς φόβῳ, ἢ καὶ ἐφ' ἱστῷ / ἑστήκει Μουσέων λάτρις ἐφαπτομένη, "she stood by the distaff in fear of her mother, she stood at the loom, handmaid of the Muses, to whom she applied herself". Such a topos, featuring a self-taught accomplished genius, is by all means too widespread[8] to allow for a literal reading – which many adopted[9] – of the glaring pun of the weaver-poetess, serving the loom and the Muses at once, almost a feminine counterpart of Archilochus soldier and poet in fr. 1 W.[2] (εἰμὶ δ' ἐγὼ θεράπων μὲν Ἐνυαλίοιο ἄνακτος / καὶ Μουσέων ἐρατὸν δῶρον ἐπιστάμενος, "I am servant of Lord Enyalios and connaisseur of the Muses' lovely gift").[10] Yet the cliché does not prove sufficient

[5] On the form ἠλακ- (Planudes, *versus* ἠλεκ- in mss **AC**, upheld by Neri 2003, 191), some cogent remarks can be found in Zel'čenko 2005, 159–160, who argues for its authenticity; see also Chantraine, *DELG* 409.

[6] *Cf.* e.g. Antip. Sid. *AP* 7.14.6 (*HE* 11.241): see also Veniero 1905, 199 and Eitrem 1932, 2477–2488.

[7] On this epigram (authorship, text, style, *loci similes*), see Neri 2003, 194–197.

[8] On its presence – for example – in the biographic tradition concerning Clotilde Tambroni, see now Tosi 2011, 2–5.

[9] Discussion and bibliography in Neri 2003, 47–48.

[10] "Λάτρις kann nicht zu Μουσέων gezogen werden, da dieses von ἐφαπτομένη abhängt", wrote Scheidweiler (1956, 42 n. 1), quoted *verbatim* by Page (1981, 346): "the reader has an irresistible impulse to take Μουσέων λάτρις together (...) and then finds that ἐφαπτομένη makes this impossible". Yet this elaborate *ordo verborum* – λάτρις placed between Μουσέων and ἐφαπτομένη as possible *apo koinou* and at once as if it were to thwart the 'craved contact' – accredits the attempt of the composer to both fulfil the "irresistible impulse" and produce a feminine version of what Archilochus said about himself, ensuring the accomplished transformation of λάτρις ἠλακάτης (feminine translation of θεράπων Ἐνυαλίοιο ἄνακτος) in λάτρις ἐφαπτομένη (the participle being in assonance, still not in competition, with the Archilochean ἐπιστάμενος) Μουσέων (see 'Ion' *AP* 7.44.5f. = *FGE* 2.574f. λάτρις / Πιερίδων). The verse seems to be mimicking and mocking the anon. *AP* 5.135.3 (*HE* 57.3904) Βάκχου καὶ Μουσέων ἱλαρὴ λάτρι καὶ Κυθερείης, "cheerful waitress of Bacchus and the Muses and the Cytherea". A Μουσάων θεράπαινα is found also in *GVI* 479.1 (Berea, after 150 AD), looking at the

to support, even though well-grounded, West's thesis (1977, 117–118, 1996, 24–25), according to which this cute *Cinderella*, an exquisite blending of *ingenium* and *ars*, would be nothing but the miserable disguise of a forgery. Distaff and spindle were common to women of all ages and social ranks;[11] it is not surprising that an 'upper-class woman' – as Erinna had to be – was able to compose masterpieces between the 5th and the 4th century BC.[12]

Here again, notwithstanding the persistence of topic features such as demanding mother, distaff and spindle (*cf.* Sapph. fr. 102.1 V. γλύκηα μ ᾶ τ ε ϱ, οὔτοι δύναμαι κρέκην τ ὸ ν ἴ σ τ ο ν, "my sweet mother, I can no longer weave the web"), we shall reckon with the possibility of an autoschediastic genesis for the notation; the compiler might have found the frame of his biographical texture (*cf.* fr. 4.22–27, 39) precisely in the poem (on account of the recurring ἠλακάτη).

It is not certain – albeit possible – that weaving is implied in the epigram that Meleager preposed to his *Garland*, where Erinna is compared to the παρθενόχρως κρόκος (*AP* 4.1.12 [*HE* 1.3937] = test. 9 (ἐμπλέξας) / καὶ γλυκὺν Ἠρίννης παρθενόχρωτα κρόκον, ("weaving Erinna's sweet, virginal-colour crocus"), that is to say to the only flower ('saffron') well-known in antiquity for dyeing cloth: besides, κρόκος recalls the κρόκη, 'thread', 'flock of wool', and the verb κρέκω, 'to spin'.[13] Yet a reference to weaving is explicit in Christhodorus' *ekphrasis* of the statues that were still in the Zeuxippos of Constantinople in 532 AD and included an Erinna, allegedly in bronze (*AP* 2.108–110 = test. 12): παρθενικὴ δ' Ἥριννα λιγύθροος ἕζετο κούρη, / οὐ μίτον ἀμφαφόωσα πολύπλοκον, ἀλλ' ἐνὶ σιγῇ / Πιερικῆς ῥαθάμιγγας ἀποσταλάουσα μελίσσης, "Erinna sat, clear-voiced maiden, not handling plots with many twists, but quietly distilling drops of Pierian bee's honey".[14] The dualism of weaver and poetess, already fixed in 'Archilochaean' terms by *AP* 9.190 (= test. 7), leans here decidedly towards the second value, up to the point that we might consider the possibility for Christodorus' image to be *entirely* poetic and literary. After v. 108 (where, as in *AP* 9.190,7f., Erinna is set beside Sappho, portrayed at v. 69 by Christodorus as Πιερικὴ δὲ μέλισσα λιγύθροος ἕζετο Σαπφώ, "Sappho sat, clear-voiced Pierian bee"), the counterbalance (vv. 109f.)

Μουσάων θεράπων in *Marg.* fr. 1.2 W.[2] In Eva Stehle's opinion (2001, 198), these words, intertwining weave and poetry, summarise the contents of the poem in which Erinna, well-behaved "good daughter" fond of the 'flamboyant' Baucis, sublimates her clashing feelings "by quietly weaving/composing a text that could go out into the world in her place and speak publicly of her love for Baucis."

[11] See e.g. *Il.* 1.31, 3.125, 6.490–492 = *Od.* 2.356–358 = 21.350–352, 5.62, 6.52f., 7.105, 18.315, 19.138–140, Hes. *Op.* 779, Eur. *El.* 542, *Or.* 1431–1433, Xen. *Mem.* 2.7.5, *Oec.* 7. 6.36, Leonid. *AP* 7.726 (*HE* 72), Antip. Sid. *AP* 7.424.6 (*HE* 29.375), Diog. Laert. 6.97f., Sen. *Ph.* 103f., Quint. Smyrn. 6.481. Bibl. in Neri 2003, 47–48 n. 52.

[12] *Cf.* Neri 2003, 48 n. 53.

[13] *Cf.* Chantraine, *DELG* 580–581, 585–586. When Meleager decided to include also the three epigrams attributed to Erinna in antiquity (*AP* 6.354, 7.710, 712 = frr. °5–7: *cf.* Neri 2003, 85–88, 431–440) in the collection, he was certainly influenced by the celebratory themes of such epigrammatic *elogia*, from Asclepiades' γλυκύς (test. 4.1), to the depiction of the παρθένος-poetess (testt. 4–7) and the grieving strain of her poetry (test. 5): in fact κρόκος likely refers to both virginity and death (*cf.* Neri 2003, 201–202).

[14] The passage seems to be heavily indebted to Leonid. *AP* 7.13 (*HE* 98) = test. 6 Παρθενικὰν νεάοιδον ἐν ὑμνοπόλοισι μέλισσαν / Ἤρινναν Μουσέων ἄνθεα δρεπτομέναν / Ἅιδας εἰς ὑμέναιον ἀνάρπασεν. ἦ ῥα τόδ' ἔμφρων / εἶπ' ἐτύμως ἁ παῖς· "βάσκανος ἔσσ' Ἀίδα", "As Erinna, maiden composing first-fruits songs and among the poets bee gathering the flowers of the Muses, Hades kidnapped her for his wedding. Judicious, she said it rightfully, the child: "Hades, you are envious", which – perhaps – described a statue portraying Erinna: for a bronze statue of the poetess by Naucides, see Tatian 33.2 = test. 13. *Cf.* Neri 2003, 42–45, 192, 209–210.

between μίτος πολύπλοκος which Erinna does *not* "handle",[15] and the "suave droplets"[16] which she instead "distils"[17] hints at the familiar (and Callimachean) opposition between the slight, slender and graceful poetess and the knotty "tangled texture" of the Homeric epic,[18] just as in *AP* 9.190.1–4 her slender three hundred verses kept up with Homer.[19] Here we have a synthesis *cum variatione* of the notorious epigram (on which *Suda* and Eustathius, testt. 16a-b, are based too) and of his daring *synkriseis*, even though deprived of any allusion to the 'distaff'.

Spinning in Erinna

Epigrammatists refer to spinning in many ways. Their images focus on either everyday life (the work at the loom: testt. 7, 12), or poetic inspiration (framing lines: testt. 7, 12 and perhaps 9), or Fate (the thread of life untimely broken: test. 5). Such references recur in Erinna's own words, as preserved by *PSI* IX 1090, the papyrus containing the longest of her fragment (*SH* 401 = fr. 4). The autoschediastic genesis of these *testimonia* proves the importance of the theme in Erinna's poetry. In particular, three passages of hers seem to refer directly to the weave:

First: at vv. 5–17, Erinna allusively paraphrases the game of χελιχελώνη (*PMG* 876c χελιχελώνη, τί ποιεῖς ἐν τῷ μέσῳ; / ἔρια μαρύομαι καὶ κρόκην Μιλησίαν. / ὁ δ' ἔκγονός σου τί ποιῶν ἀπώλετο; / λευκᾶν ἀφ' ἵππων εἰς θάλασσαν ἅλατο, "'Torty-tortoise, what are you doing in the middle?' 'I'm winding off wool and a Milesian thread'. 'But that baby of yours, how did he ruin?' 'Off white horses into the sea he (…) jumped!'"):[20] she also reproduces the second line of the related jingle – which is the first answer of the tortoise's 'mother' (ἔρια μαρύομαι καὶ κρόκην Μιλησίαν) – at v. 13, where one should probably read ἀμ]νίδα πέξα[ς, "you sheared the ewe lamb".[21]

The way Erinna evokes words and gestures of this feminine play (mentioned by Poll. 9.122.125 and Eust. *Od.* 1914.56–58 = Suet. *Παιδ.* *19, p. 72 Taill.), with their initiatic implications (*e.g.* the 'dive into the sea'-motif; its masculine counterpart is the game of χύτρα or χυτρίνδα described in Poll. 9.113f., whose jingle evokes the 'stewing in a cauldron'-motif),[22] and the nocturnal setting (*cf.* vv. 6, 12) with its ritual connotation (*cf.* v. 8) lead us to think that the poetess is pacing here through the steps of a maiden ritual

[15] Here first linked, as it appears, to the act of weaving, the verb usually refers to the Homeric act of 'handling' a bow (*cf. e.g. Od.* 4.227, 8.196, 215, 15.461, 19.475, 586).

[16] The noun ῥαθάμιγξ designates a 'spatter' of mud or blood in epic war scenes (*cf. e.g. Il.* 11.536, 23.502, Hes. *Th.* 183: see Bulloch 1985, 118–119), though it appears to be already transferred to a poetic context in Pind. fr. 52g M. (*cf.* Rutherford 2001, 342 and n. 11). In the *ekphrasis* it depicts "the glimmers of the splendour" of Cypris (v. 79) and the "suave droplet" of Xenophon's *ars dicendi* (v. 392).

[17] The verb (although without preverb) occurs in Christodorus at v. 161 (referring to Andromache). The joint presence of a *diektasis*- and a loose form of feminine participles at a short distance is noteworthy, even if justified by epic precedents.

[18] This meaning is attested for the term μίτος already in Pherecr. fr. 156.6 K.-A. (*cf.* LSJ⁹ 1138), whereas πολύπλοκος (*cf.* LSJ⁹ 1441) is "un tecnicismo della retorica", as observed by Conti Bizzarro (2001, 105–106), who indulges on μίτος too, regarding it as "un'allusione anche alla *Conocchia*".

[19] According to Zel'čenko (2003, 37–43), ἀλλά is concessive, while ἐν σιγῇ is the focal point of the synthesis: "Erinna non filava la lana, tuttavia componeva versi, oggi in silenzio"; in this (problematic) perspective, the expression would allude to the title of the poem.

[20] *Cf.* Neri 1998, 19–24; 2003, 242–253, 270–294.

[21] For such a reading (and a reconstruction of the *status quaestionis*), see Neri 2002, 61–68 and 2003, 270–277.

[22] On the *Leitmotiv* of the initiatic 'dive into the sea', see Gallini 1963, 61–70, 85–88; Brelich 1969, 376–377; Burkert 1985, 260–264; Versnel 1990, II 48–60; Moreau 1992; Graf 1993; Bremmer 1994, 44–54, 69–72. On 'brewing in a cauldron', *cf.* Gallini 1963, 66; Halm-Tisserant 1993.

feast, where girls abandoned their dolls (*cf.* vv. 3f., 20–22), played the χελιχελώνη (*cf.* vv. 5–17) and rose up from κόραι to νύμφαι (*cf.* vv. 3f.), "marriageable girls": the poem describes an initiation to life, grimly turned for Baucis – dead just after the marriage – into a tragic initiation to death, liable to be revived and lamented in the poem (*cf.* v. 18).[23]

Erinna 'translates' the rhyme with a puzzling nexus, where ἀμνίς stands for ἔρια,[24] πέκειν for μαρύεσθαι[25] and "I" of the χελιχελώνη switches to "you" for Baucis: this suggests that she might have intended to reinvent rather than to repeat – as apparent in the second answer of the 'tortoise' (λευκᾶν ἀφ' ἵππων εἰς θάλασσαν ἅλατο, the starting line of the game) implied at vv. 14f. (with a dramatic ἐς β]ạ[θ]ὺ κῦμα[/ λε]υκᾶν μαινομέν[οισι(ν) ἐσάλαο π]ọσσ̣ὶṿ ἀφ' ἵ̣[π]πω̣[ν, "into the deep flood, off white mares, by crazy feet, you jumped down").[26] Since the memory of such a maiden game (vv. 5–17) is sealed, at v. 18, by the poetess' lament over her unfortunate friend (τα]ῦ̣τα τυ Βαυκὶ τάλαι[να βαρὺ στονά]χεισα γόημι, "this I'm sobbing, crying deeply for you, miserable Baucide", where the epanalectic ταῦτα suggests that the frame of memory is itself weaving the 'canvas' of the *goos* for Baucis), we can cautiously assume that this grief is precisely the tragic key by which Erinna herself reads past words and acts. An indicative aorist is used in place of the present, fixing the event in that very instant of the past, the persistent act of weaving thus transformed in a one-act play. 'Saying', as in the content of the rhyme, hereby slides into 'doing', a deed or an action in a precise moment in the childhood of two friends. By shifting ἔρια μαρύομαι from a literary to a metaphorical meaning, Erinna enhances the symbolic strength of her vocabulary, giving the new image ("shearing the ewe lamb") an added value in respect of the original one ("spinning the wool"). "Shearing the ewe lamb"– unusual at the least, if not premature act – constitutes in fact the exact counterpart of a proverb, ἀμνὴν (or ἀμνία) θερίζειν,[27] "mowing the ewe lamb", ἐπὶ τῶν χρόνῳ ὕστερον τιμωρουμένων, "said of those who

[23] On the presence of young boys and girls at feasts overnight, and on the role of the παρθένοι in rites underlying 'passages' in life by means of ancient initiatic procedures, *cf.* Brelich 1969, 38–39, 58–60, 64–65, 138, 145–147, 152–153, 162–164, 181, 192–193, 231–238, 436. On playing and dancing as essential components of such rites, *cf.* Brelich 1969, 122–126; Lonsdale 1993, 169–193.

[24] The term ἀμνίς occurs only in Theocritus (5.3 and 138f.), meaning 'ewe lamb', as a perfect synonym of ἡ ἀμνός (*cf. e.g.* Theocr. 5.144, 149). A metaphoric value of 'wool', 'fleece' is attested for the latter only by Hesych. α 3726 L., although similar examples of synecdoche *a maiore* (the animal name being used to refer to the material obtained from it) can be found in *Il.* 7.238f., 12.105, 137, up to Verg. *Aen.* X 785 and later (*cf.* Lausberg, *HR* 296). It is therefore possible to establish an equation such as ἔρια = ἀμνίς.

[25] The verb πέκειν has three peculiar meanings (already described by Ap. Soph. 129.30–32 Bekk., *schol.* **Vat** Theocr. 5,98, Hesych. π 1418–1420 H., Phot. 411.18f. P., *Suda* π 918–922, 986 A.; see Forbes 1964, 20, 70 nn. 165–167; Gow & Page 1965, II 284; Bulloch 1985, 142–143): 'to tease' (see e.g. *Od.* 18.315f., Hdn. *GG* III/1 313.23f., III/2 502.34f., 902.16f.), 'to comb' (with reflexive meaning, used with a direct object meaning 'locks': *cf. e.g. Il.* 14.175f. alluded by Synes. *Calv.* 20.22, Pind. fr. 320 M., Call. *Lav. Pall.* 31f., *schol.* **bT**[il] *Il.* XIV 176a, *Suda* ε 2019 A.) and 'to shear', usually of ovins (*cf. e.g.* Hes. *Op.* 775, Theocr. 5.98f., 28.12f., Philostr. *Vit. Ap.* 1.32, Aelian. *NA* I 38, Porph. *Abst.* 1.21.4, and Eust. *Il.* 975,55–57), but exceptionally of human beings (*cf. e.g.* Simon. *PMG* 507, mocked by Ar. *Nu.* 1356 and then Euphor. *AP* 6.279.1f. [*HE* 1.1801f.], Theodorid. *AP* 6.155.1f. [*HE* 1.3506f.]). The verb is seldom explained as ξαίνειν (*cf.* Ap. Soph. 129.31f. Bekk., *schol.* Vat Theocr. 5.98, Hesych. π 1419 H., Choerob. *Orth. An. Ox.* 2.208.2–4 Cr. = Hdn. *GG* III/2 502.34f., Phot. 411.18 P. = *Suda* π 986 A.), meaning both 'teasing' (*cf. e.g. Od.* 22.423, with εἴρια) and 'weaving' (*cf. e.g.* Ar. *Av.* 827 as well as LSJ[9] 1187). The equation πέκειν = μαρύεσθαι is thus legitimate (*cf.* LSJ[9] 1129).

[26] *Cf.* Neri 2003, 277–287.

[27] See Paus. Att. α 88 E. ἀμνὴν θερίζειν· παροιμία ἐπὶ τῶν χρόνῳ ὕστερον τιμωρουμένων. ὅταν γὰρ ᾖ ἀμνίον, οὐκ ἀποκείρεται, "mow the lamb: the proverb is said of those who are going to be punished in the coming time. The lamb in fact, as such, is not sheared", whose sources are Hesych. α 3717 L., Phot. α 1225 Th. and Eust. *Od.* 1627.13f. καὶ ὅτι ἀμνὴ ἐνιαύσιον ἄρνα δηλοῖ. καὶ ὅτι παροιμία ἀμνεῖν (*scil.* ἀμνήν?) τὸ θερίζειν ἐπὶ τῶν ἐκδιαλείματος τιμωρουμένων, τὰ

are going to be punished in the coming time": the unwise and untimely action is thus a warning of a punishment to come. It is hereby apparent how the 'ewe lamb shearing' applies to Baucis' parable, as a metaphorical description of an originally innocent game, albeit set in an initiatic setting. The sudden shearing, expressed by a punctual yet absolute aorist, foreshadows a τιμωρία which is no more χρόνῳ ὕστερον as it was in the proverb, but outright and conclusive, such as death. The new phrasing, ἀμ]νίδα πέξα[(ς), entails three separate layers: first the rhyme, accompaniment of a game riddled with initiatic implications (the rite of passage by which a group of κόραι may reach the *status* of νύμφαι); then the proverb, with its allegorical deprecation of precocity, naturally doomed to retribution; last the fate of Baucis, damned to a *mors immatura* approaching right after crossing the threshold of youth. The metaphor works here as a 'melting chamber': the joke of a childish game going in, the yoke of a dire presage getting out. Needless to say, in the interplay of analogies and metaphors the unfortunate *aoros* plays both the role of the 'weaver', soon reached by an ill-fated τιμωρία, and that of the 'ewe lamb', fatefully 'mown' at dawn on her first harvest time. Considering the assonance between ἀμνίς and Βαυκίς and the extensive resort to the '(ewe) lamb' as a symbol of the young and innocent soul,[28] often doomed to self-sacrifice, one can suppose that Erinna found in games, words and rites of childhood – *a posteriori* transformed and overhauled – as many symptoms of the tragedy and tiles of a 'retrospective prophecy'. Fate came as luck would have it, 'on time' as a 'lamb shearing', unripe and ill-timed.

Second: according to the most reliable readings and supplements, vv. 22–27 run as follows: ἅ τε πὸτ ὄρθρον / μάτηρ, ἀε[ίδοισα σὺν ἁδυλό]γ̣οισιν ἐρείθοις, / τήνας ἦλθ[εν ἄγοισα, πονευμέ]να ἅμφ' ἁλίπαστον· / α̣ἱ̣ μικραὶ στ̣[ονάχησαν, ἔτ' ἃ]ν φόβον ἄγαγε Μο̣ρ̣μ̣ώ, / τᾶ]ς̣ ἐν μὲν κο̣[ρυφᾷ μεγάλ'] ὦ̣ατα, ποσσὶ δὲ φοιτῇ / τέ]τ̣ρ̣[α]σιν̣, ἐκ δ' [ἑτέραν ἑτέρας] μετεβάλλετ' ὀπωπάν, "and towards dawn, (singing with the sweet-talking) workers, mum came (and picked them up and prepared) the salted meat. If the little girls (had cried), the Bogey would have scared them, who (has large ears on her head), and runs on her four feet and was always changing her aspect".[29] A morning scene,[30] where a mother – allegedly Erinna's[31] – arranges the work ἀμφ' ἁλίπαστον while singing,

γὰρ ἄρνια ἐν τῷ ἔαρι ἄκαρπα (*scil.* ἄκαρτα?) ἐᾶται, "*amne* indicates the one-year lamb; and there is a proverb, 'mow the lamb', which is said of those who are punished after a time interval, because during the spring lambs are left untouched"; Photius adds θερίζειν δὲ τὸ διάγειν τὸ θέρος †ἄκαρπον† (*scil.* ἄκαρτον) ὡς τὰ ἀμνία· ἔαρος γὰρ γεννώμενα τοῦ θέρος ἐᾶται †ἄκαρπα† (*scil.* ἄκαρτα), "*therizein* means to spend the summer untouched, like lambs, because lambs born in the spring are left untouched during the summer", an interpretation (or rather a condensation of a broader esegetic context) which seems to neglect the proverb. See also Diogenian. 1.69 and Apostol. 2.65.

[28] See e.g. Ar. *Pax* 935, Strat. *AP* 12.250.2, Clem. Al. *Protr.* 12.119.1, Hesych. α 3713 L.

[29] *Cf.* Neri 2003, 156–158, 323–349 (with *status quaestionis* and discussion of the passage, "the most puzzling place in the entire text", according to West 1977, 106): however, despite the arguments there produced, I do not believe anymore in ἦλθ[ε δὲ at v. 24 (and δέ in third position), nor in ἔθε]ν at v. 25 (ἔτ' ἄν – which may alternatively be suggested – is validated by drama, with its three occurrences in Aeschylus, one in Soph. *OC* 1110, eight in Euripides, four in Aristophanes). At the beginning of v. 24 πήνας, "wefts", is attractive (*cf.* Maas 1934, 208, accepted by Quasimodo 1940, 101, 236 and Scheidweiler 1956, 46–47, 51), but the papyrus exhibits τήνας beyond any doubt (unless a misspelling on the part of the scribe is assumed).

[30] As for πὸτ ὄρθρον, see Theocr. 5.126 (τὸ πότορθρον), and 4.3, 5.113 (τὰ ποθέσπερα), 4.33 (τὸ ποταῷιον), 18.56 (ἐς ὄρθρον in a domestic scenery, at dawn of a nuptial feast). Interchangeability between adverb and the syntagm constituted by preposition + noun is attested already in Thuc. 2.3.4 τὸ περίορθρον, 6.101.3 περὶ ὄρθρον. Even κηρύσσειν ὑπ' ὄρθρον (of the heralds) in *Batr.* 103 matches ὑπορθρίαις φωναῖς in *Anacreont.* 10.9 W. Before Erinna, πρὸς ὄρθρον appears in Aristophanes (*Lys.* 1089, *Ec.* 20; in *V.* 772 we find instead κατ' ὄρθρον).

[31] This is the opinion of most scholars: ἅ τε πὸτ ὄρθρον / μάτηρ, in fact, could hardly refer to anything different from "mother at dawn", the mother of the speaker (*cf.* Theocr. 2.145, 4.9, 11.54 and 67). Besides, the involvement of Erinna's

together with some ἔρ(ε)ιθοι: were the maidens to cry, they would have called Mormò – that is, a dreadful bogey used to deter children from tantrums and equipped with big ears, four legs and a constantly changing appearance.[32] The nature of these ἔρ(ε)ιθοι and of the epithet ἀλίπαστος (or ἀλίπαστον) are crucial to pinpoint setting and meaning of the scene.

The term ἔριθος[33] appears twice in the *Iliad* (18.550, 560) to describe a 'reaper', whereas in the *Odyssey* (6.32) Athena defines herself as Nausicaa's συνέριθος while promising her help in washing the trousseau for her future marriage; the etymology is obscure (*cf.* Chantraine, *DELG* 371), though the word seems to mean 'wage earner', 'daily worker', as possibly testified by Hesychius (ε 5840 L. ἔριθοι· οἱ γεωργοί. παρὰ τὸ τὴν ἔραν ἐργάζεσθαι, ἥτις ἐστὶ γῆ. καταχρηστικῶς δὲ καὶ οἱ ἐριουργοὶ ἢ μισθωτοί, "*erithoi* indicates the farmers, from 'working the *era*, *i.e.* the soil'. By extension also the wool worker or the wage earners") and Choeroboscus (*An. Ox.* 2.199.8–11 Cr. ἔριθος: ι· σημαίνει δὲ τὸν ἐργάτην, καὶ τὴν γυναῖκα τοῦ ἐργαζομένου τὰ ἔρια· γέγονεν δὲ τὸ ἀρσενικὸν ἀπὸ τοῦ ἔρα, ὅ ἐστιν ἡ γῆ, ἔραθος, καὶ κατὰ τροπὴν τοῦ α εἰς ι ἔριθος· καὶ ἀπὸ τοῦ ἔριον, ἔριθος, "*erithos*: by *iota*: indicates the worker, and the wife of the wool worker; the masculine derives from *era*, *i.e.* the soil, also *erathos*, and – by change of *alpha* in *iota* – *erithos*; also from wool, *erion* > *erithos*"). The term is found in Hesiod in connection with a woman (the notorious "servant" in *Op.* 602f.): further on, "à la suite d'un rapprochement par étymologie populaire avec ἔριον" (Chantraine, *l.c.*) which is still detectable in Choeroboscus, the meaning will shift to define the waged 'spinner' or 'weaver'.[34] Such figure of a textile wage earner had to be popular already during the 4th century BC, as suggested by the title *Ἔριθοι* attributed to some comedies by Alexis (frr. 177–183 K.-A.) and Amphis (frr. 17–19 K.-A.). Elements such as the references to singing,[35] the term ἀλακ̣ά̣τ̣αν at v. 39 (see below), the emphasis laid by Erinna's *testimonia* on her work of 'spindle and distaff' (testt. 5.4, 7.5f. see above, pp. 196 and 200 n. 31) and the title *Ἠλακάτη* sometimes assigned to the poem (see below pp. 206–207) support the idea that the *clausola* of v. 23 praised "weavers" rather than "harvesters".[36]

mother in the poem is confirmed by test. 7.5 (ἐπ' ἠλακάτῃ μητρὸς φόβῳ, "by the distaff in the fear of her mother") and by v. 29, where traces suggest τ'[ ]ματρός rather than τε[ᾶς]ματρός. On different interpretations (such as Baucis' mother, a game's 'mother', Erinna herself, a bride-mother, the 'mother' of a *thiasos*, the bossy mother of a family business, and so on), *cf.* Neri 2003, 325–327.

[32] On the *nursery-bogey* of Hecate's retinue, *cf.* Neri 2003, 339–340, 345–346.

[33] We are keen to maintain ἐρείθοις in the text despite a number of facts, namely that: ἔριθος is the only occurrence known by the Alexandrians in the *Iliad* (*cf. ThGL* III 2018); the spelling ερειθ is unknown in Greek; -ει- could well be a prosodic rendering for -ῑ-, for "infolge der spätern Schreibung ει für ῑ sind Formen wie ὀκτείρω statt des richtigen -ῑ́ρω auch in ältere Texte eingedrungen" (Schwyzer, *GG* 184, with bibl.); the Laconian dialect leans towards darkening sounds like ε and ει into ι (*cf.* Thumb-Kieckers, *HGD* 82–83, 84, 87). However, ει/ι alternate in a number of dialects (*cf.* Schwyzer, *GG* 192–194) including Laconian itself (*cf.* ἱράνα/εἱράνα, ϝίκατι/ϝείκατι, εἰρήν/τριτίρην, and Thumb-Kieckers, *HGD* 90, 100, 104, 123, 278–279), and the term, even though not rare, has been written by the scribe with an accent (placed between ε and ι, as usual for accents in diphthongs), thus suggesting an unusual variant, perhaps unexpected in a dialectal context which would have required ἐρίθοις instead.

[34] *Cf. e.g.* Dem. 57.45, Theocr. 15.80, *P. Hib.* 121.34 (III sec. a.C.), Eratosth. fr. 10 Pow. ἢ χερνῆτις ἔριθος (not necessarily a weaver) ἐφ' ὑψηλοῦ πυλεῶνος / δενδαλίδας τεύχουσα καλοὺς ἤειδεν ἰούλους, "where a female wage earner, in the high vestibule, sang beautiful folksongs preparing barley cakes" (a similar situation). In Soph. fr. 286 R.^{2} and Philostr. *Im.* 2.28 we are dealing with a spider. See *ThGL* III 2018.

[35] On singing as an accompaniment for (textile) work, *cf. Od.* 5.61, 10.221–228, Eur. *Hyps.* fr. 752f.9–14 K., Theocr. 24.77, *GVI* 474, Leonid. *AP* 7.726.1–5 (*HE* 72.2411–2415), Eratosth. fr. 10 Pow. (see above), Posidipp. 55.1f. A.-B. (where Μοῖρα untimely snatches away ἀθύρματα καὶ πρὸς ἑῴαν / κερκίδα Σαπφοῦς ἐξ ὀάρων ὀάρους, "delights and Sapphic conversations, one after the other, at the sound of the morning shuttle" of Nicomache: a *Stimmung* which is quite close to that of our poem), Ov. *Trist.* 4.1.13, Ath. XIV 618d, *schol.* **V** Ar. *Ra.* 1315.

[36] Despite Latte 1953, 84–85 (who prefers "reapers") and Page 1981, 344 (who cautiously states: "nothing is said in this passage about spinning or weaving. ἔριθοι are day-labourers, servants of the household; reference to weaving may be introduced by way of supplement or inference, but is not in the text", "the workers are male, not female"). On the

As regards αλῐ̣π̣αστον in *clausola* at v. 24, it will be necessary to leave aside the idea of separating ἁλί from παστόν,[37] emending the sequence[38] or providing it with a meaning never attested ("cloth interwoven with purple", West 1977, 106–107 proposed,[39] on the comparison with terms like ἁλιπόρφυρος, χρυσόπαστος, ἁλουργής, ἁλιανθής, ἁλίχλαινος: nevertheless, it is unclear why Erinna should have resorted to such a 'puzzling' expression). The simplest solution seems to be the following: ἁλίπαστος is well attested as an attribute of some sort of κρέας, as well as substantivized adjective referring to elaborate (as salted: see e.g. Ar. *Ach.* 521, 831, 1099) or rather luxurious meats, therefore unsuitable to workers, as pointed out by Bowra (1936, 331f.).[40] In the light of all this, Bowra tended to prefer an interpretation (therefore mellowed, on the basis of *Il.* 18.559f. where αἱ δὲ γυναῖκες / δεῖπνον ἐρίθοισιν λεῦκ' ἄλφιτα πολλὰ πάλυνον, "and the women spread a lot of white barley flour for the wage workers' meal") of ἁλίπαστον as a bakery product ("it therefore seems unlikely that the *meal* [italics are mine] here was meant for the woolworkers [...] Erinna's meal must be rather special").[41] Still, this concern does not seem positively relevant to the text, where we find no mention of luxurious "beverage being offered to the workers", nor of a mother calling for the ἁλίπαστον. On the contrary, the preposition ἀμφ(ί) and parallels such as *Il.* 18.559 (βοῦν δ' ἱερεύσαντες μέγαν ἄμφεπον, "they prepared a great ox that they had slained for the sacrifice") or Xen. *Cyr.* 5.5.44 (ὁ μὲν Κυαξάρης ἀμφὶ δεῖπνον εἶχεν, "Cyaxares attended to his dinner") suggest that the mother (perhaps together with "the workers") might have been "busy ἀμφ' ἁλίπαστον".[42]

Yet a mother, some weavers, some salty meat and a frightening bogey cannot be easily combined in the same context.

Bowra[43] supposed a scene of awakening and call to work: since the weavers are "waged", we would be dealing with "a mother" ("more likely to be Baucis' than Erinna's") who led a household of textile manufacturing.[44] Such an interpretation was accepted by many scholars.[45] Shifting the property from Baucis'

whole issue, see Neri 2003, 327–328.

[37] *Cf.* Vitelli 1928, 14 = 1929, 142; Orvieto 1929; Edmonds 1938, 203; Cervelli 1953, 215; Vara Donado 1973, 73–74 (even though in the value of "salt-coated"); Lloyd-Jones & Parsons 1983, 191 ("de lustratione thalami", with παστός = θάλαμος); yet "alongside the sea" is rather ἀμφ' ἅλα (*cf. e.g. Il.* 1.409). Scribes do resort to the 'double accent' elsewhere (πὸτόρθρον for πὸτ ὄρθρον at v. 22, δὲφοιτῆι for δὲ φοιτῆι at v. 26) in order to suggest the correct *distinctio* in two terms of an ambiguous sequence. Therefore, one would expect αλὶ̣π̣άστον here, or αλὶ̣π̣αστον at the very least. Yet a *breve* is noted on ι, and no accent: this prosodic mark argues for the most unusual term (ἁλίπαστον).

[38] Maas (1929, 117) suggested ἀλίπαστον ("von λιπάζω?"), Scheidweiler (1956, 46–47, 51) ἀλίπαντον (that is "slender and skinny", with reference to the bogey called by the "mother" to force the "maids" at work).

[39] As foreseen by Stella 1929, 836.

[40] Bowra 1936, 331–332. As for ἁλίπαστος, see Aristom. frr. 6.12 K.-A., Eubul. fr. 6.9f. K.-A., Archestr. fr. 57.4 O.-S. (*SH* 188.4 = fr. 57.4 Mont.), Ath. 16.658a. Substantivate adjective: *cf.* G. Kaibel *ap.* Kassel & Austin II 567. Only in Corn. Long. *AP* 6.191.3f. (*FGE* 1) the adjective refers – by a comic *oxymoron* – to a "ripe" (*i.e.* dry) olive as part of a miserable offering. On servants' food, from Hippon. fr.°194,8 Dg.² to Aesch. *Ag.* 1041, from Soph. *Ai.* 499 to Xen. *Cyr.* 1.2.11, see Neri 2003, 331 n. 399.

[41] For a survey of possible interpretations, *cf.* Neri 2003, 331 n. 400.

[42] So recently (correctly) Nasta 1995/1996, 104: "venne invitandoti ad aiutarla a preparare la carne salata".

[43] Bowra 1933, 181–182; 1936, 326 = 1953, 152.

[44] Bowra 1936, 331 = 1953, 157.

[45] *Cf.* also Maas 1934, 208; Diehl 1935, 209; Collart 1944, 186; Page 1950, 487 (who assumed, though, "that the 'Mother' [...] is Erinna herself, playing 'Mothers and Children' with Baucis: the 'attendant toilers' would be a row of dolls, or imaginary"; see also Page *ap.* Bowra 1953, 157 n. 1 and 1981, 344, and further on Barnard 1978, 205; Bernabé 1994, 143, 258 n. 4; Balmer 1996, 60: "as girls / we played weddings with our dolls, brides in our soft beds, / or sometimes I was mother allotting dawn wool / to the women, calling for you to help spin out / the thread"); Latte 1953, 80, 83–85 (with M. Pohlenz), Scheidweiler 1956, 44–47, 51 (according to whom a manageress-mother evokes a skinny and ἀλίπαντος bogey in order to compel her young ἔριθοι to everyday work), West 1977, 105–107, 112; on different hypotheses (such as Erinna's farewell to her household chores in Edmonds 1938, 203; the *ubi sunt?* of Quasimodo 1940, 101, who translates "le bambole stanno riverse / sui letti nuziali; e presso il mattino / la madre cantando più non reca / il filo sulla rocca e

to Erinna's mother, others have implicitly accepted the identification of such a household with a wool mill.[46] Indeed ἔριθος can hardly refer to slaves, thus dismissing the idea that mother and workers were merely busy with homework:[47] Latte rightly stated that "für Herstellung für den Hausbedarf ist eine Mehrzahl freier Arbeiterinnen ausgeschlossen".[48] Nor does Scheidweiler' idea of Erinna and Baucis being subject to child labour by Erinna's mother (self-interested tutor of Baucis who was parentless from early childhood), evidently unwaged, seem plausible.[49] Yet considering test. 7.5f. (where Erinna is tied-up to "distaff and loom in fear of her mother" and appears as λάτρις) and the papyrus, where the "workers" are followed by "young girls" and a "dreadful Mormò" (v. 25), one can figure Erinna and Baucis, although not as direct ἔριθοι to the mother of one of them,[50] as more or less compulsively trained in such activity since early childhood.[51] The scenario of a family-owned *Betrieb* led by a matron is not sufficiently supported by evidence: textile manufacturing in antiquity allowed women to work at home, in order to provide their οἶκος with clothes,[52] whereas men were normally in charge of manufacturing the apparel destined for sale: W. Thompson's essay[53] has the merit of proving that, even if women cannot be ruled out as possible participants in the commercial phase,[54] men played the leading role in directing a workshop, especially in final stages of the manufacturing, namely the weaving[55] (albeit it seems not implausible that half-processed products prepared by waged teasers and spinners were used too). In other words, even if it does not prove uncommon to employ ἔριθοι in 4th century BC, it is still peculiar, as remarked by Latte, "daß die Hausfrau mit ihnen zu tun hat".[56]

i dolci cosparsi di sale"; the *thiasos* supposed by Vysoký 1942, 97–98; a bride-mother bringing dolls to her children at dawn according to Cervelli 1953, 215; a nuptial context in Michelazzo Magrini 1975, 235, or rather a gastronomic one in Cavallini 1980, 83 and 1991, 125: "le bambole, [nostre compagne?] / [giacevano in cerchio?] sui letti; e la madre, verso il mattino, / cantando venne [con le?] tessitrici; / [...?] e preparando le focacce col sale") and the subsequently implied textual reconstructions, *cf.* Neri 2003, 325–329.

[46] *Cf.* Cervelli 1953, 215, then also Latte 1953, 84 (who returned the workshop to Baucis' mum) and West 1977, 106, who gave the leading role back to Erinna's mother: "Erinna's mother seems to have a number of ξυνέριθοι, and apparently ran a small business". Antip. Sid. *AP* 6.174 (*HE* 5) is implied, according to whom "a dedication" is involved, though not "by a consortium" of weavers, defined as just ἅλικες. West's reconstruction on this point has been accepted by Vara Donado 1980, 307.

[47] *Cf.* Latte 1953, 84; Collart 1944, 187; Levin 1962, 199; Pomeroy 1978, 20; Page 1981, 344; Rauk 1989, 113; so, as it appears, Bernabé 1994, 151: "criadas".

[48] Latte 1953, 84.

[49] Scheidweiler 1956, 47. Yet, the examples of "nicht für Lohn"-ἔριθοι collected by the scholar are limited to Cypris Philaenion. *AP* 6.284, who supplies the nimble hetaera Flenio with a smuggled coat, proving to be a skilled spinner and worker. He had better to mention Athena συνέριθος for Nausicaa in *Od.* 6.34, where the term appears in the mere value of 'helper' (*cf. e.g.* Ar. *Pax* 786, Plat. *Resp.* 533d, *Leg.* 889d, Ap. Rh. 3.942).

[50] The trade of ἔριθος was proper to lower classes, though not to slaves (*cf.* Dem. 57.45, and Dover 1993, 358, Pomeroy *et al.* 1994, 109). Such origin is then highly unlikely for a polished and learned poetess such as Erinna (the epithet λάτρις, in test. 7.6, is deliberately paradoxical: see above pp. 196–197).

[51] So Vara Donado 1980, 307. That weaving, not intended as wage labour, was not exclusive to lower classes, has been repeatedly affirmed by Pomeroy 1975, 30, 40, 149, 199–200; 1978, 20, Pomeroy *et al.* 1994, 34, 103–104. See also Bérard 1986, 79–81 and Dorati 1998, 41–42. On weaving, *cf.* Forbes 1964, 153–174 (especially 164–167).

[52] *Cf. e.g.* Theocr. 28.1f., where ἀλακάτα is the principal tool of οἰκωφελία (see Papadopoulou in this volume).

[53] Thompson 1982, 217–222.

[54] Yet the main character of Aeschylus' monody in Ar. *Ra.* 1331–1363 says (vv. 1346–1351): ἐγὼ δ' ἁ τάλαινα / προσέχουσ' ἔτυχον ἐμαυτῆς / ἔργοισι, λίνου μεστὸν ἄτρακτον / εἱειειειλίσσουσα χεροῖν / κλωστῆρα ποιοῦσ', ὅπως / κνεφαῖος εἰς ἀγοράν / φέρουσ' ἀποδοίμαν, "But I, the wretched one, happened to be performing my work, the spindle full of flax thread tu-tu-tu-turning in my hands and making a clew, so that early in the morning to the market I could bring it to sell". See Hagborg 1827, 13 n. 4.

[55] See also Möller 1976, 72–82. Too cut-and-dried seem then Dorati's remarks (1998), depicting weaving as an exclusively feminine activity. On wool-manufacturing as a chiefly domestic activity see Whitehorne 1995, 70: "the woollen industry [...] was the domestic industry *par excellence* of the ancient Greek world". On wool-processing in Ancient Greece see Forbes 1964, 11–16 (on textile factories, pp. 196–211, 235–236). On feminine labour see also Brock 1994, 336–346 (p. 338 on wool-working). On women's properties within the οἶκος, *cf.* Wolff 1944, 43–95; Ste. Croix 1970, 273–278; Schaps 1975, 53–57; 1977, 323–330; 1979; Hirschon 1994; Foxhall 1989, 22–44 (with further bibliography).

[56] Latte 1953, 84.

So it seems unlikely that these ἔρ(ε)ιθοι were either domestic help of a private house, or employees of a bossy mistress: they apparently do not fit in any of the settings supposed by the scholars. Nonetheless, Theocritus attests (15.80, *cf. IG* II/5 477d) at least a traditional activity pursued by waged (female) weavers: the production of (sacred) garments for a temple or religious rituals.

It is not easy to combine into a coherent framework the considerations carried out so far. Most scholars – both supporters of the 'domestic type-scene' and champions of the *Betrieb* – regarded the scene as a 'sample' of Erinna's childhood fondly spent with Baucis.[57] Yet some evidence sharply clashes with the 'domestic' thesis.

First of all, as noticed by Bowra, the "salty meat" "must be rather special, and perhaps it is meant for the children after their moonlit festival. In that case the ἄριστον would be more elaborate than usual".[58] Second, as pointed out by Michelazzo Magrini, Erinna's aorist is always used with a punctual value (*cf.* vv. 13, 15, 16), thus referring to "un avvenimento unico, ricordato con ricchezza di dati: possono essere ulteriori particolari della festa che abbiamo supposto rievocata all'inizio del testo da noi conservato?".[59] Moreover, time determinations in the papyrus (v. 12 σε]λ̣άνν̣α̣ν, v. 22 πὸτ ὄρθρον) and the neat opposition between τα]ῦ̣τα and τῆν̣[α at vv. 19f. (that is between what the poetess remembers, vv. 5–17, and what she neglects, vv. 21–27) speak against the presence, at the same feast, of the maidens playing the χελιχελώνη in the moonlight and of the mother busy with something "salty" πὸτ ὄρθρον. A ritual background is alleged also by Vysoký, who is deeply influenced by the numerous parallels he detected with Sappho, and subsequently sets the scene in a real *thiasos*, where the "mother" is actually the governess.[60] This hypothesis, though not fully acceptable, has the merit of focusing on the relationship between Erinna and Baucis without resorting to external expedients such as orphanage, adoptions and exchanges.

If the χελιχελώνη (covering a huge section of the poem, vv. 5–17) has to be framed into the lunar setting of a nuptial rite, it might well have been the end of a period of shared living (*cf.* fr. °5.7f. ἁ συνεταιρὶς / Ἤρινν(α), "the helpmate Erinna") and the final step of the educational *iter* of the two girls: during this period, they might have learned the feminine art of weaving and all the skills valuable for brides. Such *praxis* of separation from the family and attendance to rituals echoing initiatic practices is well attested in classical and postclassical times:[61] spinning and weaving had a major role in this educational process, relatives of the initiates performed special duties during the rites of separation and integration (which often involved a large part of the community), mothers cooked for the initiates.[62] Approaching the closure of the poem, Erinna recalls the full moon when her destiny and Baucis' disentangled after an obscure game, and gives all this a key-role in the weave of her *goos* (v. 18). Omitting (v. 31) what she believes Baucis already forgot (v. 28), Erinna recalls also the very beginning of their life together, portraying their dolls left in the beds, a mother singing while arranging comfort food for a feast,

[57] Along this exegetic track a major role is played by parallels such as *Il.* 18.550–560 (the peaceful everyday life blooming on Achilles' shield seems to be reproduced by Erinna in a – even lexically – minor key: Iliadic characters such as the king, the reapers, women and children, are paralleled here by the mother, the ἔρειθοι and the girls) and *Od.* 6.52f. (Nausicaa's awakening), where Arete ἐπ' ἐσχάρῃ ἧστο σὺν ἀμφιπόλοισι γυναιξίν, / ἠλάκατα στρωφῶσ' ἁλιπόρφυρα, "Arete was sitting with her maids at the hearth, turning sea-purple wool".

[58] Bowra 1936, 332 = 1953, 157–158. "Una fiesta" is suggested by Bernabé 1994, 143, who sets it "en las tareas de la casa". Bowra's remark is accepted by Nasta 1995/1996, 178: "il pasto di Erinna deve essere piuttosto speciale e forse [...] è destinato ai bambini dopo la loro festa illuminata dalla luna".

[59] Magrini 1975, 234.

[60] Vysoký 1942, 96–98.

[61] *Cf.* Brelich 1969, 436.

[62] *Cf. e.g.* Plut. *Thes.* 23.3f. See Brelich 1969, 91 n. 119, 445.

weavers joining them (some from the village, some producing cloth for the rites, some who had to help the mother in the textile training of the initiates) and a Mormò, compelling the children to surrender to their next step in life.

Following a narrative strategy that takes her back in time, in search for the 'symptoms' of Baucis' tragic death, consistently enough, Erinna ends up regarding this laborious sunrise as the dawn of a friendship, finally culminating in the night of the χελιχελώνη.

Third: at v. 39, in a section of the papyrus which preserves only (meaningful) initial words of lines (see especially v. 37 ἐννεα[και]δέκατος, v. 38 Ἡρ̣ί̣ν̣ν̣α̣ι̣) we shall read (as Maas did) ἀλακ̣ά̣τ̣αν ἐσ̣όρει̣[σα(-).[63]

The form is Aeolic and Doric, maybe already a *Vordorisches*.[64] Since no breathing, accent and diacritical mark is visible on the vowels, it is likely that the term has already occurred in the poem, and the scribe did not feel the urge to reproduce here its prosodic value.

"Eyes fixed on the distaff".[65] The subject, the meaning and the circumstances of this sequence cannot be detected, and Latte already questioned Maas' supplement: "dann müßte der Anblick der Spindel (nicht die Aufforderung, zu spinnen) irgendwelche Gefühle oder Handlungen ausgelöst haben, was schwierlich wahrscheinlich ist."[66]

"Schwierlich wahrscheinlich" were in fact Edmonds' supplements (Ἀλακάταν ἐφόρει[σ' ἐπέων σοφὰ Ἐλλείθυια]), according to which a Baucis-"Ilithyia well-informed in verses" allegedly helped Erinna in composing the *Distaff*). The same holds true for Cervelli's supplements (ἀλακάταν ἐ[σ]ορεῖ[σ' ἰνδάλλομαι εἰνὶ κραδίαι): Erinna, at the sight of the distaff, would imagine stretching her arms towards her dead friend. To be rejected is also Vysoký's attempt to join the participle to the *Moira* ἀλακάταν ἐ[π]ορεῖ̣[σα λινόκλωστον). Stehle instead believes that the term "suggests that she described (or reverted to) her mother's insistence that she remains home at her weaving".[67]

The distaff, together with the ἄτρακτος and the ἱστός, was the tool *par excellence* within feminine equipment (see above, p. 197): still it had possibly become familiar to Baucis and Erinna right before the notorious "nineteenth year", allegedly during the educational *iter* bringing them together from childhood to the threshold of marriage (*cf.* vv. 5–17, 21–24). As a residue of that friendship – just as the memories of the days spent together, the echo of the χελιχελώνη, the remains of the dolls – it could well "auflösen irgendwelche Gefühle" in Erinna, unable to

---

63 Maas 1929, 117. On possible readings of this verse, *cf.* Neri 2003, 158, 395–397. The *incipit* of vv. 37–39 contained all the elements on which later literary (autoschediastic) biographies on Erinna were built up: so ἐννεα[και]δέκατος (death at nineteen) and ἀλακ̣ά̣τ̣αν (the title of the poem, but Maas 1935, 55 rightly asserted that the occurrence should not be overestimated in the debate: "ἀλακάταν col. III ist wegen der Verstümmelung der Umgebung unverwertbar").

64 Aeolic: *cf.* Theocr. 28, title in mss **CD**, vv. 1, 22. Doric: see χρυσαλάκατος in Pind. *O.* 6.104, *N.* 5.36, 6.36, frr. 29.1, 128c.1 M., Soph. *Tr.* 637 (ἠλακάτα in Eur. *Or.* 1431 is a false Dorism: *cf.* Vitelli 1929, XIV, Chantraine, *DELG* 409). As regards the *Vordorische*, *cf.* Thumb-Kieckers, *HGD* 195–197. Modern Greek has ἠλακάτη, ἀλεκάτη and λεκάτη. The etimology is obscure: a few proposals are in Boisacq, *DELG* 318 and Frisk, *GEW* 628; even more speculative is van Windekens, *Amend* = *DECLG* 94–95. For Steen Due (1965, 9) "it seems reasonable to suppose that this was the original meaning of ἠλακάτη: 'the part of a reed which is between two joints'". On the tool, *cf.* Blümner 1912, 122–125. On ἠλάκατα, see now Magnelli 1999, 141–142.

65 According to Bernabé 1994, 144, the "distaff" is here mentioned "probablemente porque Erinna la mira sin poder trabajar".

66 Latte 1953, 89.

67 See respectively Edmonds 1938, 203; Cervelli 1953, 224; Vysoký 1942, 105; Stehle 2001, 197. Stehle 2001, 197.

attend Baucis' funerals and forced to grieve her friend through keepsakes and memories, in silent solitude (*cf.* vv. 31–34). It seems therefore reasonable to assume that Erinna is the one who turns her gaze on a distaff brimming over with memories, as she could not do with the dead Baucis: the 'vertical' connection between οὐδ̣' ἐσιδῆν, a o r i s t infinite, at v. 33, and ἐσ̣όρει̣[σα(-), p r e s e n t participle, at v. 39 (both at the beginning of the verse) is evident. The 'distaff' mirrors a shared childhood, and acts in Erinna's grief and *goos* as a substitute for her friend's corpse. This outfit of spiritual objects, this attitude to look back on past words and things for discovering their new value, make it plausible that even in antiquity the poem received (by Erinna herself?) a title linked to the object which better resumed Baucis' bitter fate: yesterday's friendship, separation of today.

The *Distaff*

According to the *Suda* (test. 16a) – an information repeated by Eustathius (test. 16b) and almost certainly derived from Hesychius' *Onomatologos* – Erinna "wrote the *Distaff*". Despite some scholars being sceptical about the title, which has been at times either rejected or questioned,[68] no substantial reason exists to cast it out *a priori*, an exception being perhaps the fact that such compositions hardly received titles in pre-Hellenistic times.[69] Nevertheless, most scholars accepted this title, sometimes – in particular Alan and Averil Cameron – explicitly defending it.[70] The note seems at any rate to be confirmed by the presence of ἠλακάτη not only in the poem (fr. 4.39), but also in two epigrams for Erinna (testt. 5,4, 7,5; "in quite different ways and surely independently"),[71] in Christodorus' and Tatian's allusions to spinning and weaving (testt. 12,109, 13), and in the manifold value of the distaff, which is tool of everyday feminine work and keepsake of the dead Baucis, device of the Parcae spinning fates and gear of the Muses weaving the poem, and symbolizes well Erinna's and Baucis' *bioi paralleloi* (as far as it can be reconstructed), the main – if not the only one – theme of the poem.[72] As regards the (feeble) claim by Crusius that the title had obviously been inferred by lexicographers from epigrammatic *testimonia* (chiefly by test. 7.5), Cameron and Cameron rightly replied: "why should anyone have picked out the one word ἠλακάτη from that poem and assumed that it was the title of Erinna's poem on Baucis? It is by no means a natural assumption".[73]

Yet not all these 'confirmations' are, in the end, utterly persuasive. In particular:

a) The ἠλακάτη, just as the ἄτρακτος and the ἱστός, was some sort of symbol of womanhood:[74] one should not overestimate its presence – in the form ἠλακάτη, different from the ἀλαχ̣ά̣τ̣αν in fr. 4.39[75] – in a couple of epigrams praising a short poem on weaving. Be it as it may, both epigrams date from a

[68] Bibliography and *status quaestionis* in Neri 2003, 94–95 n. 133.

[69] In particular, 'symbolic' or allusive titles, such as *Ἠλακάτη* would be, were apparently not widespread (perhaps except for comic poetry) before the Hellenistic age.

[70] *Cf.* Cameron & Cameron 1969. Bibliography and *status quaestionis* in Neri 2003, 95 n. 136. The authenticity of the title was recently defended by Zel'čenko 2003 and 2005.

[71] Cameron & Cameron 1969, 287.

[72] Bibliography with a survey of different interpretations in Neri 2003, 95–96.

[73] Crusius 1907, 458; Cameron & Cameron 1969, 285.

[74] *Cf.* Neri 2003, 96 n. 143 with bibliography.

[75] Although the title was *Ἠλακάτη* (*cf.* West 1970, 277 n. 3, 1977, 96 n. 3), it would still be surprising for the epigrammists to use a different form. Moreover, Asclepiades allusively provided his epigram on Erinna with a Doric shade (test. 4). On the 'mixed form' "Alakate" in modern scholarly literature, *cf.* Neri 2003, 96 n. 144.

time (plausibly 3rd/2nd century BC) when an Alexandrian *Erfindung* might well have named a poem left untitled by its authoress.

b) In chapters 33 and 34 of his *Oratio ad Graecos*, Tatian draws a distinction between Roman women – wise and always attentive to their distaffs – and Greek ἑταῖρα, including Erinna (test. 13): it is hence unclear why should he allude to her again as to a weaving poetess author of a "*Distaff*" (are we dealing with an ambivalent instance of Greek *pietas*?).[76] A bulletproof allusion to the title is not to be found in Christodorus either (test. 12). In any case, both Tatian and Christodorus wrote when the poem might have already been named *Distaff*.

c) The word ἠλακάτη can indeed receive manifold meanings, all (even too) suitable for Erinna's work and to her biographic image as appointed by Alexandrians: this could easily give birth to the title. If this is the case, Theocritus might have already composed his *Distaff* when the compilers of the *Pinakes*[77] decided to resort to the same title for a short poem dedicated by a graceful poetess to her beloved friend, two girls whose thread of life was untimely broken – according to the tradition.

In conclusion, we can save *Suda*'s and Eustathius' witnesses (testt. 16a–b), both probably drawn from Hesychius: not many the validations, no opposite evidence either. One thing rests assured: if it was not Erinna, it was indeed the history of literature that gave the poem its final title *Ἠλακάτη*.

A thread of friendship

Both the tradition concerning Erinna and *PSI* IX 1090 agree on the *compagnonnage* binding Erinna and Baucis (see fr. °5.7f.): this has sometimes been regarded as a homosexual relationship by modern scholars,[78] who tried to counterbalance the lack of information by means of reveries. The most varied settings have been envisaged, such as a *thiasos* in Sappho's style, as well as a religious order having Erinna as high priestess,[79] or rather a family factory of matriarchal inspiration.[80] Much of this relies

[76] It seems to me both speculative and improbable that Erinna is the exception and that Tatian's τῆς παρ' ὑμῖν παιδός, "of the girl with you" (33.5) refers to her (not to Sappho, mentioned right before as γύναιον πορνικὸν ἐρωτομανές, "sex-maniac little bitch"), as states Zel'čenko 1998 (see also Angelici 1956/1957, 92). The terms γύναιος and παῖς are here interchangeable, as suggested by the sentence following παιδός: τούτου χάριν αἰδέσθητε, μαθηταὶ μὲν ὑμεῖς τῶν γυναίων εὑρισκόμενοι κτλ., "therefore, you ought to be ashamed to be caught even as girls' pupils".

[77] Erinna's poem might, for instance, have been listed in the Callimachean *pinax* of poets even without title, only with an *incipit* (possibly including the word ἠλακάτη), as happened to Sappho's, Alcaeus' and Anacreon's poems and to some dramas (*cf.* Wendel 1949, 32–34; Pfeiffer 1973, 218). The poem might then have been named after an evocative image at the time of Aristophanes' editorial *krisis* (and thus preserved from Artemidorus and Theon onwards). This hypothesis would certainly untie some knots, yet it cannot be proven.

[78] Bibliography and discussion in Neri 2003, 48 n. 54. For a different opinion, see Gutzwiller 1998, 77, who thinks that the poetry of Erinna-'bee' would be marked by the "chastity of bees" and by "nonerotic honey". For Erinna's chastity (even though postulated on the basis of terms meaning 'youth' rather than 'virginity') argue also Zel'čenko 1998, 55, 57 and Stehle 2001: Erinna should be a suppressed "good daughter" allured by a maverick Baucis and finally able to settle her inner struggle "by quietly weaving/composing a text that could go out into the world in her place and speak publicly of her love for Baucis [...]. More, therefore, than lament, her poem is affirmation of Baucis" (p. 198).

[79] Possibly consecrated to Demeter, as prove – if ascribed to Erinna – the "Doric-Aeolic" hexameters in fr. °°16, in which a priestess Erinna would have devoted herself to a clearly cultual weaving. Bibliography and discussion in Neri 2003, 48 n. 55.

[80] Led by a mother (fr. 4.23 and 29), be it Baucis' (*cf.* Bowra 1936, 330 = 1953, 156 and Latte 1953, 84) or Erinna's (*cf.* Cervelli 1953, 215, Scheidweiler 1956, 47 and West 1977, 106), assisted by a crew of ἔριθοι (F 4.23). Erinna and Baucis (the second one parentless, or rather daughter of one of the ἔριθοι, therefore guest in Erinna's dwelling, according to Scheidweiler 1956, 47 and West 1977, 108) would have then be children accustomed to playing (Latte) or else prematurely working, under the threat of the Mormò (Scheidweiler and West, on the basis of fr. 4.25 and test. 7.5f.): see above pp. 200–204.

of course on the role assigned to the uncertain spots of autobiography contained in fr. 4. Like many aspects in Erinna's story, the scenery of the relationship with Baucis is partly autoschediastic and partly hypothetic: no argument can ever aim at being conclusive. The relationship, along with the weave surrounding it, belongs to an educational *iter* – in which the two girls walked at the same pace – begun at dawn, when a mother was arranging a feast (possibly evoked by the 'salty meat' of v. 24) and closed at dusk at the doorsill of nuptials, with dances and games (vv. 5–17) sealing the passage from maidens to women. By means of the insisted use of the aorist (more suitable to the memory of unique and unrepeatable moments that to mere references to everyday life) Erinna discloses her *goos* for Baucis (v. 18) and encloses their friendship within the steps of the formative path of a Laconian girl between 5th and 4th century BC. No more νηπία (this is perhaps the meaning of ṿηπιάσ̣α̣σ̣α̣ referred to Baucis at v. 29) the young girl entered puberty (the door-sill of which is sealed by a bogey like Mormò, at v. 25) and was taught the whole know-how of the perfect landlady by several figures (mothers of other initiates and cultic staff such as the ἔριθοι at v. 23). Once youth was closed in the moonlight, the initiate could await the *gamos*, the secure harbour of her future life.

Baucis did not find any future life in that harbour: she was soon deprived of her marriage by untimely death. The circumstances of the event are obscure, and the scarce details we know are shady: they emerge from one of the 'pseudo'-Erinnian epitaphs for her girlfriend (fr. °6), where Baucis' bridal torches were used for her funerary pyre, lighted by her father-in-law. The motif of the bride deceased on the day of her wedding is the feminine counterbalance to the myth of Hymenaios, *aoros* ("died untimely") son of Calliope dead ἐν γάμοισι χροϊζόμενον, "as was first touched in marriage" (Pind. fr. 128c.7 M. = 56.7 Cannatà Fera) and will constitute a recurring topos of Hellenistic poetry.[81] Given the success of the poem between 3rd century BC and 2nd AD, precisely the real story of Baucis' death as sung by Erinna might have inspired Hellenistic epigrams, filled with *pathos* for the *aoros thanatos* arranged in very many melodramatic variations. In fr. °6.5–7 Baucis is referred to as γένος Τηνία ("of Tenian family") and her father-in-law carries her to the pyre: in Bergk and Wilamowitz's opinion, this entails that Baucis died (or at least was buried) on foreign soil.[82] This hypothesis is supported by details such as the *propemptikon* composed by the poetess for her beloved friend (fr. 1) and the naval-nuptial value of πλόος (fr. 1.1) and εἴσβασις (fr. 4.28), which describe the initiation to marriage, together with the impossibility – merely legal – to attend her funeral (fr. 4.31–34). The detail of her father-in-law (not her husband) laying Baucis on the pyre, far from increasing the *pathos* of the scene, might suggest that she died together with her husband right after the wedding, and could well have been borrowed by the author of fr. °6 from the poem itself. It is easy to be seized by the alluring Sirens of autoschediastic reconstructions; yet, by postulating a shipwreck, one would shed a new light on Baucis' πλόος as well as on her "dive ἐς β]α̣[θ]ὺ κῦμα", "into the deep flood" (fr. 4.14), on her εἴσβασις, which turned out to be a sail to Hades, as well as on the "transference of Erinna's words about Baucis to Erinna herself"[83] on behalf of Alexandrian epigrammatists, who portrayed Erinna as an *aoros*, captured by Hades εἰς Ἀχέροντα, διὰ πλατὺ κῦμα καμόντων, "over the wide wave of the dead, to Acheron" (test. 5.3).[84]

[81] See the accurate study by Szepessy 1972, and recently Khan 1995/1996, 123–124 and Hunter 1996, 15, who points out that in Pindar Hymenaios "died 'in the act of first love-making'", and asks: "did Erinna use this conceit somewhere in her poem?" (n. 60).

[82] Bergk 1866, 928; 1882, 144. Wilamowitz's 1913, 230. See recently Lefkowitz & Fant 1992, 335 n. 10: "her birthplace may be mentioned because she died away from home". On ethnic epithets in epitaphs for people dead on foreign lands, *cf.* Preger 1891, 208; Lapini 1995, 123.

[83] Cameron & Cameron 1969, 287. See now Hunter 1996, 15 n. 57.

[84] Depicted as a new Persephones, as a tragic prey either of Hades' βασκανία (testt. 6, 9) or of an unrelenting Moira

The thread breaks?

An *aoros* poetess, then. Nineteen years old, to be correct. The age at which Erinna died was regarded with suspicion by scholars, although it is supported by the evidence:[85]

a) in fr. 4.37 (*incipit*) ἐννεα[και]δέκατος .[is clearly legible, and the *incipit* of the following line mentions Erinna by name;

b) according to Asclepiades (test. 4), Erinna would have *produced* her πόνος at nineteen (vv. 1f.), dying shortly afterwards (vv. 3f.);

c) the extraordinary case of a nineteen-year-old girl capable to parallel Homer is mentioned by an anonymous epigrammatist (test. 7.3f.) beyond any doubt related to Asclepiades (although it seems hard to determine who depends on whom);

d) *Suda* (test. 16a), probably following the *Onomatologos* by Hesychius Milesius, explicitly states that Erinna τελευτᾷ παρθένος ἐννεακαιδεκέτις, "she died when she was just a nineteen-year-old maiden", an information further carried on by Eustathius (test. 16b: ἐτελεύτησε παρθένος (...) ἀπῆλθε δὲ ἐννεακαιδεκέτις, "and she came to the end of her days in her youth (...) yet she left when she was nineteen years old").

Indeed, one might quibble about the details. The age ἐννεα[και]δέκατος in fr. 4.37 results from a supplement, and nothing guarantees for it to be Erinna's, even though her name appears in the following line. Moreover, allusions to her death at nineteen do not, *a rigore*, appear anywhere earlier than the *Suda*, for Asclepiades and the anonymous epigrammatist want just to enhance the poetess' early talent.[86] However, Asclepiades asserts that Erinna died shortly after that age, an element which is fair to relate with the papyrus' ἐννεα[και]δέκατος. This is even more legit if we are dealing, as it seems probable, with autoschediasms. An educated guess is that "nineteen"(-year-old) in fr. 4.37, in a context of doleful bridal death, referred to Erinna's and Baucis' age as *agentes*[87] – the age at which was usual for a Laconian girl to get married.[88] If so, Asclepiades, by composing an epigram on an *aoros thanatos*' poem, might have transferred Baucis' fate to Erinna. As a safeguard stands Asclepiades' ὡς ἄν,[89] a precautionary measure of which the anonymous epigrammatist of test. 7 did not feel the need. Such transfer would explain the youthful zest of this poetry, the absence of other *ponoi* by Erinna as well as the sullen silence on the poetess' biographic details, probably opaque already at the time.[90] As a result, the process could have been the following:

a) the poem, which portrays both Erinna and Baucis *agentes* as 19 years old at the moment of Baucis' death, is allegedly "not long";

b) young poets (as Asclepiades himself, cf. *AP* 12.46 [*HE* 15]) produce *ponoi* marked by *oligostichie* and *brevitas*;

c) Erinna *agens* is the same age as Erinna *auctrix*.

(test. 5), as a swan singing not for long and yet sublimely (testt. 5, 8), the poetess became a challenging pathetic theme capturing the attention of many epigrammatists. Bibliography in Neri 2003, 50–51 n. 64.

[85] *Status quaestionis* (with bibliography, discussion and a survey on the reception of the topos) in Neri 2003, 51 nn. 65–66.

[86] *Status quaestionis* and bibliography in Neri 2003, 52 nn. 69–70.

[87] If the image in fr. 4.45f. (καὶ χαιτα.α̣ν̣.[/ πραϋλόγοι πο̣λ̣ιαί, ταὶ γήραος ἄνθεα θνατοῖς⌋, "and the hair (...) soft-speaking hoary temples, that are flowers of old age for the mortals") refers to Erinna (most certainly not to Baucis), it is difficult to figure her abducted by Hades at nineteen. Cavallini (1991, 128–129, sceptical as well on the poetess' *immatura mors*) supposed a projection into the future (see also Skinner 1982, 267 and Bernabé 1994, 144) whereas Arthur (1980, 56, followed by Stehle 2001, 186–187, who persistently writes "Erinna" in her study on fr. 4) in a rather intriguing fashion suggested that Erinna *auctrix* is here recalling the nineteen-year-old Erinna *agens*. On this passage, see Neri 2009.

[88] *Cf.* Cartledge 1981, 94–95.

[89] The conjunction has a hypothetical value: *cf.* Kühner-Gerth, *GG* 1.243f.; Schwyzer-Debrunner, *GG* 669.

[90] So Bowra (1936, 335 = 1953, 161), who charges the *Suda* with this deliberate autoschediasm.

Baucis' bitter destiny (lamented in fr. 4 and resumed in an epitaph in frr. °5f.) encouraged such an autoschediasm: in fact, is well-attested in Hellenistic epigrams "la tendenza ad identificare il poeta con la propria opera, creando uno stereotipo per ogni genere: Saffo e Anacreonte, poeti d'amore, sono considerati passionali ed edonisti; Ipponatte e Archiloco, come giambografi, saranno irascibili e violenti; Pindaro è cantore della gloria, umana e divina: quindi esempio di moralità e *pietas*. L'epigramma, nel suo breve giro, si presta facilmente a queste schematizzazioni, anzi, spesso le esige".[91] By dedicating her brief poem to an *aoros*, Erinna, or better her literary counterpart, ended up in sharing – which caused her success – Baucis' fateful doom.

And it is no doubt significant that all this was done through textile imagery, in which all aspects of Erinna's biography and poetics find, so to say, their common thread: the concrete work of the girls, probably as part of an educational path that led them together to the *status* of *nubendae*; the relentless action of the Moirai, who spin the fates and soon severed Baucis' thread of life; and finally the title – *Distaff* – of that poem, which was able to deliver a minor and private story to the immortal fame, and in which already the first Alexandrians appreciated a verbal texture of a rare poetic finesse. A *ponos* on a *ponos*: but sweet, and therefore immortal.

[91] Barbantani 1993, 66.

Appendix

Test. 4 (Asclep. *AP* 7.11 [*HE* 28]) Ὁ γλυκὺς Ἠρίννας οὗτος πόνος, οὐχὶ πολὺς μέν, ὡς ἂν παρθενικᾶς ἐννεακαιδεκέτευς, ἀλλ᾽ ἑτέρων πολλῶν δυνατώτερος· εἰ δ᾽ Ἀίδας μοι μὴ ταχὺς ἦλθε, τίς ἂν ταλίκον ἔσχ᾽ ὄνομα;	This is the sweet labour of Erinna, not very long, as being that of a nineteen-year-old girl, but much more effective than that of many others; if Hades had not kidnapped me so early, who would have enjoyed such a fame?
Test. 5 (anon. *AP* 7.12 [*FGE* 39]) Ἄρτι λοχευομένην σε μελισσοτόκων ἔαρ ὕμνων, ἄρτι δὲ κυκνείῳ φθεγγομένην στόματι ἤλασεν εἰς Ἀχέροντα διὰ πλατὺ κῦμα καμόντων Μοῖρα, λινοκλώστου δεσπότις ἠλακάτη· σὸς δ᾽ ἐπέων, Ἤριννα, καλὸς πόνος οὔ σε γεγωνεῖ φθίσθαι, ἔχειν δὲ χοροὺς ἄμμιγα Πιερίσιν.	Soon, just as you bleed a spring of poems created by the bees, soon, just as you uttered sweet sounds with swan-like voice and accent, the Fate – mistress of flax-spinning distaff – pushed you down, over the wide wave of the dead, to Acheron. But the beautiful labour in verses proclaims, Erinna, that you will not perish, and that you have choirs with the Pierides
Test. 7 (anon. *AP* 9.190 [*FGE* 38]) Λέσβιον Ἠρίννης τόδε κηρίον· ἁδύ τι μικρόν, ἀλλ᾽ ὅλον ἐκ Μουσέων κιρνάμενον μέλιτι. οἱ δὲ τριηκόσιοι ταύτης στίχοι ἶσοι Ὁμήρῳ, τῆς καὶ παρθενικῆς ἐννεακαιδεκέτευς· ἣ καὶ ἐπ᾽ ἠλακάτῃ μητρὸς φόβῳ, ἣ καὶ ἐφ᾽ ἱστῷ ἑστήκει Μουσέων λάτρις ἐφαπτομένη. Ψάπφω δ᾽ Ἠρίννης ὅσσον μελέεσσιν ἀμείνων, Ἤριννα Ψάπφους τόσσον ἐν ἑξαμέτροις.	Here is Erinna's Lesbian honey-comb: sweet, small, of course, and yet all coated in the Muses' honey. Her three hundred verses are equal to Homer, though she was just a girl of nineteen years. She stood by the distaff in fear of her mother, she stood at the loom, handmaid of the Muses, to whom she applied herself. And as much as Sappho excels Erinna in lyrics, so much Erinna excels Sappho in hexameters.
Test. 9 (Meleag. *AP* 4.1.12 [*HE* 1.3937]) (ἐμπλέξας) 5 καὶ γλυκὺν Ἠρίννης παρθενόχρωτα κρόκον 12	(weaving) Erinna's sweet, virginal colour crocus
Test. 12 (Christod. *AP* 2.108-110) Παρθενικὴ δ᾽ Ἤριννα λιγύθροος ἕζετο κούρη, οὐ μίτον ἀμφαφόωσα πολύπλοκον, ἀλλ᾽ ἐνὶ σιγῇ Πιερικῆς ῥαθάμιγγας ἀποσταλάουσα μελίσσης	Erinna sat, clear-voiced maiden, not handling plots with many twists, but quietly distilling drops of Pierian bee's honey
Test. 16a (*Suda* η 521 A.) Ἤριννα, Τεΐα ἢ Λεσβία, ὡς δὲ ἄλλοι Τηλία· Τῆλος δέ ἐστι νησίδιον ἐγγὺς Κνίδου· τινὲς δὲ καὶ Ῥοδίαν αὐτὴν ἐδόξασαν. ἦν δὲ ἐποποιός. ἔγραψεν Ἠλακάτην· ποίημά δ᾽ ἐστιν Αἰολικῇ καὶ Δωρίδι διαλέκτῳ, ἐπῶν τ'. ἐποίησε δὲ ἐπιγράμματα. τελευτᾷ παρθένος ἐννεακαιδεκέτις. οἱ δὲ στίχοι αὐτῆς ἐκρίθησαν ἶσοι Ὁμήρῳ. ἦν δὲ ἑταίρα Σαπφοῦς καὶ ὁμόχρονος.	Erinna: from Teos or Lesbos, according to others from Telos: Telos is a small island near Cnidus; there were also some who thought that she was from Rhodes. She was a poetess in hexameters. She wrote the *Distaff*: this is a poem in Aeolic and Doric dialect, in three hundred hexameters. She also composed epigrams. She died as a maiden just nineteen years old. Her verses, however, were judged as equal to Homer. She was contemporary and companion of Sappho.

Test. 16b (Eust. *Il.* 326.43-327.9) ὅτι δὲ καὶ γυναῖκες ἦσαν οὐ μόνον ἄλλως σοφαί, ἀλλὰ καὶ μελοποιοί, οὐ μόνον ἐκ τῆς ῥηθείσης Πραξίλλης δῆλον, ἀλλὰ καὶ Σαπφοῦς καὶ Κορίννης καὶ Ἠρίννης (...) ἰστέον δὲ ὡς ἡ ῥηθεῖσα Ἤριννα Λεσβία μὲν ἦν ἢ Ῥοδία ἢ Τεΐα ἢ Τηλία, ἐκ Τήλου νησιδίου ἐγγίζοντος τῇ Κνίδῳ. ποιήτρια δὲ ἦν καὶ ἔγραψε ποίημα Ἠλακάτην Αἰολίδι γλώσσῃ καὶ Δωρίδι ἐν ἔπεσι τριακοσίοις. ἦν δὲ ἑταίρα Σαπφοῦς καὶ ἐτελεύτησε παρθένος. οἱ δὲ στίχοι αὐτῆς ἐκρίθησαν ἐνάμιλλοι τοῖς Ὁμήρου. ἀπῆλθε δὲ ἐννεακαιδεκέτις. δήλωσις δὲ τῶν ῥηθέντων καὶ τὸ εἰς αὐτὴν ἐπίγραμμα τοῦτο· [test. 7]. καὶ τοῦτο μὲν τοιοῦτον.	The fact that women were skillful, not only in other fields, but also in lyric poetry, can be seen clearly not only from Prassilla mentioned above, but also from Sappho, Corinna and Erinna (...) We must also know that the above-mentioned Erinna was from Lesbos or from Rhodes or from Teos or from Telos – which is a small island near Cnidus. She was a poetess and wrote the poem *Distaff*, written in Aeolic and Doric dialect, in three hundred hexameters. She was also companion of Sappho, and came to the end of her days in her youth. But her verses were regarded as worthy to compete with those of Homer. Yet she left when she was nineteen years old. A demonstration of what has been said, finally, is given also by this epigram for her [T 7]. And that's all.
Fr. 4.5-17]χελύνναν σ]ελάννα· χε]λ̣ύννα· (-)]τ̣ελῆσ[-]ω̣π̣ει· -]α φύλλοις[ ].λᾶσσει· σε]λάνν̣α̣ν ἀμ]νίδα πέξα[(-) -]α̣[.]υ κῦμα[λε]υκᾶν μαινομέν[οισι(ν) π]ο̣σσὶ̣ν̣ ἀφ᾽ ἵ̣[π]πω̣[ν· αἰ]α̣ῖ ε̣.ω μέγ᾽ ἄϋσα· φ[ ]χελύννα ἀλ]λομένα μεγάλας[ ]χορτίον αὐλᾶς·	(...) 'tortoise' (...) moon (...) 'tortoise' (...) accomplish (?) (...) saw (?) (...) leaves (...) shakes (?) (...) moon (...) (you) sheared the ewe lamb (...) deep (?) flood off white mares, by crazy feet, (you jumped down). "Alas, (I'll stop)", I yelled loudly; and 'tortoise' (in my turn), jumping out (I began to chase) (through) the fence of the great court.
Fr. 4.22-27 (...)ἅ τε πὸτ ὄρθρον μάτηρ αε[- (-)].οισιν ἐ̣ρείθοις τ̣ήνασ ἠλθ[-]ν̣α ἀμφ᾽ ἀλί̣π̣αστον· α̣ἰμικραιστ[(-) -]ν φόβον ἄγαγε Μορ̣μ̣ώ -]. ἐν μὲν κο̣[-].ατα· ποσσὶ δὲ φοιτῇ -]..[.]σιν̣· ἐκ δ᾽[ ]μετεβάλλετ᾽ ὀπωπάν·	and towards dawn, (singing with the sweet-talking) workers, mum came (and picked them up and prepared) the salted meat. If the little girls (were to cry), the Bogey would have scared them, who (has large ears on her head), and runs on her four feet and was always changing her aspect.
Fr. 4.37-39 ἐννεα[και]δέκατος .[Ἠρίνν̣α̣ι̣[.]ε φίλαι π.[ἀλακ̣ά̣ταν ἐ̣σορει̣[σα(-)	nineteenth (...) to your Erinna (?) (...) eyes fixed on the distaff

Bibliography

Angelici, M. 1956/1957 Erinna. Diss. (typescript) Bologna.

Arthur, M. B. A. [Katz] 1980 The Tortoise and the Mirror: Erinna PSI 1090. *CW* 74, 53–66.

Balmer, J. 1996 *Classical Women Poets.* Newcastle-upon-Tyne, 57–63.

Barbantani, S.1993 I poeti lirici del canone alessandrino nell'epigrammistica. *Aevum(ant)* 6, 5–97.

Barber, E. J. W. 1991 *Prehistoric Textiles. The Development of Cloth in the Neolithic and Bronze Ages.* Princeton NJ.

Barnard, S. 1978 Hellenistic Women Poets. *CJ* 73, 204–213.

Bérard, C. 1986 La condizione delle donne. In A.A. V.V., *La città delle immagini. Religione e società nella Grecia antica.* Trad. it. Modena (ed. or. L'ordre des femmes. In A.A. V.V., *La cité des images. Religion et société en Grèce antique.* Lausanne–Paris 1984, 79–96).

Bergk, T. 1843[1], 1853[2], (1866-)1867[3], 1882[4] (repr. 1914) *Poetae lyrici Graeci.* Lipsiae, 632–634[1], 702–704[2], 925–929[3], 141–146[4].

Bernabé, A. 1994 Erinna. In A. B. Pajares & Helena Rodríguez Somolinos (eds), *Poetisas griegas.* Madrid, 141–157.

Blümner, H. 1912[2] *Technologie und Terminologie der Gewerbe und Künste bei Griechen und Römern* I. Leipzig–Berlin.

Bowra, C. M. 1921, 1933 Erinna. In J. U. Powell & E. A. Barber (eds), *New Chapters in the History of Greek Literature* I, III. Oxford, I 53–54, III 180–185.

Bowra, C. M. 1936, 1953 Erinna's Lament for Baucis. In A.A. V.V., *Greek Poetry and Life. Essays Presented to G. Murray on His 70th Birthday January 2, 1936.* Oxford 1936, 325–342 (repr. with corrections and additional notes in *Problems in Greek Poetry.* Oxford 1953, 151–168).

Brelich, A. 1969 *Paides e parthenoi* I. Rome.

Bremmer, J. N. 1994 *Greek Religion.* Oxford.

Brock, R. B. 1994 The Labour of Women in Classical Athens. *CQ* n.s. 44, 336–346.

Bulloch, A. W. 1985 *Callimachus: The Fifth Hymn.* Cambridge.

Burkert, W. B. 1985 *Greek Religion.* Oxford.

Cameron, A. & Cameron, A. 1969 Erinna's Distaff. *CQ* n.s. 19, 285–288.

Cartledge, P. 1981 Spartan Wives: Liberation or Licence? *CQ* n.s. 31, 84–105.

Cavallini, E. 1980 *Poetesse greche e romane* [*Erinna*]. Venice–Rome, 77–94.

Cavallini, E. 1991 Due poetesse greche. [Erinna]. In De Martino, 117–135.

Cervelli, M. 1953 Erinna e l'età che fu sua. *Annali dell'Istituto Superiore di Scienze e Lettere "S. Chiara" di Napoli* 4, 195–246.

Chávez Reino, A. L. 2010 Ecos de Teopompo en la Suda. In Vanotti, 207–266.

Collart, P. 1944 La poétesse Erinna. *CRAI* 183–199.

Conti Bizzarro, F. 2001 Note alle testimonianze su Erinna. *Vichiana* s. 4ª 3, 101–106.

Costa, V. 2010 Esichio di Mileto, Johannes Flach e le fonti biografiche della Suda. In Vanotti, 43–55.

Crusius, O. 1907 Erinna. *RE* 6/1, 455–458.

De Martino, F. (ed.) 1991 *Rose di Pieria.* Bari.

Diehl, E. 1925[1], 1935[2] *Anthologia Lyrica.* I/4. Lipsiae, 486–488[1], 207–213[2].

Dorati, M. 1998 Lisistrata e la tessitura. *QUCC* n.s. 58, 41–56.

Dover, K. J. 1993 *Aristophanes. Frogs.* Oxford.

Edmonds, J. M. 1938 Erinna P.S.I. 1090. *Mnemosyne* s. 3ª 6, 195–203.

Eitrem, S. 1932 Moira. *RE* 15/2, 2477–2488.

Flach, H. 1883 *Geschichte der griechischen Lyrik nach den Quellen dargestellt* I. Tübingen (518–522).

Forbes, R. J. 1964[2] *Studies in Ancient Technology* IV. Leiden.

Foxhall, L. 1989 Household, Gender and Property in Classical Athens. *CQ* n.s. 39, 22–44.

Gallini, C. 1963 Katapontismós. *SMSR* 34, 61–90.

Gow, A. S. F. & Page, D. L. 1965 *The Greek Anthology* 1, *Hellenistic Epigrams* I–II. Cambridge (I 97f. II 281–284).

Gow, A. S. F. & Page, D. L. 1968 *The Greek Anthology* 2, *The Garland of Philip and Some Contemporary Epigrams* I–II. Cambridge.

Graf, F. 1993 Initiationsriten in der antiken Mittelmeerwelt. *AU* 36/2, 29–40.

Gutzwiller, K. J. 1998 *Poetic Garlands. Hellenistic Epigrams in Context.* Berkeley CA–Los Angeles CA–London.

Hagborg, N. 1827 Veterum poëtarum in Erinnam, Graecanicam vatem, encomiastica carmina. Diss. Upsaliae 1827. In Tranér 1827, VII 9–16.

Halm-Tisserant, M. 1993 *Cannibalisme et immortalité. L'enfant dans le chaudron en Grèce ancienne*. Paris.

Hirschon, R. (ed.) 1984 *Women and Property: Women as Property*. London–Canberra.

Hunter, R. 1996 *Theocritus and the Archaeology of Greek Poetry.* Cambridge.

Kaldellis, A. 2005 The Works and Days of Hesychios the Illustrios of Miletos. *GRBS* 45, 381–403.

Kassel, R. & Austin, C. *Poetae Comici Graeci* (*PCG*) I, II, III/2, IV, V, VI/2, VII, VIII, Berolini-Novi Eboraci 1983 (IV), 1984 (III/2), 1986 (V), 1989 (VII), 1991 (II), 1995 (VIII), 1998 (VI/2), 2001 (I).

Khan, H. A. 1995/1996 The Robe and the Crown: Medea's Celebration of Jason's Wedding. *MCr* 30/31, 119–125.

Lapini, W. 1995 I frammenti alcibiadei di Crizia: Crizia amico di Alcibiade? *Prometheus* 21, 1–14, 111–130.

Lardinois, A. & McClure, L. (eds) 2001 *Making Silence Speak: Womens' Voices in Ancient Greek Literature and Society*. Princeton.

Latte, K. 1953 Erinna. *NAWG* 79–94 (= *Kleine Schriften zu Religion, Recht, Literatur und Sprache der Griechen und Römer*. München 1968, 508–525).

Lefkowitz, M. R. & Fant, M. B. (eds) 1992[2] *Women's Life in Greece & Rome*. London (1982[1]).

Levin, D. N. 1962 Quaestiones Erinneanae. *HSPh* 62, 193–204.

Lloyd-Jones, H. & Parsons, P. J. 1983 *Supplementum Hellenisticum.* Berolini–Novi Eboraci.

Lonsdale, S. H. 1993 *Dance and Ritual Play in Greek Religion.* Baltimore MD–London.

Maas, P. 1929 Erinna. *DLZ* n.s. 6, 116f.

Maas, P. 1934 Erinnae in Baucidem nenia. *Hermes* 59, 206–209.

Maas, P. 1935 Erinna. *RE* Suppl. 6, 54–56.

Magnelli, E. 1999 *Alexandri Aetoli testimonia et fragmenta.* Florence.

Magnelli, E. 2004 App. Anth. II 400 C.: Oppiano come Erinna? *Prometheus* 30/3, 269–274.

Michelazzo Magrini, M. 1975 Una nuova linea interpretativa della Conocchia di Erinna. *Prometheus* 1, 225–236.

Möller, W. O. 1976 *The Wool Trade of Ancient Pompeii*. Leiden.

Moreau, A. 1992 Initiation en Grèce antique. *DHA* 18, 191–244.

Nasta, L. 1995/1996 Erinna. Diss. (typescript) Salerno.

Neri, C. 1998 Cambio di ritmo (Erinna: SH 401,14f. ~ Carm. pop. fr. 30(c) PMG 876). *Prometheus* 24, 19–24.

Neri, C. 2002 L'agnella e il rasoio (Erinna SH 401,13). In María José García Soler (ed.), *Τιμῆς χάριν. Homenaje al Profesor P. A. Gainzarain*. Vitoria-Gasteiz, 61–68.

Neri, C. 2003 *Erinna. Testimonianze e frammenti*. Bologna.

Neri, C. 2009 Die weißen, sanftredenden Schläfen (SH 401, 46 = Herinn. fr. 4,46 N.). *RhM* n.F. 152, 113–123.

Nosch, M.-L. 2010 In M. Del Freo *et al.*, The Terminology of Textiles in the Linear B Tablets, including Some Considerations on Linear A Logograms and Abbreviations. In Cécile Michel & M.-L. N. (eds), *Textile Terminologies in the Ancient Near East and Mediterranean from the Third to the First Millennia BC*. Oxford, 338–373.

Orvieto, A. 1929 Erinna e Bauci. Una poetessa diciannovenne di 23 secoli fa. *Il Marzocco* 34/8 (24.2.1929), 1.

Page, D. L. 1950[3] *Greek Literary Papyri* I. London–Cambridge, MA. (120–123) (= *Select Papyri* III, London-Cambridge MA 1950[3] [486–489], quoted here; 1941[1], 1942[2]).

Page, D. L. 1981 *Further Greek Epigrams*. Cambridge.

Pfeiffer, R. 1973 *Storia della filologia classica dalle origini alla fine dell'età ellenistica*: Ed. it. a c. di M. Gigante & S. Cerasuolo. Napoli 1973 (ed. or. *History of the Classical Scholarship. From the Beginnings to the End of the Hellenistic Age.* Oxford 1968).

Pomeroy, S. B. 1978 Supplementary Notes on Erinna. *ZPE* 32, 17–22.

Pomeroy, S. B. *et al.* 1994 Women in the Greek World. In Elaine Fantham *et al.* (eds), *Women in the Classical World*. New York–Oxford (164f.).

Preger, T. 1891 *Inscriptiones Graecae Metricae ex Scriptoribus Praeter Anthologiam Collectae*. Lipsiae.

Rauk, J. 1989 Erinna's Distaff and Sappho Fr.94. *GRBS* 30, 101–116.

Rutherford, I. 2001 *Pindar's Paeans.* Oxford.

Schamp, J. 1987 = J. S., *Photios historien des lettres. La Bibliothèque et ses notices biographiques*. Paris.

Schaps, D. M. 1975 Women in Greek Inheritance Law. *CW* n.s. 25, 53–57.

Schaps, D. M. 1977 The Woman Least Mentioned. *CQ* n.s. 27, 323–330.
Schaps, D. M. 1979 *Economic Rights of Women in Ancient Greece*. Edinburgh.
Scheidweiler, F. 1956 Erinnas Klage um Baucis. *Philologus* 100, 40–51.
Scholz, U. W. 1973 Erinna. *A&A* 18, 15–40.
Skinner, M. B. 1982 Briseis, the Troian Women and Erinna. *CW* 75, 273–299.
Ste. Croix, G. E. M. de 1970 Some Observations on the Property Rights of Athenian Women. *CR* n.s. 84, 273–278.
Steen Due, O. 1962 The Meaning of the Homeric Formula χρυσηλάκατος κελαδεινή. *C&M* 26, 1–9.
Stehle, E. 2001 The Good Daughter. Mother's Tutelage in Erinna's Distaff and Fourth-Century Epitaphs. In Lardinois & McClure, 179–200.
Stella, L. A. 1929 Intorno ai nuovi frammenti di Erinna. *RIL* 62, 827–838.
Szepessy, T. 1972 The Story of the Girl Who Died on the Day of Her Wedding. *AAntHung* 20, 341–357.
Thompson, W. 1982 Weaving: A Man's Work. *CW* 75, 217–222.
Tosi, R. 2011 *I carmi greci di Clotilde Tambroni*. Bologna.
Tranér, J. 1827 *Novem Graecarum poëtridum, quae supersunt, carmina*. Upsaliae (VII: *Erinnae, Graecanicae vatis, quae exstant residua*).
Vanotti, G. (ed.) 2010 *Il lessico Suda e gli storici greci in frammenti. Atti dell'incontro internazionale. Vercelli, 6–7 novembre 2008*. Rome.
Vara Donado, J. 1973 Obra de Erinna y algunas reconstrucciones textuales. *Habis* 4, 41–79.
Vara Donado, J. 1980 Demos a Erinna lo que de Erinna es. *SPhS* 5, 303–308.
Veniero, A. 1905 *I poeti de l'Antologia Palatina* (…) I/1 (…). Ascoli Piceno.
Versnel, H. S. 1990 *Inconsistencies in Greek and Roman Religion* II. Leiden.
Vitelli, G. 1928, 1929 Frammenti della Conocchia di Erinna. *BSAA* 24 (1928), 9–16 (repr. with corrections and additions in *PSI* IX 1090 [1929], XI, XIII-XIV, 137–144).
Vysoký, Z. K. 1942 Erinna. *LF* 69, 85–113.
Wendel, C. 1949 *Die griechisch-römische Buchschreibung verglichen mit der des vorderen Orients*. Halle.
West, M. L. 1977 Erinna. *ZPE* 25, 95–119.
West, M. L. 1996 *Die griechische Dichterin. Bild und Rolle* (Lectio Teubneriana 5 [29.3.1996]). Stuttgart–Leipzig, 24–26, 43.
Whitehorne, J. 1995 Women's Work in Theocritus, Idyll 15. *Hermes* 123, 63–75.
Wilamowitz, U v. W.-Moellendorff 1913 *Sappho und Simonides*. Berlin, 228–230.
Wolff, H. J. 1944 Marriage Law and Family Organization in Ancient Athens. *Traditio* 2, 43–95.
Zel'čenko, V. V. 2003 Le silence d'Erinna (AP II, 108–110). *Hyperboreus* 9/1, 37–43.
Zel'čenko, V. V. 2005 review of Neri 2003. *Hyperboreus* 11, 149–167.

Further bibliography on Erinna (items not present in Neri 2003)

Allroggen-Bedel, A. 2008 Frauen unter den "homines illustri": eine Gruppe griechischer Dichterinnen in der Villa Albani. *Hephaistos* 26, 71–91.

Bowman, L. 2004 The "women's tradition" in Greek poetry. *Phoenix* 58, 1–27.

Braungart, W. 2007 Tod und Kunst, Geist und Bewußtsein: Zu Eduard Mörikes "Erinna an Sappho". *Oxford German Studies* 36, 76–96.

Capellà, M. C. 2004 *Poetes gregues antigues.* Barcelona, 107–129.

Downes, J. M. 2010 *The Female Homer. An Exploration of Women's Epic Poetry*. Newark NJ.

Falchi, F. M. 2009 Quasimodo traduttore di Erinna, Lamento a Bauci. *RPL* 32, 161–179.

Greene, E. 2005 *Women Poets in Ancient Greece and Rome*. Norman OK.

Gutzwiller, K. J. 2005 Genre development and gendered voices in Erinna and Nossis. In Greene, 202–222.

Keith, A. 2006 Critical trends in interpreting Sulpicia. *CW* 100, 3–10.

Kost, K. 2005 Erinna. In H. H. Schmitt & E. Vogt (eds), *Lexikon des Hellenismus*. Wiesbaden, 305f.

Levaniouk, Olga 2008 Lament and Hymenaios in Erinna's Distaff. in Ann Suter (ed.), *Lament. Studies in the Ancient Mediterranean and Beyond*. Oxford, 200–232.

Magnelli, E. 2004 (see above).

Manwell, E. 2005 dico ergo sum. Erinna's Voice and Poetic Reality. In Greene, 72–90.

Mayor Ferrándiz, T. M. 2000 Erina, el llanto por el paraíso perdido. *Epos* 16, 395–405.

Murray, J. & Rowland, J.M. 2007 Gendered Voices in Hellenistic Epigram. In P. Bing & J. S. Bruss (eds), *Brill's Companion to Hellenistic Epigram Down to Philip*. Leiden, 211–232.

Neri, C. 2009 (see above).

Rayor, D. 2005 The Power of Memory in Erinna and Sappho. In Greene, 59–71.

Rowland, J. M. 2012 Footnotes to Sappho. An examination of the female poets of Greece. Diss. Michigan 2012.

Sens, A. 2003 Asclepiades, Erinna and the Poetics of Labor. In P. Thibodeau & H. Haskell (eds), *Being There Together. Essays in Honor of M.C.J. Putnam on the Occasion of his Seventieth Birthday*. Afton MN, 78–87.

Spanoudakis, K. 2007 review of Neri 2003. *Gnomon* 79, 204–209.

Torné, R. T. 2014 La mort i l'inframón en els lírics grecs. In J. Redondo & T. Ramon (eds), *Apocalipsi, catàbasi i mil·lenarisme a les literatures antigues i la seua recepció.* Amsterdam, 293–305.

Vos, M. J. de 2012 Negen aardse muzen. Gender en de receptie van dichteressen in het oude Griekenland en Rome. Diss. Nijmegen 2012.

Waack-Erdmann, K. 2011 Erinna – Wer ist denn das? *AU* 54, 20–23.

Zel'čenko, V. V. 2003, 2005 (see above).

12. Textile and Textual Poetics in Context: Callimachus' 4th *Iamb* and Theocritus' *Idyll* 28

Maria Papadopoulou[*]

331 BC is a landmark year in the long history of Egypt, which itself goes back well into prehistoric times. It is the year Alexander of Macedonia took a decisive step towards establishing his own empire: following his visit to Memphis, where he was crowned Pharaoh,[1] he decided to found a city on the Mediterranean coast of Egypt across from the Island of Pharos (which is Greek for 'lighthouse').[2] According to Alexander's historian, Arrian, and his biographer, Plutarch, it was Homer who appeared to him in his sleep and inspired him to build a city there. This city, "the capital city of Asiatic Europe, if such a thing could exist" as Lawrence Durrell aptly put it, would become a cultural beacon projecting its light onto Western civilization.[3]

After his death, Alexander's body was brought to Egypt by the first Ptolemy, Sotēr, the founder of the legendary Library and the *Mouseion*.[4] The intellectual activity at Alexandria started under Ptolemy Sotēr and was continued by his successor, Ptolemy II Philadelphus. The Ptolemies' focus on Alexander, as well as other divine parentages, was important for both internal and foreign policy: it provided legitimacy and continuity with the pharaonic past and authoritative precedence over the other *epigonoi*. Born a human but deified before his death, Alexander became a hero in both literature and legend. The Library and the *Mouseion*, both Ptolemaic institutions situated within the palace complex, not only provided a cultural link to Greek heritage but were also politically significant. Ptolemaic ambitions for political dominance came hand–in–hand with their aspirations to cultural superiority. Religious as well as literary manifestations provided the cultural space that served as a vehicle for the political ideological

[*] I am grateful to Professors Marie-Louise Nosch and Mary Harlow, and to Dr Giovanni Fanfani for their insightful comments and unfailing support.

[1] During his lifetime Alexander had considered himself a descendant of the Homeric heroes. Soon after entering Egypt he received the typical fivefold title of the pharaohs consisting of 1) The Horus name; 2) The Nebti name; 3) The Golden Horus name; 4) The Throne name; and 5) The Birth name.

[2] The Pharos lighthouse, one of the Seven Wonders of the ancient world, was designed by the Rhodian Sostratos from Knidos and erected in the 3rd century BC; it stood there until the 14th century AD, when it was destroyed by an earthquake.

[3] Durrell 1961, 130.

[4] The *Mouseion* was the closest ancient parallel to a modern university. Until the time of the Ptolemies the word '*Mouseion*' literally meant 'shrine of the Muses'. In emulation of the peripatetics, the first two Ptolemies dedicated it to the Muses, not only because these were the deities protecting the cultivation of arts and sciences, but also because Aristotle's *Lykeion* at Athens had a shrine of the Muses. See Erskine 1995, 40.

programme of the Ptolemies to ascertain their control over Egypt and widen their sphere of influence in the rest of the world. Science, too, especially astronomy served the dynastic programme of the Ptolemies.[5] The influence of this political agenda on the literary products of the period cannot be overlooked.

Spinning the yarn of a cultural empire: Callimachus and Theocritus in Ptolemaic Alexandria[6]

Alexandrian scholarship aimed at continuing and reinterpreting the literary and scientific advancements achieved in the classical period. Scholars from all over the Greek world swarmed to Alexandria; they saw it as their major task to hand down the classical heritage to posterity, not by simply collecting and copying works, but by subjecting these texts to linguistic, philological and literary criticism. Two of the most learned representatives of the Alexandrian school of poetry were Callimachus and Theocritus, who constantly invited their readers to delve into their multi-layered poetic discourse with equal erudition.

Greek poets conceived of poetic composition as analogous to the process of producing cloth from yarn.[7] The metaphor of weaving poetry is found in most Indo-European languages.[8] While the classical theory of metaphor views it as a figure of speech, *i.e.* a linguistic/poetic device, the advocates of metaphor as a cognitive device view it as a conceptual process that shapes both language and thought. According to this view, metaphor is perceived as a kind of semantic 'abnormality' in linguistic usage, which requires an alternative mapping of meaning outside the typical domain of reference. To illustrate: what is typically conceived as spun is yarn and as woven is cloth. If a poem is described in terms of spinning or weaving, which is often the case in Greek literature, this mapping is metaphorical and leads to conceiving texts in terms of textiles. In both Greek and Latin, the words used to describe the process of turning yarn and cloth have cognitive contiguity to the process of producing a story, often in verse, *i.e.* a poem. The semantic contiguity of the terms *text* and *textile* in modern English owes this metaphorical mapping to Latin. It should be noted that, while in modern English there is no apparent semantical contiguity between the terms *text* and *textile*, their etymology can be traced back to the same latin word (*textus*) which is a calque from Greek *huphos*, which denotes both 'weaving' and 'literary style'.[9] In the cognitive theory of metaphor, inaugurated by Lakoff & Johnson (1980), metaphors are not exclusive to the realm of language, namely literary language, as purported in the classical theory of metaphor, which goes back to Aristotle; they shape thought and everyday linguistic interactions to map one conceptual

[5] *Cf.* the naming of the new star after the queen Berenike II of Cyrene, who got married to Ptolemy III Euergetes in 246 BC. See Gutzwiller 1992; Clayman 2011.

[6] All the passages quoted from Callimachus are from Pfeiffer (in two volumes) Oxford 1949, 1953. The *scholia* to Callimachus are poorly preserved; the *diegeseis* (summaries) of the poems are preserved in papyri dated to the Roman period. All the passages from Theocritus are from Gow (in two volumes) Cambridge 1952 and all reference to the *scholia* are from Wendel 1914(= TLG). All the translations are my own, unless otherwise stated.

[7] Wool was probably the most popular fibre in Greece during the 1st millennium; it was warm, soft and easy to dye in vivid colours. Both Sicily and Miletus, Theocritus' and Theugenis' homelands respectively were famous centres of wool trade.

[8] Indo-Iranian, Avestan, Greek, Celtic, Germanic, Russian, Irish are just a few of them, see West 2007. Textile metaphors shape the concepts of literary composition also in English, e.g. "to spin a yarn" metaphorizes the actual spun fibre to a story; to "weave x into a story" means "to compose a story".

[9] *Textus* (*i.e.* 'woven') is the past participle of the Latin verb *texere* ('to weave'), *cf.* Lewis & Short *s.v.*

domain onto another, seemingly unrelated, domain.[10] Terms such as *rhapsōdeō* 'to stitch a song', (LSJ *s.v.* ῥαψῳδέω; hence *rhapsōdos* 'song-stitcher') and expressions such as 'to weave a song/ hymn' (see LSJ *s.v.* ὑφαίνω III.2, ὕφος 'web' as in 'spider web' and 'web of words'; πλέκω II.2, πλοκή 'plot of a story', also 'secret scheme or plan') show that there is a systematic conceptual link between textile production and poetic composition manifest in the Greek language.[11]

Textiles had been central to Egyptian economy since dynastic times.[12] In Ptolemaic times two important textile traditions intersect in Alexandria. Nevertheless, the role of textile production in relation to textile imagery in Alexandrian poetry has been largely overlooked. My hypothesis that textile imagery plays a prominent role in the major works of Alexandrian poets is here exemplified in Callimachus' *Iamb* 4 and Theocritus' *Idyll* 28. In this chapter I will focus on the role of textile imagery in the content and structure of these two poems. What I hope to show is that, while textile imagery functions differently in each poem (it is subtle in Callimachus – it is central in Theocritus), it remains a major tool used in order to serve the poet's purpose in each poem.

On how (not) to dress for an iamb: the laurel-olive quarrel and the *exōmis* insult (Call. 194.22ff. Pf.)

The details of Callimachus' life facts are contradictory or completely unknown to us.[13] He is usually dated between 320–240 BC, a period coinciding with the golden era of the Ptolemaic age, the reign of Ptolemy II Philadelphus. What is certain is that his family came from Cyrene, a colony of the island of Thera in Libya and that he moved to Alexandria where he became one of the most prominent intellectuals of his time. He frequented the Ptolemaic court and was assigned the project of cataloguing the Library holdings.[14] Being a poet who cultivated all sorts of poetic genres (hymns, aetia, elegies, epyllia, iambs and epigrams), a critic, and a grammarian, he also composed prose works.[15] According to Callimachus, Apollo himself gave him precepts on what his poetry should be like: thin (*leptos*), not overweight, following untrodden paths, not taking the road where heavy carts travel, singing like a cicada not an ass (fr. 1.22–30 Pf.).[16] In his famous elegy in honour of queen Berenice celebrating the queen's chariot victory in the Nemean games (*Aetia* book 3 fr. 254–69 *SH* = 383.15 Pf.) Callimachus refers to the textiles woven by Egyptian women with the adjective *leptaleos* ('very fine', 'of very fine quality').[17]

[10] Lakoff & Johnson 1980 introduced a powerful theory for the study of the conceptual role of metaphor. This theory has been further developed by Johnson 1987, Lakoff 1987, Lakoff & Turner 1989, Gibbs 1990, 1992, Kövecses 2002. "Metaphor is a cross-domain mapping in the conceptual system" Lakoff 1993, 203. See also Tay 2014. *Cf.* Aristotle's definition of metaphor in his *Poetics* 1457b7 transl. Loeb "Metaphor is the application of a strange term either transferred from the genus and applied to the species or from the species and applied to the genus, or from one species to another or else by analogy." See also Guilleux, this volume.

[11] McIntosh Snyder 1981; Nagy 2002; Tuck 2009; Nosch 2014; Fanfani & Harlizius-Klück (this volume).

[12] See Allgrove-MacDowell 2003, 30–37; Wild 2003, 48–51.

[13] Callimachus' self-presentation in his epigram (ep. 21 Pf.) is noteworthy: he presents his father as a man who excelled in arms and himself as the singer of songs louder than envy. "It's only fair, he concludes, for the Muses not to abandon in old age someone whom they favoured since his younger years."

[14] He produced the famous *Pinakes* ('Tablets') (*c.* 260 BC), a bibliographical survey and catalogue of the titles of the Library books consisting of 120 books. See Blum 1991; Regenbogen 1950.

[15] Krevans 2004.

[16] The Callimachean predilection for Aesopic fable is evident here, *cf.* fable 184 in the Perry index (featuring the cicada and the ass).

[17] Pfeiffer's suggests that Callimachus refers to a τελαμών (a kind of hand-woven band or belt); see further discussion in Thomas 1983, 106–107.

Egyptian flax and textiles were renowned for their excellent quality; *leptotēs* was indeed a key concept in Callimachus' poetics. The political agenda dictated by patronage was not the only difficulty Callimachus had to overcome to please both his divine and royal patrons. He was also faced with the double challenge of coming to terms with the present by alluding to the past. Fulfilling both goals at once appears to have been Callimachus' favorite pursuit.

Although recognized as the sequel to classical literature, Hellenistic poetry often emphasizes the prequel, the *aetiology*, the reason before the fact, the beginnings of a story or a custom, the least known version of a myth. Literary quarreling is the *Leitmotiv* of the whole collection of *Iambi*, the aim paradoxically being to make those involved put an end to their disputes. The book begins with Hipponax[18] coming back from the dead. It moves on to Aesopic-like allegorical style fused with the abusive, coarse vocabulary of the iamb, while it retains the Callimachean predilection for *leptotēs*. Aesop and Hipponax are picked as authoritative figures in their respective genres, the fable and the iamb. Callimachus surpasses himself in Hipponactean and Aesopic devices: didactism and derision on the one hand, plant and animal voices on the other, form a Callimachean eclectic blend of archaic-inspired material reshaped and manipulated according to the poet's needs. Animal and plant fable flourished not only in Greek, but also in Babylonian and Egyptian literature.[19] In Acosta-Hughes' terms (2002, 198) "Callimachus has incorporated into this traditional and popular structure a debate on aesthetics that is scholarly, detailed, and poetically self-referential." This self-referentiality is Callimachus' innovative addition. At the same time Callimachus is winking at Hipponax while recounting his Aesopic *exemplum*. This is apparent not only in the diction,[20] but also in the satire and venomous acuteness.

In the first iamb Callimachus resurrects Hipponax of Ephesus from the dead not for him to face old enemies, but to address Callimachus' *ephamilloi* ('rivals', 'adversaries'). Callimachus' *Iamb* 4, is a semi-decipherable text combining the traditions of the iamb and the fable.[21] In *Iamb* 4 the invective meets the admonishing poet, the polemicist meets the moralist, and all of them meet the fabulist.[22] It had not been Callimachus' idea to introduce Aesopic fable to the circles of the *Mouseion*,[23] he cannot get the credit for the use of choliambic metres for his Aesopic material either.[24] His revisionism is apparent in the admirable way he handled his material. The Callimachean

[18] Hipponax of Ephesus is an archaic iambic poet and the inventor of *choliamb* or *scazōn*. Callimachus claims to revive his style. The proverbial animosity between Hipponax and the Chian sculptor Bupalus is a point of reference for Callimachus (*Iamb* 4.3–4), even though scholars agree that Callimachean Hipponax is a lot more subtle, *contra* Rosen 2007, 177 who believes that Callimachus is no less invective than Hipponax.

[19] On animal fable in Egyptian literature see Lichtheim 2006, 157. On the similarities with a Babylonian tale recounting a dispute between two trees West 1969, 119.

[20] The more usual feminine form ὕδρα (= water serpent, feminine noun) is supplanted by the masculine ὕδρος (= water serpent, masculine noun), a Homeric *hapax* (*Il.* 2.723) also used by Hipponax (fr. 102.10 W = 105.10 D).

[21] On the content and structure of this iamb see Kerkhecker 1999, 111–115; Edmunds 2001; Acosta-Hughes 2002, 190–204; Scodel 2011, 368–383; Giuseppetti 2012, 10.

[22] Gutzwiller 1991: 148: Hesiodic Callimachus who prefers shorter poems and chooses didactic over cyclic epics is Hipponactean in his iambics and Aesopicus in resorting to *fabula* in this iamb.

[23] Demetrius of Phaleron compiled a collection of fables assigning them to Aesop around 300 BC (Perry 1962).

[24] Adrados 1999, xv: "The Cynics (…) translated [the fable] into choliambic metre, they partially modified its topics to bring it into line with their philosophy and they collected many new fables. For them the fable was a weapon for instruction and attack at the same time, a mixture of the serious and the joke".

choliambic *ainos*[25] is modelled on the Aesop fable about a pomegranate and an apple tree;[26] Callimachus chooses new protagonists (the laurel and the olive), except for the bramble,[27] which provides the linking thread between the archaic fable winning a well-deserved place in the Aesopic corpus and the Hellenistic meta-fable. Callimachus does not simply narrate the *ainos*, he provides a wider narrative into which the *fabula* is embedded.

The Callimachean skill in the handling of the poet's multiple *personae* is revealed in all its magnitude. Two speaking shrubs, the olive and the laurel, are arguing: each one aims to prove her superiority over the other, just like the quarreling Alexandrian *Mouseion* scholars; a third one, the bramble, intervenes to restore the peace. While weaving his poetic concerns into his poetic fabric, Callimachus distances his own voice, not assuming responsibility for what the 'characters' of his story say. It would be pointless to try to uncover the poet behind the mask of the boastful laurel or the one behind the mask of the self-effacing, yet 'dissembler' olive, who puts her own praise into the mouth of birds, or of the bramble who is told off by the angry laurel at the end of the extant poem.[28] Ironically, reconciliation being the aim, the actual *agōn*,[29] the longest surviving part of the iamb, is but the bickering between the laurel and the olive with regard to their worth as trees; and it ends in rather an abusive than a reconciliatory way.

The narrative, more extensively discussed in the *diēgēsis*, sets the dialogue between talking shrubs within a narrative addressed to a certain Simos (probably a derogatory *nomen loquens*)[30] as an instructive parable, which aims to teach Simos not to interrupt people who are not his equals. Callimachus (or rather the poet's *persona*) was debating with someone he considered of equal status, when Simos interrupted them. After rebuking him the poet narrates the story of laurel's quarrel for supremacy over olive at mount Tmōlus of Lydia:[31] the laurel attacks the olive first. The text is badly damaged (vv. 1–20); about 9 verses are missing. The olive's self-characterization as *phaulē* (v. 12) and her false modesty are in strong contrast to the blatant arrogance of the laurel. Now, it is the laurel who takes the stand. The beginning of her speech is lost, but what remains of it seems to me to be the centrepiece of the *ainos* and the iamb. The abusive attack on the olive reaches a climax at verses 22–23, when the laurel launches a double insult against the olive's external appearance:

> ὡριστερὸς μὲν λευκὸς ὡς ὕδρου γαστήρ,
> ὁ δ' ἡλιοπλὴξ ὃς τὰ [π]ολλὰ γυμνοῦται.

[25] *Ainos* as 'story' appears in Homer, Hesiod, Archilochus; *ainos* as 'proverb' in Euripides, Theocritus XVI, 43. Callimachus calls the exchange of insults between the laurel and the olive *mythos* (*Iamb* 4.6) and Babrius, his Roman successor, calls his collection of Aesopic fables *mythiambi*. On the meanings of *ainos* in ancient Greek literature see Van Dijk 1997.

[26] Number 213 in the Perry Index.

[27] Constan & Landry 2008, 47 argue for the *persona* of the poet represented by the bramble in this iamb.

[28] See Lelli (2004, 51–57) on the difficulties encountered by scholars in their effort to interpret *Iamb* 4 as alluding to a literary quarrel and unmask its protagonists.

[29] Gutzwiller sees in Theocritus' *Idyll* 5 (the musical *agōn* between two herds, Lacon and Comatas, with a third one as their judge) a close parallel to *Iamb* 4; Gutzwiller 1991, 146. The term brings classical drama to mind as well as the Platonic dialogues.

[30] In *Iamb* 1.33 Callimachus uses a play on words with the meaning snub-nose of Simias' name asking his interlocutor not to snub his nose. This not only demonstrates the unity the book of *Iambi*; it also proves that every single detail in it fits the Callimachean meta-poetic programme.

[31] Tmōlus was the king of Lydia. Callimachus alludes to the circumstances of Tmōlus' death (v. 101) and conveniently sets this contest on the mountain which bore his name. His metamorphosis is told by Ovid (*Metamorphoses* XI). Tmōlus is also linked to the old veneration of Dionysus, another patron god of the Ptolemies (Eur. *Bacch.* 424–464).

the left one (*sc.* shoulder) is white just like the belly of the water-snake hydros,
the right one is sunburnt as it is often exposed (*sc.* to the sun).

An example of Callimachean attention to *leptotēs* in imagery is the fact that the dichromy of the slave's shoulders and the dichromy of the olive's leaf (dark greyish-green above and hoary beneath) are not of the same hue; they were both attributed to sunburn by the Greeks. Also, the *exōmis* was considered *phaulē* (φαύλη, of low quality). This is exploited by the poet in making a play on words with *phaulia* (φαυλία, coarse), a type of olive mentioned in Callimachus' epyllion *Hecalē* as a sign of poverty.[32] Scholarship generally accepts that the garment indirectly referred to by Callimachus is the *exōmis*. The indirect reference to this garment was first suggested by Gilbert Murray in P. Oxy. 1011, 75.[33] D'Alessio suggests that the garment alluded to is ἐπωμίς, which appears to be a synonym of *exōmis* and denotes the same slave garment.[34]

The *exōmis*, as the name suggests (< Gr. *ex-* 'off' + *ōmos* = shoulder; 'off one shoulder'), was a type of tunic leaving one shoulder bare. Xenophon (*Mem.* 2.7.5) associates it with artisans; in Pollux' *Onomasticon* it is described as χιτὼν λευκὸς ἄσημος, κατὰ τὴν ἀριστερὰν πλευρὰν ῥαφὴν οὐκ ἔχων, ἄγναπτος ("a white *chitōn* without any in-woven patterns, with no seams on the left side, not fulled") (4.118.7). Pollux,[35] Pausanias[36] and Hesychius *(s.v.)* define it as the garment of slaves. It may be also identified with the one-shoulder short tunic worn by women athletes:[37] e.g. a bronze statuette of a female runner from the British Museum[38] in a short *chitōn* pinned on the left shoulder, passing under her right arm to leave the right breast exposed. The *exōmis* was not only very practical for athletes,[39] men of the military, workmen, and slaves, but also readily visible; the *exōmis* along with different types of garment is used for comic effect and mockery in Aristophanic comedy;[40] it was a cheap light garment associated with the lower classes;[41] its association with the status of slaves may have been stronger in Hellenistic and Roman times than in Classical and Archaic times; so the laurel's innuendo was able to hit home immediately.

---

[32] Call. in *Hecalē* (Pf. 334 = *SH* 283) mentions a πίτυρις kind of olive (LSJ πιτυρίς) which Philemon identifies with φαυλία (in Athenaeus' *Deipnosophists* 2.47).

[33] Hunt 1910. See also Waites 1912, 38; Clayman 1980, 24 n. 36.

[34] D'Alessio 1996: 605 n. 69: "La foglia dell' olivo è descritta in termini sprezzanti dall' alloro, come se essa fosse rivestita da una ἐπωμίς, la veste che lasciava scoperta la spalla destra, tipica degli schiavi, o comunque delle classi inferior". See LSJ *s.v.* ἐπωμίς.

[35] Pollux *Onomasticon* 4.119 (2nd century AD) ἑτερομάσχαλος· χιτὼν δουλικός, ἣν ἐξωμίδα λέγουσιν. "Heteromaschalos; a type of chitōn for slaves, which is called exōmis." The etymology of the compound *heteromaschalos* shows that it came over only one shoulder as opposed to ἀμφιμάσχαλος, which had openings for both arms. See also 7.47.

[36] Pausanias the grammarian *Collection of Attic words* (*Ἀττικῶν ὀνομάτων συναγωγή*) E 46, 4: ἦν δὲ τῶν δούλων χιτών – ὁ δὲ ἀμφιμάσχαλος τῶν ἐλευθέρων καὶ ἀστῶν – οὐκ ἐπισκεπάζων τοὺς βραχίονας. οἱ δὲ νεώτεροι καὶ θοιμάτιον ἐξωμίδα τὸ μικρὸν καὶ εὐτελές. "This one was the chitōn of slaves – that one was the sleeveless chiton of free men and citizens – it did not cover the arms. The younger ones also wore the garment exōmis, short and cheap".

[37] Serwint 1993.

[38] British Museum, inv. 208; ht. 0.11 m. A red-figure *pelikē* by the Pan Painter showing a fisherman wearing the *exōmis* (Vienna 3727; ARV 555.88). It also appears on an Attic red-figure *pyxis* by the Wedding Painter, *c.* 450 BC, illustrating a scene from Aeschylus' *Diktyoulkoi* (Princeton, C. Clairmont Collection; ARV2 924.35). For a modern reconstruction see: Oakley 1982, 110–113 and 116 – Plate 12.

[39] The *exōmis* was also associated with the *apobates*, *i.e.* riders who took part in a special contest during the Panathenaic festival. (Plaque B XXV of the Parthenon frieze).

[40] Aristoph. *Lysistrata* 662–664; 1020–21; *Ecclesiazusae* 267.

[41] Plutarch, *On tranquility of mind* 470F5, Dio Chrysostom, *Orations* 7.32.3.

The laurel, which is already shaking with anger (vv. 9–10), puts the olive's inferiority into relief by means of the double imagery of the snake and of the slave, piquing on the bichromy of the colour of the olive's leaves. Callimachus expresses the idea of bichromy one time in reference to the snake and one time allusively (of the slave's shoulders). The former is literal, the latter is allusive; this is precisely why it achieves to aptly combine iambic vituperation with the subtlety of Callimachean *leptotēs*. The derogation of the olive associates her colour (again) with low descent, character, and morals: let us note that according to tradition Aesop was a slave, but neither Aesop, nor (his?) slave *exōmis* are mentioned anywhere in the extant poem. The laurel's jeer encapsulates Callimachean conjunction of bluntness with allusiveness.

The laurel's derision of the olive reaches new peaks: she mocks the olive for the difference in colour between the two sides of her leaves. She thus creates an analogy between the olive leaf and a slave's shoulders which were different in colour, too: the right shoulder was always exposed to the sun, whereas the left shoulder was covered by the *exōmis*, a garment usually worn by slaves and workmen, specially designed so as to permit full use of the right hand. The distance between the two plants is magnified and maximised by being both literal (*i.e.* the imagery of the crawling snake) and metaphorical (*i.e.* the imagery of the slave). As the snake holds the lowest possible position among fauna, and the slave holds the lowest possible rank among people, so does the olive rank the most inferior among plants. The inferiority of the olive is spatially represented, both literally (up *vs.* down) and metaphorically (high *vs.* low social standing) on the vertical axis and on the horizontal axis (left *vs.* right). The slave imagery of lowness becomes even more intense as it unfolds on two axes: literally, on the horizontal axis and metaphorically, on the vertical axis. An extra interpretative layer is that the slave may allude to the persona of Aesop, the well-known fabulist and Callimachus' literary model in his book of *Iambs*.[42] The laurel accuses the olive of slavishness. To that the olive retorts an equally indirect and sly praising of herself by means of a second *ainos* embedded within the first one. The second *ainos* presents two birds, each taking the sides of each of the trees, who finally, declare the victory of the olive.

The *exōmis* allusion is the climax of the laurel's effort to underline her own elevated position. No other abuse of the olive or self-praise can take the laurel closer to achieving her goal. The height of her argument is reached through this slave slander of the olive's external appearance. The exchange of insults between plants involves an animal and a human. No other comment to vex the olive can be more powerful and more self-referential, precisely due to its anthropomorphism. While the olive is compared to a snake and a slave, the laurel is a sacred plant: she attends sacred rites at every home, or shrine, or temple, or in the Delphic oracle, or in every city in need of purification ever since Bacchus purged the people of Miletus using her twigs. There is subtlety and complexity even in the grand aesthetics represented by the laurel, which is typically associated with Apollo.[43] And the most complex, subtle and hence most powerful of all is the imagery of the mark the *exōmis* leaves on the body of the slave.

[42] Acosta-Hughes 2002, 201.
[43] Acosta-Hughes 2002, 191.

Weaving poetry into praise: the *Distaff* of Theocritus[44]

Theocritus is an emblematic figure of pastoral poetry. Little of his life is known with certainty; some autobiographical information is inferred through his poems, but most of it is conjectures, not solid fact.[45] In the rest of this section I will attempt a close reading of Theocritus' 28th *Idyll* entitled *Distaff* (*Ἀλακάτα*), since this seems to be the best method to unweave the threads of this tightly knit text unfolding in just 25 verses.

Fig. 12.1 "The Distaff" by D. Galanis (1879–1966). Engraving from the book The Idylls of Theocritus 1953. Aquatint, 18 × 13 cm. Donated by the artist to the National Gallery, Athens, Greece. Π.1843.

ΘΕΟΚΡΙΤΟΥ ΑΛΑΚΑΤΑ[46]

Γλαύκας, ὦ φιλέριθ' ἀλακάτα,
δῶρον Ἀθανάας
γύναιξιν νόος οἰκωφελίας αἶσιν ἐπάβολος,
θέρσεισ' ἄμμιν ὑμάρτη πόλιν ἐς
Νείλεος ἀγλάαν,
ὄππα Κύπριδος ἶρον καλάμω χλῶρον ὑπ'
ἀπάλω.
τυίδε γὰρ πλόον εὐάνεμον αἰτήμεθα πὰρ
Δίος,
ὄππως ξέννον ἔμον τέρψομ' ἴδων
κἀντιφιληθέω,
Νικίαν, Χαρίτων ἰμεροφώνων ἴερον φύτον,
καὶ σὲ τὰν ἐλέφαντος πολυμόχθω
γεγενημέναν[47]
δῶρον Νικιάας εἰς ἀλόχω χέρας ὀπάσσομεν,
σὺν τᾷ πόλλα μὲν ἔργ' ἐκτελέσῃς ἀνδρεΐοις
πέπλοις,[48]
πόλλα δ' οἶα γύναικες φορέοισ' ὐδάτινα

[44] The only paper solely concerned with Theocritus' *Idyll* 28 is Cairns 1977, which argues that the poem is an *anathēmatikon*.

[45] Little is known of his life. He spent some time in his homeland, and some time in Alexandria; at some point he probably sought the patronage of Hiero II in Syracuse and of Ptolemy II in Alexandria; he probably visited Cos and Miletus, although it cannot be proven whether his literary journeys recount actual ones. He is commonly thought to author a collection of 31 poems, known as *idylls*, characterized by diversity in both form (language, metre) and content (genre, subject matter). The ones commonly thought to claim no authenticity are 8, 9, 19–21, 23, 27 of his *Idylls*. The numbers follow Gow's edition (1950).

[46] The text is from TLG = Gow A. S. F. 1950 *Theocritus* Vol. 1. Introduction, Text and Translation. Cambridge.

[47] Ivory spindles must have been quite unusual and were probably imported from the East. See Barber 1992, 62–64, especially image 2.30 on p. 64 depicting a late Bronze Age ivory spindle from the Attica, Greece, from the collection of the National Museum in Athens (see C. Sauvage 2014 on ivory distaffs).

[48] On the garment term *peplos* see Cleland *et al.* 2007, 143 "although commonly used for female garments – and long dresses worn by men, particularly Persians – in drama, in wider literature peplos most commonly refers specifically to the Panathenaic peplos".

βράκη.[49]
δὶς γὰρ μάτερες ἄρνων μαλάκοις ἐν βοτάνᾳ πόκοις
πέξαιντ' αὐτοέτει, Θευγένιδός γ' ἔννεκ' ἐυσφύρω·
οὕτως ἀνυσίεργος, φιλέει δ' ὄσσα σαόφρονες.
οὐ γὰρ εἰς ἀκίρας οὐδ' ἐς ἀέργω κεν ἐβολλόμαν
ὄπασσαί σε δόμοις, ἀμμετέρας ἔσσαν ἀπὺ χθόνος.
καὶ γάρ τοι πάτρις ἂν ὠξ Ἐφύρας κτίσσε ποτ' Ἀρχίας,
νάσω Τρινακρίας μύελον, ἄνδρων δοκίμων πόλιν.
νῦν μὰν οἶκον ἔχοισ' ἄνερος ὃς πόλλ' ἐδάη σόφα
ἀνθρώποισι νόσοις φάρμακα λύγραις ἀπαλάλκεμεν,
οἰκήσῃς κατὰ Μίλλατον ἐράνναν πεδ' Ἰαόνων,
ὡς εὐαλάκατος Θεύγενις ἐν δαμότισιν πέλῃ,
καί οἱ μνᾶστιν ἄει τὼ φιλαοίδω παρέχῃς ξένω.
κῆνο γάρ τις ἔρει τὦπος ἴδων σ' 'ἦ μεγάλα χάρις
δώρῳ σὺν ὀλίγῳ·πάντα δὲ τίματα τὰ πὰρ φίλων.'

THEOCRITUS' DISTAFF[50]

Distaff, fond of wool-spinning, bright-eyed Athena's gift
to women who are capable of increasing their household,
accompany me to the splendid *polis* of Neileus,
where lies the precinct of Cypris green amidst the soft reed.
Towards this place we ask Zeus to give us fair voyage,
so as to have the pleasure to see my friend and host Nicias,
beloved of the Graces who have beautiful voices,
and return his welcome with you, my gift of wrought ivory,
which I will place in the hands of his wife,
with whom you will accomplish many products of women's toil, men's long robes,
and many finely worked transparent tunics that women wear,
for twice in the year the mothers of the lambs are shorn of their soft fleeces in the pastures,
for Theugenis with the beautiful ankles;
so industrious is she, she is fond of what people of sound mind pay attention to.
I would not give you to the home of a weak or idle woman
because you come from my own land.
And your country is the one that Archias of Ephyra founded of old,
the very marrow of theTrinacria island, the polis of excellent men.
Now in the home of a man who knows a lot
about many a drug to ward off baneful diseases,
you shall dwell in lovely Miletus with Ionian folk,
so that Theugenis may be famed for her distaff among her fellow country-people,
and so that you may always serve as a reminder to her of her poet-friend.
And one would say this upon seeing you: "indeed, great grace and gratitude goes with a small gift;
precious is all which comes from friends."

[49] On the garment term *brakos* (neuter) see Cleland *et al.* 2007: 23: "Generic term for 'cloth'; a long robe or tunic. Sappho, 70; Theocritus, 28.11."

[50] This is a blank verse translation of the poem, *i.e.* the emphasis is placed on rendering the content, not the form (*e.g.* rhythm, metre) of the poem. An effort has been made, however, to preserve the visual organisation of the text in lines, *i.e.* the lines of the translated text roughly correspond to the verses of the original poem, wherever this has been possible.

Fig. 12.2 White-ground funerary lekythos. Made in Athens. and attributed to the Brygos painter, c. 490 BC. British Museum GR 1873.8–20.304 (Vases D 13). It depicts a finely dressed woman spinning wool. Spinning wool and weaving cloth for the family were the principal activities of Athenian women and girls. This vase is said to have been found in a tomb, probably of a woman as tribute to her domestic skills.

Fig. 12.3 The Flemish Proverbs (Die niederländischen Sprichwörter also known as the Blue Cloak or the Topsy Turvy World) (1559, 117.2 × 163.8 cm) by Pieter Bruegel the Elder depicts a scene in which humans, animals and objects offer literal illustrations of Dutch proverbs and idioms. In the lower middle of the painting two elderly women (one seated, the other one standing) spin wool using a spindle and a distaff (the object the seated woman holds in place with her legs).
The scene ironically remarks on the absurdity of having two people do the job of one. It is an illustration of a Dutch proverb which literally translates: "one winds on the distaff what the other spins" (i.e. it takes two people for gossip to spread). Ironically, using a distaff and a spindle to spin is typically performed by a single spinner. Courtesy of Gemäldegalerie, Staatlichen Museen zu Berlin, Preußischer Kulturbesitz, Kat. no. 1720. Photographer: Volker H. Schneider

The title of the poem (*ἀλακάτα*), which denotes a tool for spinning, has been traditionally translated as 'distaff'.[51] An equally accurate term would be 'spindle'.[52] The spindle is a tapered stick of wood, bone or bronze, weighted with spindle whorls. A distaff consists of a shaft of wood or metal shaped at one end into a spike for holding the wool. The spinner takes a few inches of wool fibre and attaches it to her spindle. Holding the spindle in one hand she then flicks the spindle to set it spinning. The distaff holds the fibre to be spun in an orderly fashion and keeps the fibre in a safe place. This allows the spinner to easily draw fibre out of the bundle, draw it together with her fingers and spin it to thread with the help of the rotary motion of the spindle regulated by the weight of the whorl.

[51] *Cf.* Cleland *et al.* 2007, 56 *s.v. ēlakatē*; 2007, 175–176 *s.v. spindle*; 2007, 48–49 *s.v. distaff.*

[52] The rendering of the Greek term by means of the English term 'distaff', however, does not fail to strongly connote the conventional perception of femininity present in Theocritus' poem (*cf.* the expression the distaff side, *i.e.* the female side). Similarly, the French standard translation 'la quenouille' (distaff) suffers in terms of accuracy (the more appropriate term would be 'le fuseau'), but connotatively manages to render the allusion to femininity (*cf.* the French expression 'tomber en quenouille' usually used of political power fallen into the hands of women).

She then winds it on to the shaft of the spindle. As it was the use of the spindle which required agile hands, this is a more probable gift to Theugenis, who is described as an extremely competent spinner (v. 22 *eualakatos*). While ἄτρακτος and ἠλακάτη could both designate the spindle, which is the most necessary tool for spinning, one cannot be certain whether Theocritus' ἠλακάτη should be translated as 'distaff' or 'spindle'. This could be, in fact, an instance corroborating the long tradition of decorated distaffs which archaeologists interpret as votive offerings. Can this Greek term designate two different types of objects or is it Theocritus' choice not to be more precise? Neither archaeological nor textual testimonies can help in elucidating this matter any further. In what follows the traditional title of the *idyll* is used to refer to the poem, but both terms designating two different spinning tools are used to designate the actual gift Theocritus meant as a gift to his friend's wife.

A textile reading of the poem is not only pointed at by the title and the theme; by favouring tactile diction over visual or auditory Theocritus surely intended to present his heart-felt poem so that its recipients would be able to 'feel' it, too.[53] The traditional motif of poetry as a gift is reworked by Theocritus through the meta-poetic image of the distaff/spindle. The occasion of the poem: the poem accompanied a gift for Theocritus' friend Nicias' wife;[54] the gift is an ivory distaff or spindle (v. 8 ἐλέφαντος πολυμόχθω γεγενημέναν). The word *ālakatā* is mentioned nowhere else except in the opening line of the poem. Its omnipresence, though, is suggested by the poet's device to address the whole *Distaff* (poem) to the distaff/spindle (object), hence the second singular person verbs (v. 3 θέρσεισ', v. 10 ἐκτελέσῃς, v. 21 οἰκήσῃς, v. 23 παρέχῃς) and pronouns (vv. 8, 16, 24 σ'), and most importantly the vocative at the beginning of the poem (v. 1 ὦ φιλέριθ' ἀλακάτα). Both the distaff/spindle (object) and the *Distaff* (poem) will be accompanied by the poet, who will take the boat trip to Miletus to see his friend (v. 3 ἄμμιν ὑμάρτῃ) and deliver his present to Theugenis in person (v. 9 εἰς ἀλόχω χέρρας ὀπάσσομεν). A dominant theme in the poem is the reference to the gods, the cities, and the three persons involved in the gift-giving:

First, the gods invoked by the poet are (in the order their names occur in the poem): Athena, Aphrodite and Zeus. Goddess Athena (v. 1) loves spinning and is, undoubtedly, an accomplished weaver; the distaff/spindle was her gift to women.[55] Kypris (v. 4) is Aphrodite *eu-ploia* (*eu* + *ploos* = a good voyage) who was connected with Arsinoe. Her (and Arsinoe's) holy precinct mentioned in line 4 must be the temple of the goddess at Miletus; the poet also alludes to her temple at cape Zephyrion (named after the wind which blows from the West – since this is an eastbound journey). The reeds mentioned evoke both the image of the Nile and an idyllic natural setting where everything is smooth. The noun πλόον (v. 5) alludes to Arsinoe-Aphrodite *euploia*, the protectress of Ptolemy's navy which helped him expand his sphere of influence in the Aegean.[56] The poem opens with a

[53] Tactile images evoking softness: v. 4 ὄππα Κύπριδος ἶρον καλάμω χλῶρον ὑπ ἀπάλω ("where the sanctuary of Kypris Aphrodite lies fresh and green among the soft reeds"); v. 9 (εἰς ἀλόχω χέρρας ὀπάσσομεν ("will place it into the hands of the spouse"); v. 11 γύναικες φορέοισ' ὐδάτινα βράκη ("women donning long gowns as transparent and supple as water"; v. 12 μαλάκοις (…) πόκοις ("soft wool"); auditory image: v. 7 Χαρίτων ἰμεροφώνων ("of the Charites who have lovely voices").

[54] The wife status denoted in v. 9 could also be deduced by the gift itself, as the spindle is a typical attribute of domesticity and wife, not bride-to-be (Marinatos 2002, 155 fig. 2).

[55] Apart from the *aegis* and the spear, the spindle is also an attribute of Athena; *cf.* image 18. *Cf. Anthologia Palatina* 6. 47 and 48, two epigrams on a widow, Bitto, who at the age of 40 retired from the production of textiles, dedicated her spindle to Athena and became a *hetaira* (McKechnie 2005, 70).

[56] In the first *Iamb* of Callimachus the cup Bathycles bequeathed to the wisest man was passed on from one sage to another and finally returned to the hands of Thales of Miletus who dedicated it to the lord of the Neileos people (*Iamb* 1.57 τῷ μεδεύντι Νείλεω δήμου). Moreover, Callimachus *Hymn to Artemis* (*H.* 3.227 Χιτώνη Μιλήτῳ ἐπίδημε-Κεκροπίηθεν) refers to the founding of the city of Miletus by Neleus, son of the Attic king Codrus, following the goddesses'

Fig. 12.4 Tetradrachm (35mm, 16.72g, 12h) depicting the head of Athena Ilias standing right, holding distaff and filleted spear; at her feet to right, owl standing right; to inner left, monogram above star; It reads Α-ΘΗΝΑΣ ΙΛΙΑΔΟS, and ΦΟΙΝΙΚΟΣ in exergue. ca 188–133 BC.

reference to Athena *glaukōpis*; it goes on with a reference to the first *oikist* of the city, Neileos, who was an Athenian (v. 3). Zeus (v. 5) (not Poseidon, nor Dioskouroi or any other gods) is asked to provide a favourable and pleasant voyage. Theocritus associates Zeus with Ptolemy II in his *Encomium to Ptolemy* II (*Idyll* 17.1–8) which opens with the admonition to always begin and end with Zeus, the best among gods, and goes on to refer to Ptolemy, the best among men.[57]

Second, the cities and their founders: Miletus' *oikist* (founder) was Neileos, son of Codrus (3, 21). By mentioning the founder, Theocritus stresses the link between Ionian Miletus and the city of Athens.[58] Miletus was renowned for its fine wool and textiles (*cf. Id.* 15.125).

Syracuse (vv. 16, 17, 18) was the poet's homeland; reference to Corinth (Ephyra is an older name of Corinth)[59] as an integral part of his Syracusan identity is not a first in Theocritus; in idyll 15.91–93 two Syracusan women proudly refer to themselves as Corinthian: Πασάμενος ἐπίτασσε· Συρακοσίαις ἐπιτάσσεις. / Ὡς εἰδῇς καὶ τοῦτο, Κορίνθιαι εἰμὲς ἄνωθεν, / ὡς καὶ ὁ Βελλεροφῶν.

will. Again the choice of *Neileōs* cannot be random; it was probably picked due to its phonetic proximity to *Neilos* (the Nile) which constitutes one more allusion to Alexandria and the Ptolemies.

[57] For connections between Zeus and Ptolemy in *Idyll* 17 see Henkel 2011, 16 n. 17. Callimachus devotes his first *hymn* to Zeus and ties it to allusions to Ptolemy (*h.* 1.79).

[58] Ehrhardt 2003, 269–289.

[59] *Ephyra* is the place name of at least three different sites: Corinth (Hom. *Il.* 7.152), a village in Sicyon (Strabo 8.338) and an island in the Corithian Gulf (Pliny 4.56). The *Iliad* mentions another *Ephyra* near *Elis*, by the river *Sellēes* (Hom. *Il.* 2.659, Strabo 8.338). Plutach in his *Amatoriae narrationes* (*ἐρωτικαὶ διηγήσεις*) (*Moralia* 772–3) tells us how Corinthian Archias founded the city of Syracuse. Archias fell in love with Actaion, son of Melissos, tyrant of Corinth. When he tried to kidnap Actaion his family resisted and the event resulted in a tug-of-war in which Actaion was torn apart. So the ensuing famine (λιμός) and the Delphic oracle led Archias to sail to Sicily where he founded Syracuse.

Πελοποννασιστὶ λαλεῦμες, Δωρίσδειν δ' ἔξεστι, δοκῶ, τοῖς Δωριέεσσι; which roughly translates: "buy slaves and order them around. And you must know that we are Corinthian originally, as was Bellerophon. And we speak Peloponnesian; Dorians, I guess, have the right to speak Dorian". The merging of Syracusan, Corinthian and Dorian identity into a common one is made as explicit as possible.

Third, the persons involved, *i.e.* the person who gives and the one who receives: Theocritus' *vox loquens* alternates between the 1st person singular and the 1st person plural: ἄμμιν v. 3, ἔμον v. 6, ὀπάσσομεν v. 9, ἐβολόμμαν v. 15, ἀπαλάλκεμεν v. 20. Theocritus' connection with Nicias is reflected in ἀντιφιληθέω (v. 6), φίλων (v. 25), ξένος (vv. 6, 23), χάρις (v. 24). *Philia, xenia,* and *charis* are the bonds between the two men. But what does this present have to do with them? Obviously, the distaff/spindle, an emblem of female competence, is connected to Nicias only indirectly, through his wife; Theugenis will be the one to put it to good use. The homonymous poem, though, will be of use to both Nicias and Theocritus.[60]

The poem begins with the distaff/spindle and ends with *philia*. Theocritus does not like to repeat himself or others; he does so only if he doesn't have to: the *Cyclops* is addressed to Nicias, so this one has to be different. *Idyll* 11 is considered to belong to his early period, when Theocritus was a young man, interested in love matters and so was Nicias and Theugenis. The woman depicted in the *Distaff* is a tireless wife, who wove more than any other woman in Miletus.[61] Perhaps this points to a dating of the *Distaff* belongs to his early period, too. However one can only guess as to the sequence of events in Theocritus' life. One fact is more likely true than not, though: Nicias and Theocritus had already met before the poet engaged in composing the *Distaff*. Nicias is *xenos* to Theocritus (v. 6) as is Theocritus to Nicias (v. 22); Nicias is praised as poet (sacred plant of the Charites v. 7), a wise man (v. 19), and a doctor (v. 20). Beside *xenia*, what bonds the two men is also *philia* (*philoi* v. 25). The themes of *philia* and *xenia* are the motifs in-woven on this part of the poetic fabric of the poem.

Theugenis, ἡ ἄλοχος (the wife) is never addressed directly; the genitive Θευγένιδος γ' ἔννεκ' seems to underline the fact that the distaff/spindle will be hers, since it is situated in the very middle of the poem (12 lines from the beginning and 12 lines from the ending of the poem) as a central panel to the overall set of scenes woven here; she is portrayed as a fine woman, just what a woman should be like, as fine as the fine garments (vv.10–11) she will produce with the distaff/spindle in her competent hands: fine textiles define a fine home,[62] and a woman's place is in the centre of it, the exact position Theugenis' name holds in the poem. Her praise is the centrepiece of this poetic fabric and it is not exaggerated: she is both laborious and σαόφρων (judicious), the utmost female quality, which also denotes the ability to run the *oikos* well. She is a perfect match for her husband who is a doctor and "knows how to ward off many baneful diseases from men" (vv. 19–20); the distaff/spindle's new home (δόμος) is a proper match to both its mistress and its previous abode.

I will now turn to the motifs woven into the poem:

The toil (ponos) theme: Hellenistic poets use toil (*ponos*) as a metonym for poem, especially the subtle kind of poems Callimachus and Theocritus loved. Compare the use of the *ponos* theme

[60] As shown by the fact that Theocritus addresses Nicias in *Idyll* 11, entitled *Cyclops*, the Miletian and the Sicilian often engaged in poetic dialogue, over topics such as how to alleviate lovesickness.

[61] West 1967, 82–84.

[62] The opposite of Nicias' and Theugenis' *oikos* is the *oikos* of Diocleidas and Gorgo in *Idyll* 15.18–20.

in Thalysia (*Idyll* 7.51 and 7.137) a poem fraught with poetological allusion.[63] Doubtlessly the poet draws an analogy between his own poetic toil and Theugenis' toil with the distaff/spindle.

The excellence (εὐ/ἐυ) theme underlined by the presence of the following composite with *eu-* adjectives spanning the beginning: (v. 5 πλόον εὐάνεμον "a fair voyage", *i.e.* with favourable wind), middle (v. 13 Θευγένιδος (...) ἐυσφύρω "of fair Theugenis" *i.e.* literally, with beautiful ankles), and end of the poem v. 22 (εὐαλάκατος Θεύγενις "of Theugenis who is a competent spinner").

The praise theme revolves around:

1) The praise of Nicias as gifted poet (v. 7) and doctor-healer of many a disease (vv. 19–20);
2) The praise of Theugenis extends over 6 continuous verses (vv.10–16) and is repeated in verse 22, since it is her tireless and competent hands that will receive the present and produce ἔργα distinguished both by their quantity and quality: her apt spinning and weaving is a metaphor of female agency affecting the quality of her *oikos* and reflecting the quality of her *polis*; her *noos* and *sōphrosynē* are qualities that single out a woman and are perfectly matched to those possessed by her husband. Apart from being beautiful (v. 13 ἔϋσφυρος) she is εὐαλάκατος (v. 22 "laborious"), as a good woman should be:[64] she spins and weaves clothes, the wealth of her city (v. 22 ὄλβος "happiness, wealth") twice as fast as other women of Miletus (v. 12–3). Compare *Idyll* 15.20 where an explicit connection is made between bad fabrics and a bad home (*oikos*). Some lines down (*Id.* 15.122–126) the wealth of the clothes, along with the gold and the ebony is admired as a display of Ptolemy's vast riches and power; Miletus and Samos are particularly mentioned as cities producing "purple rugs that are softer than sleep" (v. 125–126 πορφύρεοι δὲ τάπητες ἄνω μαλακώτεροι ὕπνω· ἁ Μίλατος ἐρεῖ χὠ τὰν Σαμίαν καταβόσκων);
3) The praise of the distaff/spindle: it is made of wrought ivory (v. 8) and is addressed in the second person singular by the voice of the poetic *persona.* Thus the distaff/spindle is not represented as an inanimate object and is brought to life. The deixis of voice change in the final two lines: the gift is deliberately belittled by being referred to as an *oligon* ("small") gift (v. 25) to emphasize more strongly the praise of Nicias, Theugenis and Miletus;
4) The praise of the cities of Miletus (v. 3 πόλιν (...) ἀγλάαν, v. 21 Μίλλατον ἐράνναν), and Syracuse (v. 18 νάσω Τρινακρίας μύελον);
5) The praise of Theocritus: φιλαοίδω ξένω (v. 23).

The oikos and polis theme: The *oikos* is the realm of women; a good woman has a good *oikos* (v. 2). But private life is not seen as separate from public life; the etymological link between *oikos* and *oikizō* (to found a colony) and the noun *oikist* (the founder of a colony) shows the continuum between home as in private abode and home as one's own country (*patris*, v. 17). This justifies Theocritus' mention of Syracuse, his home town, and Archias, its founder: the poet addresses the distaff/spindle to say: "I would never entrust you to the hands of an incompetent woman; your new home (*oikos*) is of equal worth to your previous one (*patris*)". As the distaff/spindle will find a new home in Miletus weaving new textiles (vv. 21–23), so Theocritus will probably find a new milieu in Miletus weaving new poetic texts. The *Charites* (not the Muses) may spin a propitious

---

[63] i) 7.51 τὸ μελύδριον ἐξεπόνασα ("this ditty I composed") is a phrase the poet uses in reference to composing poetry, μελύδριον being a smaller-scale *melos* (a word alluding to *aiolikon melos*, the poems-songs composed by Sappho and Alcaeus); ii) 7.137 σκιαραῖς ὀροδαμνίσιν αἰθαλίωνες/τέττιγες λαλαγεῦντες ἔχον πόνον ("while the ash-brown cricket chirped laboriously amidst the shady leaves"); the co-reference to *ponos* (labor) and *tettix* (cicada) is a poetological allusion to Callimachean *leptotēs*. The cicada, a symbol of delicate songs becomes a symbol of Callimachean poetics which opposes the "heavy" sounding poetry of the *onos* (ass). Ambühl 1995, 209–213.

[64] *Cf.* how John Chrysostom (*Quales ducendae sint uxores = Encomium ad Maximum = PG 52,* 4th–5th century AD) describes female tasks more than five centuries later emphasizing that whereas women cannot throw the javelin, nor the arrow, they are very good with the spindle.

fate[65] for him in the *oikos* and *polis* of Nicias. Theocritus assigns anthropomorphic characteristics to this inanimate object in the hope that it will serve as a go-between.[66]

*The friendship theme (*φίλος "friend, beloved"–φιλέω "to befriend, to love") is woven around three poles: i. the distaff/spindle-object (v. 1 φιλέριθ'), ii. Nicias' reaction as he receives his friend (v. 6 ἀντιφιληθέω, *i.e.* "I will be welcomed in return") and iii. Theugenis (v. 14 φιλέει "she loves").

The xenia theme: v. 6 (Nicias to Theocritus), v. 23 (Theocritus to Theugenis and Nicias).

The gift theme: δῶρον "present" vv. 1, 9, 25. The gift is double: a valuable object from the poet to his friend's wife and a poem from a poet to another.[67]

The charis theme: v. 7 Χαρίτων, v. 24 χάρις.[68] Aphrodite and the *Charites* are intimately linked with the heart of the *polis*.[69]

Lastly, *the song as memory (mnēmē) theme*: the double gift (object and poem) will always remind Nicias and Theugenis of their poet-friend: (v. 23 μνᾶστιν (…) παρέχῃ φιλαοίδω).[70]

The king, a friend and I[71]

In this section I will endeavour to show how the concepts of *philia-xenia-charis* are pivotal in this Theocritean *Idyll*. Within the Hellenistic context the terms *philia* and *xenia*[72] were also used to describe bonds between a chosen few and Alexander the Great[73] (his best friends were called *hetairoi*) or the Lagid court; as a scholar frequenting the Alexandrian court Theocritus would certainly aspire to attain the status of *philos* and *xenos* in Alexandria, as well as Miletus.[74] In the Theocritean encomium to Ptolemy II (*Id.* 17.110) we get a depiction of Ptolemy as very generous, giving out lots of gold (riches) to kings (βασιλεῦσι), *poleis* (πτολίεσσι) and friends (ἑταίροις); these three words epitomize the focal axes of Ptolemaic foreign policy. *Philos* ("friend"), may be regarded as the technical term for courtier in the Hellenistic age. *Xenia* (or *philoxenia*) denotes a ritualized friendship, characterized by gift exchange. The friends of the kings were for the most part Greeks and Macedonians.

Friendship between two people is based on reciprocity and the mutual will to help each other in the case of danger. No such mention is made in the *Distaff*. Theocritus asks for nothing in particular, yet πάντα τίματα "anything (of value) that a friend may give" is welcome. When it

[65] The traditional spin-ster is Fate (Moira *Μοῖρα*) not Charis: *AP* 7.12 Μοῖρα λινοκλώστου δεσπότις ἠλακάτης. "Fate, mistress of the spindle which spins flax fibre into linen thread".

[66] The use of θέρσεισ' (v. 3) is interesting: θέρσεις(α) (…) ὑμάρτη (= attend me without fear). Why would the spindle be afraid? Because of parting with her master and she has to serve a new one. But this new one is a really good master.

[67] Nicias was a poet, too. *Cf.* Idyll 11, a *consolatio amoris* Theocritus addresses to Nicias in order to argue that poetry is a better cure than drugs in certain cases, as in the case of the lovesick Cyclops.

[68] *Cf.* Charites-Muses-bucolic Muses in *Idyll* 1.

[69] In the Athenian market there was a district consecrated to Demos and the Charites and an altar bearing the inscription ΑΦΡΟΔΙΤΕΙ ΗΓΕΜΟΝΕΙ ΤΟΥ ΔΗΜΟΥ ΚΑΙ ΧΑΡΙΣΙΝ ("to Aphrodite hegemon of the *deme* and to the Charites"). For Theocritus' reference to the Charites of Eteoclēs and Orchomenos as Chtonian Charites in the final verses of his encomium of Hiero II (*Id.* 16.104) see Zielinski 1924. Also note the ending line of encomium to Hiero II where the poet expresses a wish: "may I always be with the Charites" (Ἀεὶ Χαρίτεσσιν ἅμ' εἴην.)

[70] *Cf.* καλάμω "reed" (v. 4), a metonymical allusion to the pipe (*syrinx*) which was made of reeds (*cf.* Theocritus' epigram 2).

[71] For issues of patronship and Theocritus see Burton 1995; Griffiths 1979; Stephens 2003.

[72] Belfiore 1993/1994.

[73] The *etairoi* (= friends) were Alexander the Great' s circle of closest companions, his generals and comrades, the men he fought side by side with during battle.

[74] Clauss & Cuipers 2010, 38.

Fig. 12.5 Stained Glass Design for a Married Couple by Hieronymus Lang (Switzerland, 1553). This design for a circular stained-glass window commemorates the marriage between Bitzius Haller and Barbara Fluomann, members of prominent Swiss families from Bern. An inscription around the base of the roundel details the commission, while an empty scroll at the top provides room for other lettering. On the left side, above one coat of arms, a beehive supports the figure of a man covered with the family emblems and carrying a flail. The female figure on the right rests on another beehive above her coat of arms; she carries a distaff. To produce a stained-glass window, patrons commissioned an artist to make a cartoon for the design. Once the patrons had approved this image, another specialist craftsman cut the glass in shapes and colors according to the design or painted the scene directly onto a colorless glass panel.

comes to Ptolemy, Theocritus clearly perceives his role as the bard who will immortalize the patron through his song. But he does not state this humbly: his hymnic *idyll* to Ptolemy begins with an admonition ("let us sing of the one who is the best of all") and ends with a wish, that Ptolemy will ask this ἀρετήν (right to praise) of the gods. The final two verses of the *Spindle* conclude the poem by positing the notions of *charis*,[75] *gift* (δῶρον) and *timē* (both honour and remuneration)[76] at the centre of Theocritus' idea of poetic means to an end: poetic weaving is not an end in itself; it has to do with *timē*, *i.e.* respect and riches.

[75] *Charis* induces a social softening between strangers, friends, lovers, mortals and gods (MacLachlan 1993, 148). MacLachlan's analysis of this concept in early Greek poetry emphasizes the element of reciprocity between the giver and the charis receiver of charis. I thus translate it using two synonymic equivalents in English: grace (on the part of the giver) and gratitude (from the point of view of the receiver). (See my translation of the penultimate line of the text in the appendix.)

[76] Theocritus' expresses a similar preference of τιμή and φιλότητα over material things in another poem with encomiastic content, his *idyll* in honor of Hiero II of Syracuse (*Id.* 16.66–67).

Nicias' was Theocritus' and maybe also Ptolemy's friend and ally in his effort to control the city of Miletus.[77] In 262/1, faced with the danger of losing control of the city, Ptolemy II wrote a letter addressed to the *boule* (council) and *demos* (people) of Miletus. In this letter he praises their friendship and goodwill and mentions that his son and other friends who were at Miletus wrote to inform him of this. If Ptolemy II was indeed the poet's patron, then the poet's visit to Miletus and his oblique allusions to Alexandria may also reflect his *philia* and *xenia* with Ptolemy whom he elsewhere addresses as 'most excellent of men' (*Id.* 17.4 προφερέστατος ἀνδρῶν).[78]

Theocritus' Aeolic side

The Aeolic language of the poem betrays its literary affiliations. Theocritus composed two more poems in Aeolic dialect entitled *Paidika* (*Idylls* 29 and 30) which are variations on the theme of homoerotic love poetry. Hunter interprets Theocritus' exercises in Aeolic lyric as engaging not Sappho but Alcaeus, due primarily to their pederastic context (*paidika*),[79] and their authoritative figures of Aeolic *melos*. Idyll 2 is thematically closer to Sappho's *Hymn to Aphrodite* than *Idyll* 28, even though it is written in Doric dialect.

There was another poem in 300 dactylic hexameter verses, by Erinna[80] written in a literary mixture of Doric and Aeolic, an epitaph to a young girl who died, the poet's friend, Baucis. Her biography has been the source of much controversy as the Suda considers her a contemporary of Sappho, but most scholars agree that she was a 4th century BC female poet. She was held in high esteem by Alexandrian poets, if we judge by the epigrams that mention or allude to her in the *Greek Anthology* (or *Anthologia Palatina*) (*i.e.* Asclepiadēs, Antipater, Leonidas) comparing her to Homer, and paralleling her sweet verses with honey. According to epigram *AP* 9.190 she came from the island of Lesbos. *AP* 7.710 informs us that she came from Mytilene, which may only imply her artistic affinities with Sappho. The Suda tells us that she came from Telos, a Dorian island close to Rhodes. Did Theocritus have Erinna in mind, when composing in Aeolic? The date of Erinna has been the object of scholarly debate, though Neri has convincingly proposed the first half of the fourth century.[81]

An Alexandrian papyrus roll opening with poems in Aeolic like the one containing the *Spindle* evokes a particular authoritative literary past context.[82] The choice to conjure up Aeolic poetic

[77] The death and deification of Arsinoe in 270/269 BC have been suggested as *terminus post quem* of the poem (Marquaille 2008, 59). Votive plaques bearing the name of Arsinoe Philadelphus in the genitive were found in Lesbos, Samos, and Miletus among other coastal cities attesting to Arsinoe's worship, both in private and public cults, merged with elements from the cults of Aphrodite Euploia. A good guess for a *terminus ante quem* is the year Antigonos Gonatas recaptures Miletus and Samos (259/9). Ptolemy II dies in 246 BC.

[78] The opening word of the poem is a pun on the female name Glaukē. She was a contemporary and probably the mistress of Ptolemy II Philadelphus. The *scholia* to Theocritus *Idyll* 4 (*Id.* 4.31b) cite Theophrastus for the story of a ram, which fell in love with her, and Aelian (*On the nature of animals* 8.11) considers the ram Ptolemy II's rival. Whether she was really Ptolemy's concubine or not, is not of import here; what is, is that there this was rumoured, and the opening of the idyll would have brought to Theocritus' contemporaries Ptolemy II to mind.

[79] Hunter 1999.

[80] West 1977; Cameron & Cameron 1969.

[81] Most scholars agree on a date between 400 (Levin 1962) and 350 BC (West 1977); Neri (2003) dates Erinna in the first half of the fourth century BC. For extensive discussion of the 'biographical' epigrams on Erinna see Neri's chapter in this volume.

[82] Although I do not wish to challenge the fact that "the poem [is] a Theocritean evolution of Sappho. The appropriation

tradition cannot merely be a poetic one, since this would not fit the occasion of the poem: the addressees were Ionians; of course, Doric could have a complex web of associations for an Alexandrian audience (*cf. Id.* 15.93), so can Aeolic for an Ionian one. The direct reference or indirect allusion to Corinth and Syracuse, one the one hand, and Miletus and Athens, on the other, all emblematic cities of the Dorian and the Ionian world respectively is in contrast to the formalistic aspects of the poem, *i.e.* language and metre. Even though the use of both these elements, informed by Alexandrian poetics which prefer re-invention over imitation, are consonant with Hellenistic cosmopolitanism, I would argue for something else being the reason why Theocritus used Sappho and Alcaeus on a non-(homo)erotic topic.

So what could be the reason for Dorian Theocritus visiting his Ionian friend to be writing his poem-gift in Aeolic language and verse? Was this due to Alexandrian eclecticism and cosmopolitanism or does it serve historical circumstances to do with the poet's relationship not only with Nicias but also with Ptolemy II? Lesvos is mentioned one more time in an idyll important in terms of Theocritus' poetological quests. In his *Idyll* 7 entitled *Thalysia* a certain Ageanax, friend of the narrator, wishes to sail to Mytilene and the narrator composes a *propemptikon* (a poem wishing one a safe journey) to accompany him. The poem has multiple embedded narratives and is set on the island of Kos, birthplace of Ptolemy II. The names of the personae appearing in the poem have all been said to allude to poets (Hellenistic or older); but only Ageanax has a poem composed in his honour evoke a regal countenance and the song is a *propemptikon* wishing Ageanax a safe voyage.[83]

The Aeolic *Spindle* and the *propempticon* wish for a voyage to Mytilene are, I propose, elements of Theocritus' secret mission to spread the political message on behalf of Ptolemy II that his sphere of influence will include Mytilene; by doing that he can also gain the monarch's goodwill towards him. Theocritus was probably weaving new spheres of Ptolemaic influence not only in Miletus, a city with strong affiliations with the Ptolemies, but also in the Aeolic isles, notably Lesvos, which was the apple of Ptolemy's eye. On the other hand, Theocritus was a poet, not a diplomat or a cultural attaché; he certainly admired Sappho and probably saw the affinities between her wealthy, aristocratic and elitist poetic circle and the Alexandrian *Mouseion*. Her leisured life and the pursuit of culture in a beautiful island setting was probably very attractive to Theocritus and reminded him of his Syracuse and Sicily.

Conclusion: Callimachus' and Theocritus' complex model of poetic weaving

To finish the unravelling of the textile poetics in Callimachus' 4th *Iamb* and Theocritus' *Spindle* let me summarize the central pivots around which these two poems revolve, namely the genres, the personae, and the *poleis* (cities):

of the distaff/spindle as a subject of male Alexandrian discourse neatly highlights this evolution" (Acosta-Hughes 2010, 108) I am not convinced that the reasons for this poetic choice were merely artistic.

[83] *Cf.* the adjective *agēnōr* (ἀγήνωρ 'manly', 'heroic') the poet uses in 17.85 with reference to the Ptolemy as master of the Cyclades; Lesbos was not one of his territories.

Callimachus' 4th *Iamb*		
archaic iamb Hipponax Ephesus	Ainos Aesop Samos	Callimachean iamb poets of the Mouseion Alexandria
Theocritus' *Idyll* 28		
Athena Nicias Miletus	Theugenis Theocritus Syracuse	Arsinoe-Aphrodite Ptolemy Alexandria

Their poetics consists of a complex web of choices and oscillations between past and present, between traditional poetic genres and their re-invention. To paraphrase what Barchiesi claims about Horace,[84] Theocritus' approach to his distaff poetics is not just Theocritus' poetics, but also the poetry of praise within Theocritus' poetics, plus Callimachus in Theocritus, plus a prismatic view of the cultural past in order to accommodate present political goals; in the same vein, Callimachus' disposition toward his 4th *lamb* is formed by his own poetic choices, plus the poetry of censure in Hipponax, plus Hipponax in Callimachus, plus Aesop, plus Callimachus' Aesop, plus Callimachus' critics, plus those who agreed with his poetics, like Theocritus.

In Iamb 13, the closing iamb of the book, Callimachus' self-irony reaches new peaks by setting himself up by his critics as the weaver (lit. 'plaiter' πλόκος) of a host of seemingly disparate elements Ionic, Doric and τὸ σύμμεικτον[85] as it was more often than not the case the elevated poetry from Homer through to the poetry produced in the circles of the Alexandrian *literati*. The poet emphatically calls for originality in *Iamb* 12 (fr. 202 vv. 9–11 Pf.):

> ὔμμες ὦ κάλλιστα νήθουσαι (…) πᾶσιν εὐρείης ἐφ' οἱ ζῶσι χθονός
> ἔλθετ' οὐχ *ὑμήν ἀείδουσαι*
> you,[86] who weave exquisitely (...) from all of those who live on the vast inhabited land
> come, but do not sing a *wedding song*

What is more, this iamb is uniquely themed;[87] its topic is a girls' birthday party. No such poem exists in Greek literature. What can one gather from its fragmentary verses is that the poet addresses some female deities who spin (songs?) pleading with them not to sing 'the wedding song' (*humenaios* - ὑμέναιος). Obviously wedding songs were abundant, but this occasion was a little girl's birthday, so a new kind of song was needed. In this iamb Callimachus defends his poetic choices, while imagining those made by his polemicists. These lines express, in my view, the meta-poetic urgency for originality strongly felt by Callimachus. It is as if the poet admonishes his counterparts to weave poems befitting the Hellenistic context.

What Callimachus was strongly urging for Theocritus was able to put into practice with his

[84] Barchiesi 1996.
[85] This was probably the term designating the artificial language of epic poetry.
[86] Through the second person singular the poetic voice directly invokes the Fates or the Muses (D'Alessio 1997, 635 n. 136); *cf.* Kerkhecker 1999, 226; Durbec 2006, 176 n. 465.
[87] It is perhaps the only surviving example of a genre. There are, however no parallels in poetry, only in rhetoric (Acosta-Hughes 2002, 121).

Spindle. In this *idyll* Theocritus does not identify his poetic work with that of his doctor friend; the poet identifies with Theugenis, the laborious spinner who will weave λεπτὰ (delicate, flowing) stuffs for men and women with the distaff/spindle she will receive as a gift by the poet. Apparently, his aim is to produce texts which will be λεπτὰ καὶ χαρίεντα (fine and elegant) just like the textiles, whose life-like realism Gorgo and Praxinoa so ardently admire in Theocritus' *Adoniazusae* (*Women at the festival of Adōnis*) (*Id.* 15.79–82):

{ΓΟ.} Πραξινόα, πόταγ' ὧδε. τὰ ποικίλα πρᾶτον ἄθρησον,
λεπτὰ καὶ ὡς χαρίεντα· θεῶν περονάματα φασεῖς
{ΠΡ.} πότνι' Ἀθαναία, ποῖαί σφ' ἐπόνασαν ἔριθοι,
ποῖοι ζωογράφοι τἀκριβέα γράμματ' ἔγραψαν.
ὡς ἔτυμ' ἑστάκαντι καὶ ὡς ἔτυμ' ἐνδινεῦντι,
ἔμψυχ', οὐκ ἐνυφαντά. σοφόν τι χρῆμ' ἄνθρωπος.

{Gorgo}

Praxinoa, come here. See the multi-coloured textiles,
how fine and elegant they are! You 'd say they 're the garments of gods

{Praxinoa}

Oh! patroness Athena! What excellent wool-weavers toiled,
and wove these letters in so accurately - as if painters have drawn them.
How realistically they stand and roll in the textile!
They look alive, not in-woven! Humans are wise, indeed!

In Theocritus' view being an apt weaver equals being wise. Indubitably, the *Spindle* poem provides ample proof of both his poetic aptness and wisdom.

Poetry requires craftsmanship and so does spinning or weaving. Callimachus' and Theocritus'[88] seamless sewing of songs of praise with myth custom-made to meet the needs of the rising cultural empire was ambitious enough to go beyond the authoritative literary traditions that had dominated Greece until that time, such as Homeric epics and classical drama; under the pretence of imitating what was *archaic*, *i.e.* what belonged to the *archē*, the beginnings, the poetic fruit of their loom encapsulated the double need to be truly modern, on the one hand, and foster links with the Greek cultural heritage, on the other; if their skill elevated them to the status of cultural ambassadors of Ptolemaic political ideology, they surely promoted it with outspokenness and style in the same way they interwove innovative variants with existing poetic diction (*oppositio in imitando*). The intertextual relations in their texts weave a fabric of distinguished archaic 'colour' patterned by their own special and unique designs, the end product being a highly sophisticated, eclectic, and, paradoxically, original material.

Bibliography

Acosta-Hughes, B. 2002 *Polyeidia. The Iambi of Callimachus and the Archaic Iambic Tradition.* Berkeley and Los Angeles CA.

Acosta-Hughes, B. 2010 *Arion's Lyre. Archaic Lyric into Hellenistic Poetry.* New Jersey.

Adrados, F. R. 1999 *History of the Graeco-Latin Fable: Introduction from the origins to the Hellenistic Age.* Leiden.

[88] On whether it is Theocritus who followed Callimachus, or vice versa, or whether they both followed Philitas of Cos see Van Sickle 1975, n. 18.

Algrove-MacDowell, J. 2003 Ancient Egypt, 5000–332 BC. In D. T. Jenkins (ed.), *The Cambridge History of Western Textiles*, vol. 1. Cambridge, 30–37.

Ambühl, A. 1995 Callimachus and the Arcadian Asses: The Aitia Prologue and a Lemma in the London Scholion. *ZPE* 105, 209–213.

Barber, E. 1992 [1991] *Prehistoric Textiles: The Development of Cloth in the Neolithic and Bronze Ages with a special reference to the Aegean*. Princeton NJ.

Barchiesi, A. 1996 Poetry, Praise, and Patronage: Simonides in Book 4 of Horace's *Odes. ClAnt 15.1, 5–47.*

Belfiore, E. 1993/1994 Xenia in Sophocles' Philoctetes. *CJ* 89 2, 113–129.

Blum, R. 1991 [1977] *Kallimachos: The Alexandrian Library and the Origins of Bibliography*, trans. H. H. Wellisch. Madison WI.

Burton, J. B. 1995 *Theocritus' Urban Mimes. Mobility, Gender, Patronage.* Berkeley CA–Los Angeles CA–London.

Cairns, F. 1976 The distaff of Theugenis: Theocritus Idyll 28. In F. Cairns (ed.), *Papers of the Liverpool Latin Seminar*. Liverpool, 293–305.

Cameron, A. & Cameron, A 1969 Erinna's Distaff. *CQ* 19.1, 285–288.

Clayman, D. L. 1980 *Callimachus' Iambi.* Leiden.

Clayman, D. L. 2011 Berenice and her Lock. *TAPhA*, 229–246.

Clauss, J. J. & Cuipers M. 2010 *Companion to Hellenistic Literature*. Leiden.

Cleland, L., Davis G. & Llewellyn-Jones L. 2007 *Greek and Roman Dress from A to Z.* London–New York.

Constan, D. & Landry L. 2008 Callimachus and the Bush in 'Iamb' 4. *CW* 102.1, 47–49.

D'Alessio, G. B. 2007 *Callimachus: Inni, Epigrammi, Ecale, Aitia, Giambi e altri frammenti. 4th edn, revised and enlarged.* 2 vols. Milan.

Durbec, Y. 2006 *Fragments poétiques. Origines, Iambes, Hécalè, fragments de poèmes épiques et élégiaques, fragments de place incertaine*. Paris.

Durell, L. 1961 (1959) *Alexandria Quartet. Mountolive*. New York.

Edmunds, L. 2001 Callimachus Iamb 4: From Performance to Writing. In A. Cavarzere, A. Aloni & A. Barchiesi (eds), *Iambic Ideas. Essays on a Poetic Tradition from Archaic Greece to the Late Roman Empire*. Lanham MD–Boulder CO–New York NY–Oxford, 77–98.

Ehrhardt, N. 2003 Poliskulte bei Theokrit und Kallimachos: das Beispiel Milet. *Hermes*, 131. Jahrg., 3, 269–289.

Erskine, A. 1995 Culture and Power in Ptolemaic Egypt: The Museum and Library of Alexandria. *G & R*, 2nd Series, 42, 1, 38–48.

Gibbs, R. W. 1990 The Process of Understanding Literary Metaphor. *The Journal of Literary Semantics*, 19.2, 65–79.

Gibbs, R. W. 1992 When is Metaphor? The Idea of Understanding in Theories of Metaphor. *Poetics Today* 13, 4, 574–606.

Giuseppetti, M. 2012 Poetry in the Iron Age: interplay of voices in Callimachus' *Iambi*. *Aitia* 2.

Griffiths, F. 1979 *Theocritus at Court.* Leiden.

Gutzwiller, K. 1992 Callimachus' Lock of Berenice: fantasy, romance and propaganda. *AJPh* 113, 3, 359–385.

Gutzwiller, K. 1991 *Theocritus' Pastoral Analogies: The Formation of a Genre*. Madison WI.

Johnson, M. 1987 *The Body in the Mind: The Bodily Basis of meaning, Imagination, and Reason*. Chicago IL.

Hunt, A. S. 1910 *The Oxyrhynchus Papyri 7*. London.

Hunter, R. 1999 (1996) *Theocritus and the Archaeology of Greek Poetry.* New York.

Kerkhecker, A. 1999 *Callimachus' Book of Iambi*. Oxford.

Kövecses, Z. 2002 *Metaphor. A Practical Introduction.* Oxford.

Krevans, N. 2004 Callimachus and the Pedestrian Muse. In M. A. Harder, R. F. Regtuit & G. C. Wakker (eds), *Callimachus II*. Hellenistica Groningana. Leuven, 173–184.

Lakoff, G. 1993 The contemporary theory of metaphor. In Ortony A. (ed.), *Metaphor and Thought,* 2nd ed. Cambridge, 202–251.

Lakoff, G. 1987 *Women, Fire, and Dangerous Things: What Categories Reveal about the Mind*. Chicago IL.

Lakoff, G. & Johnson M. 1980 *Metaphors We Live By*. Chicago IL.

Lakoff, G. & Turner, M. 1989 *More than Cool Reason: A Field Guide to Poetic Metaphor*. Chicago IL.

Lehmler, C. 2005 *Syrakus unter Agathokles und Hieron II. Die Verbindung von Kultur und Macht in einer hellenistischen Metropole.* Frankfurt am Main.

Lelli, E. 2004 *Critica e polemiche letterarie nei Giambi di Callimaco.* Italy.

Levin, D. N. 1962 Questiones Erinnaenae. *HSPh* 66, 193–204.

Lichtheim, M, 2006 (1980) *Ancient Egyptian Literature*, vol. 3. Berkeley CA.

MacLachlan, B. 1993 *The age of Grace. Charis in Early Greek Poetry*. Princeton NJ.

Marinatos, N. 2002 The Life Cycle of the archaic Greek warrior and hero. The interplay of myth and genre in imagery. In S. des Bouvrie (ed.), *Myth and Symbol I. Symbolic Phenomena in Ancient Greek Culture. Papers from the Norwegian Institute in Athens 5.* Bergen, 153–170.

McKechnie, P. 2005 Beau Monde and Demi-Monde in Alexandria, 323–116 BC. *L' antiquité classique* 74, 69–82.

Marquaille, C. 2008 "The foreign policy of Ptolemy II". In P. McKechnie & P. Guillaume (eds), *Ptolemy II Philadelphus and his world.* Leiden, 39–64.

McIntosh Snyder, J. 1981 The Web of Song: Weaving Imagery in Homer and the Lyric Poets, *CJ* 76, 193–96.

Nagy, G. 2002 *Plato's Rhapsody and Homer's Music: The Poetics of the Panathenaic Festival in Classical Athens.* Cambridge. MA.

Neils, J. 1984 The Orestes Sarcophagus and Other Classical Marbles. *The Bulletin of the Cleveland Museum of Art*, 71, 4, 102–113.

Neri, C. 2003 *Erinna. Testimonianze e Frammenti*. Bologna.

Nosch, M.-L. 2014 Voicing the Loom: Women, Weaving and Plotting. In D. Nakassis, J. Gulizio & S. A. James (eds), *KE-RA-ME-JA: Studies Presented to Cynthia W. Shelmerdine.* Philadelphia PA 91–102.

Oakley, J. H. 1982 The Anakalypteria. *AA*, 113–118.

Perry, B. E. 1962 Demetrius of Phalerum and the Aesopic Fables. *TAPhA* 93, 287–346.

Regenbogen, O. 1950 Pinax. *RE* 20B, 1408–1482.

Rice, E. E. 1983 *The Grand procession of Ptolemy Philadelphus.* Oxford.

Rosen, R. M. 2007 *Making Mockery. The Poetics of Ancient Satire. Classical Culture and Society.* Oxford.

Scodel, R. 2011 Callimachus and Fable. In B. Acosta-Huges, L. Lehnus & S. Stephens (eds), *Brill's Companion to Callimachus.* Leiden-Boston MA, 368–383.

Serwint, N. 1993. The Female Athletic Costume at the Heraia and Prenuptial Initiation Rites. 3 *AJA* 97, 403–422.

Tay, D. 2014 George Lakoff's Theory of Conceptual Metaphor. In J. Littlemore & J. R. Taylor (eds), *Cognitive Linguistics Companion.* London, 49–57.

Thomas, R. F. 1983 Callimachus, the Victoria Berenices, and Roman poetry. *CQ* 33.1, 92–113.

Tuck, A. 2006 Singing the Rug: Patterned Textiles and the Origins of Indo-European Metrical Poetry. *AJA* 110.4, 539–550.

Tuck, A. 2009 Stories at the Loom: Patterned Textiles and the Recitation of Myth in Euripides. *Arethusa* 42, 151–59.

Waites, M. C. 1912 Some Features of the Allegorical Debate in Greek Literature. *HSPh* 23, 1–46.

West, M. L. 1967 A Note on Theocritus' Aeolic Poems. *CQ*, New Series 17.1, 82–84.

West, M. L. 1969 Near Eastern Material in Hellenistic and Roman Literature. *HSCP* 73, 113–34.

West, M. L. 1977 Erinna. *ZPE* 25, 95–119.

West, M. L. 2007 *Text-textile a metaphor for poetics-on how to weave a Hellenistic poem Indoeuropean poetry and myth.* Oxford.

Wild, J.-P. 2003 *The Eastern Mediterranean.* In D. T. Jenkins (ed.), *The Cambridge History of Western Textiles*, vol. 1. Cambridge, 48–51.

Zanker, G. 2008 *Modes of Viewing in Hellenistic Poetry and Art*. Madison WI.

Zielinski, T. 1924 Charis and Charites. *CQ* 18.3/4, 158–163.

13. Patterning Poetry: Weaving and the Construction of Didactic in Lucretius' *De Rerum Natura*, Book 1

Ann Glennie[*]

The *De Rerum Natura*, written in the mid-1st century BC by the Roman poet Lucretius, is a didactic poem composed with the basic goal of instructing its reader (and addressee, Memmius) on the tenets of Epicureanism to free him from the restraints of religion and thereby alleviate his fear of death. This chapter examines weaving vocabulary and weaving metaphor in order to discuss the author's motivation for utilizing the comparison with a generally feminine daily life activity in constructing his didactic poem regarding Epicurean philosophy.[1] I argue that straightforward weaving metaphor is ideal for Lucretius' task of illuminating complex Epicurean principles to his readers.[2] Additionally, the weaving metaphor prevalent throughout Book 1 equates Lucretius' work to an elaborately woven textile, an equation which in turn glorifies Lucretius as the poet capable of producing such an intricate, well-crafted didactic text.

First, a few words on the genre of didactic poetry. Didactic poetry is poetry composed in the meter of dactylic hexameter and meant to instruct its readers, and often a specific addressee who may be a patron or family member, about a certain topic. The genre developed as a reaction to the "heavy" genre of epic poetry, written in the same meter. Instead of tackling "weighty", and perhaps inaccessible, material such as mythic heroes and their deeds, didactic poetry conveyed potentially difficult material in such a way that it was easily understood by the work's audience. Because of the manner of its inception and its metrical relationship to epic poetry, didactic poetry

* This paper is limited to a close study of the weaving metaphor in only Book 1 of the *De Rerum Natura*. For a more complete discussion of weaving metaphor throughout the *De Rerum Natura*, see Matthew Johncock's chapter in this volume. I extend much gratitude to the Danish National Research Foundation's Centre for Textile Research as well as to the organizers, hosts, and participants of the "Spinning the Fates and the Song of the Loom" conference for the opportunity to present, discuss, and improve a prototype of this paper. Additionally, a special thanks is offered to Laurel Fulkerson for her encouragement and advisement on many occasions.

1 Metaphors for the process of poetic composition, including comparison to journey either by boat (Var. *Men.* 224; Verg. *A.* 6.1), by chariot (Lucr. 6.47; Ov. *Fasti* 2.360, 4.10, 6.585–586; Ov. *Rem.* 397–308; Prop. 3.9.57–58, 4.1.70), or by a combination of both in the same book (Ov. *Ars.* 1.39–40, 1.771–772; Verg. *G.* 2.39–46, 2.541–542) are regular in didactic poetry. Weaving, on the contrary, does not seem to be a particularly common metaphor for poetic composition in this context (*i.e.* of didactic poetry). While metaphors involving boats and chariots cannot categorically be declared "masculine" metaphors, nor weaving a "feminine" metaphor, it bears a closer examination as to why Lucretius, who uses a chariot metaphor himself in Book 6, more prevalently defers to the more "female" weaving metaphor in Book 1.

2 Previous work on weaving metaphors in Lucretius' *De Rerum Natura* has been done by McIntosch Snyder (1983) and Hendren (2012).

is not totally divorced from epic; it also often employs similar features including the invocation of a muse figure and the use of mythical exempla. But, at its core, didactic poetry is designed to teach, utilizing readily accessible metaphor and imagery to illuminate challenging concepts to its reader.

The first use of a weaving metaphor by Lucretius in Book 1 of his didactic occurs during his initial discussion about the existence and nature of atoms: *principium cuius hinc nobis exordia sumet*, "the beginning principle for our [study] will derive its warp from this place".[3] *Exordium* can mean simply a 'beginning', but additionally here means the start of the weaving, the creation of the border, and the setting of the 'warp', the vertical threads strung on the loom through which the weft, the horizontal threads, are subsequently woven to create the textile.[4] Lucretius therefore parallels the starting point of his didactic with the physical stringing of the warp threads; he initiates his writing process in the same way as the weaver initiates the weaving process.

After establishing the presences and natures of both atoms and the void,[5] Lucretius declares that everything is made up of these two substances.[6] To introduce this new section of his argument, Lucretius directly refers to weaving again, recalling the previous use of *exordium*: *sed nunc ut repetam coeptum pertexere dictis*, "but now let me resume weaving with words my task".[7] As with *exordium*, the word *pertexo* has a non-technical sense, in this case "to accomplish", but in this specific context it can only mean 'to weave'.[8] This explicit weaving diction relies on Lucretius' earlier assertion that his loom and warp are set up;[9] he is eager to get on with his weaving, a notion that signifies the poet's desire to continue with his poetic discourse on Epicureanism.

Lucretius often utilizes other words with weaving connotations, such as in the passage instructing the reader about the binding nature of atoms,

denique res omnis eadem vis causaque volgo
conficeret nisi materies aeterna teneret
inter se nexus minus aut magis indupedita
tactus enim leti satis esset causa profecto
quippe ubi nulla forent aeterno corpore quorum
contextum vis deberet dissolvere quaeque
at nunc inter se quia nexus principiorum
dissimiles constant aeternaque materies est
incolumi remanent res corpore dum satis acris
vis obeat pro textura cuiusque reperta

Finally, force and cause, without exception, would destroy everything unless eternal material preserved the interweavings having been tied together among itself either more or less, for a touch is a sufficient cause of destruction to the thing having been accomplished. Since, in fact, there are no things with an eternal body, force ought to unbind each of the parts having been woven. But as it is, because the weavings of first beginnings exist dissimilar among themselves and because the material is eternal, the things remain with an unharmed body, even while a sufficient enough force comes having hit upon the fabric.[10]

[3] Lucr. 1.149; Lucretius' Latin text and line citations after Bailey 1947, Vol. 1; translations are my own.
[4] Leonard and Smith (1942, 219 in the commentary on line 1.149) write "*exordia sumet* (…) seems to be derived from weaving."
[5] Lucr. 1.149–1.417.
[6] Lucr. 1.419–1.420.
[7] Lucr. 1.418.
[8] Leonard and Smith 1942, 245 (in the commentary on 1.418): "*coeptum pertexere dictis*: 'to finish weaving with words the web I have begun.'"
[9] Lucr. 1.149.
[10] Lucr. 1.238–1.247.

The words *nexus, textura, dissolvere, indupedita,* and *contextum* can, in other contexts, be variously translated, but collected here in relatively high frequency and with such close proximity, Lucretius must be intending to evoke their shared definitions which relate to the weaving (and unweaving) of a cloth.[11] Yet again, Lucretius uses weaving, here metaphorically, to describe the behavior of atoms and matter, as well as to illustrate Epicurean principles.

Having established that a reader might expect any amount of force to break apart and destroy matter,[12] just as weaving might be thought to be easily unraveled, Lucretius further explains that every type of matter consists entirely of the same kind of atoms. These atoms, when combined, are difficult to separate,[13] in the same way in which a tightly woven fabric stays bound together even when being acted upon by a strong force. So, having already equated his technique of writing didactic to that of a weaver setting out on her task, Lucretius now observes that Epicurean philosophic tenets themselves are similar to weaving.

Lucretius includes several other passages that have closely situated weaving words which he uses to explain Epicurean principles. He discusses the inability of force to break up matter into pieces, like the difficulty of unraveling a well-woven textile:

> *nulla vi foret usus enim quae partibus eius*
> *discidium parere et nexus exsolvere posset*

> For there would be no use for force which would be able to prepare the breaking up into parts and to unbind the weaving.[14]

Nexus, meaning 'weaving', or 'interweavings',[15] appears here first in the text and establishes the textual and textile tone for the translation of that word throughout Book 1. *Exsolvere*, a near synonym for *dissolvere*,[16] introduces the same unweaving concept.

Lucretius reiterates this theme of woven fabrics not being unwoven by force again, about 300 lines later:

> *haec neque dissolui plagis extrinsecus icta*
> *possunt nec porro penitus penetrata retexi*

> These things, having been constructed, are not able to be unbound from without by blows nor furthermore, having been penetrated, to be unwoven from within.[17]

This passage again uses *dissolvere* and, in addition, *retexere*, a variation on the previously encountered *pertexere*,[18] a term which evokes the image of "unraveling a woven fabric."[19] The use of these two words in conjunction with one another suggests their shared quality of weaving to make it clear that interwoven atoms, like interwoven threads, are sturdy.

---

[11] Leonard and Smith (1942, 229) also see weaving connotation in this section, as noted in the commentary on 1.240, but made especially clear in the commentary on 1.243: "*dissolvere*. Image clearly comes from pulling apart a woven fabric."
[12] Lucr. 1.238–244.
[13] Lucr. 1.244–251.
[14] Lucr. 1.219–1.220.
[15] Leonard and Smith 1942, 227.
[16] *Cf.* Lucr. 1.238–1.247.
[17] Lucr. 1.528–1.529.
[18] Lucr. 1.418.
[19] Leonard and Smith 1942, 255.

Two further examples utilize related weaving words to explain the acts of substances both binding and unbinding:

denique si minimas in partis cuncta resolvi
cogere consuesset rerum natura creatrix
iam nil ex illis eadem reparare valeret
propterea quia quae nullis sunt partibus aucta
non possunt ea quae debet genitalis habere
materies varios conexus pondera plagas
concursus motus per quae res quaeque geruntur

Finally, if nature, creator of things, had been accustomed to force all things to be unbound into their smallest parts, that same nature would then be able to restore nothing out of those things, therefore, because the things which are augmented by no parts would not be able to have the generative materials which they ought, various weavings, weights, blows, meetings, and motions through which things are born.[20]

and,

ne volucri ritu flammarum moenia mundi
diffugiant subito magnum per inane soluta
et ne cetera consimili ratione sequantur
neve ruant caeli tonitralia templa superne
terraque se pedibus raptim subducat et omnis
inter permixtas rerum caelique ruinas
corpora solventes abeat per inane profundum

Lest the walls of the world having been unbound suddenly flee in different directions through the void in the winged manner of flames and lest the rest follow in a similar way and the thundering space of the sky not rush upward and the earth rapidly withdraw itself beneath our feet and among all the ruins of things and of the sky having been mixed up, and unbinding bodies depart through the empty deepness.[21]

These passages reiterate Lucretius' now familiar point that atoms cannot be divided by any measure of force. His repeated use of this sort of imagery equating the strength of woven textiles to the indivisibility of atomic bonds, through the pairing of words with weaving connotations highlights Lucretius' devotion to and reliance on weaving metaphor in the construction of his didactic.

Thus far, I have treated what I consider the unambiguous uses of references to actual weaving techniques and terminology, weaving metaphor, and passages employing multiple words associated with weaving. I move now to somewhat more ambiguous cases, where context is not quite so decisive. In Book 1, *dissolvere* appears twice more in examples located about 40 lines from each other:

materies igitur solido quae corpore constat
esse aeterna potest cum cetera dissolvantur

Material, therefore, which exists in a solid body, is able to be eternal, when all the rest are unbound[22]

and

nam quidvis citius dissolvi posse videmus
quam rursus refici

For rather we see anything to be able to be unbound which by reverse process can be made again.[23]

[20] Lucr. 1.628–1.634.
[21] Lucr. 1.1102–1.1108.
[22] Lucr. 1.518–1.519.
[23] Lucr. 1.556–1.557.

Because in each of these two cases it is not paired with at least one other weaving word, however, *dissolvere* does not necessarily initially stand out as having weaving connotations.[24]

Likewise *exsolvere* appears twice more, but not in contexts which utilise any other of the previously established weaving vocabulary,

> *nisi nos cibus aridus et tener umor*
> *adiuvet amisso iam corpore vita quoque omnis*
> *omnibus e nervis atque ossibus exsoluatur*
>
> Unless solid food and soft water sustains us all, life would be unbound straightaway with a body wasted away from all muscles and bones.[25]

and

> *primum quod magnis doceo de rebus et artis*
> *religionum animum nodis exsolvere pergo*
>
> First, because I teach of great things and I continue to unbind the mind from the close packed knots of religion.[26]

Similar words, and further variants on the 'loosen' or 'unbind' verb, *solvere* and *resolvere,* also crop up throughout Book 1,[27] but again with no clear textual relation to other weaving terminology. While Lucretius may have intended all these derivatives of *solvere* to play off the obvious weaving metaphors of Book 1's didactic instruction, such a connection could not be definitively established without further analysis of the nearby words, a project currently beyond the scope of this paper. It seems likely, however, that since he previously used variants of the 'loosen' or 'unbind' word in connection with weaving references[28] Lucretius expects his audience to similarly recognize *dissolvere, exsolvere, solvere,* and *resolvere*'s weaving connotations in these passages.

Two additional words may have an even more tenuous connection to the weaving metaphor in Lucretius. First is the word *daedala* – cunning, skillful, crafty – literally meaning 'Daedalic', and derived from the name of the craftsman Daedalus, who was famous for the design and execution

---

[24] Leonard and Smith (1942, 254) in the commentary on lines 1.518–1.519 write "The exact bearing of these lines on the argument is not clear."

[25] Lucr. 1.809–1.811.

[26] Lucr. 1.931–1.932.

[27] Lucr. 1.492–1.493: *cum labefactatus rigor auri solvitur aestu / tum glacies aeris flamma devicta liquescit,* 'when firmness of gold is unbound having been destroyed by heat, then the ice of bronze having been conquered by flame melts'; Lucr. 1.1017–1.1018: *nam dispulsa suo de coetu materiai / copia ferretur magnum per inane soluta*, 'for the abundance of material having been wrenched apart out of its union would be carried through the void having been unbound; Lucr. 1.57: *quove eadem rursum natura perempta resolvat*, 'into which nature, having been broken down, unbinds again'; Lucr. 1.414–1.415: *ut verear ne tarda prius per membra senectus / serpat et in nobis vitai claustra resolvat*, 'so that I fear lest old age first creeps through the limbs, making them slow and unbinds the locks of life in us'; Lucr. 1.1053–1.1056: *dicunt omnia niti / atque ideo mundi naturam stare sine ullis / ictibus externis neque quoquam posse resolvi / summa atque ima quod in medium sint omnia nixa*, 'they say all things lean and therefore the nature of the world stands without any external blows and the highest and lowest things are not able to be unbound to any place because all things lean into the middle'.

[28] Lucr. 1.219–1.220, 1.238–1.247, 518–1.519; 1.528–1.529, 1.556–1.557, 1.628–1.634, 1.1102–1.1108.

of the labyrinth, as well as for being a recipient of Athena/Minerva's patronage.[29] There are two similar usages of *daedala* in Book 1:

> *tibi suavis daedala tellus*
> *summittit flores tibi*
>
> The cunning/skillful earth causes flowers to spring for you.[30]

and

> *redductum daedala tellus*
> *unde alit*
>
> From whence the cunning/skillful earth nourishes.[31]

There could be a connection with Athena/Minerva, a goddess of not only male craftsmen, like Daedalus (or perhaps even Lucretius), but also a goddess of weavers, which is inherent in the word *daedala* and which could have activated weaving metaphor for an ancient audience. In any case, a textile was recognized as a very complex structure in antiquity, which required skill and technique, and thus *daedala* is a suitable term for it.

In addition, the use of the perfect passive participle *comptus, -a, -um*, from *comere*, 'to bring together, comb, braid' may be meant to have textile connotations through the word's association with the binding together of strands of hair. This participle occurs in two places,

> *cui simul infula virgineos circumdata comptus*
> *ex utraque pari malarum parte profusast*
>
> For whom once the woolen band [reserved for sacrificial animals] placed around the virgin braids poured down equally on both sides of her cheeks.[32]

and

> *si tibi forte animum tali ratione tenere*
> *versibus in nostris possem dum perspicis omnem*
> *naturam rerum qua constet compta figura*
>
> If, by chance, in such a way I am able to hold your mind in my verses, until you perceive by which shape having been woven every nature of things exists.[33]

The second passage clearly draws parallels to the weaving and unweaving of matter previously referenced by Lucretius, but the connection between the passage describing Iphigenia's braids[34] and Lucretius's previous use of weaving metaphor is less evident. I think, however, the association is as simple as the principle that hair, much like threads of the weft, can be arranged into different styles, giving different appearances without being altered, cut, or destroyed; matter likewise can be differently arranged, bound, or unbound, to create new shapes without destroying it either.

I move now to the question of why Lucretius depends so heavily upon the use of weaving metaphor within Book 1. Clearly he enjoys the utility of detailing the beliefs of Epicureanism through its comparison with daily activities with which his audience would have been intimately

[29] Hyg. *Fab.* 39.
[30] Lucr. 1.7–1.8.
[31] Lucr. 1.228–1.229.
[32] Lucr. 1.87–1.88.
[33] Lucr. 1.948–1.950.
[34] For further work on connections between Iphigenia and weaving, see Emmanuela Bakola's chapter in this volume.

familiar.[35] Lucretius relies on these familiar tropes to illustrate his poem in order to make the subject matter being taught comprehensible to his audience. West notes that these tropes are the same sort of stock images present in Hellenistic philosophy, and may not be far from the types of tropes used in Lucretius' original source.[36] In fact, Greek atomists such as Leucippus and Democritus in the late 5th century BC had already used weaving terminology as an analogy to describe atomic bonding; Epicurus, Lucretius' source, drew on those philosophers when building his theory of atoms.[37] But this body of evidence begs the question: why is a "woman's task" (or art) so prevalent in a philosophical didactic poem which was presumably largely directed at a male audience,[38] in lieu of more "masculine" exempla?[39]

Weaving, especially in domestic, mythological, and allegorical contexts to which Lucretius refers, has long been considered 'women's work.' The issue of textile production in antiquity is more complicated than that. Though household women and female slaves certainly devoted a portion of their daily lives to weaving, the manufacture of textiles, especially of those for commercial consumption, was often also conducted by male specialists. Despite the fact that in actuality both male and female craftsmen worked wool in the ancient world, it is unlikely that Lucretius' supposed elite, male audience wove themselves. His readers must have, however, understood the nuances of weaving in order for them to access the metaphors through which Lucretius illuminated Epicurean philosophy.

Weaving imagery and metaphor was certainly prevalent and multifunctional in ancient literature. At its simplest, through the intertwining of threads, weaving represents unity—including marital, familial, civil, social, and cultural unity.[40] But at the same time, there is a differentiation between the statuses of the workers and weavers of wool—from mythological to actual, from male to female, from freeborn to slave. The above innate imageries evoked by weaving, however, are not particularly integral to the didactic message of Lucretius' Book 1, where he predominantly uses weaving as a metaphor either for his writing process or for the nature of atoms being held together and the tendency for matter to be indivisible.

What can an examination of the identities and purposes of the women who are depicted weaving in ancient sources lend to the analysis of Lucretius' use of weaving metaphor in Book 1? Most examples of weaving females come from the realm of myth, presumably because weaving is thematized as a domestic activity rather than a public one. The most prominent and early references to mythic

[35] Snyder 1981, 193.

[36] West 1994, 14.

[37] Scheid & Svenbro 1996, 166.

[38] The readership of ancient literature is a very controversial topic. Who was literate? Who had access to what kinds of texts? This paper does not seek to tackle this highly contentious issue, but traditional opinion suggests that a more philosophical work like the *De Rerum Natura* may have been more widely read by a male symposiastic audience than a female audience, though this opinion may also be somewhat archaic and unfounded. Epicurus was known in antiquity as the first philosopher to open his school to women (Nussbaum 1986, 497), so perhaps weaving metaphors are original to his work and thereby imitated by Lucretius, or otherwise thought by Lucretius in his time to be especially relevant in light of his possible female audience.

[39] Masculine exempla could include, for instance, metaphors such as the driving of a chariot or steering of a ship already mentioned above (Lucr. 6.47; Ov. *Ars*.1.39–40, 1.771–772; Ov. *Fasti* 2.360, 4.10, 6.585–586; Ov. *Rem*. 397–308; Prop. 3.9.57–58, 4.1.70; Var. *Men*. 224; Verg. *A*. 6.1; Verg. *G*. 2.39–46, 2.541–542).

[40] Scheid & Svenbro 1996, 9–10, 19, and 85.

female weavers—including Helen,[41] Andromache,[42] Penelope,[43] Calypso,[44] and Circe[45]—appear in the *Iliad* and *Odyssey*, canonized six to seven hundred years before Lucretius' *De Rerum Natura*. In a Greek myth[46] adapted by a fellow Latin author, Philomela uses her thread to weave the tale of her rape by Tereus, the husband of her sister, Procne.[47] Arachne is punished by Athena for her insolent weaving which depicts several of the Olympian gods' sexual indiscretions with mortals.[48] In Roman history, the fall of the monarchy and creation of the republic stems from the suicide of the faithful spinner, Lucretia.[49] While the majority of these instances are inconsequential to (and often postdate) Lucretius' work, I believe two of them can provide a tack from which to approach the Lucretian didactic use of weaving metaphor.

Both of the Homeric goddesses, Calypso and Circe, are importantly described as simultaneously singing while they weave.[50] This connection between singing and weaving has garnered much interest in recent scholarship.[51] Heath argues that women must have told and sung mythic narratives in antiquity, as exhibited by the work of another poet known to dabble in didactic, Ovid.[52] Ovid's *Metamorphoses* includes (his versions of) the tales of Philomela and Arachne,[53] and furthermore that work provides a unique opportunity for some of the other women of myth to tell their stories first hand.[54] Though impossible to prove, Heath tantalizingly suggests that this opportunity afforded by Ovid to his female characters reflects the actual practice of ancient women who were not prohibited from telling or singing myths, especially while weaving,[55] though they were not publically or literarily displayed doing so because weaving was a private activity which took place within the home.

Who was the audience for such myths told during the weaving process? While Calypso and Circe sing and weave alone, ancient women of a household and their female slaves most likely carded, spun, and wove in groups, not unlike modern knitting circles or sewing bees. The communal nature of textile processing was necessary not only to complete the initial, labor-intensive steps of the textile process including carding and spinning, but also to use large upright looms which could accommodate a weaver working in tandem with at least one other woman to speed along the process.[56] Additionally, within this utilitarian practice lies a socially mediated construct in which older women act as mentors and models to younger ones. Practically speaking, processing wool

[41] Hom. *Il.* 3.125–3.127.
[42] Hom. *Il.* 22.437–22.441.
[43] Hom. *Od.* 2.84–2.114.
[44] Hom. *Od.* 5.58–5.74.
[45] Hom. *Od.* 10.220–223.
[46] Soph. *Tereus* (*TrGF* 581–595b).
[47] Ov. *Met.* 6.576–6.577; it should be noted that Ovid's writings post-date Lucretius'.
[48] Ov. *Met.* 6.129–6.145; again, Ovid and his works post-date Lucretius.
[49] Liv. *AUC* 1.57.8–1.58.12; a contemporary of Ovid, Livy likewise writes shortly after Lucretius.
[50] Hom. *Od.* 5.58–5.74 (Calypso); Hom. *Od.* 10.220–10.223 (Circe).
[51] Heath 2011; McIntosh Snyder 1981; Tuck 2006.
[52] Heath 2011, 70.
[53] Ov. *Met.* 6.576–577 (Philomela); Ov. *Met.* 6.129–145 (Arachne); and Magdalena Ohrman's chapter in this volume.
[54] There are many examples in the *Metamorphoses* of women who either tell stories or narrate their own accounts first hand, such as the weaving daughters of Minyas (Ov. *Met.* 4.31–4.415), the stories of the Muses (Ov. *Met.* 5.294–5.678) including the embedded narrative of Arethusa in Calliope's voice (Ov. *Met.* 5.572–5.641), and Alcmena's conversation with Iole about giving birth to Hercules (Ov. *Met.* 9.273–9.323), among others.
[55] Heath 2011, 73.
[56] Foxhall 2012, 198, 201; Gillis and Nosch 2007, 174; Heath 2011, 73; Pantelia 1993, 494.

and working on a large loom cannot be done by one woman alone, so the task is simultaneously a legitimate female duty and an opportunity for women to gather, learn, impart wisdom, socialize, and weave stories.

Tuck surmises that singing was utilized by weavers as a mechanism for reproducing the intricate patterns of woven pieces, and the complexity of the song reflected the complexity of the pattern.[57] The close relationship between singing and weaving lends itself to a discussion of the connection between poetry and weaving. A physical link between poetry and weaving is elucidated by McIntosh Snyder, who notes the "mechanical parallels between the weaver's loom and the poet's lyre (…) between striking the 'strings' of the loom with the shuttle and striking the strings of the lyre with the plektron".[58] This link is also extremely evident in the Latin word *carmen* which means both 'poetry' and 'song'.

Therefore, the common shared element is song: poetry is work equivalent to song, while weaving is work put to song. Tuck adds, "indeed, virtually any pattern or design that is incorporated into the weave of a textile can be reduced to numeric sequences, given the gridlike structure of warp and weft threads of a piece of cloth."[59] The numeric sequence of song and of weaving which he describes could just as easily be applied to the numeric sequence of poetry, expressed through its meter.[60] Because all three of these media – poetry, song, and weaving – are composed from numeric elements they must have been related in the ancient mind.

I want to return to Tuck's earlier mentioned conclusion that weavers appropriated singing as a didactic tool to reproduce patterns.[61] Drawing on this concept, it can be inferred that as the difficulty level of the pattern increased, so too did the complexity of the song. A more complex song would then result in a more difficult to weave textile, as well as in a more elaborate textile. The interchangeability of song with poem affords through analogy the suggestion that the most complex poems would also be the most difficult for a poet to construct and ultimately the most elaborate works. Hence, by using weaving metaphor, Lucretius is not only detailing Epicureanism, he is highlighting his ability to create finely-woven didactic poetry.

It would not be unheard of for a male poet to assume the role of a weaver in the realm of song and poetry. After all, Lucretius has already shown himself as a weaver, through the use of vocabulary as discussed above, especially including *exordium* and *nexus*. Additionally, Achilles, the epitome of maleness, was immediately prior to the events of the Trojan War disguised as a woman and relegated to female tasks including weaving.[62] In *Iliad* Book 9, Achilles, formerly a weaver, now sings *klea andrōn*, "the glories of men",[63] just like Lucretius in *De Rerum Natura* Book 1 sings the accomplishments of Epicurus. Pantelia compares Achilles' singing of heroic deeds in Book 9 to Helen's weaving of the events of the Trojan War in Book 4 of the *Iliad*, and furthermore casts her as

[57] Tuck 2006, 542.

[58] McIntosh Snyder 1981, 193.

[59] Tuck 2006, 540.

[60] The parallel between poetry and weaving may be further strengthened by the ambiguity of whether Philomela's tapestry is a literal or, more likely, a pictorial (and possibly a pictographic) depiction of her rape by Tereus. Textiles have been observed cross-culturally to incorporate complex symbolic systems in order to convey messages and stories pictorially – either in patterns or in figures – rather than literally; if this practice was common in Roman culture, as Ovid's account of Philomela may indicate, Lucretius may likewise imagine himself a poet-weaver who encodes his didactic text with allusive and symbolic meaning.

[61] Tuck 2006, 542.

[62] Ov. *Met.* 13.162ff; Pseudo-Apollodorus, *Bib.* 3.13.8; Statius, *Ac.* 2.67ff.

[63] Hom. *Il.* 9.189.

an epic poet.[64] Just like an epic poet, Helen tells the story of heroes in a work that will survive her and continually tell the story woven upon it. By likewise weaving his poetry so carefully, Lucretius will similarly create a work like Achilles', Helen's, or Homer's which outlasts him and moreover ensures the survival of his name and of his glory through their association with the finely wrought poem.

Having suggested that we can consider Lucretius as a poet-weaver, I want to return to his calculated use of weaving metaphor to reinforce the Epicurean principles in his didactic. In Book 1 of the *De Rerum Natura*, Lucretius mocks several famous schools of philosophy,[65] including the teaching of Anaxagoras that every object can be broken down into a combination of the four basic elements, most famously through Lucretius' pun on the word *lignum*, 'wood', which includes the same letters as the word *ignis*, 'fire', thereby signifying that on its most basic level wood consists of the element fire.[66] McIntosh Snyder observes that this concept regarding the contents of substances, communicated verbally through a rearrangement of letters, reflects Lucretius' didactic equivalency between the letters and atoms as smaller elements which make up larger objects.[67] I would further postulate that weaving metaphor plays upon this same concept because individual, smaller threads are also woven together in order to form a larger object.

This connection between letters of a word and threads of a fabric is explicit in the ancient philosophies of Plato and his pupil, Aristotle.[68] Though Lucretius never mentions thread when he equates letters and atoms, the association of all three, which might have been overlooked were a reader not thinking of Lucretius as a weaver in light of his extensive use of weaving imagery, can be implied by the other obvious weaving metaphors in the didactic composition of Book 1 where weaving metaphors directly describe other concepts of Epicureanism.

Lucretius tells his reader outright how difficult the task of converting this Greek Epicurean philosophy into Latin poetry is, but adds that he accepts the undertaking regardless of that difficulty:

> *nec me animi fallit Graiorum obscura reperta*
> *difficile inlustrare Latinis versibus esse*
> *multa novis verbis praesertim cum sit agendum*
> *propter egestatem linguae et rerum novitatem*
> *sed tua me virtus tamen et sperata voluptas*
> *suavis amicitiae quemvis efferre laborem*
> *suadet et inducit noctes vigilare serenas*
> *quaerentem dictis quibus et quo carmine demum*
> *clara tuae possim praepandere lumina menti*
> *res quibus occultas penitus convisere possis*

> Nor does it escape me that it is difficult to illuminate the obscure findings of the Greeks with Latin verse, more so especially when it must be accomplished in new words because of the poverty of the language and the newness of the topic, but, however, your virtue and hope for pleasure of your sweet friendship persuades me to carry out however much labor and lead me to stay awake calm nights seeking precisely by which words and by which poetry I might be able to open clear eyes of your mind so that you are able examine thoroughly the things having been deeply hidden in those things.[69]

[64] Pantelia 1993, 495.
[65] Lucr. 1.705–1.920.
[66] Lucr. 1.891–1.892: *postremo in lignis cinerem fumumque videri / cum praefracta forent ignisque latere minutos*, "at last when, having been broken, ash and smoke can be seen in wood, and fire is able to lurk in those things divided into small pieces".
[67] McIntosh Snyder 1980, 133.
[68] Pope 1949, 77; McIntosh Snyder 1980, 35; Scheid & Svenbro 1996, 24.
[69] Lucr. 1.136–1.145.

This claim underlines Lucretius' confidence, which derives from his poetic prowess, and which he illustrates throughout Book 1 by his use and mastery of weaving metaphors to construct his didactic. Furthermore, Lucretius does not write low-level didactic which simply teaches; rather, his didactic poem has epic elements which elevate it, including the high epic meter of dactylic hexameter and the replacement of the typical epic muse with the god responsible for Lucretius' poetic creation – Epicurus.[70] Denoted in antiquity by Cicero as a craftsman,[71] Lucretius utilizes weaving implements and weaving metaphors as poetic devices to explain and communicate Epicurean philosophical principles. But on a second level, these metaphors equate the poetic work of Lucretius to an elaborately woven textile, an equation which in turn glorifies him as the writer capable of producing such intricate, well-designed, and well-executed didactic poetry.

[70] See Gale (1994, 201) and Thomas (1968, 43) for more information about the transition between the focus of the *De Rerum Natura* from Venus to Epicurus, and, thereby, the use of Epicurus as Lucretius' muse figure.
[71] Cic. *Q. Fr.* 2.9.3: *Lucreti poemata ut scribis ita sunt multis luminibus ingeni inultae tamen artis.*

Bibliography

Bailey, C. 1947 *De Rerum Natura* vol. 1, Prolegomena, Text, Critical Apparatus, and Translation. Oxford.

Bailey, C. 1947 *De Rerum Natura* vol. 2, Commentary Books I-III. Oxford.

Foxhall, L. 2012 Family Time: Temporality, Gender, and Materiality in Ancient Greece. In J. Marincola, L. Llewellyn-Jones, & C. A. Maciver (eds), *Greek Notions of the Past in the Archaic and Classical Eras: History without Histories*. Edinburgh, 183–206.

Gale, M. 1994 *Myth and Poetry in Lucretius*. Cambridge.

Gillis, C. & Nosch, M.-L. 2007 *Ancient Textiles: Production, Craft and Society; Proceedings of the First International Conference on Ancient Textiles, Held at Lund, Sweden, and Copenhagen, Denmark, on March 19–23, 2003*. Oxford.

Heath, J. 2011 Women's Work: Female Transmission of Mythical Narrative. *TAPhA* 141, 69–104.

Hendren, G. 2012 Woven Alliteration in the *De Rerum Natura*. *CJ* 107, 409–421.

Leonard, W. E. & Smith S. B. 1942 *T. Lucreti Cari De Rerum Natura Libri Sex*. Madison WI.

McIntosh Snyder, J. 1980 *Puns and Poetry in Lucretius' De Rerum Natura*. Amsterdam.

McIntosh Snyder, J. 1981 The Web of Song: Weaving Imagery in Homer and the Lyric Poets. *CJ* 76, 193–196.

McIntosh Snyder, J. 1983 The Warp and Woof of the Universe in Lucretius' 'De Rerum Natura'. *ICS* 8, 37–43.

Minadeo, R. 1969 *The Lyre of Science: Form and Meaning in Lucretius' De Rerum Natura*. Detroit MI.

Nussbaum, M. C. 1986 *The Fragility of Goodness: Luck and Ethics in Greek Tragedy and Philosophy*. Cambridge.

Pantelia, M. C. 1993 Spinning and Weaving: Ideas of Domestic Order in Homer. *AJPh* 114, 493–501.

Pope, S. R. 1949 The Imagery of Lucretius. *G&R* 18: 70–79.

Scheid, J. & Svenbro, J. 1996 *The Craft of Zeus: Myths of Weaving and Fabric*, trans. C. Volk. Cambridge MA.

Sedley, D. N. 1998 *Lucretius and the Transformation of Greek Wisdom*. New York.

Thomas, R. K. 1968 A Study of Venus' Symbolism in De Rerum Natura. M.A. thesis, Florida State University.

Tuck, A. 2006 Singing the Rug: Patterned Textiles and the Origins of Indo-European Metrical Poetry. *AJA* 110, 539–550.

West, D. A. 1994 *The Imagery and Poetry of Lucretius*. Norman OK.

14. Life Hanging by a Thread: The Weaving Metaphor in Lucretius

Matthew Johncock

Throughout the Epicurean didactic epic *De Rerum Natura* Lucretius employs a small group of gestalt metaphors,[1] which serve to explain his principal scientific theories, and to support the pillars of his moral philosophy – in turn enhancing his reader's understanding and setting them on the path towards Epicurean philosophical enlightenment and the associated state of calm detachment, ἀταραξία. By tracking the employment and development of these metaphors, we can increase our understanding of Lucretius' methodology of argumentation, and advance solutions to certain passages considered contradictory or otherwise problematic.[2] The focus of this chapter will be the extensive application of Lucretius' weaving metaphor in its primary application of explaining atomic compound structure.[3] The choice of the weaving metaphor for this task is of particular interest, since it is a conspicuous anomaly alongside Lucretius' other persistent metaphors, which are all deeply rooted in human experience, whether internal, or external from our interaction with fundamental laws and processes.[4] Nevertheless, it would be uncharacteristic for Lucretius to be misguided in his choice of metaphor, contradicting the carefully constructed analogies and metaphors that colour the whole *De Rerum Natura*. Instead, therefore, we might assume that weaving was sufficiently central to Roman life to be appropriate for Lucretius' explanatory cause.

[1] A gestalt metaphor is a complex compound metaphor consisting of multiple parts, and can be applied as a whole to explain a complex concept, or its individual parts applied to individual parts of the complex concept. For the gestalt metaphor to aid our comprehension of the complex concept, it must be rooted in our experience. Lakoff & Johnson 1980 consider these gestalt metaphors, which they term 'experiential gestalts', at length; *cf.* esp. 77–86 on the 'Argument is war' gestalt.

[2] This has been the topic of my doctoral research, which considers how Lucretius employs gestalt metaphors of life-cycle, weaving, liquids, boundaries and pleasure to enhance his explanation and the reader's subsequent understanding of the *processes* underlying his principal scientific doctrines. It is in this way that my approach differs from Cabisius 1984, and Murley 1947 and Gale 1994, 117–119, 121–122 (both in relation to the epic genre of the *De Rerum Natura*), who have considered Lucretius' socio-political and military metaphors respectively.

[3] Garani 2007, 156–187 considers Lucretius' weaving metaphor alongside other metaphors of 'fitting together', in particular in relation to his reception of Empedocles. Of these construction metaphors, weaving is by far the most persistent and broadly applied.

[4] Lucretius' gestalt metaphors of life-cycle, pleasure and liquids are clearly rooted in human experience. The gestalt metaphor of boundaries, meanwhile, is rooted in our perception and interaction with our surroundings. We as humans are individual entities bound off from all other things, and this experience is applied metaphorically to the laws of property or land division, and to the natural boundaries (rivers, mountains etc.) that separate nations or tribes – concepts exploited by Lucretius in his poem. *Cf.* Lakoff & Johnson 1980, 25–31.

Snyder[5] considers this assumption in relation to Lucretius' description of the loom (5.1350–3):

> *nexilis ante fuit vestis quam textile tegmen.*
> *textile post ferrumst, quia ferro tela paratur,*
> *nec ratione alia possunt tam levia gigni*
> *insilia ac fusi radii scapique sonantes.*[6]
>
> 5.1350–1353
>
> Plaited clothes came before woven garments. Woven garments came after iron, because the loom is made using iron tools, and such smooth heddle rods (?)[7] and noisy spindles, shuttles and bobbins (?)[8] cannot be created in any other way.

The terms in this passage are problematic since – to the despair of the modern translator – they are not clarified, and *insilia* and *scapi* are not found elsewhere in relation to parts of the loom.[9] Nevertheless, as Snyder notes, Lucretius' technical explanation signifies both his interest in, and his reader's knowledge of the loom and the weaving process. Snyder then recalls Crowfoot's observation of weaving terms in ancient literature:

> weaving and spinning were such common features of daily life that poets and playwrights expected their hearers to pick up any witty or fanciful allusion – a pun, the merest hint – to any tool or operation connected to them.[10]

This incisive general observation applies perfectly to Lucretius' employment of the weaving metaphor in his poem, as shall be explored in this chapter.[11] Weaving, as an everyday task, is an entirely appropriate metaphor for Lucretius' purpose, while its complex technology provides a broad range of specific processes applicable to multiple scientific contexts. That Lucretius chose to employ a persistent weaving metaphor and expected it to be comprehensively understood by his reader, makes

---

[5] Snyder 1983, 38–39.

[6] Latin quotations are from Rouse & Smith 1975 (Loeb edition), with some minor alterations.

[7] It is unclear what *insilia* are, but an etymological connection with *insilio* 'to jump' suggests a moving part. I follow Snyder's suggestion 'heddle rods', which were moved onto a bracket to separate the warp threads (as if, at a stretch, made to jump).Bailey and Gale's 'treddles' are unattested as part of the Roman warp-weighted loom; Bailey 1947, n. *ad loc*; Gale 2009, n. *ad loc*.

[8] It has been assumed that *scapi* refers to some kind of beam or rod, since the term was otherwise used of comparable long, thin structures such as plant stems and column shafts (OLD, *scapus*, 1, 2a). Bailey 1947, n. *ad loc.* and Gale 2009, n. *ad loc.*, 'leash rods', must, however, be incorrect since these were a medieval addition to the loom. Cloth-beam is a possible reading, since, if it was able to revolve in its place to roll up a finished portion of cloth (*cf.* Barber 1991, 106; Wild 1970, 61), this might have been noisy. Alternatively, the shed rod might, at a stretch, have been considered noisy as the warp threads struck it when moved. However, an altogether different suggestion may be advanced. Since the three terms *fusi*, *radii* and *scapi* are grouped together, perhaps *scapi* might have been another tool used to hold the weft while it was passed through the warp-threads. 'Bobbins' might, therefore, be a reasonable translation. In fact, Barber 1991, 107 notes that such an item was found in Troy II, and recognised as being made from 'the stem of a very young tree' – *i.e.* a *scapus*.

[9] Furthermore, Lucretius' knowledge of the techniques of weaving was perhaps not matched by his knowledge of its history, since weaving was in fact possible long before the invention of iron tools (Barber 1991, 79 places weaving in the 7th millennium BC, some 5,000 years before the invention of iron tools).

[10] Crowfoot 1936, 36.

[11] We should also heed Crowfoot's observation more generally and approach Lucretius' other principal metaphors, rooted as they are in human experience, in this way. Pope 1949, 79 later offered a similar exhortation to attend more assiduously to Lucretius' imagery in general: "It is possible that some of the problems in Lucretius will be resolved by a consideration of his imagery in its richness and variety, the reality which it gives to theoretic processes, and the clarity with which abstractions are realized". The later work of the likes of West, Gale and Garani has proved the prescience of this recommendation.

clear weaving's central position in Roman life. In reading the *De Rerum Natura* we should be especially alert to collocations of weaving terms, and to weaving terms repeated in similar contexts.

By this approach, we shall be able to unpick the intricacies of Lucretius' expansive weaving metaphor. Such is its extent that a full examination is impossible in a paper of this length. Therefore, special attention shall be paid to its primary applications in explaining the structure and dissolution of atomic compounds, and to the consequences these correspondences have for Lucretius' moral philosophy.[12] Exploring key examples within these doctrines will exemplify the breadth of Lucretius' weaving metaphor.

Atomic Bonding

It has been frequently discussed[13] how a philosophy that considers sensory evidence as the touchstone of truth (4.379–521; Ep. *ad Her.* 38) can be built upon a foundation of invisible bodies. The answer lies in the Epicurean theory that 'true reasoning' (*vera ratio*) must be applied to the evidence received by our senses, which are themselves unimpeachable (4.462–521; Ep. *ad Her.* 50–52). This *vera ratio* is informed by evidence we know from sensory experience to be true. Therefore, we must find a metaphor or analogy from our experience in order to explain processes hidden to our senses.[14] Metaphor – specifically, visible processes standing for the invisible – applied to the evidence we can perceive provides the truth when conclusive sensory evidence is unavailable.[15] Therefore, since atoms cannot be sensed by humans, Lucretius employs metaphors to depict them and the compounds they form, and how these compounds come apart again. The principal metaphors chosen for this purpose are life-cycle and weaving, which together explain respectively the birth-death cycle of things, and atomic bonding[16] and dissolution, accounting for the vast majority of physical processes in the *De Rerum Natura*.[17]

We first find metaphors of weaving and birth (part of the life-cycle metaphor) applied to atoms in the prologue:

quod superest, vacuas auris animumque sagacem
semotum a curis adhibe veram ad rationem,
ne mea dona tibi studio ***disposta*** *fideli,*
intellecta prius quam sint, contempta relinquas.
nam tibi de summa caeli ratione deumque

[12] As a result, I shall overlook its application to certain processes, such as sensory perception, weather and changes of state, and its role in explaining Lucretius' methodology or argumentation and composition (for which see Ann Glennie's chapter in this volume).

[13] Most recently by Asmis 1999, 2009; Schrijvers 2007; and Lehoux 2013, who considers how Lucretius' appeals to vision aid his explanations of things unseen.

[14] This approach is well summarised by Bailey 1947, 643–644.

[15] Epicurus (albeit not referring to metaphor) states that the senses and judgment must be applied together 'in order that we may have indications whereby we may judge both the problem of sense-perception and the unseen' (*ad Her.* 38; Bailey's translation).

[16] A weaving metaphor notably still present in modern scientific explanations of atomic interactions. Weaving metaphors in ancient atomism are also found, with less breadth and frequency, in Epicurus (διαλύω, of atomic dissolution; περιπλοκή and πλεκτικός, of atomic bonding), perhaps originating from Empedocles and Leucippus (to whom Aristotle attributes διαλύω and διάλυσις in *de gen. et corr.* A8, 325a2; KRS 545). Weaving metaphors had broad applications in philosophy and science in general. For an overview, *cf.* Harlizius-Klück 2011.

[17] Not all, however, since the concepts of *voluptas* and void, principally portrayed by the metaphors of pleasure and liquids, are central to Lucretius' theories of initiation and movement.

disserere *incipiam, et rerum primordia* ***pandam,***
unde omnis Natura creet res auctet alatque
quove eadem rursum Natura perempta resolvat,
quae nos ***materiem*** *et* ***genitalia corpora*** *rebus*
reddunda in ratione vocare et ***semina*** *rerum*
appellare suëmus et haec eadem usurpare
corpora prima, *quod ex illis sunt omnia* ***primis.***

1.50–61

What is more, apply open ears and keen judgment, far removed from worries, to true reasoning, lest you abandon my gifts, **laid out** before you with loyal devotion, disregarded before they have been understood. For I shall begin to **separate the threads** of the highest logical array of the heavens and the gods, and I shall **unfold** the first-threads of things, from which Nature gives birth, makes grow and nurtures everything, and into which the same Nature, spent, unties them again – these, when we are explaining the reasons for things, we have become accustomed to call the "**mother substance** and **procreative bodies** of things", and to name "the **seeds** of things", and to call habitually these same things "**first bodies**", because all things consist of these **first things**.

The metaphors are evident in Lucretius' atomic terminology itself, and the surrounding brief explanations of the atoms' qualities. The birth metaphor is the more noticeable of the two, evident in the terms *materies* (etymologically linked to *mater*, 'mother', and thus best translated as 'mother substance'),[18] *genitalia corpora, semina, corpora prima* and *prima*, and the striking description of *Natura* (herself the embodiment of birth, cognate with *nascor*, 'to be born')[19], who 'gives birth to, makes grow and nurtures everything' utilising these creative bodies.

The appropriateness of this metaphor for describing the creation of things has led many to suppose[20] that *primordia* carries a 'creative' meaning of 'first-beginnings' (from *ordior*, 'to begin')[21]. However, a different metaphorical sense can be garnered from the root *ordo*, which carries a weaving meaning evident in its cognates *exordium*, 'the warp set up on the loom before the web is started',[22] and one sense of *ordior*, 'to lay the warp of (a web)'.[23] With this in mind the *primordia* would be the 'first threads' – i.e. the warp threads – set on the loom to initiate weaving.[24] This reading is eased by

[18] A translation supported by the close proximity of *mater* (1.168) and *materies* (171) in Lucretius' argument that if nothing could come from nothing, then all things could come from all, and thus no fixed 'mother' or 'mother-substance' would be required (experience tells us otherwise). Bailey 1947, n. *ad* 1.168 and Brown 1984, n. *ad* 1.170–171 note the etymological figure, but nevertheless translate *materies* as 'matter'.

[19] Ernout-Meillet 1939, *s.v. nascor* give the first meaning of *natura* as "action de faire naître, naissance".

[20] Among them Bailey 1947, Leonard & Smith 1965, Brown 1984 (all n. *ad* 1.55); Sedley 1998, 38–39; Schrijvers 2007, 258.

[21] OLD, *ordior*, 2, a sense also carried in the other definitions offered by the OLD. This usage would be consistent with occurrences in Pacuvius (fr. 52.3, Warmington), and Cicero (*Leg.* 2.7, twice; *Part.* 7.21).

[22] OLD, *exordium*, 1; *cf.* esp. *Inc. trag.* 181, *te ab summo iam detexam exordio* 'I shall now weave out a web for you from the start of the warp'.

[23] OLD, *ordior*, 1.

[24] A reading advocated by Snyder 1983, 40–41, who suggests 'first warp-threads'. The apparent conflict between atoms as essentially round seeds (*cf.* 3.186–207, where soul atoms are compared to poppy seeds and contrasted with wheat ears) is never confronted or explained in the poem. It should not, however, prove too problematic, since each metaphor serves a different purpose, one explaining atoms' creative powers, the other their ability to form and be untied from compounds. Each highlights one aspect of atoms while suppressing others, a function of metaphor explored by Lakoff & Johnson 1980, 159–184, esp. 164–166.

the accompanying explanation in the prologue that Nature 'unties' (*resolvat* 1.57) things back into their *primordia*, or 'first threads'[25]. Even Lucretius' description of his composition emphasises the weaving metaphor, as proofs are 'separated and laid out' (*disposta*, like warp threads on a loom), and he undertakes to 'separate the threads' of (*disserere*, *i.e.* 'explain') the heavens and gods, and 'unfold the first-threads of things' (*rerum primordia pandam*).[26] Lucretius' atomic weaving metaphor is consistent with his methodology: he will unravel the truths of the universe by unravelling things into their 'first-threads' (*i.e.* atoms), and then set them up on the loom in order to weave them together in logical order (expressed by *exordia* in 1.149) for the benefit of his reader's understanding[27].

'Atoms are the first-threads of things'

These atomic threads are 'first' because they are the original threads into which Nature unties things, and cannot be split further into constituent threads. If this could occur, re-creation would be impossible, just as a cloth unravelled too far cannot be woven back together. Lucretius explains this sense of *primordia* in his doctrine of smallest parts (1.599–634), which are said to cling together so they can neither be 'torn apart' (*revelli* – 1.608) nor have anything 'torn away' (*avelli* – 613) from them. Lucretius then expands further:

> *denique si minimas in partes cuncta* ***resolvi***
> *cogere consuesset rerum Natura creatrix,*
> *iam nil ex illis eadem reparare valeret* (…)
>
> 1.628–630

> Lastly, if Nature the creator had been accustomed to force everything to be **untied** into their smallest parts, now the same Nature would have no power to restore anything from them (…)

Therefore, atoms are not only the threads that create compound structures, but also the smallest point beyond which unweaving cannot occur without the destruction of everything.[28]

Any doubt about this meaning of *primordia* can be dispelled further by considering Lucretius' choice of other atomic terms rooted in weaving metaphor. Firstly, Lucretius employs the term *exordia*, which shares the same root as *primordia* and carries similar base meanings of 'beginning' and 'threads'.[29] In Lucretius, *exordia* occurs eleven times, of which five replace *primordia* in the sense 'atoms' (2.333; 3.31, 380; 4.45, 114), and a further four pertain to either '(atomic) beginnings'

[25] The translation 'first-threads' seems preferable to 'warp' or 'warp-threads', principally owing to the importance of primacy to Lucretius' atomic vocabulary, and to highlight the semantic connection between *primordia* and *principia*, which are used interchangeably when one would be unmetrical (*e.g. principia*, *primordium*).

[26] *dispono* formed from the prefix *dis-*, which pertains to separation, and *pono*, 'to lay out', suggesting the separating and laying out of (threads); OLD, *pando* 2a, 'to spread or stretch out cloth'; *dissero* formed of *dis-* and OLD, *sero*[2], 'to entwine, interlace', thus 'to separate the threads'.

[27] Despite the assertion of Scheid & Svenbro 1996, 165 that Lucretius does not explicitly state that he is 'weaving' his poem or that it constitutes a metaphorical 'fabric', our reading here suggests it is heavily implied.

[28] For the atoms' inability to be unwoven, and the effects caused if they could be, *cf.* 1.518–519, 528–529 (note the collocation of *dissolui* ('to be untied' – 528) and *retexi* ('to be unwoven' – 529)), 556–557, 559.

[29] OLD, *exordium*, 2 and 2b refer to beginnings and *exordium*, 1 to the start of a web, meanings which combine in *exordium*, 3, 'the primal or rudimentary stage in the development of things' (used by Lucretius of his atomic 'first-threads' and general '(atomic) beginnings') and *exordium*, 4, 'the introductory part of a speech or book', which suggests the beginning of the 'thread' of a narrative or argument.

(5.331, 471) or simultaneously to atomic beginnings and 'congregations of atoms' (2.1062; 5.430).[30] Another usage is in the sense of 'first arguments' (1.149), embodying the favoured ancient metaphor 'rhetorical arguments are woven fabrics'.[31] The established weaving metaphor of *exordia* in Latin and its close relationship with *primordia* in the *De Rerum Natura* strengthen our translation of *primordia*, and deepen our understanding of atoms as 'first threads' from which all things are made.

Secondly, let us consider Lucretius' employment of *filum* ('thread') as an atomic term.[32] Among proofs that atoms are colourless, Lucretius provides an analogy from weaving: just as 'when [purple cloth] is pulled apart **thread by thread**' (***filatim*** *cum distractum est* – 2.831) its colour becomes fainter (and eventually disappears) as its threads become smaller, so can coloured things be made of colourless atoms. Or, to bring Lucretius' *primordia* more clearly into view, a coloured item can be torn until only colourless 'first-threads' remain. In another example, Lucretius explains the emission of sensory films from things by an analogy with light cast from a theatre awning (4.75–89). Since light emanates from awnings, Lucretius argues, so too must all things emit *effigias tenvis* ('fine images' – 4.85) endowed with *subtili filo* ('fine-woven [atomic] thread' – 4.88). The weaving sense of *filum* here is enhanced by the semantically-related adjectives *tenvis* ('fine')[33] and *subtilis* ('fine-woven')[34], and also by the woven nature of the awning with which the analogy is made.[35] Both examples emphasise the woven nature of atomic compounds, and strengthen our readings of *primordia* and *exordia*.

Thirdly, a particularly important term in the context of Lucretius' 'atomic bonding is weaving' metaphor is the atomic attribute *ordo* ('order', '(linear) array'). The word pertains to things set out in rows or levels, or to physical or logical order,[36] and thus appropriately to the regimented pattern of (straight) threads in a woven item. This is the weaving meaning shared with the atomic terms *primordia* and *exordia*, whose suffixes are from the same root. This shared etymology is emphasised by three collocations of *ordo* and *primordia* in the *De Rerum Natura* (1.1021–1022; 2.883–884; 5.184–194).[37] These passages, in addition to emphasising a shared weaving meaning by etymological figure, also explain the importance of changed *ordo* among the *primordia*. In particular, among a discussion of different things arising from others (worms from the ground; different animals from the same grass; fire from wood – 2.871–885), Lucretius asks, do you see now, therefore, that it matters greatly in which **array** the **first-threads** of things are, and how they are placed (...)? (*iamne vides igitur magni* ***primordia*** *rerum / referre in quali sint* ***ordine*** *quaeque*

[30] Four occurrences of *exordia* are in the phrase *exordia rerum*, a clear verbal echo of *primordia rerum*, frequent in the opening two books (32 occurrences), while another is in the phrase *exordia prima*. Thus, *primordia* and *exordia* are more closely associated, strengthening our reading of a shared weaving metaphor between them.

[31] For weaving terminology in rhetorical contexts, *cf. e.g. implicitas* ('tied together') in Cic. *Inv.* 1.97 and *contextus* ('interwoven') in Cic. *Fin.* 5.83, Snyder 1981, 193–194, and Nagy 2002, 70–98.

[32] The word occurs six times in the *DRN* (and once in its derived form *filatim*), of which four are in its alternative meaning of 'form', 'figure' or 'shape' (2.341; 5.572, 581, 589). A fifth, though, is used of the 'fine threads of a spider's web' (*aranei tenvia fila* – 3.383), nature's paradigm of weaving.

[33] Used of woven garments and cloth in *e.g.* Pl. *Men.* 518; Verg. *A.* 4.264, 8.33; Tib. 2.3.52; Ov. *Ars.* 3.445.

[34] Cognate with *tela*, 'web (on the loom)' or specifically 'the warp threads', as noted by Ernout-Meillet, *s.v. subtilis*: *sans doute terme de tisserand: de *sub tela "qui passé sous la chaîne", cf.* subtemen, *qui désigne les fils les plus fins de la trame.*

[35] A weaving analogy strengthened by the preceding analogy, in which 'cicadas take off their snug tunics' (*teretis ponunt tunicas* (...) *cicadae* – 4.58) and 'the slippery snake takes off its clothing' (*lubrica serpens / exuit* (...) *vestem* – 4.60–61). The sensory films also consist of a woven texture, albeit far finer.

[36] Evident in the various definitions given by the OLD.

[37] Note also *dispositura* (5.192) nearby, which carries a similar sense appropriate to 'laid out threads' as its cognate *disposta* in 1.52.

locata (…)? – 2.883–884, *cf.* 1.908–909; 2.894–896). Changed *ordo* of the 'first-threads' causes variety in things made of similar atoms, just as changed order of the warp threads in weaving creates different patterns of cloth from the same threads[38]. This analogy emphasises still further the range of scientific applications implicit in the term *primordia.*

This doctrine is emphasised in a frequently repeated, occasionally modified list of atomic attributes:

> (…) *sunt quaedam corpora quorum*
> *concursus motus* ***ordo positura*** *figurae*
> *efficiunt ignis, mutatoque* ***ordine*** *mutant*
> *naturam* (…)
>
> 1.684–687
>
> (…) there are certain bodies whose collisions, motions, **array**,
> **placements**, shapes make up fire, and by changed **array** they
> change its nature (…)

> (…) *concursus motus* ***ordo positura*** *figurae*
> *cum permutantur, mutari res quoque debent.*
>
> 2.1021–1022
>
> (…) when the collisions, motions, **array**, **placements**, shapes are
> now changed, things must also be changed.

We note the repetition of 1.685 in 1.1021, and the polyptoton *ordo (...) ordine* in 1.685–686, and also the vivid juxtaposition *ordo positura,* a near tautology embodying the fundamental premise of atomic order. A strongly reminiscent asyndetic list occurs shortly before our first example above, with *ordo positura* replaced by *conexus*:[39]

> (…) *varios* ***conexus*** *pondera plagas*
> *concursus motus, per quae res quaeque geruntur.*
>
> 1.633–634
>
> (…) the various **interweavings**, weights, blows, collisions,
> motions by which everything is carried out.

Again Lucretius emphasises the importance of collision (here expressed also by *plagas*), motion and form (here *pondera* instead of *figurae*), but we now find the concept of position or order supplied by *conexus*, a weaving term from *necto* 'to weave', 'to twine together'.[40] That *ordo* and *positura* can be interchanged with *conexus* in this way emphasises their weaving meaning and also the weaving metaphor applied to Lucretius' atoms.

'Compounds are woven structures'

With the intrinsic weaving metaphor of Lucretius' atoms confirmed, we can now expand upon the concept of bonding expressed by *ordo* and *conexus* above. Atomic compounds in Lucretius are consistently explained in terms of interweaving. This is particularly clear in Lucretius' second argument, that nothing dies to nothing (1.215–264), in which compounds are said to be

[38] For the various patterns created by the arrangement of the starting border, *cf.* Harlizius-Klück 2016.

[39] The term seldom appears outside Lucretius. TLL, *conexus* lists eight other occurrences, of which the earliest is Vitr. 10.1.5 (also in a weaving context). Its cognate *conexio* is far more common.

[40] These, and several other weaving definitions, are offered by OLD, *necto*.

formed by atomic *nexus* ('bindings' – 1.220, 240, 244), which give compounds *textura* ('(woven) texture' – 1.247), and *contextum* ('interweaving' – 1.243). If these woven compounds could be destroyed to nothing, no special force would be needed to 'untie' (*exsolvere* – 220; *dissolvere* – 243) them. In fact, things cannot be destroyed unless a special force, sufficient to break them apart, gets in and 'unties' (*dissoluat* – 223) them. Things are made from interwoven atoms and can be untied back into them, but no further (*cf.* the smallest parts theory, discussed above).

The nature of these interweavings, and the *ordo* and *positura* of the first-threads within, accounts for the varied qualities of things, just as it does in weaving on the warp-weighted loom. The reason different blows are needed to 'untie' different things is that their 'mother-substance' (*materies* – 1.239) is 'less or more entangled among itself in bonds' (*inter se nexus minus aut magis indupedita* – 1.240). Thus, a given thing's ability to withstand impact directly correlates to how tightly its first-threads are interwoven. This is emphasised later among a discussion of atoms coming together and apart by constant movement (2.95–111), preceding the famous analogy of dust motes in the sunlight (2.112–124). Lucretius describes the formation of dense, heavy substances:

> *et quaecumque magis condenso conciliatu*
> *exiguis intervallis convecta resultant,*
> ***indupedita** suis **perplexis** ipsa figuris,*
> *haec validas saxi radices et fera ferri*
> *corpora constituunt et cetera de genere horum.*
>
> 2.100–104
>
> And those which, by being more tightly-packed in union, jump back and are carried along by small intervals, themselves **entangled** by their own **intertwined** shapes, these constitute the strong roots of stone and the cruel bodies of iron, and others of their kind.

Here the ability of atoms in iron and stone to be entangled explains both their density (and therefore weight) and their creation through entangled atomic compounds that cannot be easily separated. The density (or rather, viscosity) of liquids is then explained in similar terms. Lucretius contrasts the flow of wine and olive oil through a colander:

> *et quamvis subito per colum vina videmus*
> *perfluere, at contra tardum cunctatur olivom,*
> *aut quia nimirum maioribus est elementis*
> *aut magis hamatis inter se* ***per**que **plicatis**,* (…)
>
> 2.391–394
>
> And we see wine flow through a colander as quickly as you like, but contrarily slow olive oil lingers, either because no doubt it is of larger particles, or those more hooked and **entwined** among themselves, (…)

The difficulty of *maioribus elementis* aside,[41] an explanation given for the slow movement of oil is that its *elementa* are conjoined by hooks entangled with the atoms' thread-like

[41] This phrase has lead commentators, among them Giussani 1923, Bailey 1947 and Smith 1975 (all n. *ad loc*), to assume confusion on Lucretius' part, because he cannot mean that each olive oil atom (assuming *elementa* = *primordia*) cannot fit easily through a hole in the colander – for they would then be visible. Instead *elementa* might specifically mean 'atomic congregations' or 'particles', a translation supported by the regular meaning of *elementa*, 'alphabet' (*i.e.* 'a group of letters' – *cf.* Lucretius' analogy between shared atoms in things and letters in words, with the example of the shared *elementa* (*i.e.* the grouping *IGN*) of *lignum* and *ignis* – 1.907–914). With this interpretation, Lucretius' two possible explanations of the oil's slow passage would be: i) that the oil's *primordia* form large congregations (= *elementa*) that pass with difficulty through a small hole, or ii) that these congregations are entangled by hooks into larger groups, which get caught trying to pass simultaneously through multiple holes.

shapes.[42] As a result, oil atoms cannot flow singularly through the colander's single holes but, we infer, pass through with difficulty in large chains of atomic groups.

The weaving metaphor explains both the density and cohesion of substances. This is entirely appropriate, as woven fabrics can differ in strength, weight or density by the degree to which their threads are entwined: the greater the gaps between them, the softer, lighter or weaker the item becomes.[43] Again, a woven item's individual strands are more difficult to separate the more closely interwoven they are, and if we try to draw a thread from a densely-woven cloth, many might come out with it.[44]

'Destruction is unweaving'

With the woven nature of compounds confirmed, we can now consider the commonest facet of Lucretius' atomic threads metaphor: 'the destruction of atomic compounds is unweaving'. We have seen glimpses of this metaphor already in 1.57 (*Natura* untying things into their *primordia*), in Lucretius' second argument, and subsequent proofs, that 'Nature **unties** everything back into its bodies, and does not **destroy** things to nothing' (*quidque in sua corpora rursum / **dissoluat** Natura neque ad nilum **interemat** res* – 1.215–216), and among his explanations that atoms themselves cannot be 'untied' or 'torn apart' (1.599–634). Lucretius expands this metaphor across the six books of the *De Rerum Natura*, extending its application to a wide variety of theories.

Woven atomic congregations hold all things together, from the smallest groups of atoms to the world itself. For things to disintegrate, unweaving (signified by *solvo* and its compounds, or synonyms such as *retexo*) must be initiated, whether by an external blow (the noun or participle *ictus* – 1.222, 528, 1055; 2.952, 954; 6.514; or *plaga* – 1.528; 3.818; 5.363) or penetration inside from without (*penetro* – 1.223, 529; 3.471). A third cause, also frequently driven by a blow (2.99, 136, 241 etc.), is movement, embodied in the analogy for random atomic movement of dust motes dancing in the sunlight (2.112–124), which are 'driven along by repeated unions and **tearings apart**' (*conciliis et **discidiis** exercita crebris* – 2.120). Unweaving is the result of each process, and appropriately so, since a woven item can be untied by an external blow, by being penetrated and unravelled, or by being torn apart by opposite movements. In each case initial unravelling can result in complete dissolution down to the first-threads.

The dust motes analogy depicts the process of unweaving at the smallest level. It is equally applicable, however, on a macroscopic level, as seen among Lucretius' consideration of the infinite universe at the end of Book 1. Matter and void are infinite (1.958–1051) and the universe has no middle, so atoms fall downwards from all directions and never lie at permanent rest (1.1052–1113). Were matter finite, 'it would be carried **untied** through the great void' (*ferretur magnum per inane **soluta*** – 1.1018) or rather would never have been brought together to create anything (1.1019–1020). Again, infinite matter is required to keep the world alive and, as an animal deprived of food dies, 'so all things must be **untied** as soon as mother-substance has ceased to be supplied' (*sic omniadebent / **dissolui** simul ac defecit suppeditare / materies* – 1.1039–1041). Finally, Lucretius argues against

[42] The process of felting brings a similar result, and Lucretius may have this in mind here. It is difficult for the modern reader, however, not to think of Velcro.

[43] A concept echoing Aristotle's use of the cognates of *solvo*, λύω and διαλύω, 'to loosen' for the rarefication of the four elements into one another (esp. in *Cael.* 3.6), and for changes of state (frequently in *Meteor.* 4).

[44] In the case of the oil analogy we might also think of pulling a clump of hair from a plughole (with each hair representing an *elementa* or congregation of atoms). This is made difficult both by the hairs being intertwined into bigger congregations, and by each congregation passing through multiple holes at once.

assumptions that the universe has a middle. While some argue the earth cannot 'be untied' (*resolvi* – 1.1055) because all presses upon it to the middle, in fact there is no middle and thus the earth is susceptible to future untying. Similarly, matter does not tend the other way, since then the world would be '**untied** instantly through the great void' (*subito magnum per inane* ***soluta*** – 1.1103, repeated almost verbatim from 1.1018; *cf.* also *solventes* in 1.1108). The metaphor of unweaving explains the destruction of all things into their first-threads.

Unweaving and Death

This destruction is expressed by the important recurrent metaphor of death, an integral part of Lucretius' gestalt 'life-cycle' metaphor.[45] This is emphatically introduced in Lucretius' second argument, that nothing is destroyed to nothing (1.215–70), as seen in his opening gambit:

> *huc accedit uti quidque in sua corpore rursum*
> *dissoluat Natura neque ad nilum* ***interemat*** *res.*
> *nam si quid* ***mortale*** *e cunctis partibus esset,*
> *ex oculis res quaeque repente erepta* ***periret****;*
> *nulla vi foret usus enim quae partibus eius*
> *discidium parere et nexus exsolvere posset.*
>
> 1.215–220

> Added to this is that Nature unties each thing into its own bodies again and does not **destroy** things to nothing. For if something were **mortal** in all its parts, each thing would **perish** and be snatched from our sight in an instant; for there would be no need of force, which could bring about the tearing up of its parts and untie its bonds.

Again it is a collocation of terms that alerts the reader to an important metaphorical correspondence, an effect aided by the animism (*i.e.* the attachment of animate qualities to inanimate objects) of *periret* and, especially, *mortale*. Mortality is applied to all things (*quidque*), although not their constituent parts (*e cunctis partis* here stands for 'atoms'). Most importantly for Lucretius' explanatory purpose the metaphor of death is synonymous with the metaphor of untying: a thing would die (*periret*) unless a specific force were required to cause 'tearing apart' (*discidium*) and to 'untie the bonds' (*nexus exsolvere*) of the object. Later in the argument, the same force would 'kill' (*conficeret* – 239) all things, unless their *materies* held them 'less or more entangled among themselves in bonds' (*inter se nexus minus aut magis indupedita* – 1.240). A touch would be sufficient cause of 'death' (*leti* – 240), since there would be no 'fabric' (*contextum* – 1.243) of everlasting particles requiring a force 'to untie' (*dissolvere* – 1.243) it. The two metaphors are interchangeable: the unweaving of a thing into its first-threads is its moment of death, and *vice versa*. Conversely, something immortal cannot be untied, as we discover among Lucretius' smallest parts theory: 'the **first-threads** of things must be of **immortal** body, into which each thing might be **untied** at the last moment' (*esse* ***inmortali primordia*** *corpore debent, /* ***dissolui*** *quo quaeque supremo tempore possint* – 1.545–546). In short, for the keen reader of Lucretius a term of unweaving should bring death to mind, a term of death unweaving.

[45] Death and birth are equal partners in the cycle of life, an intrinsic part of Lucretius' physical and cosmological science. This duality is symbolised in particular by *Natura*, who (despite the birth connotations of its name) both creates and destroys things (1.56–57), and also by the complementary opening arguments that nothing comes from nothing (1.149–214) and nothing is destroyed to nothing (1.215–270).

Unweaving soul and body

The interchangeability of these synonymous metaphors ('unweaving is death', 'death is unweaving') assumes a special relevance in Book 3 as Lucretius explores the nature of the soul and its interaction with the body. The soul is bodily, Lucretius argues (3.161–176), and therefore consists of a woven structure. Specifically it is 'exceedingly fine-woven' (*persubtilem* – 179) in nature – a quality applied elsewhere (in its non-intensive form *subtilis*) to other atomic compounds.[46] Lucretius then expands his description:

haec quoque res etiam naturam dedicat eius,
quam ***tenui*** *constet* ***textura*** *quamque loco se*
contineat parvo, si possit ***conglomerari****,* (…)

3.208–210

This proof besides also makes its nature clear, of how **fine** a **texture** it consists and in how small a space it might hold itself together, if it could be **gathered into a ball**, (…)

Two complementary characteristics are applied to the soul here. It is *tenuis* like fine-woven cloth, but can also be gathered into a small ball, like yarn in a clew or wool gathered for teasing. The soul's constituent parts are also woven in nature: the breath leaving the dying man is 'fine' (*tenvis* – 3.232), and the accompanying heat 'loose-knit' (*rara* – 3.235). As with other compounds, several of the soul's qualities can be expressed by how loosely its threads are intertwined.

Like other compounds the soul can also be woven together with the body, albeit with an important distinction. Whereas other compounds are woven together only on the outer surface, the soul is '**interwoven** through the blood vessels, organs and flesh' (***nexam*** *per venas viscera nervos* – 3.217) such that they exist as one, 'because the first-things have been **intertwined** in this way from first birth' (***implexis*** *ita principiis ab origine prima* – 3.331). Body and soul could not, Lucretius asserts (3.688–697), 'be woven together to such an extent' (*tanto opere adnecti* – 3.688) if the soul entered whole from outside because it is 'so **interwoven** through the blood vessels, organs, flesh and bones' (*ita* ***conexa*** *est per venas viscera nervos / ossaque* – 3.691–692), and we see they are so 'interwoven' (*contextae* – 3.695) that they cannot 'untie' (*exsolvere* – 3.696) themselves safely from the organs of the body.

This metaphor contrasts the weaving together, either end-to-end or surface-to-surface, of two pieces of cloth, which depicts the usual process of compounds cohering, and the interweaving of two types of thread (here those of soul and of various bodily organs) to make a patterned or textured cloth from scratch. The metaphor is still more specific in an earlier denial of Democritus' theory of how body and soul atoms are combined (3.370–95). Contrary to Democritus' 'first-threads' (*primordia* - 3.372) of body and soul, which are 'laid out' (*adposita* – 3.373) alternately one-by-one to 'weave together' (*nectere* – 373) the limbs, Lucretius asserts the soul atoms are in fact smaller and fewer (3.374–7), and therefore 'set out **loose-knit** throughout the limbs' (***rara*** *per artus / dissita sunt* – 3.376–377). We are to think of two types of thread – one thick, the other fine – being set at intervals on the loom.[47] These are woven together not in equal quantities, but with

[46] Lightning (2.385; 6.225), vision films (4.88), mental image films (4.743, 747), and wind (4.901). Atoms themselves (specifically the weaving term *exordia*) are described as *subtilis* in 4.114–115.
[47] A process captured by *adposita* and *dissita*, which together recall *disposta* in 1.52, discussed above.

the thick thread forming the main structure, and the thin inlaid at intervals. Unlike two compounds stitched or woven together surface-to-surface, which can be disentangled without destroying either of them, body and soul are 'finely interwoven' (*subtiliter* (…) *conexae* – 3.739–740) from birth, and cannot be disentangled without the destruction of both.

This concept, and the earlier assertion that souls cannot 'untie' (*exsolvere* – 3.696) themselves from bodies and remain 'unharmed' (*incolumes* – 3.696), together embody the theory that the unweaving of body and soul symbolises the point of death for living things. In fact this has already been introduced among arguments against the Empedoclean theory that the four elements are required for growth (1.803–829). Instead, Lucretius asserts, we need atoms, which also make up the four elements, for growth, or else life 'would also be **untied** from all our flesh and bones' (*omnibus e nervis atque ossibus* ***exsoluatur*** – 1.811). A description of a blow to the body extrapolates this metaphor further:

> ***dissoluuntur*** *enim* ***positurae*** *principiorum*
> *et penitus motus vitales* ***inpediuntur****,*
> *donec materies, omnis concussa per artus,*
> *vitalis animae* ***nodos*** *a corpore* ***solvit***
> *dispersamque foras per caulas eiecit omnis.*
> *nam quid praeterea facere ictum posse reamur*
> *oblatum, nisi discutere ac* ***dissolvere*** *quaeque?*

2.947–953

For the **arrangements** of first-things are **untied** and the vital motions are deeply **entangled**, until the mother substance, shaken up throughout all the limbs, **unties** the vital **knots** of the soul from the body, and, scattered outside, casts it out through the pores. For what else do we reckon the inflicted blow could do, except to shatter and **untie** everything?

The weaving metaphor here is undeniable, and – particularly in the phrase *vitalis animae nodos* – introduces correspondences, extrapolated extensively in Book 3, between mortality and the unweaving of body and soul.

A particularly emphatic example of Lucretius' intertwining of death and unweaving occurs at the beginning of his arguments that body and soul are inextricably entwined and cannot live separated (3.323–369). They ensure each other's mutual safety:

> *nam communibus inter se radicibus haerent*
> *nec sine pernicie* ***divelli*** *posse videntur.*
> *quod genus e thuris glaebis* ***evellere*** *odorem*
> *haud facile est, quin intereat natura quoque eius,*
> *sic animi atque animae naturam corpore toto*
> ***extrahere*** *haud facile est, quin omnia* ***dissoluantur****:*
> ***inplexis*** *ita principiis ab origine prima*
> *inter se fiunt consorti praedita vita;* (…)

3.325–332

for they cling among themselves by shared roots, and it is seen that they cannot be **torn apart** without destruction. In which class it is not easy to **tear** odour **away** from lumps of frankincense without its nature also dying, thus it is not easy to **draw** the nature of mind and soul **out of** the whole body without everything being **untied**: with first things **interwoven** in this way from first birth they exist among themselves endowed with shared life; (…)

Body and soul both live intertwined and die when untied.[48] The soul's destruction is caused by being torn apart and dragged away, just as the nature of frankincense 'dies' when its odour is torn away (*i.e.* by burning). This concept is strengthened by the collocation of two words expressing dying, and five expressing unweaving.[49]

Further intertwining of the weaving and death metaphors occurs in Lucretius' discourse on the soul's mortality (3.417–829). In his first proof, Lucretius compares the soul leaving the body with water pouring from a shattered vessel, and smoke and mist dispersing into the air. Since soul atoms are even smaller and swifter than those of these substances (3.425–433), he urges his reader, 'believe that the spirit is also poured out and **dies** much more quickly and is **untied** into its first bodies much more swiftly' (*crede animam quoque diffundi multoque* ***perire*** */ ocius et citius* ***dissolvi*** *in corpora prima* – 3.437–438).[50] There is a near tautology in this description, since death and untying are simultaneous and inextricably linked, as emphasised by the chiasmus *perire ocius et citius dissolvi*. Elsewhere, Lucretius argues that a gradually dying man's soul dies gradually too (3.526–547), since the soul is 'torn up' (*scinditur* – 531) and so 'it must possess a mortal nature' (*natura* (…) *mortalis habendast* – 3.531–532).[51] Finally, Lucretius explains that the soul cannot exist outside its fixed place in the body, just as trees are not found in the sky, nor blood in sticks, etc. (3.784–799), 'for which reason, when the body **dies** you must confess that the soul, **drawn apart** throughout the whole body, **perishes** too' (*quare corpus ubi* ***interiit, periisse*** *necessest / confiteare animam* ***distractam*** *in corpore toto*' – 3.798–799). The weaving term *distractam* (*cf.* 2.831) depicts the threads of the soul 'drawn apart' in the body, like a cloth being unravelled in all directions. The simultaneous death of body and soul is emphasised by the juxtaposition *interiit, periisse*, which can scarcely be captured in translation. Lucretius emphasises repeatedly that death cannot occur without the unweaving of body and soul, nor the unweaving of body and soul without death.

The consolation of the *De Rerum Natura*

We must now consider these persistent weaving and death correspondences in the broader context of the *De Rerum Natura*. Throughout the first two books Lucretius has carefully intertwined the metaphors of unweaving and death to explain the dissolution of things. Therefore, when unweaving and death terms are applied to body and soul in Book 3, the keen reader should map the atomic processes that cause the destruction of things onto the death of living beings. The inevitable

[48] For this concept *cf.* esp. 3.576–579: 'when all the body's covering is **untied** […] you must confess that the spirit and the sensations of the mind are **untied**' (***resoluto*** *corporis omni / tegmine* […] ***dissolui*** *sensus animi fateare necessest / atque animam* – 3.376–378). The emphatic etymological figure *resoluto* (…) *dissolui* emphasises these co-dependent processes.

[49] A metaphor is aided by a subsidiary horticultural metaphor, expressed by *divelli*, *evellere* and *extrahere*, of a plant being uprooted. This image is picked up by the polyptoton *convellitur* (3.340) (…) *convulsi* (3.343) in the contrasting exemplum of heat departing from water (3.339–349). The metaphor complements the concurrent weaving metaphor of *dissoluantur* and *inplexis*, since roots themselves are also threadlike and entangled. Just as uprooting a plant damages the soil in which it is planted, so pulling out threads damages a cloth, and so the departure of the soul damages the body.

[50] Note the liquids metaphor *diffundi* ('poured out') applied to the soul's departure, and the literal *diffluere* ('flow away' – 3.435) describing the water leaving the vessel. This technique of employing terms appropriate to a literal description also to its accompanying analogy is typically Lucretian, and West 1994 (*cf.* 43–48 for examples) labelled it "transfusion of terms". This rhetorical technique is also employed by Lucretius to join multiple connected arguments together, in which context West 1975 labelled the shared terms 'responsions'.

[51] For a similar correspondence, *cf.* 3.640–641: *at quod* ***scinditur*** *et partis discedit in ullas, scilicet* ***aeternam*** *sibi* ***naturam*** *abnuit esse* ('but because it is **torn up** and departs in all directions, rest assured it does not allow itself to possess **eternal nature**'), with *scinditur* picking up from *discissa* ('torn apart') in 639.

metaphorical 'death' of inanimate objects anticipates the inevitable literal death of body and soul. To draw this link the reader must identify the transition of the *metaphorical* 'death' of things to the *literal* death of living beings. The unweaving metaphor applied to both guides the way. As a result, the reader who has understood and accepted that things perish by being unwoven into their constituent first-threads is, assuming the efficacy of Lucretius' explanatory method, partway towards accepting their own impending death by the same means.

This sets up arguments, towards the end of Book 3, that fear of death and the accompanying endless grief for dying or dead loved ones are unnecessary. In this section, friendly consolation gives way to hard logic (introduced by 'death is nothing to us' – 3.830)[52] and a hectoring, withering tone against those fearing their own death or grieving for dying loved ones.[53] Lucretius' approach perhaps provides insufficient reassurance that death is not to be feared or excessively lamented, but we may find consolation elsewhere in the epic. Lucretius has stated that *naturae species ratioque* (*i.e.* the outward appearance and reasoning of Nature, and our reasoning in interpreting it)[54] alone can dispel darkness and fear of the mind (1.148; 2.61; 3.93; also later in 6.41). He has applied this methodological approach to lay bare the process of atomic unweaving that causes death. Perhaps, therefore, Lucretius believes this reasoning should provide sufficient consolation. Through the persistent unweaving/death metaphorical correspondence Lucretius has not only strengthened his arguments, but has perhaps guided the keen reader towards philosophical enlightenment. Furthermore, since the death and unweaving metaphors have been so inextricably intertwined throughout the *De Rerum Natura*, when we reach the plague at the end of Book 6, despite the lack of explicitly stated unweaving metaphors, the very mention of death alone should suggest the unweaving with which it is synonymous. Lucretius' extrapolated metaphors have helped to guide his reader to this point, at which they must apply what they have learnt in a final test of their understanding and acceptance of Epicureanism.[55]

Weaving and Cares

This tentative association between Lucretius' weaving metaphor and the philosophical consolation of his reader is enhanced by a further correspondence. Frequently in the *De Rerum Natura* Lucretius describes the cares 'binding' our minds, and considers how we might untie them. Such a correspondence should not be unexpected, since our minds are bodily, and, like all bodily things, consist of threads susceptible to interweaving, tying and untying. Therefore, we might infer, understanding the weaving process underlying fears and cares should help the reader to

[52] The passage is mirrored well by the matter-of-fact reality of the Athenian plague at the end of Book 6, An appropriately bleak ending according to Leonard & Smith 1942, n. *ad* 6.1286: 'Such a conclusion would probably have appealed with especial force to a poet whose temperament is marked by a profound and brooding melancholy'.

[53] Konstan 2013, 206–207, acknowledges Lucretius' tone as "a tad gruff" and "brusquely" dismissive, but argues that Lucretius takes issue not with grief in general, as others have assumed, but specifically with eternal (*aeternus* – 3.907, 911) grief. As Konstan notes, this view, far from being peculiarly Epicurean, is a commonplace to consolatory literature. Nevertheless, regardless of whether Lucretius' opinion of grief is less dismissive than first assumed, he certainly does not offer an explicit, heartfelt consolation.

[54] The ambiguity of Lucretius' syntax is reflected in my translation, and also highlighted by Brown 1984, n. *ad loc.*: "[The phrase] implies an antithesis between nature's outward aspect and her inner (atomic) realities, at the same time suggesting the two main instruments of Epicurean scientific procedure, observation and reasoning on the basis of it."

[55] In other words, the reader must make the logical connection for themselves, based on their understanding of the unweaving and death metaphors because, as Fowler 2007, 232 has it: "There are certain arguments and conclusions at which pupils have to arrive by themselves." Fowler, following Clay 1983, 250–66, also supports the idea of the plague scene as "a kind of test for the reader"; *cf.* also Segal (1990) 43–5, 234; Gale (2000) 23; Müller (2007) 251–4; Morrison (2013) 223–32.

dispel them. This approach may seem rather subtle, but is explicitly outlined, albeit briefly, in 1.921–950 (the majority of which is repeated almost verbatim at 4.1–25), where Lucretius explains his motivation and methodological approach. Driven by love of the Muses (1.924–925), Lucretius seeks novelty (1.925–928) and the accompanying prize from the Muses of a hitherto unawarded garland (928–930). The reason Lucretius might be crowned before all others is then given:

> *primum quod magnis doceo de rebus et* ***artis***
> *religionum animum* ***nodis exsolvere*** *pergo,*
> *deinde quod obscura de re tam lucida pango*
> *carmina, Musaeo contingens cuncta lepore.*
>
> 1.931–934

> firstly because I teach great subjects and proceed **to untie** the mind from the **tight knots** of *religio*, and then because I compose so bright a song about a dark subject, infusing everything with the charm of the Muses.

The Muses are expected to crown poets for their literary aptitude (the second of these two reasons), but not for their moral impact (freeing men from *religio*), yet here the latter is given prominence (*primum*...), while the former is introduced secondarily (*deinde*...). Nevertheless, moral didacticism impels Lucretius' poetic purpose, and the prize he seeks echoes his praise for Epicurus' contribution to man's enlightenment (1.62–79; 3.1–30; 5.1–54). However, while the *religio* conquered by Epicurus in 1.62–79 was an oppressive force 'looming over mortals with horrible appearance' (*horribili super aspectu mortalibus instans* – 1.65), here it is expressed by the weaving metaphor as a knotted structure constricting the mind in bonds, which must be untied by teaching. Thus Lucretius draws an implicit etymological link between *religio* and *religo* 'to bind',[56] creating a subsidiary meaning perhaps underlying each usage of *religio* in the *De Rerum Natura*.

This facet of the weaving gestalt metaphor is recurrently applied to other cares elsewhere in the epic. Like *religio*, which to Lucretius is simply a specific form of fear, the fear of death also binds the mind in knots. Therefore, those lamenting that a dying man will be robbed of life's rewards would, if they could also see that he will not miss them in death, '**untie** themselves from great anguish and fear of mind' (***dissoluant*** *animi magno se angore metuque* – 3.903). More general cares can also be untied. Looking down upon military manoeuvres on the Campus, Lucretius argues, cannot make fear flee 'and the terrors of death leave the breast empty and **untied** from care' (*mortisque timores / tum vacuum pectus linquunt curaque* ***solutum*** – 2.45–46)[57] – for *ratio* alone has this power. Finally sexual desire is similarly harmful (4.1073–1170) and is to be avoided as vigorously as the emotional reactions of fear and *religio*. It is best to avoid the beginnings:

> *nam vitare, plagas in amoris ne laciamur,*
> *non ita difficile est quam captum retibus ipsis*
> *exire et validos Veneris perrumpere* ***nodos****.*
> *et tamen* ***implicitus*** *quoque possis* ***inque peditus***
> *effugere infestum,* (...)
>
> 4.1146–1150

[56] A reading advocated vigorously by Springer 1977, after Smith 1975, n. *ad loc*, and also Brown 1984, n. *ad loc*. The implied etymological figure echoes that of *super* (...) *instans* in 1.65, which surely recalls *superstitio* (a negative synonym for *religio* roughly equivalent to English 'superstition'), as noted by Smith 1975; Brown 1984 (both n. *ad loc.*)

[57] Echoed later in Lucretius' exploration of sleep, which '**unties** cares from the breast' (*curas e pectore* ***solvat*** – 4.908), albeit only temporarily.

> For to avoid being lured into the nets of love is not as difficult as it is to get out, having been captured in the nets themselves, and to break through the strong **knots** of Venus. And nevertheless you can also, having become **entwined** and **entangled**, escape from danger, (….)

This is of course a metaphor from hunting or fishing, but, in addition to the obvious similarities between a net and a woven item, the specifics are rooted in Lucretius' weaving metaphor, expressed by terms used elsewhere of atomic bonding[58] and, in the case of *nodos*, of *religio* (1.932 = 4.7).

From these instances an important conclusion can be drawn regarding the methodology of Lucretius' moral philosophy, accounting for the apparent lack of consolation in the *De Rerum Natura*. The keen reader would notice the metaphor of 'unweaving the emotions' and recall the unweaving of compound structures into their first-threads. By these shared applications Lucretius simultaneously reemphasises the mind's bodily nature and guides the reader along the path to understanding our mind's constriction by fear, love and *religio*. More crucially, just as a special force is required to unweave compounds, so cares can only be untied by *ratio*. Indeed, we have the power to do this, by applying the *vera ratio* of our burgeoning knowledge of Epicureanism, which lays bare the truth of *rerum natura*. This is a remarkably subtle technique of consolation, but is carefully constructed around a consistently extrapolated weaving metaphor, and Lucretius' exhortations for the reader to 'pay attention to true reasoning' (*adhibe veram ad rationem* – 1.51; 2.1023),[59] and his reiterations of the importance of *naturae species ratioque*. Just as understanding of the atomic unweaving process responsible for the death of body and soul might soothe our fear of death, so might we untie that fear itself if we understand how it ties the mind in bonds. Regardless of our opinions on the efficacy of such a consolation, we might at least marvel at its construction.

Concluding remarks

This brief exploration of certain applications of one of Lucretius' principal metaphorical gestalts has many implications. Firstly, we have seen the depths to which Lucretius' metaphor extends, and just how specific certain applications are. This approach can be applied to any of Lucretius' principal metaphors with similarly fruitful results, which further strengthens our instinct that metaphor is essential to Lucretius' methodology.[60] Secondly, certain applications of Lucretius' principal metaphors can be found in reference to Lucretius' composition,[61] such as the weaving-together of arguments, the pleasure of poetry and the birth of his epic poem. These shared correspondences form a mirror in which the *De Rerum Natura* reflects *rerum natura* themselves as an embodiment of *naturae species ratioque.*[62] Thirdly, this approach of reading persistent metaphors can be applied more broadly to other scientific or philosophical texts, both ancient and modern, an approach one suspects would be particularly fruitful for the works of Darwin and Dawkins, highlighting suspected overlaps with Lucretius in terms of not just content, but also methodology.

[58] *nodos* of melting metal and ice (6.356, 878) and of the soul untied from the body (2.950); *impeditus* (= *indupeditus*) of entangled elements (1.240; 2.102, 459; 6.453, 1010).

[59] The phrase *vera ratio* itself occurs 17 times in all (1.51, 498, 513, 623, 637, 880; 2.82, 176, 229, 645, 1023; 4.446; 5.23, 406, 1117; 6.767, 853).

[60] As I argue in my doctoral thesis, currently in preparation.

[61] See Glennie's chapter in this volume.

[62] Many have compared Lucretius' text with the universe it describes, at times stretching the analogy towards breaking point. For the sounds and word-plays of the *DRN* as mirroring atomism, *cf.* Friedländer 1941 and Snyder 1980; for the *DRN* as a *simulacrum*, relying on the reader's interpretation, of the universe, *cf.* Thury 1987.

Bibliography

Asmis, E. 1999 Epicurean Epistemology. In K. Algra, J. Barnes, J. Mansfeld & M. Schofield (eds), *The Cambridge History of Hellenistic Philosophy*. Cambridge, 260–294.

Asmis, E. 2009 Epicurean Empiricism. In J. Warren (ed.), *The Cambridge Companion to Epicureanism*. Cambridge, 84–104.

Bailey, C. 1926 *Epicurus: The Extant Remains*. Oxford.

Bailey, C. 1947 *Titi Lucreti Cari:* De Rerum Natura *Libri Sex* (3 vols). Oxford.

Barber, E. J. W. 1991 *Prehistoric Textiles: The Development of Cloth in the Neolithic and Bronze Ages*. Princeton NJ.

Brown, P. M. 1984 *Lucretius* De Rerum Natura *I*. Bristol.

Cabisius, G. 1984 Social Metaphor and the Atomic Cycle in Lucretiu. *CJ* 80 No. 2, 109–120.

Clay, D. 1983 *Lucretius and Epicurus*. Ithaca NY.

Crowfoot, G. M. 1936 Of the Warp-Weighted Loom. *Annual of the British School at Athens* 37, 36–47.

Ernout, A. & Meillet, A. 1939 *Dictionnaire Étymologique de la Langue Latine*. Paris.

Fowler, P. 2007 Lucretian Conclusions. In M. R. Gale (ed.), *Oxford Readings in Classical Studies: Lucretius*. Oxford, 199–234.

Friedländer, P. 1941 Pattern of Sound and Atomistic Theory in Lucretius. *AJPh* 62, 16–34.

Gale, M. R. 1994 *Myth and Poetry in Lucretius*. Cambridge.

Gale, M. R. 2000 *Virgil on the Nature of Things: The Georgics, Lucretius, and the Didactic Tradition*. Cambridge.

Gale, M. R. 2009 *Lucretius:* De Rerum Natura *V*. Bristol.

Garani, M. 2007 *Empedocles* Redivivus*: Poetry and Analogy in Lucretius*. London.

Giussani, C. 1896–8 *T. Lucreti Cari* De Rerum Natura *Libri Sex* (4 vols). Torino.

Glare, P. G. W. (ed.) 1982 *Oxford Latin Dictionary*. Oxford.

Harlizius-Klück, E. 2011 Weben, Spinnen. In R. Konersmann (ed.), *Lexicon philosophischer Metaphern,* 3rd revised edn (1st edn published 2007). Darmstadt, 504–524.

Harlizius-Klück, E. 2016 Textile Technology. In G. L. Irby (ed.), *A Companion to Science, Technology, and Medicine in Ancient Greece and Rome*. Oxford.

Kirk, G. S., Raven, J. E. & Schofield, M. 1983 *The Presocratic Philosophers*, 2nd edn (1st edn published 1957). Cambridge.

Konstan, D. 2013 Lucretius and the Epicurean Attitude to Grief. In D. Lehoux, A. D. Morrison & A. Sharrock (eds), *Lucretius: Poetry, Philosophy, Science*. Oxford, 193–209.

Lakoff, G. & Johnson, M. 1980 *Metaphors We Live By*. Chicago.

Lehoux, D. 2013 Seeing and Unseeing, Seen and Unseen. In D. Lehoux, A. D. Morrison & A. Sharrock (eds), *Lucretius: Poetry, Philosophy, Science*. Oxford, 131–151.

Leonard, W. E. & Smith, S. B. 1942 *Lucretius* De Rerum Natura *Libri Sex*. Madison WI.

Morrison, A. D. 2013 *Nil igitur mors est ad nos?* Iphianassa, the Athenian Plague, and Epicurean Views of Death. In D. Lehoux, A. D. Morrison & A. Sharrock (eds), *Lucretius: Poetry, Philosophy, Science*. Oxford, 211–232.

Müller, G. 2007 The Conclusions of the Six Books of Lucretius (trans. Bettina Reitz). In M. R. Gale (ed.), *Oxford Readings in Classical Studies: Lucretius*. Oxford, 234–54. (First published (1978) as Die Finalia der sechs Bücher des Lucrez. In O. Gigon (ed.), *Lucrèce,* Fondation Hardt Entretiens sur l'antiquité classique 24. Geneva).

Murley, C. 1947 Lucretius, *De Rerum Natura*, Viewed as Epic. *TAPhA* 78, 336–346.

Nagy, G. 2002 *Plato's Rhapsody and Homer's Music: The Poetics of the Panathenaic Festival in Classical Athens*. Washington DC.

Pope, S. 1949 The Imagery of Lucretius. *G&R* 18 No. 53, 70–79.

Rouse, W. H. D. 1975 *Lucretius:* De Rerum Natura. rev. M. F. Smith (1st edn published 1924). Cambridge MA.

Scheid, J. & Svenbro, J. 1996 *The Craft of Zeus: Myths of Weaving and Fabric* (trans. C. Volk, originally published 1994 as *Le métier de Zeus: Mythe du tissage et du tissu dans le monde gréco-romain*. Paris). Cambridge MA.

Schrijvers, P. H. 2007 Seeing the Invisible: A Study of Lucretius' Use of Analogy in *De Rerum Natura*. In M. R. Gale (ed.), *Oxford Readings in Classical Studies: Lucretius*. Oxford, 255–288.

Sedley, D. 1998 *Lucretius and the Transformation of Greek Wisdom*. Cambridge.

Segal, C. P. 1990 *Lucretius on Death and Anxiety: Poetry and Philosophy in De rerum natura*. Princeton NJ.

Snyder, J. M. 1980 *Puns and Poetry in Lucretius'* De Rerum Natura. Amsterdam.

Snyder, J. M. 1983 The Warp and Woof of the Universe in Lucretius' *De Rerum Natura*. *ICS* 8 No. 1, 37–43.
Springer, L. A. 1977 The Role of *Religio, Solvo* and *Ratio* in Lucretius. *CW* 71 No. 1, 55–61.
Thury, J. M. 1987 Lucretius' Poem as a *Simulacrum* of the Rerum Natura. *AJPh* 108 No. 2, 270–294.
West, D. 1975 Lucretius' Methods of Argument (3.417–614). *CQ* 25 No. 1, 94–116.
West, D. 1994 *The Imagery and Poetry of Lucretius* (Revised edn, first published 1969). Bristol.
Wild, J. P. 1970 *Textile Manufacture in the Northern Roman Provinces.* Cambridge.

15. Colchis, Wool and the Spinning Fates in Catullus *Carmen 64*

*C. Alfaro**

The subject of one of the longest and most interesting poems by Catullus, *Carmen* 64 - also known as 'The Wedding of Peleus and Thetis'[1] - falls squarely into the theme of this volume: we find ourselves dealing with an epyllion,[2] a very 'Catullan' *topos,* in spite of its baroque Hellenistic inspiration. In any case, the Catullus version is completely original.[3] The Hellenistic or Alexandrian origins, structure, precedents and underlying intention of the poem have all been discussed at length in the existing bibliography.[4]

However, this chapter will focus on the metaphors related to fate and destiny (symbolised by the Spinning *Parcae*), and their origin in the actual crafts of spinning and weaving.[5] These two crafts appear to have been conceived as indispensable elements for graphically describing the reality of life and the social environment.[6] This might be attributed to the fact that one of the greatest

* I wish to thank the editors G. Fanfani, M. Harlow and M.-L. Nosch, The Danish National Research Foundation's Centre for Textile Research (CTR), Saxo Institute, University of Copenhagen for the invitation to participate in the Symposium and in this book, and for the help with the revision of the translation. My gratitude also to the colleagues and friends who have provided images for the text: Henning Nielsen; Prof. Dr. Regine Schulz and Dr. Antje Spiekermann (Director and Technical Coordinator of the Roemer- und Pelizaeus-Museum Hildesheim), and Dr. Michael Tellenbach (Director Archaeology and World Cultures rem. Curt-Engelhorn-Stiftung für die Reiss-Engelhorn-Museen) for his friendship and his generous help. This work has been carried out under the project I + D + I of the Ministry of Education and Science of Spain (University of Valencia) "Wool in Roman Ibiza. Aspects of pastoralism around purple production workshops (*Timaeus Project*. Ref. HAR2010–21065), and DressID Project, Clothing and Identities. New perspectives on Textiles in the Roman World (Ref. 2007–1765/001–001 CTU COOPMU). This project has been funded with support from the European Commission. The publication reflects the views of the author, and the Commission cannot be held responsible for any use which may be made of the information contained therein.

1 The English translation of Catullus 64 is adapted from the most recent Loeb (F.W. Cornish 1913, revised by G.P. Goold), unless otherwise stated.

2 Or love story with a mythological background which aims at describing human feelings.

3 Perrotta 1931; Boucher 1956; Curran 1969; Flobert 1976; Forsyth 1975; Giangrande 1977; Kinsey 1965; Putnam 1961; Ramain 1922.

4 Lafaye 1894, 101; Pasquali 1920; Ramain 1922; Willamowitz-Moellendorf 1940; Perrotta 1931; Waltz 1945; Boucher 1956; Putnam 1961; Kinsey 1965; Curran 1969; Flobert 1976; Forsyth 1975; Giangrande 1977.

5 Alfaro 1986; Andrés 2010.

6 This aspect can be appreciated at the beginning of the poem (line 10) when the poet refers to the hull of the ship Argo, built (or better 'woven/sewn', if we remember the *technē* for constructing archaic ships) with pines from Mount Pelion, resembling a wooden (*pinea*) fabric (*texta*): *pinea coniungens inflexae texta carinae*: see Nieto-Prieto 2008, 26–29; Alfaro 2008, 244–245).

revolutions in human history must have been made thousands of years ago, namely the development of spinning and the difficult craft of creating, with the spun threads, flat surfaces that could be used for covering and protecting the body (during life and after death).[7] Metaphors connected to spinning and weaving had a big advantage for the ancients: they were observed from infancy and therefore universally understood, since they were common activities in their societies. The poem is full of allusions to the usefulness of threads and the origin of the fibres that make up the wool yarn (traditionally, the raw material most often mentioned in classical literature since Homer). It also makes reference to the part played by the thread in the *Parcae*'s hands as an instrument for showing mankind, through its strength, but also its fragility, the meaning of fate.

Working with wool, the invisible *labor matronalis*, marked the lives of Roman women as much as it marked the lives of Greek women. It somehow characterized their destiny as controlled beings, tied by 'invisible', but secure threads to their domestic duties, represented by *lanificium*.[8] This was a way of ensuring that women, eternally suspected of hatching troublesome plots, had their hands occupied and were conveniently located for easy surveillance in the *gynaeceum*.[9] In the Republic weddings represented the transfer of monitoring responsibilities (*manus tutela*) from the father to the husband.[10]

Nevertheless, the myths engendered by ancestral work with wool spread beyond the *gynaeceum* limits and affected the lives of men too. Were women, perhaps, the creators of these myths? Or should we call them stories? The fact is that it was precisely around the task of spinning that a whole, specific metaphorical language originated; a language derived from daily routine that sought to explain the brevity of life and the role of Fate/Chance in each person's future. Favourable or adverse, chance, destiny or fate was fixed at the transcendental moments in a person's life: birth, procreation and death. The birth of a new being represented hope and confidence in the future; a wedding symbolized maturity and the possibility of creating the basic unit of society: a family. As the life cycle ended, the passage of death had to be correctly accomplished at the appropriate time and with the proper conditions. The three stages, however, did not always develop as hoped.[11] A person's fate was unpredictable at each stage. The responsibility lay with the prudent, silent, but implacable *Moira*/*Fata*/*Parca*.[12]

Leaving aside the fact that the goddesses of destiny possessed the power to control people's lives from the supreme god (Zeus/Jupiter), the personification of Fate as a spinner-woman should not be unexpected. The deities with similar functions such as Demeter, Kore, Artemis, Eileithyia

[7] See Douglas Cairns, this volume.

[8] In this context, spinning represented the control and seclusion of women, while thread was a symbol of strength, in opposition to their weak nature. Perhaps, that is why wool yarn and knots are frequently used in ancient magic when tying material things to the person, but also when controlling their destiny. See the interesting text in a papyrus collected by Preisendanz 1973–1974, 108–109: "And take a lead tablet and write the same spell and recite it. And tie the lead leaf to the figurines with thread from the loom after making 365 knots while saying as you have learned: *Abrasax, hold her fast!* You place it, as the sun is setting, beside the grave of one who has died untimely or violently, placing beside it also the seasonal flowers".

[9] Orestano 1951; Cantarella 1991; Cantarella 1996.

[10] The Romans always tried to keep women, even in adulthood, under tutelage because of their superficial nature: *propter anime leuitatem* (Leg. XII. Tab., Tabula IV, 1). For the textile symbolism connected with the *manu captio* in iconographical sources see Scheid & Svenbro 1996, 84–88.

[11] Van Gennep 1909; Turner 1969; Eliade 1975; La Fontaine 1985.

[12] Hom. *Il.* 5.613; 18.119; 19.87; *Od.* 3.269; 11.292; 19.592f., etc. Often with the sense of misfortune and death (μόρος = destiny, death); in Hesiod *Th.* 903–906, the Moirai are carriers of good and bad things; Krause 1949; S. de Angeli 1992.

or Athena herself were also women. Among them all, the actual executor of fate is Parca/Moira. First one, and over time a group (the *Parcae/Moirai*): a pair, a trio and, on rare occasions, a quartet, although this was not a linear, chronological development.[13]

Scholars who devote attention to the field of ancient textile technology and economy have always been particularly interested in the *Moira* Clotho, the one whose function is represented by women's most emblematic work: spinning.[14] She is normally accompanied by two,[15] three[16] or four companions of similar character. Starting with Hesiod, their names seem to stay the same, Atropos and Lachesis; and in the case of the *Parcae/Fatae*, Nona and Decuma.[17] All of them are intelligent and fair personalities (unrelated to the blind Fortuna, as in the case of the destiny deities of the Etruscans and Romans).[18] According to the different versions of the myth, they have one function or another, always close to the fate of people. Plato makes them three spinners who run the major axes of the world and the cosmos moved by the threads of destiny and identifies all this with the spindle to improve the understanding of the image of the universe that he wants to present.[19] Music (organizing the universe) and the work of spinning and weaving go hand in hand. So the Fates sing their eternal song, Lachesis sings past events, Clotho present and Atropos future ones. It is possible to identify a parallelism between the three moments in the life course of a person in terms of the spinning work process if we focus on both the three sisters, and on the spinning technique itself.[20] In terms of the spinning process, the present should be considered as the thread coming out at the moment of the twist of the fibres thanks to rotating spindle (universal character); the past would be the thread already rolled around the spindle which signifies the lived experiences of the individual; and the future is represented by the unworked wool fibres that, tied around the distaff, are waiting to be arranged and twisted as ideas symbolizing the real present, i.e. the spun thread. The representation of the spinning fate in the plastic arts varies tremendously. In the case of the Etruscan and Roman *Fatae*, the idea of destiny is simplified using a good/bad conception of the destiny.

Like the *Moirai,* the *Parcae* are perceived ambivalently; somewhere in between good fairies and implacable enforcers of what they had devised for each mortal at the crucial moment of their birth. In *Carmen* 64, Catullus always uses the term Parca in the plural, Parcae.

Few texts have approached human destiny as beautifully as the long *Carmen* 64. This paper will not address the question of whether the whole poem should be interpreted as an epithalamium for Thetis and Peleus, or, on the contrary, whether this idea should only be applicable to the final fragment, where the *Parcae* perform their chants.[21] However, we agree with those who identify two

[13] Hesiod, *Th.* 211–220, identifies them as a trio. Greek vases, and the ancient authors who indiscriminately use the singular or the plural forms, are the basis of this diversity.

[14] According to De Angelis (1992, 636): "All' interno dell'epica omerica (…) solo in un caso, *Il.* 24.49, esse (Moirai) compaiono come un gruppo."

[15] Plu. *Mor.* 385c.

[16] Plu. *Mor.* 385c; Paus. 10. 24. 4.

[17] These names come from the most important months of pregnancy; *i.e.*, the ninth and the tenth. De Angelis 1992, 637; Hild 1896, 1018. Gell. III, 16, 9. According to Aulus Gell. 3.16.9 (*Attic Nights* 3.16.11) Caesellius Vindex in his *Ancient Readings* says: The names of the Fates are three: Nona, Decuma, and Morta.

[18] Hild 1896, 1019: in Etruria: Miura-Mnerva/Afrodita Urania, Nortia-Fortuna/Tyche (blind) and Athrapa. In Rome: *Fatum/Moira* (*Tria Fata:* Parca, Nona and Decuma).

[19] *Rep. Rep.* 616b–c.*Cf.* Andrés 2010, 65–76.

[20] Alfaro 1986, 177.

[21] Waltz 1945, 93–94; see also Leathers Daniels 1972–1973 with analysis.

kinds of sexual union: legitimate and illegitimate, in the poem's two main episodes.[22] A legitimate union, or *conubium*, was regarded as a binding stable relationship in Catullus' time, although divorce did exist.[23] This was the normal practice among the upper classes, to whom the poem was addressed. A union accomplished through the so-called 'marriage by *confarreatio*', including *sponsaliae*, was difficult to dissolve. On the other hand, an illegitimate union was the result of blind passion (Ariadne) or selfish interest (Theseus). This was thought to end only in a disastrous outcome such as the abandonment of Minos' daughter on Naxos and her eventual straying into Bacchus's arms. Marriage could be the first symbol or "metaphor for the ties of life" in *Carmen* 64.[24] Its purpose is procreation and Thetis will fulfil this aim. Nonetheless, the ties that bind mortals do not apply to deities. This would explain the goddess' escape, after fulfilling her mission as a woman and giving birth to her son Achilles, to the paternal sea in search of freedom. The final deviation from the norm eliminates the moralistic character that the poem would have had if Thetis and Peleus had grown old together.

Structure of the poem

Before continuing, let us recall the two episodes recounted in those 408 magnificent lines. The first is divided into two parts: one at the beginning and the other at the end of the poem, leaving the central part for the second episode, the story of Theseus and Ariadne. This central part constitutes the bulk of the whole narrative. We might see a sign of Catullus' Hellenistic influences in the poet's wish to frame Ariadne's story (most popular in his time) within the main episode that gives the work its name, 'The Wedding of Thetis and Peleus'. Like Velázquez in *Las Hilanderas* (*The Spinners*) centuries later, Catullus brings the secondary plot into the foreground.

First Episode

The first episode is divided for us into three 'scenes'; two before the story of Ariadne and Theseus, and a third that ends the poem (Thetis and Peleus's wedding).

1) The beginning includes a flashback to the perfectly cinematographic encounter between Peleus and Thetis. Almost flying, he "swept the deep blue expanse" (*caerula aequora*) on board a swift ship, the Argo, the super-ship that Athena had built. Beautiful Thetis proves to be an agile swimmer; for twice she reached the fast ship and met Peleus's eyes; he was already in love with her. Peleus is sailing from his present homeland, Iolcus, towards the waters of the river Phasis in the kingdom of Aeëtes, Colchis. Together with his companions the Argonauts, especially Jason, he has to accomplish his mission, *i.e.* the rescue of the Golden Fleece (Fig. 15.1). On his way through Thessaly, his second home, he succeeds in concluding a traditional *sponsaliae* contract for marriage to Thetis, barely sketched by the poet, but, apparently, properly sealed by Jupiter himself and, curiously, ratified by the young goddess's grandmother and by Oceanus (ll. 28–30).

[22] Ramain 1922, 143, 145.
[23] Gardner 1986; Treggiari 1991; Gourevitch & Raepsaet-Charlier 2001; Cenerini 2002.
[24] On the ties of life and the divinities who preside over them, see Deonna 1918, 141–152. Many of the magic rings with ties of initials that the author collects might be engagement or marriage rings.

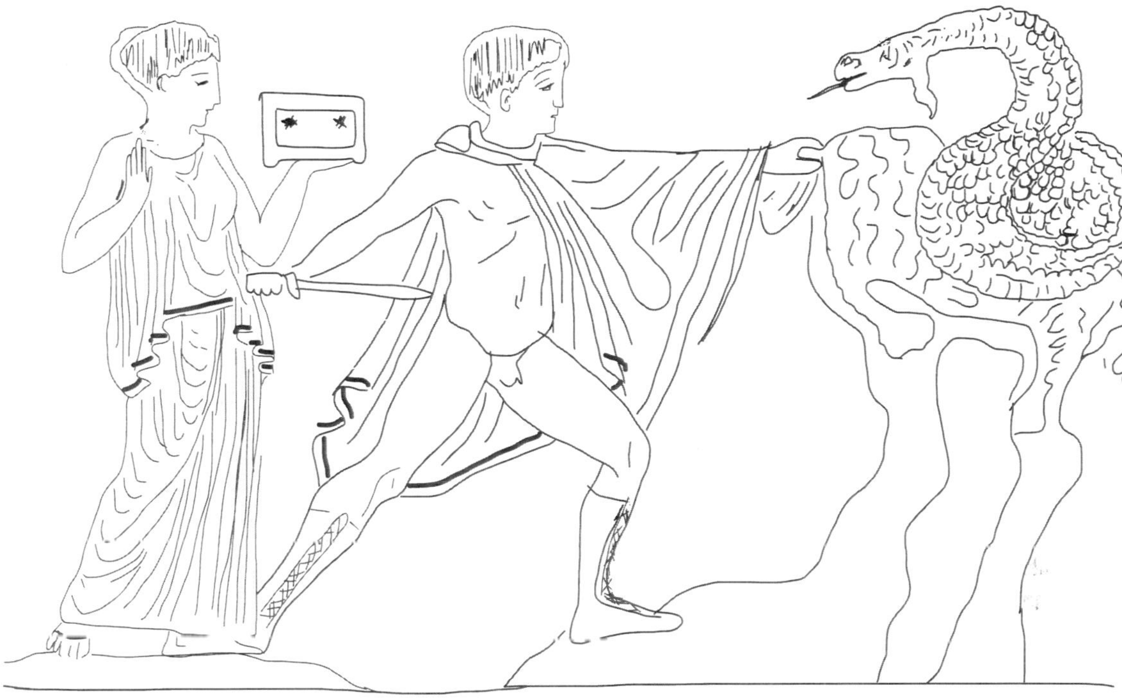

Fig. 15.1 Jason, Medea, the dragon and the golden-fleece. Drawing by the author after DS I, *fig. 507.*

2) The second scene, which begins by briefly confirming that the marriage promise had been made in accordance with human norms (*finito tempore,* line 31), consists of the celebration at Peleus' house with friends and neighbours. They leave their homes throughout Thessaly in order to bring their presents to the king while the final preparations for the wedding are being made.
3) The last scene takes place after the long *excursus* on Ariadne. In fact, it concludes this interrupted episode (ll. 267–408). It includes the arrival of the gods at the palace and the nuptials, accompanied by the *Parcae*'s songs. As they spin, they determine the couple's fate and the future of their heroic son, Achilles. With a frenetic rhythm recalling the songs of spinners and weavers, this is the crucial moment in the poem, which explains the name given to the work since the Renaissance: 'The Wedding of Thetis and Peleus'.

Second Episode

The second episode is a long digression (ll. 50–278) that describes the coverlet covering the bride's bed (*lectus genialis*), installed in the middle of the palace. The bedspread shows the whole story of Theseus and Ariadne in its rich tapestry weave and hypnotises the guests, and also the narrator, with its colours and beauty (Fig. 15.2).

Fig. 15.2 Coptic tapestry of sleeping Ariadne in Naxos. Illustration fom RPM Hildesheim.

Symbols and metaphors related to fate

Apart from the first metaphor for the ties of life, represented by marriage, we find patent symbolism directly connected to fibres, spinning, weaving and fine tapestry work of the kind that resembles painting. Though less highlighted, it is worth mentioning other themes present, such as the quality of being dressed or naked in a symbolic sense and the many references to colour.

The presence of elements of the textile process in both episodes might help create the connection between the two: all these metaphors associated with the colour of the wool, spinning and weaving form a common thread of the poetic discourse.

The symbolic importance of colour[25]

In line with Homeric aesthetics, Catullus's sea is blue, but also silver, purple, golden and white, by reason of its foam. In some versions of the mythical voyage, the ram that flies Helle and Phrixus to Colchis is also purple (Fig. 15.3), of a dark bluish shade like Tyrian purple, since, this way, it can fly unnoticed in the stormy sky. By the time Phrixus reaches Aeëtes' kingdom, fate has

[25] Colour in the world of aristocracy and royal power: Reinhold 1070; Heuzey & Heuzey 1935, 83–85; Briant 1982, 374; Casartelli 1998, 115–116; Alfaro 2013; X. *Cyr.* 7.1; Plb. 6.53; Heliod. *Etiop.* 3.3.5 and 3.4.2.

Fig. 15.3 Phrixus, Helle and the purple ram. Drawing by the author, after DS I, fig. 503.

been unkind to Helle, who has fallen off the ram into the sea, known as the Hellespont, the Sea of Helle, since then.[26] According to this version, the ram's fleece turns golden after the sacrifice.

We can imagine that, in Antiquity, ship sails were dyed with those colours related to the local raw materials normally used for dress and household textiles. The sails were external signs for

[26] The death and the fleece could also have been symbolically united in a pastoral culture, such Pontus Euxinus as shown by Pindar (*P.* 4.158–159). He describes the habit of burying the dead wrapped in a protective fleece, where their most precious possession, the soul, would be deposited: see Harrauer & Hunger 2008, 91. In the fictionalised version of *The Golden Fleece* by Robert Graves there is an interesting historical point made that an anonymous Vatican 5th-century AD mythographer had access to a very old source of the legend that the fleece was the one used by Zeus himself to go up to heaven and come down to earth. There is a symbolic connection between the Golden Fleece, royalty and power. Many current monarchies still hold a *toison d'or* in high esteem (for example, the Spanish King). The topic of the journeys undertaken by gods, sorcerers, fakirs, etc. between the earth and heaven thanks to fibrous elements such as strings, lianas, threads, etc. has been studied by Eliade 1962, 233–277. *Cf.* also Deonna 1918.

Fig. 15.4 Parcae *group of the so-called* "Puteal *of the Moncloa*" *(I c. A.D.), MAN, Madrid. Drawing by R. Schneider, 1880.*

identifying the ship at a distance. Cleopatra VII's ship had purple sails (a symbol of royalty).[27] But the symbolic/practical meaning of the sails in Catullus' poem emerges at a dramatic moment (ll. 215–237): a predicted death. It was agreed by Theseus and his father that the ship would hoist white sails as it approached Athens if the mission with the Minotaur had been successful and black ones if it had failed. Perhaps, in this the way fate punished the Argives for their arrogance: Theseus carelessly forgot to change his sails and on seeing the black sails, his father committed suicide. The sails are described as *carbasus obscurata decet ferrugine Hibera* (l. 227); that is, dyed with a black, ferruginous dye in the style of Iberia, the kingdom east of Colchis.

Ariadne, in her long lament, regrets not having been taken by Theseus, at least, as a simple household slave to spread his bed "with purple coverlet" like the one that covered the marriage bed of Thetis and Peleus (l. 163: *purpureaue tuum consternens ueste cubile*). The *Parcae* are described as follows (ll. 307–309; see also fig. 15.4): "A white garment cloaked all round their trembling frame, and bound their ankles in purple edging; snow-white ribbons adorned their rosy head" – a metaphor alluding to the fact that, like other deities, they had a halo over their head.[28]

[27] Alfaro 2013.

[28] Translation by Sara T. Nalle.

Raw materials: gold, fur, flax, hemp

At the start of the poem, the destination of the voyage is established: Colchis, Aeëtes' kingdom, particularly the river Phasis.[29] Nowadays it is called the Rioni, rich in gold dust and one of the largest rivers in the area. There is a town with the same name. Today the importance of gold to Colchis is well known thanks to archaeological finds.[30] In Antiquity, other gold-bearing rivers that flowed into the sea near the Phasis may also have been exploited. This may well explain a colonial situation and the hidden interest behind the myth.

The key text describing the way the gold was obtained is Strabo.[31] People today still use a system quite similar to the wooden troughs in which the ancients put the fleeces, using sections of fleece or even fragments of felt to retain the gold nuggets.[32] The account offers information about Colchis's wealth, not only regarding gold, but also cattle and sheep, which must have been the greatest riches of the powerful in these nomadic and semi-nomadic societies that travelled every year to the north.[33] "The country is fertile and its produce is good (...). It furnishes all materials for ship-building. It produces them in great plenty, and they are conveyed down by its rivers. It supplies flax, hemp, wax, and pitch in great abundance", says Strabo, and continues: "Its linen manufacture is celebrated, for it is exported to foreign parts".

The symbolism of spinning

The symbolism of spinning is most evident in the poem thanks to the presence of the *Parcae* and their eternal task. Catullus describes the process with absolute perfection. The slight trembling of their body, the distaff held in the left hand full of soft wool and closely observed gestures: those made with the hands, the teeth polishing the thread, moistening it with the mouth (dry from rubbing the wool to even up the threads), as the balls of wool gradually piled up in the wicker baskets. The threads made by the three sisters come out with apparent ease, almost on their own.

---

[29] The Glaucus and the Hippos are its tributaries, according to Strabo XI, 2.17.

[30] Especially at the sanctuary in Vani, a town in the Caucasus that was never looted. David Lordkipanidze continued the work of his father, Prof. Otar Lordkipanidze, there.

[31] *Geog.* 9.2.19: "In their country the winter torrents are said to bring down even gold, which the Barbarians collect in troughs pierced with holes, and lined with fleeces (*mallotais dorais*); *and hence the fable of the golden fleece.*" This description is almost more realistic than the reports that we have access to about the mountainous regions in this part of modern Georgia, around the ancient Dioscurias mentioned by Strabo. It should be pointed out that Catullus, living in 87–54 BC, as an educated man of his time, follower of the Alexandrian-Hellenistic aesthetics, describes a mythical world immersed in the Greek tradition. Links between mainland Greece and Black Sea colonies, which were very rich zones of the Colchis, was undoubtedly a subject of interest for Rome from a political and economic point of view.

[32] I am grateful for the information provided by Prof. Manuel Viladevall Solé (UAB), who clearly explained the gold extraction system in rivers, which is still used nowadays. The water is full of metallic gold particles, as well as dielectric particles as the siliciclastic ganga which allows the gold flakes to adhere to the grease of the fleece, thanks to its electric conductivity (http://www.publicacions.ub.es/refs/pub_dig/pros_aluv.pdf). In personal communication Prof. Viladevall Solé told us that: "los romanos utilizaban plantas del tipo "jara" y probablemente tomillo, capaces de adsorber las partículas de oro y el tema que encontré apasionante es el uso de plumas de ganso por parte de los egipcios para recuperar el oro. Estas, al igual que la piel de lechales, son muy grasientas y las partículas de oro tienen tendencia a ser adsorbidas por la grasa y en consecuencia concentrarse en el vello o entre las plumas".

[33] Tanais, in the north, in the Bosphoran Kingdom, was founded by the Greeks, as Strabo (11.2.2) and Pliny the Elder (*NH* 6.20) inform us. Pliny adds that the Greeks were from Panticapeaum. Strabo 11.2.3, write: "It was an important interchange market both for Asian and European nomads and for seamen coming from the Bosphorus to trade in Maeotis. The first provided slaves, skins and what nomads could offer; the second brought, in exchange, dresses, wine, and all that pertained to a civilized life."

It is the spindles that pull the wool, although it is the *Fatae* who make the spindles turn. The most interesting point is that as they pull the fibres, they also pull the fates (*sed uos, quae fata secuntur, currite ducentes subtegmina, currite fusi,* ll. 326–327, "but run you on, drawing the woof-threads which the Fates follow, run, spindles!"). The fates, destiny, become interwoven in human lives, represented ultimately by the thread. Each strophe, in this final part of the poem describing the wedding, is like a chorus monotonously repeated like the very work of spinning.

As I have previously argued elsewhere,[34] the past life is contained in the thread rolled on the spindle, an instrument that has many iconographic representations. The present is symbolised by the Parcae spinning; and the future by the still shapeless fibres which remain on the distaff as projects waiting to be developed. So the spindle-distaff tandem must be read from the bottom up. The most dangerous moment is the present; because the thread might break. The break might be accidental, as often happens during spinning, or by the will of Fate, because the spindle where the past (or the person's memory) is stored is already full and there is no other possibility than to rest-die.

The accidental rupture, the fatality programmed by the *Parcae*, is always troubling and is what drives people to display good behaviour and pay due attention to the gods. It might even be suggested that Ariadne, in recalling the struggle between Theseus and Minos, alludes to the symbolic relation between spinning and death. In lines 149 and 150, referring to the fight between Theseus and her father-brother Minos, she says: *certe ego te in medio uersantem turbine leti / eripui, et potius germanum amittere creui* ("I, when you were tossing in the very whirl of death [we can imagine the *Parca*'s rapidly turning spindle], saved you, and set my heart rather to let my brother go than to fail you").

This act of spinning also conceals an interesting metaphorical element: the multiple symbolic meanings of the spindle with regard to women's personality and attitudes. In this poem it refers to the woman who is going to take the step of entering married life. The act of spinning symbolises, in the one sense, the woman's docility in marriage and the home. It must not be forgotten that in Greece the spindle is associated with the *Moirai*, but also with Demeter and Kore, goddesses closely linked to the idea of fecundity and survival after death.[35]

The quilt covering the marriage bed

Like the use of certain festive attire, covering the *thalamos* with rich bedspreads depicting figured scenes must have been a tradition among the upper classes. Perhaps the extension of this part of the poem parallels the fact that all the friends and inhabitants of Thessaly were meant to have a close look at it, since it was a family jewel, a narrative piece that told stories of the heroes and gods. Delicately-woven, complex images were patiently made in daily life by women but also, in myths, by some goddesses. This covering, perhaps, was a present from Athena Ergane, rather than the work of unstable and volatile Thetis, who escaped from the demanding fate of marriage as soon as she had given birth to her son and returned to the freedom of the immense sea with her sisters, the Nereids. Theseus, a hero comparable to Hercules for the amount of exemplary adventures he was involved in, must have been also remembered in Colchis and territories further north as the

[34] Alfaro 1986.

[35] Recently it has been argued that in the High Empire period the spindle had erotic connotations that can be seen, with a large dose of imagination, in the inscriptions on some of the spindle whorls that have survived (*e.g. AVE DOMINA SITIIO* = salve, mistress, I am thirsty; *IMPLE ME SIC VERSA ME* = embellish or fill me and make me turn; or *GENETA VIS CARA* = dear young woman, do you want? *Cf.* Dondin-Payre 2005 and 2007). This same erotic sense of the spindle has been interpreted in other languages of the same period: Gallic, Iberian or Celtiberian (De Bernardo-Stempel *et al.* 2010).

Fig. 15.5 The artist Henning Holm Nielsen's interpretation of the labyrinth episode. © Nielsen.

one who defeated the Amazons, led by Hippolyta; these women did not fit the *labor matronalis* model and everything it symbolised.[36]

With his description of the purple bedcover, Catullus intentionally marks both the new family and Peleus's royalty, especially now that he is marrying a woman from a divine family. The *lectus genialis* where Achilles will be conceived and, for that night, placed in the middle of the palace, is described as follows:

Puluinar uero diuae genial locatur
sedibus in mediis, Indo quod dente politum
tincta tegit roseo conchyli purpura fuco.

ll. 47– 49

But see, the royal marriage bed is being set for the goddess in the midst of the palace, smoothly fashioned of Indian tusk, covered with purple dyed with the rosy stain of the shell.

Artistic and full of colour, the spread depicts the story of the Minotaur, king Minos' daughter Ariadne, and Theseus, whose sole aim is to put an end to the demands imposed on the Argives by Minos, the evil king of Crete. Passionately in love with Theseus, Ariadne loses her prudence. Confused

[36] Mossé 1983.

by Daedalus, she follows his advice and offers the man who will kill her own father a humble, but very valuable instrument to prevent him from getting lost in the labyrinth: a ball of wool, one of the many she must have spun while dreaming of a different fate from the one awaiting her. In its humility, the thread would be the passport to escape from the enormous maze, Theseus's fate (Fig. 15.5). However, the ball/fate she had made, in somebody else's hands, was going to be cruel to her: she had simply been a tool for Theseus, who would abandon his aide to her fate on Naxos, perhaps by divine command.

The symbolism of nakedness as opposed to being dressed

This appears at two moments that might represent approach to and abandonment of marriage. On the one hand, we encounter it connected to Thetis and the Nereids and the freedom in their life and bodily movements. These deities ignore life in society and conventions of modesty until, as in this case, they come into contact with civilization through marriage imposed by *pudicitia*.

On the other hand, we find the symbol with the opposite meaning in the episode of Ariadne on Naxos, when she is abandoned by Theseus. Her appearance, with her soft clothes fallen down onto the shore of the algae-filled, tempestuous sea symbolizes her pain, her lack of interest in everything around her (ll. 63–66):

> Nor does she still keep the delicate headband on her golden head, nor has her breast veiled by the covering of her light raiment, nor her milk-white bosom bound with the smooth girdle; all these, as they slipt off around her whole body, before her very feet the salt waves lapped.

The poet is narrating the end of the process of change in Ariadne: abandoned, she definitively turns her back on social norms and enters the prejudice-free world of the person who is going to console her: Bacchus. It is a symbol of deification, and somehow represents an antagonistic contrast to Thetis's humanisation.

Bibliography

Alfaro, C. 1986 El hilado y el tejido antiguos en el simbolismo del Puteal de la Moncloa. In R. Olmos (ed.), *Coloquio sobre el Puteal de la Moncloa. Estudios de iconografía II*. Madrid, 171–180.

Alfaro, C. 2008 Las cuerdas y los materiales de cestería: su interpretación histórica en el ámbito de la colonización griega. In X. Nieto and M. Santos (eds), *El vaixell grec arcaic de Cala Sant Vicenç*. Girona, 244–253.

Alfaro, C. 2013 Purple and Aristocracy: Colour, blood, and luxury as social identifiers in Antiquity. In C. Alfaro, J. Ortiz, & M. J. Martinez (eds) *Luxury and dress. Political power and appearance in the Roman Empire and the Provinces*. Valencia.

Andrés, M. A. 2010 *El hilo de la vida y el lazo de la muerte en la tradición indoirania*. Valencia.

Blum, H. 1998 *Purpur als Statussymbol in der Griechischen Welt*. Bonn.

Boucher, J.-P. 1956 A propos du Carmen 64 de Catulle. *REL* 34, 194–218.

Briant, P. 1982 Conquête territorial et stratégie idéologique: Alexandre le Grand et l'idéologie monarchique achéménide. In *Rois, tributs et paysans. Études sur les formations tributaires du Moyen-Orient ancien*, Annales littéraires de l'Université de Besançon 269. Paris, 357–401.

Brunello, F. 1973 *The Art of Dyeing in the History of Mankind*. Vicenza.

Cantarella, E. 1991 *La calamidad ambigua*. Madrid.

Cantarella, E. 1996 La sexualidad de la mujer romana. In A. Pérez Jiménez & G. Cruz-Andreotti (eds), *Hijas de Afrodita: la sexualidad femenina en los pueblos mediterráneos*. Madrid, 115–130.

Casartelli, A. 1998 La funzione distintiva del colore nell'abbigliamento romano della prima età imperiale. *Aevum* 72, 109–125.

Cenerini, F. 2002 *La donna romana*. Bologna.

Curran, L. 1969 Catullus 64 and the Heroic Age. *YClS* 21, 186–204.

De Angeli, S. 1992 Moirai. *LIMC* VI, 1, 636–648.

De Bernardo-Stempel *et al*. 2010 Nueva fusayola con inscripción en signario celtibérico de la Necrópolis vaccea de las Ruedas de *Pintia* (Padilla de Duero-Peñafiel, Valladolid). *Paleohispanica* 10, 405–426.

Deonna, W. 1918 Le nœud gordien. *REG* 31, 141–184.

Dondin-Payre, M. 2005 Épigraphie et acculturation: l'apport des fusaïoles inscrites. In *Le monde romain à travers l'Épigraphie: méthodes et pratiques*. Lille, 133–146.

Dondin-Payre, M. 2007 *Instrumentum* et Épigraphie lapidaire: scission ou complémentarité? L'exemple des fusaïoles inscrites. In *Acta XII Congressus Internationalis Epigraphiae Graecae et Latinae*. Paris, 425–432.

Eliade, M. 1962 *Méphistophèles et l'androgyne*. Paris.

Eliade, M. 1975 *Iniciaciones místicas*. Madrid.

Flobert, P. 1976 Catulle et les cheveux des Parques (LXIV, 309). *REL* 51, 140–157.

Forsyth, P. Y. 1975 Catullus 64: the descent of man. *Antichthon* 9, 35–55.

Gardner, J. F. 1986 *Women in Roman Law and Society*. London.

Giangrande, G. 1977 Catullus 64. *LCM* 21, 229–231.

Gourevitch, D. & Raepsaet-Charlier, M.-Th. 2001 *La femme dans la Rome antique*. Paris.

Harrauer, Ch. & Hunger, H. 2008 *Diccionario de Mitología griega y romana* (ed. española revisada por J. Fernández & A. Martínez, translation J. A. Molina). Barcelona.

Heuzey, L. & J. 1935 *Histoire du costume dans l'antiquité classique. L'Orient, Égypte, Mésopotamie, Syrie, Phénicie*. Paris.

Hild, J. H. 1896 *DS* II, 2 1016–1021, *s.v. Fatum*, Αἶσα, Μοῖρα, *Parca*.

Kinsey, T. E. 1965 Irony and structure in Catullus 64. *Latomus* 24, 912–925.

Krause, W. 1949 Zeus und Moira bei Homer. *Wiener Studien* 64, 10–52.

Lafaye, G. 1894 *Catulle et ses modèles*. Paris.

La Fontaine, J. 1985 *Initation. Ritual drama and secret knowledge across the world*. London.

Leather Daniels, M. 1972–1973 The Song of the Fates in Catullus 64: Epithalamium or Dirge? *The Classical Journal* 68, 97–101.

Lordkipanidze, O. & Mikéladze, T. 1990 La Colchide aux VIIème-Vème siècles. Sources écrites antiques et archéologie. In O. Lordkipanidze and P. Lévêque (eds), *Le Pont-Euxin vu par les Grecs, Symposium de Vani (Colchide), sept.-oct. 1987.* Paris, 167–187.

Lordkipanidze, O. 1990 Vani dans la structure du royaume Colchidien. In O. Lordkipanidze and P. Lévêque, *Le Pont-Euxin vu par les Grecs, Symposium de Vani (Colchide), sept.-oct. 1987.* Paris, 289–318.

Mosée, Cl. 1983 *La Femme dans la Grèce antique*. Paris.

Mosée, Cl. 1999 Le travail des femmes dans l'Athènes de l'époque classique. In C. Alfaro (coord.) *Más allá de la labor matronalis: aspectos del trabajo profesional femenino en el mundo antiguo. Saitabi* 49, 223–227.

Nieto-Prieto, X. 2008 L'Arquitectura del vaixell. In X. Nieto and M. Santos (eds), *El vaixell grec arcaic de Cala Sant Vicenç*. Girona, 23–64.

Orestano, R. 1951 *La struttura giuridica del matrimnio romano. Dal dirito classico al diritto giustinianeo*. Milan.

Pasquali, G. 1920 Il carmen 64 di Catullo. *Stud. It. di fil. Class*. 1.1, 1–22.

Perrotta, G. 1931 Il carmen 64 di Catullo e i suoi pretesi originali ellenistici. *Athenaeum* 9, 177–409.

Preisendanz, K. 1973–1974 *Papyri Graecae Magicae*, IV. Stuttgart.

Putnam, M. 1961 The art of Catullus 64. *HSCP* 65, 165–205.

Ramain, G. 1922 Catulle. Sur la signification et la composition du poème 64. *RPLHA* 46, 135–153.

Reinhold, M. 1970 *History of Purple as a Status Symbol in Antiquity. Latomus* 116, Brussels.

Squicciarino, N. 2012 *El vestido habla*. Madrid (Ital. ed. Rome 1986).

Treggiari, S. 1991 *Roman Marriage. Iusta coniuges from the Time of Cicero to the Time of Ulpian.* New York.

Turner, V. 1969 *The Ritual Process: Structure and Anti-Structure*. New York.

Van Gennep, A. 1909 *Les rites de passage*. Illinois.

Waltz, R. 1945 Caractère, sens et composition du poème LXIV de Catulle. *REL* 23, 92–109.

Willamowitz-Moellendorf, U. 1940 Hellenistische Dichtung in der Zeit des Kallimachos II. reprint *Hermes* 75, 298–325.

16. From *Calathos* to *Carmen*: Metapoetics in the Story of the Daughters of Minyas (Ovid *Metamorphoses* 4)

Magdalena Ohrman

In this chapter, I will revisit two passages from Ovid's *Metamorphoses*, both of which are often read as using weaving as a metaphor for Ovid's own writing. I am interested in potential influences of elegiac verse on Ovid's story of the stubborn daughters of Minyas, and will consider the different methods by which Ovid dramatises the conflict between welcoming of Bacchus as a new god and the traditional worship of Minerva in this story.[1]

Ovid's description of the Minyeides is very different from that of the more famous weaving contest between the mortal Arachne and Minerva later in the *Metamorphoses* (Ov. *Met.* 6.1–145). In the sixth book of Ovid's *Metamorphoses*, Arachne and Minerva set up their looms and prepare in the same way, with an initial emphasis on the twin looms and later, the description of one loom as well as how the contestants perform the same actions in tandem:

haud mora, constituunt diuersis partibus ambae
et gracili geminas intendunt stamine telas;
tela iugo uincta est, stamen secernit harundo,
inseritur medium radiis subtemen acutis,
quod digiti expediunt, atque inter stamina ductum
percusso pauiunt insecti pectine dentes.
utraque festinant cinctaeque ad pectora uestes
bracchia docta mouent, studio fallente laborem.
illic et Tyrium quae purpura sensit aenum
texitur et tenues parui discriminis umbrae,
(…)
illic et lentum filis immittitur aurum
et uetus in tela deducitur argumentum.

Ov. *Met.* 6.53–69

They both set up the looms in different places without delay and they stretch the fine warp upon them. The web is bound upon the beam, the reed separates the threads of the warp, the woof is threaded through them by the sharp rods which their busy fingers ply, and when shot through the threads of the warp, the notched teeth of the hammering slay tap it into place. They speed on the work with their mantles close girt about their breasts and move back and forth their well-trained hands, their eager zeal

[1] Translations from Ovid's *Metamorphoses* are based on Miller's translation in Goold 1999, with some adaptations. Other translations are my own. For the Latin text of the *Metamorphoses,* I use the *OCT* of Tarrant 2004.

> beguiling their toil. There are inwoven the purple threads dyed in Tyrian kettles, and lighter colours insensibly shading off from these. (…) There, too, they weave in pliant threads of gold, and trace in the weft some ancient tale.

We might say that in the Arachne episode as a whole, the focus is on the artistic composition, on the choices made by two artists using the same tools. The practicalities of weaving are less important, too, as both the image by Minerva and that by Arachne are presented to the audience as finished works of art (Ov. *Met.* 6.70–128). This is supported through verbal reminiscences of other *ekphrases*, for instance that of Thetis' coverlet in Cat. 64, or that of Aeneas' shield in Verg. *Aen.* 8.[2] This means, as Oliensis has shown, that the difference between them is developed only through the images they create within a given medium.[3] The differences between the two designs have often been read as outlining two different ways of writing epic poetry – epic poetry with different artistic programmes, but epic poetry nonetheless.[4]

When weaving is used in a metapoetic context earlier in the *Metamorphoses*, this is not yet the case. To my mind, Ovid uses the story of the daughters of Minyas and their weaving to consider not so much the product created as the tools used for the composition of the *Metamorphoses*. Throughout this episode, Ovid emphasises the sequence and different aspects of wool work (roving, spinning and weaving, *Met.* 4.34f; spinning, 36 and 54; passing the weft the through the loom, 275).[5] The reader is prompted to consider individual elements of a weaver's work as the sisters work and tell their tales, and later as the transformation of the sisters into bats is first heralded by the metamorphosis of their looms into vines and plants: loom, weft, yarn, and warp change in turn.[6] The creation of fabric, stressed through the large incidence of technical vocabulary in the episode,[7] itself implies the combination of different elements into a new, coherent whole, and the most recent critical commentary on the episode points to Ovid's use of the metapoetically significant phrase *leui pollice deducens* ("drawing out the thread with light thumb", 4.36), connected to the programmatic *carmen deductum* (lit. "a drawn-out song"), as a crucial starting point for such readings.[8] The position of the Minyeides as the first human internal narrators of the *Metamorphoses* might be sufficient to suggest that their weaving (and their stories) should indeed be taken to illustrate the methods or aims of the external author or the text as a whole.[9] This allows the work of the Minyeides to function as a parallel for Ovid's *Metamorphoses*

2 Harries 1990, 65–69. See also the chapter by C. Alfaro in this volume on Cat. 64.

3 Oliensis 2004, 293.

4 *E.g.* Leach 1974, 103–104.

5 Rosati 1999, 243. Ov. *Met.* 4.33–37 are cited in full below; spinning in 4.54 is described as the wool following the spinner's thread (*lana sua fila sequente*). In 4.54, Ovid also uses the verb *orior*, connected to the setting up of the loom, see n. 7 below. Ov. *Met.* 4.275 reads *quae radio stantis percurrens stamina telae* (as she [Alcithoe] ran through the warp of the standing loom with a rod. *Radius* is often translated as 'shuttle' (*cf. OLD* s.v. *radius* 3b) but probably rather refers to an elongated, rod-like narrow implement put to diverse uses. *Cf.* Ov. *Met.* 6.56 (cited above) and *ThLL* s.v. *radius* II C 3.

6 Ov. *Met.* 4.394–398 with the following stages: *telae* (weft, 394), *uestis* (cloth, 395), *fila* (threads, 396), *stamen* (warp, 397). *Cf.* Anderson 1997, 457.

7 Heath 2011, 89–93 offers a most valuable discussion of the high incidence of technical vocabulary and its metapoetic connotations, indicating also the often-overlooked technical meaning of *orsa est* ("she began", *Met.* 4.54; 167), referencing the setting up of a loom, which closely links the Minyeides' storytelling with their weaving. *Cf.* also Rosati 1999, 244–245.

8 The phrase alludes to the proem's promise of a *carmen deductum*, which is commonly held to indicate the text's commitment to Callimachean aesthetics. *Cf. e.g.* Barchiesi and Rosati 2007, 252; Jouteur 2001, 71; Myers 1994, 79–80, Leach 1974; Rosati 2002, 275f, and, in passing, Liveley 2011, 56. On the proem specifically, Wheeler 1999, 8–30.

9 Barchiesi and Rosati 2007, 242–243; Heath 2011, 86–88; Janan 1994, 427; Rosati 1999.

at large, incorporating stylistic features, topics and themes from different literary genres[10] into an innovative, new type of epic.

I would like to add to the already extensive discussion of metapoetics in this passage by suggesting that Ovid not only uses the Minyeides to sign-post his willingness to engage with other genres, but that he has also included within this episode a concrete example of one of the many ways in which this happens across the *Metamorphoses*. My concern, however, is not initially with the adoption of elegiac themes in the sisters' stories, nor with their use of elegiac vocabulary. Instead, I am interested in stylistic features, supported by but not inherently dependent on the medium of elegiac verse or metre. One such stylistic convention found in elegy is what I will loosely call stichic organisation of content,[11] which to my mind is present in Ov. *Met.* 4.32–41.

Use of the elegiac distich

I need to preface the discussion of the transference of such stylistic features to the *Metamorphoses* and the detailed analysis of the Minyeides episode with a few comments on their use in erotic elegy. I am particularly interested in two features of elegiac verse: first, the often-observed tendency of the elegiac distich to prompt the containment of syntactical units within each distich,[12] which places additional emphasis on the opposition between the metrically contrasting hexameter and pentameter lines, and secondly, the employment of metrical *aptum* within the elegiac distich. The latter merits consideration in some detail.

Because of the alternation of hexameter and pentameter lines in elegiac verse, the elegiac distich affords authors the opportunity to create and utilise tensions arising from expectations of content based on metrical *aptum*. It has been suggested that content in elegy may be organised according to a principle where different content is presented in different lines, agreeing with (or breaking) metrical *aptum*.[13] For example, erotic elegy may use such differences between hexameter and pentameter content to underscore the specific perspective of the elegiac lover: This is the case when Penelope describes rumours from Troy in Ov. *Epist.* 1.13–22, where the hexameter lines recount events familiar from Homer but the pentameter lines detail Penelope's emotional reaction.[14] Propertius 2.22a.29–34 provides another example, which manipulates, rather than accepts outright, the notion that certain content should be presented in a particular metre or in a particular part of the elegiac distich. Propertius' playful allusions to notions of metrical *aptum* in this passage shows that the ancient reader could be expected to recognise and appreciate this type of stichic organisation of content in a range of different realisations. The Propertian narrator compares his own prowess in lovemaking with the powers of Hector and Achilles in warfare, but ostensibly introduces the mythological *exempla* to the elegiac sphere as lovers as well as soldiers: their valour was not diminished by their prowess as lovers.[15]

---

[10] See below for allusions to and interaction with erotic elegy. Keith 2002, 256–263 highlights the interactions with tragedy in the Minyeides episode.

[11] *Cf.* Morgan 2012, 205, stressing the literary effects of continual manipulation and transgression of such literary or indeed metrical conventions.

[12] Barsby 1973; Heinze 1919, 75–76; Kenney and Clausen 1982, 455; Morgan 2010, 347; Morgan 2012, 207.

[13] Jacobson 1974, 252; Spoth 1992, 38–39; Barchiesi 1997, 21–23; Miller 1997; Morgan 2000; Morgan 2010; Morgan 2012. For the importance of alternation between perspectives, *cf.* also Sharrock 1990.

[14] Jacobson 1974; Spoth 1992, 38.

[15] Whitaker 1983, 117–118, Neumeister 1983, 70. *Cf.* also Dué 2002, 110–111.

quid? cum e complexu Briseidos iret Achilles,
num fugere minus Thessala tela Phryges?
quid? ferus Andromachae lecto cum surgeret Hector,
bella Mycenaeae non timuere rates?
ille uel hic classis poterant uel perdere muros:
hic ego Pelides, hic ferus Hector ego.

Prop. 2.22a.29–34

Did the Trojans flee any less before the spears of Thessaly because Achilles came from the embrace of Briseis? Did the fleet from Mycenae fear the war any less because the ferocious Hector rose from Andromache's bed? The former could destroy navies, the latter city walls: here, I shall be Achilles, here, I shall be the ferocious Hector.

The metre reinforces the impression that love and war are intertwined as the metrical *aptum* is overturned: the traditionally heroic hexameters describe lovers arising from the beds of their beloveds, the soft pentameters their performance as warriors.[16] The narrator proceeds to place the heroes of old in a sphere distinctly different from his own, describing first their military tasks of overthrowing city walls and naval forces in a hexameter line (2.22a.33), but in the pentameter line (2.22a.34), he locates his own intention to emulate their prowess exclusively within the area of elegiac love through the emphasis placed on the anaphor of *hic ego* ("here, I"). Lines 33 and 34 restore the metrical *aptum* and the division of *militia* ("warfare") and *amor* ("love"), thus ensuring the effectiveness of the metaphor of *militia amoris* ("lover's warfare").

I have highlighted examples from erotic elegy. However, when considering the potential influence of elegiac conventions on Ovid's *Metamorphoses*, it is helpful to bear in mind that the type of content organisation outlined in the two instances above is by no means exclusively connected with erotic elegy; indeed, a recent discussion by Llewelyn Morgan cites further examples from more broadly themed works, such as Propertius' fourth book, Ovid's *Fasti*, and the *Corpus Tibullianum*.[17]

Thus far, my concern has been with content organised in a pattern, which alternates in agreement with the distinct parts of the elegiac distich. The final half of this paper will suggest that this pattern has exercised at least some influence on Ovid's *Metamorphoses*, and that, although the pattern is realised in the *Metamorphoses* without the motivation or support of explicit metrical alternation, it is available to Ovid as one of many ways in which he might engage with elegiac intertexts within the epic framework of the *Metamorphoses*. Similar to the function of such stichic content organisation in elegiac verse, a stichic pattern is employed in *Met.* 4.32–41 in order to emphasise differences between two opposing value systems. We return now to Ovid's narrative of the daughters of Minyas, who stubbornly reject Bacchus in favour of their old goddess, Minerva.

Stylistic patterns in Ovid's tale of the daughters of Minyas

In telling the story of the daughters of Minyas in *Met.* 4 immediately after the cautionary tale of Pentheus in *Met.* 3, Ovid creates a tension between two competing ways of life and worship. He varies the description of the Minyeides by adapting a stichic pattern, which expresses positive and negative value judgements on their behaviour in alternate lines in *Met.* 4.32–38. This pattern is effective in making

[16] Ovid refers repeatedly to the rules of metrical aptum, *cf.* programatically in *Am.* 1.1.17–28; *Rem.* 371–386.
[17] Morgan 2012, 212–215.

the reader consider the situation of the Minyeides more carefully despite a strong initial condemnation of their impiety, which I will indicate presently, and the reader's own knowledge of the eventual punishment of the sisters by Bacchus. The first lines on the Minyeides point clearly to their impiety:

at non Alcithoe Minyeias orgia censet
accipienda dei, sed adhuc temeraria Bacchum
progeniem negat esse Iouis sociasque sorores
inpietatis habet. (…)

Ov. *Met.* 4.1–4

But not Minyas' daughter Alcithoe; she will not have the god's holy revels admitted; nay, so bold is she that she denies Bacchus to be Jove's son! And her sisters are with her in the impious deed. (…)

The alliteration of *at* ("but") and *Alcithoe* (1) is combined with an emphasised contrastive conjunction (*at*)[18] in order to highlight Alcithoe and her sisters' deviation from the behaviour expected of them.[19] Subsequently, both *adhuc temeraria* ("so bold", 2) and *inpietas* ("impious deed", 4) indicate that the subsequent punishment of the sisters is justified. Underlining this, a similar phrase recurs in 4.389 just prior to the transformation of the sisters and the conclusion of their tale.[20] Thus, obvious expressions of condemnation bracket the story of the transformation of the Minyeides, which in turn brackets the stories presented by the sisters themselves:

finis erat dictis, et adhuc Minyeia proles
urget opus spernitque deum festumque profanat (…)

Ov. *Met.* 4.389f

There was an end to her story; but still did the offspring of Minyas ply their tasks, despising the god and profaning his holy day (…)

The two passages are linked not only through the return to the framing situation but also through the placement of *adhuc* ("still") in the same position in both lines. Furthermore, common to both framing passages is the reference to ancestry and familial ties across generations. Alcithoe questions Bacchus' divine ancestry at the opening of the book (2f), and the sisters together are described as *proles* ("offspring") of Minyas in 389.[21] Ovid's ring composition supports the teleological drive of the overarching narrative of the emerging Bacchus worship by serving to emphasise the condemnation of the sisters. The references to continuity of family lines, which thus bracket the Minyeides episode, further contribute to this emphasis.

However, the intervening passage gradually puts forward a more nuanced view of the daughters of Minyas. In 4.5 the focus shifts to the Theban women, then to their performance of Bacchus' rites. Newlands has shown that there are unsettling hints in Ovid's description of the Theban women. They are described as *matresque nurusque* ("mothers and young wives", 4.9) immediately before

[18] Bömer 1976, 14.
[19] Barchiesi and Rosati 2007, 244–245.
[20] But *cf.* Leach 1974, 111, noting that the punishment imposed on the Minyeides in Ovid's version is less harsh than those exacted in parallel versions of Nicander and Aelian. *Cf.* Ael. *VH* 3.42 and Nic. 55.
[21] With the exception of the patronymicon *Minyeides*, occuring in Ov. *Met.* 4.32 (*cf.* Anderson 1997, 410 and Barchiesi and Rosati 2007, 245), Ovid consistently refers to the Minyeides as *sorores* throughout the intervening passage. This might serve to highlight their affinity and loyalty to Minerva by placing them in a context of female social networks. But *cf.* also Nikolopoulos 2004, 146–147, suggesting that the Minyeides through their actions 'align themselves with a particular masculine construction of the feminine ideal'.

abandoning the tasks evoked by these value-laden words in order to take up the thyrsus-staff and the *tympanum*.[22] Ovid, however, does not exploit this uneasiness at once. From 4.11 through to 4.31 Ovid illustrates the Theban acceptance of Bacchus by alluding to, and finally adopting fully, the form of invocation and hymn to the god (4.18–31).[23] Only when Ovid returns to description of the Minyeides themselves in *Met*. 4.32–38 does he address fully the ambiguity of the moral and religious issues inherent in such a ready acceptance of the new god.

In the following, I will suggest that the sense of ambiguity surrounding the Minyeides' actions (whether they illustrate piety or impiety) is heightened by a stichic pattern, which relies on content division. Several comments are made below on metrical features, which place further emphasis on particular aspects of the content. Importantly, this is not intended to suggest that there are any purely metrical similarities with elegiac verse. Instead, the pattern of content organisation here consists of a pronounced alternation between negative and positive modes of describing the Minyeides present in the narrative of 4.33–37:

> *"Placatus mitisque" rogant Ismenides "adsis",*
> *iussaque sacra colunt; solae Minyeides intus*
> *intempestiua turbantes festa Minerva*
> *aut ducunt lanas aut stamina pollice uersant*
> *aut haerent telae famulasque laboribus urgent.*
> *E quibus una leui deducens pollice filum*
> *"dum cessant aliae commentaque sacra frequentant,*
> *nos quoque, quas Pallas, melior dea, detinet" inquit*

Ov. *Met*. 33–37

> "Oh, be thou with us, merciful and mild!" the Theban women cry; and perform the sacred rites as the priest bids them. The daughters of Minyas alone stay within, marring the festival by toiling out of time, plying their household tasks, either spinning wool, or thumbing the turning threads, or hang on to the loom, and press their maidens with work. Then one of them, drawing out the thread the while with light thumb, says: "While the other women are deserting their tasks and thronging this so-called festival, let us also, whom Pallas, a truer goddess, keeps"

While on the whole the alternation between positive and negative descriptions of the Minyeides' situation is organised in a stichic pattern, the first sense unit focussed on the Minyeides does contain an enjambment; it commences in 4.32 and ends in 4.33: *solae Minyeides intus / intempestiua turbantes festa Minerva* (...) ("the daughters of Minyas alone stay within, marring the festival by toiling out of time (…)"). The words associated with the Minyeides in 33 are exclusively negative. First, the Minyeides' isolation is illustrated through the metre by the placement of *solae* ("alone") after the penthemimeral caesura in 32, which in terms of word placement separates the sisters from the women of Thebes, the subject of the immediately preceding *colunt* ("perform").[24] This isolation and their insistence on their craft are described in negative terms, with the adjective *intempestiua* ("out of time") emphasised both due to its initial position in the line and due to the ensuing penthemimeral caesura in 33. Metrical parallels also suggest condemnation of the sisters: Bömer notes that series of spondees such as *intempestiua turbantes* commonly occur in passages describing

[22] Newlands 1986, 147.
[23] Bömer 1976, 16; Danielewicz 1990; Anderson 1997, 411.
[24] Barchiesi and Rosati 2007, 251 suggests that Ov. *Met*. 4.31–32 re-establish the opposition between the Theban women and the Minyeides first seen in Ov. *Met*. 3.773 and Ov. *Met*. 4.1.

doubt and deliberation.[25] As a result, the Minyeides appear in stark contrast to the women of Thebes who attend to the rites ordered, *iussaque sacra colunt* ("they perform the sacred rites as the priest bids them"), in the first half of 4.32. There, the word order suggests that the Theban women follow a logical progression from the god's demand to their fulfilment. The reader might expect Bacchus' vengeance on the Minyeides to follow swiftly. Instead, line 34 defers the reader's condemnation of the Minyeides; the sense of pause is strengthened by the spondees prevalent in the first half of the line: *aut ducunt lanas aut stamina pollice uersant* ("either spinning wool, or thumbing the turning threads"). As the rhythm of the line shifts to dactyls in the second half of the line, it comes to reflect the pattern of the women's work: the slow and careful extending of the wool, followed by the quick turning of the thread and the spindle.[26] Far from condemning the sisters, the line connects them to ideal paradigms of matronly virtue.[27]

The pattern of alternation between praise and condemnation of the Minyeides emerges fully when Ovid returns to describing the sisters in negative terms in 4.35. Here, the use of *haero* ("hang on to") and *urgeo* ("press") give a strong negative colour to the Minyeides' work, as both words imply resistance.[28]) Both words are emphasised by their position in the hexameter. The sisters' over-persistence on their craft, and, through the mention of the harassed slaves, its detrimental effect on others emerges clearly.[29]

Line 36 continues the pattern of alternation, presenting the reader once more with the image of the Minyeides as quietly working their wool in the fashion of ideal *matronae*. Moreover, here a sense of community amongst the sisters is implied by *e quibus una* ("one of them"). In this line, the verse once more runs lightly, perhaps suggestive of the women's pleasure and light-handedness at their work, thus reinforcing their positive portrayal.[30]

The pattern of alternation is sustained in line 37. While the unnamed daughter of Minyas reproaches the women of Thebes for lacking devotion to duty,[31] she also singles out herself and her sisters as non-participants in the religious ritual celebrated by the others. The isolation first made apparent through *solae* ("alone", 4.32) implicitly recurs in this line, reminding the reader of the condemnation of the Minyeides initially suggested in 4.1–4.

[25] Bömer 1976, 265 and *cf.* Ov. *Met.* 5.171.

[26] Such mimicking of textile working processes through metre and sound-play occurs surprisingly often in Latin poetry. For example, when used in hexameter verse to mean 'with the thumb' in the context of spinning, the word *pollice* is consistently placed in the fifth foot dactyl – the most reliably fast-moving part of the line – in order to illustrate the speed of the spinner's hand motion as she twists the spindle. When used in other contexts, the placement of *pollice* across the hexameter verse is considerably more varied (*e.g.* Ov. *Met.* 11.17 on music; Prud. *Apoth.* 1026 on pottery). This applies to hexameter texts from the Republic through to Late Antiquity.

[27] For wool work as a paragdigmatic activity, *cf.* Barchiesi and Rosati 2007; Bömer 1976, 25–26.

[28] On the connotations of *haereo* ("hang on to"), Barchiesi and Rosati 2007, 252, who emphasise the voluntary seclusion of the sisters indicated through *intus* ("within", 32). Bömer 1976, 27, also notes the similarities with Ov. *Met.* 4.390 (quoted above), which further strengthens the negative associations of *urgeo* ("press") here. *Cf.* also *ThLL* s.v. *haereo* 1 B; *OLD* s.v. *urgeo*.

[29] *Cf.* also Barchiesi and Rosati 2007, 252, suggesting an opposition between the Minyeides' treatment of their servants and the egalitarian character of the Bacchus worship described in Ov. *Met.* 4.4–9.

[30] The metapoetic associations of *leui pollice* ("light thumb", 37) are repeatedly noted, connecting the Minyeides to Callimachean or elegiac aesthetics, *cf.* Barchiesi and Rosati 2007, 252; Jouteur 2001, 71; Myers 1994, 79–80, and, in passing, Liveley 2011, 56.

[31] The reproach is heightened by her phrasing: *cesso* ("to desert") may imply the abandonment of an appointed task (*cf. ThLL* s.v. *cesso* 1; Bömer 1976, 27, thus connecting to the problematic description of the Theban women as *matres nurusque* ("mothers and young wives") in *Met.* 4.9. Furthermore, though it is often used with *sacra* (*e.g.* Ov. *Met.* 3.581; 691; 732; 10.436), *frequento* ("to throng") may refer to the movements of an unorganised throng (*cf. ThLL* s.v. *frequento* D and Ov. *Fast.* 3.251; Sil. 13.783; Tac. *Ann.* 5.10).

Line 38 describes the Minyeides as faithful to their goddess alone and, in contrast to line 37, provides a positive motivation for their behaviour. However, negative undertones intrude: first, through *detinet* ("keeps") with a semantic reach from the innocuous to the negative.[32] Similarly, while the internal speaker surely intends *melior dea* ("a truer goddess") as praise of Minerva, the comparative firmly establishes Minerva and Bacchus as contestants,[33] bringing to mind of stories of Minerva not taking lightly challenges to her superiority, a topic to be exemplified by Arachne in *Met.* 6.[34] Again, the text forebodes the tragic outcome for Minerva's loyal followers here.

As the focus shifts from the description of the sisters and their impending misfortune to the story-telling situation in 4.39, the alternating pattern seen in lines 33–38 is abandoned. However, in the following lines the text retrospectively reinforces the reader's awareness of its presence by drawing attention specifically to modes of discourse: The reader's interest in the tales told by the sisters is built up gradually through a series of lines establishing the premise of their narration (4.39–41), and a tantalising *praeteritio* of rejected stories (4.42–54) before the tale of Pyramus and Thisbe is introduced in 4.55. Interestingly, lines 4.39–41 may be understood metapoetically as discussing differing types of discourse:

> *"utile opus manuum uario sermone leuemus*
> *perque uices aliquid, quod tempora longa uideri*
> *non sinat, in medium uacuas referamus in aures."*

Ov. *Met.* 4.39–41

> "[let us also] lighten with various talk the serviceable work of our hands, and take turns to bring something, meanwhile, to ears unoccupied, so as to beguile the time."

Exploration of the metapoetic allusions of this passage will retrospectively support the presence of an altern ating pattern borrowed from elegy in the description of the Minyeides. It furthermore places the stories the sisters tell in an elegiac context. It is heralded by the phrase *leui pollice* ("with light thumb") in 4.36, which – as mentioned – alludes to Callimachean aesthetic preferences.[35] The use of *leuemus* ("lighten") in 4.39 might be taken as a reference to epic discourse being lightened to suit the *Musa leuis* ("light-hearted Muse") of elegiac verse[36] in correspondence with the elegiac themes developed in the stories told by the Minyeides.[37] The *utile opus* ("serviceable work", 41), which the sisters wish to alleviate through their story telling, is on one level their weaving. On another, the phrase references the didactic thrust of epic poetry,[38] which is to be broken up by *uario sermone* ("various talk"), referring to a varied discourse. This phrase is often used neutrally to refer to a discussion or narrative combining many different topics,[39] but in a stricter, technical sense,

[32] *Cf. ThLL* s.v. *detineo* A and B respectively. Bömer 1976, 27, notes *detinet* (keeps) as unpoetic.
[33] But *cf.* also Barchiesi and Rosati 2007, 253.
[34] *Cf.* also Newlands 1986, 150f.
[35] Barchiesi and Rosati 2007, 252.
[36] *Cf.* Prop. 2.12.22.
[37] Readings highlighting elegiac elements in the Minyeides' tales include Due 1974, 126–127; Keith 2001; Perraud 1983; Robinson 1999, 222.
[38] *Cf. ThLL* s.v. *opus* 1.IV for the suggestion that *opus* ("work") commonly refers to epic. But *cf.* also *ThLL* s.v. *opus* 1.IV.B.2 for references to a variety of genres.
[39] Barchiesi and Rosati 2007, but *cf.* Anderson 1997, 415, who relates it to the Alexandrinian demand for novelty and variety.

it is also applicable to elegiac verse.[40] The emphasis on taking turns (*per uices* ("take turns") in 4.40) supports this interpretation.[41] Thus, the sustained pattern of alternation in lines 4.32–38 is preceded by the intrusion of a different form into the epic narrative through the hymn to Bacchus (4.18–31), and succeeded by allusions to a different form or discourse in 4.39–41, both features, which draw the reader's attention to Ovid's manipulation of epic discourse.

Through its allusions to elegiac organisation of content, the passage on the Minyeides draws the reader's attention to Ovid's play with the conventions of the elegiac genre. These factors combine to make the daughters of Minyas better placed as internal narrators of tales with strong elegiac associations, some of which have already been explored in scholarship.[42] Moreover, several recent readings of their narratives emphasise the presence of paradox and contradiction in the tales told.[43] The sequence of stichic alternation between extremes in *Met.* 4.33–38 heightens the reader's perception of such dichotomies. More importantly, the alternating structure of 4.33–38 mirrors the ambiguity with which Ovid describes Bacchus-worship in *Met.* 3 and 4,[44] and offers the possibility to understand the Minyeides as victims undeserving of the punishment described in *Met.* 4.389–415. The allusions to a form of alternation between options familiar to his readership from elegy allow Ovid to 'pause' temporarily the teleological drive of his epic verse and examine in detail their specific situation.

Conclusions

To conclude: we have seen that Ovid in his story of the Minyeides uses tools associated not only with metapoetic discourse generally but specifically with other literary genres. I hope that I have added to the ongoing discussion of the inclusion and transformation of non-epic elements in the *Metamorphoses* and how they contribute to Ovid's shaping of his own, post-Vergilian, epic. Comparing the story of the Minyeides to that of Arachne, the Minyeides episode in my reading prompts a discussion of formal features of literary genres and the potential transferability of such features to the new text. The Arachne episode, with its twin looms, is concerned with the artistic disposition of a material within a given framework, that is, with the shaping of content rather than the selection of form. By considering not only the content but the sequence of Ovid's descriptions of weaving women, we can extend our understanding not only of individual passages and their potential interpretations, but also of how Ovid gradually expands on the tantalising statements from the proem, where the combination of epic themes with the aesthetics of the elegiac *carmen deductum* is first suggested. It is only when both episodes are considered together that we trace the full process of the weaving of the *Metamorphoses,* from *calathos* to *carmen*.

[40] Admittedly, few metapoetic associations can be detected in Ovid's use of *sermo* (talk) elsewhere. Exchanges between lovers in Ovid's amatory works are sometimes loosely described as *sermo*, *e.g.* Ov. *Am.* 2.5.19; *Ars* 1.467; 569; 3.480. More helpful is in Verg. *Aen.* 1.748f: *nec non et uario noctem sermone trahebat / infelix Dido longumque bibebat amorem*. This passage introduces a similar intrusion of elegiac themes in epic. Jouteur 2001, 72 understands *uario sermone* as referring not to elegiac verse itself but to Ovid's combination of elegiac themes and epic form in the tales subsequently told by the Minyeides.

[41] I also wonder whether the placement of the (syntactically unconnected) words *in medium uacuas* (4.41) on either side of the penthemimeral caesura might not be suggestive of the emptiness or pause arising from the strong diaeresis of the elegiac pentameter line in precisely this position.

[42] *Cf.* Due 1974, 126–127; Keith 2001; Perraud 1983; Robinson 1999, 222; Knox 1986, 35–37.

[43] Janan 1994, focusses on the contradiction between female desire and normative society; Shorrock 2003, on diverging versions of the story simultaneously accommodated through its metaphors. *Cf.* also Leach 1974, 110 for the stories told by the Minyeides as illustrating both 'a vicarious experience of passion and a justification of withdrawal'.

[44] For which *cf.* Newlands 1986, 147; 150–151 and, in passing, Leach 1974, 107–108.

Bibliography

Anderson, W. S. 1997 *Ovid's Metamorphoses Books 1–5. Edited, with an introduction and commentary.* Norman OK.
Barchiesi, A. 1997 *The Poet and the Prince. Ovid and Augustan Discourse.* Berkley CA *et al.*
Barchiesi, A. & Rosati G. 2007 *Ovidio Metamorfosi. Volume II (Libri III–IV).* Oxford.
Barsby, J. 1973 *Ovid's Amores: Book 1.* Oxford.
Bömer, F. 1976 *P. Ovidius Naso Metamorphosen. Kommentar von Franz Bömer. Buch IV–V, Wissenschaftliche Kommentare zu Griechischen und Lateinischen Schriftstellern.* Heidelberg.
Danielewicz, J. 1990 Ovid's Hymn to Bacchus (Met. 4.11ff): Tradition and Originality. *Euphrosyne* 18, 73–84.
Dué, C. 2002 *Homeric Variations on a Lament by Briseis.* Lanhamn MY *et al.*
Due, O. S. 1974 *Changing Forms. Studies in the Metamorphoses of Ovid.* Copenhagen.
Goold, G. P. 1999 *Ovid Metamorphoses. Books I–VIII with an English Translation by F. J. Miller.* Cambridge MA.
Harries, B. 1990 The Spinner and the Poet: Arachne in Ovid's *Metamorphoses. Proceedings of the Cambridge Philological Society* 216, 64–82.
Heath, J. 2011 Women's Work: Female Transmission of Mythical Narrative. *Transactions of the American Philologcial Association* 141, 69–104.
Heinze, R. 1919 *Ovids Elegische Erzählung.* Leipzig.
Jacobson, H. 1974 *Ovid's Heroides.* Princeton NJ.
Janan, M. 1994 'There beneath the Roman Ruin Where the Purple Flowers Grow': Ovid's Minyeides and the Feminine Imagination. *AJPh* 115, 427–448.
Jouteur, I. 2001 *Jeux de genre dans les Métamorphoses d'Ovide*. Paris.
Keith, A. M. 2001 Etymological Wordplay in Ovid's 'Pyramus and Thisbe' (*Met.* 4.55–166). *Classical Quarterly* 51, 309–312.
Keith, A. M. 2002 Sources and Genres in Ovid's *Metamorphoses 1–5.* In B. Weiden Boyd (ed.) *Brill's Companion to Ovid.* Leiden 235–269.
Kenney, E. J., & Clausen W. V. 1982 *The Cambridge History of Classical Literature II: Latin Literature.* Cambridge.
Knox, P. 1986 *Ovid's Metamorphoses and the Traditions of Augustan Poetry.* Cambridge.
Leach, E. W. 1974 Ekphrasis and the Theme of Artistic Failure in Ovid's Metamorphoses. *Ramus* 3, 102–142.
Liveley, G. 2011 *Ovid's Metamorphoses. A Reader's Guide.* London.
Miller, J. F. 1997 Meter, Matter, and Manner in Ovid, Ars amatoria 1.89–100. *The Classical World* 90, 333–339.
Morgan, L. 2000 Metre Matters – Some High-Level Metrical Play in Latin Poetry, *Proceedings of the Cambridge Philological Society* 46, 99–120.
Morgan, L. 2010 *Musa Pedestris. Metre and Meaning in Roman Verse.* Oxford.
Morgan, L. 2012 Elegiac Meter: Opposites Attract. In B. K. Gold (ed.), *A Companion to Roman Love Elegy. Oxford* 204–218.
Myers, S. 1994 *Ovid's Causes. Cosmogony and Aetiology in the Metamorphoses.* Ann Arbor MI.
Neumeister, K. 1983 *Die Überwindung der elegischen Liebe bei Properz (Buch I-III), Studien zur klassischen Philologie.* Frankfurt am Main.
Newlands, C. 1986 The Simile of the Fractured Pipe in Ovid's Metamorphoses 4. *Ramus* 15, 143–153.
Nikolopoulos, A. 2004 *Ovidius Polytropos. Metanarrative in Ovid's Metamorphoses, Spudasmata.* Hildesheim.
Oliensis, E. 2004 The Power of Image-Makers: Representation and Revenge in Ovid *Metamorphoses* 6 and *Tristia* 4. *Classical Antiquity* 23, 285–321.
Perraud, L. A. 1983 Amatores Exclusi: Apostrophe and Separation in the Pyramus and Thisbe Episode. *The Classical Journal* 79, 135–139.
Robinson, M. 1999 Salmacis and Hermaphroditus: When Two Become One: (Ovid, Met. 4.285–388). *Classical Quarterly* 49, 212–223.
Rosati, G. 1999 Form in Motion: Weaving the Text in the *Metamorphoses*. In P. Hardie, A. Barchiesi & S. Hinds (eds) *Ovidian Transformations. Essays on the Metamorphoses and its Reception*. Cambridge Philological Society, Cambridge, 240–253.
Rosati, G. 2002 Narrative Techniques and Narrative Structures in the *Metamorphoses.* In B. Weiden Boyd (ed.) *Brill's Companion to Ovid.* Leiden, 271–304.

Sharrock, A. 1990 Alternae Voces-Again. *Classical Quarterly* 40, 570–571.

Shorrock, R. 2003 Ovidian Plumbing in "Metamorphoses" 4. *Classical Quarterly* 53, 624–627.

Spoth, F. 1992 *Ovids Heroides als Elegien, Zetemata. Monographien zur Klassischen Altertumswissenschaft.* München.

Wheeler, S. M. 1999 *A Discourse of Wonders. Audience and Performance in Ovid's Metamorphoses.* Philadelphia PA.

Whitaker, R. 1983 *Myth and Personal Experience in Roman Love-Elegy: A Study in Poetic Technique, Hypomnemata. Untersuchungen zur Antike und zu ihrem Nachleben.* Göttingen.

17. Redressing Epic: Blood and Gold on the Cloak of Polyneices in Statius' *Thebaid*

Danielle Frisby[*]

This chapter addresses the generic and metapoetic significance of cloaks in the epic tradition. It examines cloaks which are described by the external narrator in such a way as to evoke their manufacture, which connects them to the weaving metaphor. More specifically it examines a particular passage from the subversive Flavian epic *Thebaid* (11.396–402) where one of the pivotal characters, Polyneices, rejects a cloak marked out by its generic associations.[1] Intertextuality between the cloak of Polyneices and other robes and cloaks woven in epic will be examined, cloaks which all build up an impression that there is a pile of 'epic robes' representing their genre.[2] The description, comportment and spurning of the epic robe by the *Thebaid's* hero will add clarity to the nature of the weaving metaphor in epic, turning over ideas about how epic is represented by such metaphors.

The *Thebaid* is positioned in the epic genre in such a way as to render it a fruitful text for intertextual study.[3] The poem alludes to multiple and various other texts, often in a self–aware manner, highlighting self–referentiality or metapoetics as a level on which meaning operates.[4] It frequently blends its content with poetics in quite a disturbing way, creating antagonism with the previous literary tradition. A consequence of this is that intertextual study of the text is both rewarding and often very dense. The few lines I have chosen to treat here exemplify this. Weaving which evokes the construction of poetry is a pivotal generic marker for discussing how allusion and intertextuality generate meaning in the poetics of the *Thebaid.* In turn, the consciously intertextual nature of the *Thebaid* makes it a rewarding lens through which to study the dynamics of the weaving metaphor being reworked through the literary tradition.[5]

* With many thanks to the editors of this volume and hosts of the Spinning Fates workshop, Marie–Louise Nosch, Mary Harlow and Giovanni Fanfani and to those present on the day whose comments and suggestions were gratefully received.

1 On these lines see Venini 1970, 108–10 *ad* 396–400.

2 Faber 2008, 208–209 outlines the use of the weaving metaphor in Latin literature.

3 On intertextuality and its pervasiveness in Latin literature see Conte 1986, 23–95; Farrell 1991, 3–25; Hinds 1998; Edmunds 2001. Particularly useful on Statian intertextuality in the *Thebaid* see Smolenaars 1994, xxvi–xlii; Pollmann 2004, 53–57; Ganiban 2007, 6–9; McNelis 2007, 21–24; Parkes 2012, xxix–xxxiii, and for the literary sources relevant to Book 11 in particular Venini 1970, xii–xxvi, wherein she notes the tragic influence is particularly strong.

4 On distinctions between 'intertextuality' and 'allusion' see Fowler 1997, 13–34; Gale 2000, 4–6; Edmunds 2001, 153–154, although the terms do seem to be used interchangeably by some, including I have to admit, myself. For the original ideas of intertextuality see Kristeva 1969.

5 Is it important to distinguish between the 'motif' of the weaving metaphor and its function as a potential allusion in the text? Verbal echoes cannot be ignored, nor can the potential connection in the motif to a way of representing the genre through it, and if there is something Statius does not shy away from it is playing with genre stereotypes, so I do not see this as an issue.

Part of the way in which the *Thebaid* creates multiple layers of meaning lies in its int*ra*textuality, not just its int*er*textuality. Looking at the internal links in the text amplifies the resonance a motif such as the weaving metaphor might have in the poem. This chapter will begin by looking at how connections created within the *Thebaid* between Polyneices' cloak and one woven for Juno earlier in the text links the *Thebaid* to the *Iliad,* to show the line of connectedness from Polyneices' cloak to previous epic robes. From there we will move onto allusions in our passage to cloaks and robes given to other epic heroes, considering also how they connect to the intertextual links to robes for goddesses and building up the picture of the epic robe. We will close the analysis with the cloak as identifier of the dead hero on the battlefield and finally re–assess the significance of Polyneices rejecting his cloak, a metaphorically activated garment spun from the threads of previous literary robes from the epic tradition.

Polyneices' robe

sic hostile tuens fratrem; namque uritur alto
corde quod innumeri comites, quod regia cassis
instratusque ostro sonipes, quod fulua metallo
parma micet, quamquam haud armis inhonorus et ipse
nec palla uulgare nitens: opus ipsa nouarat
Maeoniis Argia modis ac pollice docto
stamina purpureae sociauerat aurea telae.

Theb. 11.396–402

Thus he (Polyneices) spoke, regarding his brother (Eteocles) with hostility. Polyneices is set ablaze in his high heart because of how many comrades his brother has, because of the regal helm, and the horse decked out in purple, because the bronzed shield gleams with precious metal. He himself, however, is not dishonoured by his own weaponry, resplendent in a cloak which was not commonplace at all: Argia herself had made it from scratch, in Maeonian fashion, and with clever fingers had integrated golden threads with purple ones.[6]

At this juncture in the epic the Seven against Thebes have fought, in the main died, and have left their founder member, the exiled Theban prince Polyneices, to his fate at his brother Eteocles' hands. The two sons of Oedipus are on the battlefield, with Eteocles having just been pushed out of the Theban gates by the Fury Tisiphone (11.388), and we come to 11.396–402 as Polyneices spots his brother in his regal finery.

Who is looking at whom becomes instrumental to what is going on here, for we start with everything from Polyneices' point of view; *tuens* ("regarding", 396) signals the act of looking, and the look is not a friendly one, indicated by *hostile* ("with hostility", 396). Polyneices' response to his brother is something we might be ready for. The enmity of the sons of Oedipus is what they are famous for in tragedy, after all. Statius puts a different spin on it by focussing on the visual aspect of this enmity, and highlighting, in doing so, the potential for epic association which is contaminated by tragic association.[7]

Polyneices' emotional reaction is couched in terms of fire, *uritur alto / corde* ("he is set ablaze in his high heart", 396– 397). The root of this is not longstanding fraternal antagonism, as we might

[6] The critical edition of the *Thebaid* used throughout is Hill (1996); of the *Aeneid* Mynors (1969); of Homer' *Iliad* Wilcock (1978; 1984). The editions used of the *Argonautica*, both that of Valerius Flaccus and Apollonius Rhodius are the Loebs of Mozley (1934) and Race (2009) respectively.

[7] On the gaze as assaultive see Lovatt 2013, 310–347, on the final dual of Polyneices and Eteocles see 343–344.

expect to be told, but based entirely on the material manifestation of Eteocles' position. Unnecessary repetition of *quod* ("because", 397–8) builds an impression of Eteocles having so much, the sight of which stimulates Polyneices' envy. He starts with Eteocles' *comites* ("comrades", 397), a sore point as Polyneices has managed to put all of his comrades in the way of danger and most have died; he has no–one to blame but himself on this score.[8] There is also the helmet which designates Eteocles as king, the horse, notably clad *ostro* ("in purple", 398), and finally the shine on the shield.[9] It is at this point that we have an abrupt change in point of view. The external narrator now turns the audience's gaze, previously directed through Polyneices' envious eyes, back onto the character and proceeds to respond to the character's issues. We are swiftly appraised of Polyneices' own position, and left to conclude that he is not so badly suited up himself, our estimation hinging on the tantalising and allusive mini–ekphrasis of Polyneices' cloak.

Polyneices has both the richness of purple *purpureae telae* ("purple threads") and the glitter of gold *stamina aurea* ("golden threads", 402) about him, according to the narrator. Moreover, this cloak has been woven for him by a princess: his Argive wife Argia, and we are given an insight into her craftsmanship since she did it in the Maeonian fashion, but notably *pollice docto* ("with clever fingers [lit. thumb]", 401). Line 402 is a golden line, with the patterning of noun and adjective reflecting the interwoven threads, something which signals to us that the weaving metaphor has been activated here. This cloak responds in every way to what Polyneices found so enviable in his brother's garb. In it the exile is resplendent, with *nitens* ("shining", 400) answering the metallic shine of *micet* ("gleams", 399) on Eteocles' shield, which in Polyneices' eyes rendered him so kingly. Then the narrator offers a brief description of the garment, evoking its construction in a terse, but rich two and a half lines. The *purpureae* ("purple", 402) of the threads reply to the *ostro* ("purple") which the horse is decked out in, with the *aurea* ("golden", 402) answering the precious metal (*metallo* "metal", 398) which decorates Eteocles' shield.[10]

The maker of this fabulous robe, the Argive princess Argia, reminds the audience of Polyneices' potential claim to the Argive throne, and this is reinforced by Adrastus (Argia's father and king of Argos) in his offer of the throne of Argos to Polyneices (*Theb.* 11.433–5).[11] The verb applied to her assembly of the garment *novarat* ("she had made from scratch/ made anew", 400) should alert us to possible repetition and renewal of something which has been done before on a poetic level, as well as in the context of the metaphor. The emphasis on her part in the cloak's manufacture, and the alignment of the poetic arrangement of words with the very threads makes plain the potential to read the poetics of weaving and ekphrasis into the passage.

[8] Only Adrastus escapes at 11.441–446, after ineffectually trying to persuade Polyneices to stop (429–435). Venini 1970, 109 *ad* 397 notes that *innumeri* is hyperbole, and to do with Polyneices' state of mind. This is all part of Polyneices' focalisation of the scene.

[9] Bradley 2009, 189 notes that 'the way [murex] crystals sat upon the surface of the fabric caused it to refract light so that the garment appeared to shimmer and glow', which I understand to refer to the particulates of the pigment on the wool. This too then could add to the lustrous effect.

[10] On purple in the ancient world see Reinhold 1969, 301–3; 1970; Bradley 2009, 189–211, on *ostrum* see 192–193.

[11] The theme of looking and directing of the gaze is also important to how Statius works Polyneices and Argia's relationship in the *Thebaid.* Augoustakis 2010, 82 and Parkes 2012, 91 *ad* 4.89–92 note Argia urging Polyneices at 12.325–8, after he has died, to look at her. In our passage, rather ominously, he is similarly incapable of 'looking' at his wife, or even thinking of her.

Internal links: Juno's *peplos*

The passage from Book 11 alludes to numerous models, in a typically Statian manner.[12] An important link is to another passage within the *Thebaid,* which describes the dedication of a robe to the goddess Juno by the Argive women (10.56–64). These women are from the same community as Argia, and the robe they dedicate shares commonalities with Polyneices' cloak further linking the two garments. The alliterative interlacing of gold and purple threads which take the items down to the level of fabrication is one shared aspect: *variis ubi plurima floret / purpura picta modis mixtoque incenditur auro* ("it blooms in various ways with much depicted purple and is set ablaze with interlaced gold").[13]

There is also a verbal echo at 11.401 which references the method of the garment's manufacture in *modis* ("ways"), which reinforces a connection between Argia and her fellow countrywomen, as well as connecting the ekphrastic elements of the two passages. Alongside verbal echoes, in Statius' work conceptual or thematic echoes often work to create connections in the text, and for our analysis of this passage the marital theme becomes an important one. At 10.56–64 the women, who are all involved in successful marriages (56–7), dedicate their robe to the goddess who is associated with marriage and the unity of the family.

Upon the robe is the image of the goddess herself being wed to Jupiter (61–64). All the optimism and potential of that union is embodied in the image, described in a brief ekphrasis by the narrator. This optimism looks forward to a less pleasant future, for the narrator finishes the ekphrasis with a snap back to the present in *nondum furtis offensa mariti* ("not yet offended by the affairs of her husband", 64). This is fundamental to the persuasive tactics of the Argive women, who also use Juno's wrath against Jupiter's Theban lovers to great effect in their prayer (10.67–69). Juno's youth, innocence and the fragility of the situation are brought out by the description which precedes the cynical comment of 64: *lumine demisso pueri Iovis oscula libat / simplex* ("with eye downcast the innocent girl tastes the kisses of boy–Jupiter", 62–3). This scene is evoked on a variety of levels by the cloak which Argia weaves for her husband. It cannot help but recall the couple's recent marriage, recent, at least in terms of the narrative, how she first appeared to him likewise modest and virginal, with face downcast at 2.230–235, and how their marriage is also destined for sorrow.

The fire imagery which is part of the dedicative robe's description (*incenditur* "it is set ablaze", 60) connects to Polyneices' negative emotional reaction. In Polyneices the fire is internal feeling, unlike the external shine which the "blazing" thread intimates. In turn we are prompted to think of Juno experiencing a fiery feeling at *Theb.* 10.76–78:

> *non saevius arsit*
> *Herculeae cum matris onus geminosque Tonantis*
> *secubitus vacuis indignaretur in astris.*
>
> not more savagely did she burn when she became
> indignant at the burden of Hercules' mother and
> doubled affair of the Thunderer, with her left abandoned
> among the stars.[14]

[12] Though this is something notable in other poets' works, see Pollmann 2004, 53–54.

[13] See Nagy 2010, 278–308 on the importance of purple in robes dedicated to goddesses.

[14] Here Williams 1972, 44 *ad* 76f. notes *geminos* must refer to the prolonged night which Jupiter spent with Alcmene, rather than his multiple affairs with various Theban women due to the mention of Alcmene within the line. Given the narrator's fondness for doubling in the *Thebaid* and the fact that this comes as part of repeated reminders to Juno of

Jupiter and Juno's traditional antagonism, something which becomes long established throughout epic, is proleptically flagged up in the robe ekphrasis. Juno's response to the Argive petition further emphasises the downward spiral which the relationship of Juno and Jupiter took from that first moment. The robe dedicated by the Argive women works with Juno's habitual anger and jealousy through the image depicted upon it to achieve their goal: divine assistance. Thus the robe dedication foregrounds both the potential unity and discord encompassed by the theme of marriage. The fiery antagonism of Juno's marital strife is a far cry from the innocent nuptials in which we saw her involved on the robe, yet is clearly connected to them. The alignment of Juno and Polyneices in their fiery jealously casts an ominous shadow over the potential for emblematising unity which the garment carries. Rather, it reinforces its status as an item which sows the seeds of discord, as the connections to a traditionally difficult divine relationship shape ideas of Polyneices and Argia being doomed to a dysfunctional relationship.

Robes for goddesses

Thebaid 10.56–64 alludes to *Iliad* 6.286–296, where a *peplos* is brought for Athene by a party of Trojan women seeking for her to act on their behalf.[15] The dedication contrasts, in its emphasis on unity, to that of the *Iliad*, where we saw Theano alone approach the goddess Athene (6.298–310). In the *Thebaid* it is *all* of the women and their children who beseech Juno, in a scene which emphasises social cohesion (10.49–53).

The Statian robe picks up on the Homeric robe in some very obvious ways, but also deviates meaningfully from the Homeric model at key points.[16] The robes are alike in craftsmanship and quality, and comparable in the way they shine. Homer compares his robe to a star (6.295), and Statius picks up on the shining in his thread which blazes through the garment (10.60). However, the Statian robe dedicated to Juno, through its deviation from the Homeric model, highlights certain issues in the Iliadic robe. The image on Statius' robe is given a fuller description; it is clearly made for purpose and is therefore, down to the last thread, entirely appropriate for its recipient. The same cannot be said of the Trojan robe.[17] It comes from those robes which Hecuba already owns, and its decoration is not fully described but instead the weight goes on its provenance. It was not, we learn, specifically woven for dedication to Athene, but was in fact made by Sidonian slaves.[18] These slaves were brought back by Paris in the same voyage on which he brought Helen, therefore not at all likely to please the goddess snubbed in the golden apple fiasco by Paris' very choice of Aphrodite and Helen over what Athene had to offer.

This is not to say, however, that the Iliadic robe is not beautiful or that it is selected without care; it is at the bottom of the pile (ἔκειτο δὲ νείατος αλλων "it lay at the bottom of the rest", 6.295) as the narrator points out. This implies it was kept for best, or indeed, not worn at all.[19]

that affair, I wonder if there is not a sense in which the Argive women, by the very nature of their appeal are in some way doubling the effect here.

[15] On the dedicating of the *peplos* in the *Iliad* see Graziosi & Haubold 2010, 162 *ad* 6.297–311, Kirk 1990, 199–200 *ad loc.* On its dedication in the *Panatheneia* see Scheid & Svenbro 1996, 18–22; Nagy 2009, 550–551.

[16] See Juhnke 1972, 143; 178; 363 on this connection.

[17] See Graziosi & Haubold 2010, 162 *ad* 6.297–311 on other aspects which contribute to the inappropriate nature of Theano's prayer and the Trojan women's dedication to Athene.

[18] Scheid & Svenbro 1996, 17–18 note the dedication of the robe as inappropriate due to the Sidonian slaves who made it.

[19] The adjective *kallistos* (very beautiful) used to describe it, though, differs from the more goddess-specific term *khariestatos* (most delightful) which Helenus used in his instructions, as noted by Graziosi & Haubold 2010, 159 *ad*

Its inappropriateness for dedication becomes especially noticeable only in the very direct comparison Statius invites through allusion to the robe dedicated by the Argive women to Juno. Statius' robe improves on the Iliadic situation by being bespoke, constructed by the right kind of people (56–58) and having upon it an image highly appropriate to its recipient (61–64). The narrator devotes time in the narrative to describe this image, a different tack to the Iliadic narrator who points to his robe's star–like qualities, purple decoration and *evokes* its manufacture in his description of work as παμποίκιλα (rich and varied), but says no more on the actual images depicted.[20]

Statius begins his description apparently picking up on the conventional floral decoration (59–60) we might imagine being upon the Iliadic robe, but by continuing into the details of an image goes further than Homer in what can be construed as a competitive way. In this ekphrasis, which has already aligned skill in poetics and textile weaving with words woven like threads, the poet makes a metapoetic statement about the *Thebaid's* status relevant to the *Iliad* as he asserts the qualities of his robe in such a way as to trump the Trojan dedication.[21] This becomes especially pointed when we consider that Juno grants the request of the Argive women (*Theb.*10.71–72) while Athena coldly turns away (*Il.*6.311).[22] In 11.396–402 the evocation of the Argive Robe from Book 10 fills in the gaps as we read the brief account of Argia's handiwork.

The skilled interweaving present in the robe woven for Juno, its bespoke nature and the ominous knell it sounds for newly made marriage is picked up in Polyneices' cloak. The object's beauty and appropriateness is rendered partially through reference to his new wife Argia's skill in making it, partially through the narrator's almost antagonistic attitude in presenting it to the character as comparable to the trappings of Eteocles. A point is therefore made not only about Juno's and Athene's acceptance or rejection of woven gifts, but also what it means for Polyneices' to overlook his. The ekphrastic element and the connection of Argia to her fellow countrywomen through the overarching theme of family and marriage intimate that Polyneices has something bespoke, socially important and loaded with poetic meaning; he is unaware of what he is missing. This feeds into his privileging of Thebes over everything else, and his obsession with power at Thebes.

The Andromache complex

The way in which Statius presents Polyneices and Argia as a potential reprisal of Hector and Andromache lends importance to Andromache as an interlocutor in the intertextual dialogue in this context.[23] Polyneices' robe was gifted to him by his wife, and the way in which the *Thebaid's*

6.288–295. Kirk 1990, 199 *ad* 294–5 however notes that "these vv. are more detailed and striking than the equivalent description in Helenos' and then Hektor's instructions at 90f. ~ 271f., where the dress was simply 'the loveliest and the largest in the palace, and the one she liked best.'" Although I remain unconvinced by the differentiation between embroidery and weaving which he subsequently makes in this note, the following point about something being at the extremity or just out of reach as a poetic shorthand for value and beauty seems eminently true in this case.

[20] Reitz 1996, 15–19 discusses the famous simile from the *Argonautica*, noting that comparisons between men and stars in Homer are often accompanied by negative connotations; of course with Jason wearing *that* cloak when he is so compared on arriving at Lemnos (*Arg.*1.774–781), Reitz, 16, points to the foreboding nature of the simile, effected by the proximity between this and the cloak ekphrasis. This then invites the negative association of astral comparanda, especially stars, to bleed into the shining cloak motif as well. Bulloch 2006, 49 n. 9 also points to this, and takes analysis of the cloak further 57–68.

[21] Weaving itself of course can be a competitive occupation, see Minerva and Arachne in Ovid *Metamorposes* 6.1–145.

[22] See Barchiesi, 1998 130–132. Graziosi & Haubold 2010, 158 *ad* 6.311 call Athena's response 'chilling'.

[23] See Lovatt 2013, 256 on Argia as Andromache and also Achilles at the same time as she is driven on by visions of Polyneices in her search for his corpse (12.187–193).

narrator uses it to redress Polyneices' envy of his brother's kingly appearance suggests it functions as a way of asserting status.[24] Polyneices shows a lack of awareness about this status, which is partially literary, partially to do with failing to self–identify as an Argive, as connected to her, rather than his brother. Through his lack of awareness Polyneices shows his unwillingness or inability to fully inhabit the role of the epic hero.[25]

The motif of the epic hero's association with weaving and robes goes back to the *Iliad.* At *Il.* 22.440–442 Andromache is weaving while Hector is dying. The ideal epic wife does her duty as the epitome of epic heroism does his. The craftsmanship involved in Andromache's labour is made clear to us (ἐν δὲ θρόνα ποικίλ' ἔπασσε "she wove elaborate flowers into the fabric", 441), as is a link to *Iliad* 6, where Hector, meeting her after he has been to instruct his mother about the robe dedication, tells Andromache to go back to her work. Hector balances the role of wife and husband succinctly (490–493), a balance revisited through allusion in Statius as Polyneices prepares for his final battle. In *Iliad* 22 it is a glimpse into the private world of women, afforded by the Homeric narrator at an emotive moment. The robe at this moment is unfinished, and when she suspects Hector's life has been cut short (22.448), Andromache's weaving is left so too. Unaware of his death she continues for a time; the delay isolates the new widow.

Polyneices' robe too comments on the distance between husband and wife. Where the distance in the *Iliad* is that of physicality and domain, however, which affords sympathy for the couple, for Polyneices and Argia it is not even the case that they form alternate sides of a shared epic identity. The *Thebaid's* narrator justifies Polyneices' outfit back to his hero, responding to Polyneices' envy of his brother and trying to impress upon him his epic potential. Argia has done her bit for the cause, but she is far away from the attacked city in Argos and her presence here reminds us that Polyneices has a city to which he can return, and a legitimate claim to the Argive throne.[26] This is an idea made clear in the *Thebaid* by Adrastus, when at 11.433–435 he offers Polyneices the Argive throne outright. There, as here, the Theban exile is blind to everything but enmity for his brother; there, as here, he ignores the proffered change in direction and chooses to return to his family.

Vergilian Mediation

The connection to the Homeric Andromache and her poetic associations with weaving are taken up by Vergil in the *Aeneid.* The context of Vergil's allusion to Andromache is one of gift giving, an act concerned with ideas of nation and descent, both literal in the content of the poem but also literary in the context of the epic genre. Part of the *Aeneid's* self–assertion as Homeric epic, Andromache's character and actions resonate to reinforce yet question the ability of descendants to deal with the

[24] On cloaks in epic associated with the feminine and becoming associated with gender issues see Lovatt 2013, 167–168; 182–185.

[25] Literarily self–aware characters are quite normal in the *Thebaid,* indeed for Polyneices to be so unaware of his own literary history renders him somewhat anomalous. See *e.g.* Feeney 1991, 341 on Oedipus' self–awareness in the *Thebaid* and 393 on Jupiter's. Pollmann 2004, 55 states "another sign of the conscious intertextuality of S(tatius)'s epic is the fact that some of his heroes and heroines are aware of their literary past." She points to Antigone in book 12 in particular knowing that the search for Polyneices is "traditionally her task (*cf.* 366–367)".

[26] A literary glance back to Argia could be seen as a diminished *literal* glance back to Argia; she is set in opposition to Thebes as far back as *Theb.*4.89–92, where the sight of her *dulces avertit pectore Thebas* ("turns sweet Thebes away from his heart"), something recapitulated by her at 12.190–191 (See Parkes, 2012, 91 *ad* 4.89–92). By this point however, Polyneices cannot see Argia anymore, the external narrator supplies the information to us, but in his blindness to his own dress he has not spare a thought for his wife: Thebes blots everything out. On this section, and Argia as alternative to his Theban family see Lovatt 2013, 231.

role to which they have been assigned. This moment forms part of the allusive dialogue which takes place in the Statian passage, concerning the representational potential of robes gifted by epic wives. We see Andromache giving robes to the Trojans, especially the Trojan prince Iulus, who is not quite ready for such weighty emblems of epic association.

Andromache is implicitly connected to the 'fabrication' of epic. The link in her weaving to epic narrative in the *Iliad* is exploited in the *Aeneid,* and comments on the issues arising from picking up the epic genre where Homer left off. The continuation and problematic nature of this poetic association is brought out in this passage, and is crystallised in the multiple senses in which we understand the verb *onerat* ("she burdened", 485).[27] Plurality of meaning for this verb stems directly from the context of its use being connected to Andromache's role in the weaving metaphor. The 'weight' comes as a consequence of repeated action; as the text makes clear, representing it by repeating the gift giving and increasing the number of gifts as the idea is repeated. She starts off giving the Trojan youth one item: *Phrygiam Ascanio chlamydem* ("a Phrygian chlamys to Ascanius", 3.484), but continues to pile up the textiles, and so weight of epic expectation: *textilibusque onerat donis* ("she burdened him with woven gifts", 485). Ascanius is too young for this great weight, and Vergil marks the gesture's significance by describing it as an honour (*honore,* 484). This gives a clear sense of the representational value of the gift.

This is not the only attempt at epic cloak gifting in the *Aeneid* which shows awareness of the power of such gifts but which also highlights inadequacy. In another attempt at epic identity forging of a slightly different kind, Dido gifts Aeneas a cloak at *Aen.*4.262–264.[28] This functions as a particularly important source for our Statian passage, as the same passing reference is made to construction of the garment. The very threads are exposed to the audience's view: just as at *Theb.* 11.402 *tenui telas discreverat auro* ("she separated threads out with delicate gold" *Aen.* 4.264). The position of Dido's gifted cloak undermines her attempt to consolidate her hold on Aeneas, and in fact seals her fate to be abandoned by the hero, as he leaves soon afterwards. Heroes have long been given clothing by lovers as opposed to wives.[29] Immediately upon describing the robe Dido had given Aeneas, he is called on by Mercury to leave Carthage (4.265–78), with disastrous consequences ensuing for Dido. This sounds an ominous knell for Argia and Polyneices' relationship, and indeed just as Mercury intervenes to push Aeneas onward to his epic destiny (just after Statius has described Polyneices' envy of his brother's garments and the cloak made for him by his wife), the Furies incite Polyneices and his brother to fight; *iamque in pulvereum Furiis hortantibus aequor / prosiliunt* ("and now with the Furies' encouragement they leap forward onto the plain" 11.403). Polyneices is summoned to his dire fate as Aeneas heeds the call of his destiny.

The Furies' intervention in the ekphrasis comes as part of a continual battle between Fury and narrator which plays out through Book 11. The narrator, especially in his allusion to the more elaborately described robe dedicated to Juno, may have been about to describe Argia's gift to her husband in more detail: *iamque* ("and now" 11.403) signals an abrupt transition, and the brothers' action begins. The Furies intervene in the narrator's attempt to slow down the narrative with the description of the robe; this becomes particularly obvious when viewed against the intratextual reference.

[27] Nagy 2009, 176 "Andromache's onerous gift of textiles to Ascanius is Virgil's metaphor for the burden of this epic past, which is the poet's own onerous gift to the notional descendent of Ascanius (…) Caesar Augustus".
[28] Hunter 1993, 179 notes the allusion to the cloak which Apollonius' Jason wears in Aeneas' cloak here.
[29] On cloaks and their association with female abandonment and loss see Lovatt 2013, 168.

It is the dysfunctional nature of Dido and Aeneas' relationship and the dissociative potential in the garment which Statius picks up on and inverts in his allusion to Argia's weaving technique. Where Dido *discreverat* ("had separated out") the threads with gold (*auro*), Argia *sociaverat* ("had united") purple and gold threads. The technicalities of what the women are doing on the loom may be slightly different, but working with gold and purple thread, the succinct one–line reference to their weaving and the verbal echo in *aurum/aureus* ("gold/golden"), as well as the identical tense of the verbs, make the comparison stand out plainly. The words used to describe Argia and Dido's action draw attention to the difference in the types of relationship they have with their heroic lovers. Aeneas and Dido are famously sharing a sham marriage which keeps him from his heroic destiny; Polyneices and Argia have the real thing, and by ignoring this Polyneices rejects the opportunity to become Aeneas.

Statius' inversion of the meaning of the weaving verb used reveals Argia as Polyneices' Lavinia, a new start in a different location with all the legitimising potential which comes with her.[30] This is another layer of meaning inherent in the allusive melting pot of Polyneices' robe; more repetition to add to the already weighty fabric. Caught between allusion to Andromache and Dido, the weaving act of Argia simultaneously conveys both her legitimacy as an epic wife, and her impending abandonment. Polyneices' abandonment of her plays out the same inversion, he forsakes her and goes back (something Aeneas is forbidden from doing) to disaster and death in his home town of Thebes.

Heroic emblems: shields and cloaks

At *Theb.* 11.396–402, as noted above, the narrator is at pains to make everything about Polyneices' cloak respond to Eteocles' garb and accoutrements. In particular, the gold and purple of the thread responded to the precious metal on Eteocles' shield and the tack on his horse respectively. Having established a powerful allusive link to the way Vergil constructs his line about Dido's gift of a cloak, it becomes clear that Statius' approach is as much to do with accentuating what is so enviable about Eteocles' trappings. The construction of the characters of the Seven and their counterparts in the *Thebaid* owe an intertextual debt not only to the epic tradition, but the tragic too. Shields and their devices are a stand–out feature of both the Aeschylean and Euripidean characterisation of the Seven, and in the *Thebaid* attention is likewise paid to shield devices and use, enacting a tension between the tragic and the epic in the very presentation of arms; a quintessentially epic feature.[31]

Shields and their ekphrases are an important part of defining the type of hero the protagonist is in epic, but the bleeding of shield and cloak ekphrases into one another becomes indicative of differentiation and attitudes to cosmic dominance or manipulation as the tradition progresses.[32] The *Thebaid*'s position, hanging between epic and tragedy, is encapsulated in this moment of ekphrastic allusion. Eteocles' shield has no particular design upon it, all we know is that it gleams with precious

[30] The generous father–in–law Polyneices has in Adrastus, as well as the latter's obedience to an oracle regarding his daughters' prospective husbands, evoke the Italian king Latinus. On this see Legras 1905, 220; Vessey 1973, 71, 94–95; Bernstein 2008, 76.

[31] On this see Harrison 1992, 247–252. See Ripoll 1998, 149 on the shield of Capaneus and 158 on that of Theseus. On shields in Aeschylus see Zeitlin 2009. On Achilles' shield see see Austin 1975, 115–28; Andersen 1976; Taplin 1980, 1–21; Lynn–George 1988, 174–200; Sprague–Becker 1990; 1995; Edwards 1991, 200–209; Hubbard 1992; Stanley 1993; Stansbury–O' Donnell 1995; Dubel 1997; Alden 2000, 48–73; Moog–Grünewald 2001; Nagy 2003, 72–88; Scully 2003.

[32] On the cosmic resonances of Aeneas' shield for example, see Hardie 1986, 336–375.

metal. We can assume it probably does have an image on it, in the same way that Polyneices' cloak almost certainly has decoration of some kind on it, but the pace of the narrative at this point leaves little room for description: the short delay afforded by the hint at ekphrasis is swiftly trampled by the Furies' intervention (*Theb.* 11.403–404). Nevertheless, it is significant that in choosing how to respond to his twin's garb the narrator should pick up on the cloak as the 'statement' part of Polyneices' outfit. Polyneices certainly must be armed with spear, sword and shield, as Eteocles is, he also has a horse as his brother does too.[33] The shield and cloak assimilation draws attention to the way in which both cloaks and shields form part of heroic identity. It is his cloak however on which the poet dwells. It assimilates yet also separates Polyneices and Eteocles in heroic terms, more specifically it makes Polyneices the type of hero who has an implicit link to the feminine, and a link which paradoxically encompasses abandonment at that.

The first hero who we must think of against a backdrop of cloaks and female abandonment is Jason. Both the Apollonian and the Valerian Jason have much light to shed on Polyneices' potential for epic heroism here. Apollonius shows Jason's importance as a hero can rival that of Achilles, but crucially in a different way.[34] Jason's cloak is shinier than Achilles' shield, we are told at *Arg.* 1.725–726: τῆς μὲν ῥηίτερον κεν ἐς ἠέλιον ἀνιόντα /ὄσσε βάλοις, ἢ κεῖνο μεταβλέψειας ἔρευθος ("more easily would you cast your eyes at the sun than look on that bright colour"). This comes as a direct challenge to the stars on Achilles' shield; the implicit message is that Jason challenges Achilles as an epic hero. The competitive element is particularly noticeable in the already existing competition within the comparison between the cloak and the sun. The connection between poetic composition and ekphrasis makes this competitiveness resonate on a poetic level, and the marked difference in heroic type communicated through the different items associated with each, can also be read into the type of epic Apollonius is composing in response to that of Homer.

Jason's cloak, in its comparison to the sun, also evokes the way in which the robe dedicated to the goddess Athene was compared to a star in *Iliad* 6. Jason's heroic cloak plays with imagery surrounding textiles in epic and draws attention to the antagonistic relationship which the later literary tradition has with earlier epic. With competitive metapoetics and epic redefinition foregrounded in Apollonian one–upmanship with Homer, Polyneices has options in the heroic models he could follow. In the fiery shimmer of Polyneices' cloak, and its competition with the metal–shine of Eteocles' shield, the competition in the literary tradition is brought onto the Theban plain. This plain is laden with conflicting genres and temporal uncertainty as the characters and scenarios veer between tragedy and epic, Rome and the Heroic age of Greece. Statius' game of "which brother has the shiniest suit" distils the competition in the epic literary tradition which is described in terms of astral bodies, rather disturbingly, into the motivation for mutual fratricide: a tragic turn.[35]

[33] Which features prominently in the final struggle, as both brothers crash into each other and are entangled with their horses, like sinking ships (11.509–524).

[34] By Hunter 1993, 53–59 but also by Shapiro 1980, 263–286; Clauss 1993, 120–147; Belloni 1995; Harrison 2001, 81–83; Bulloch 2006, 57–68 who takes a stance against seeing the cloak as indicative of Jason's more feminine version of heroism but has a useful survey on other literature on the subject; Lovatt 2013, 183–185.

[35] On the sexuality in Polyneices and Eteocles' final encounter see Ganiban 2007, 187, connecting incest and fratricide together in one fell swoop. For Polyneices' final dual to be so overlaid with tragic emphasis connects to his turn away from the epic potential expressed in the cloak and goes towards explaining what that potential has been turned into.

Valerian mediation

Valerius' Jason acts as another Roman intermediary between Greek and Statian epic.[36] He combines elements of Apollonius' Jason with echoes of Vergil's Aeneas, manipulating established connections of poetics and textiles in an epic context alongside Jason's cloak association, with all its connotations of female abandonment.[37] Dido and Aeneas are strong intertextual presences in this scene, when Jason is gifted a cloak on his departure from Lemnos by Hypsipyle.[38] But like Andromache from *Aeneid* 3 Hypsipyle gives a *chlamys,* highlighting Greek identity and retrospective links in that direction.[39]

Hypsipyle gives this cloak as part of a gift package (*haesuraque caro / dona duci promit* "she offered gifts which would stay with him to the leader beloved by her", *Argonautica* 2.408–409), which comprise the pluralised textiles (*textosque labores* "woven works", 409) in allusion to Andromache's gifts for Iulus from *Aeneid* 3 (*textalibusque onerat donis* 485).[40] Unlike these gifts however, the narrator describes images which Hypsipyle had woven into the garment, linking both to her fabrication of the garment and to its role as a gift for a hero. The first image, to which the more detailed description is given, is one that specifically relates to Hypsipyle's role in saving her father from the slaughter of menfolk on Lemnos, something which happened before Jason's arrival. The second is that of Ganymede's rape on Ida.[41] The identification of Ganymede as the "Phrygian" (417) evokes the Phrygian cloak Andromache gives to Iulus (*Phrygiam Ascanio chlamydem* "and a Phrygian cloak to Ascanius", *Aen.*3.484). Can a cloak depicting a quintessentially "Phrygian" myth be thought of as Phrygian? Coupled with its more immediate content relating to Hypsipyle's father, identity issues seem to be foregrounded in the allusiveness of Jason's cloak here. The two events may seem baffling when presented together, but illicit actions, fear of disturbance and savagery in the natural world, the woods, removal of a figure from their proper context and most particularly furtive activity connect Thoas, Ganymede, and indeed Jason in the image. In alluding to Andromache, Valerius retreats to a figure who represents the *Iliadic* and Trojan past, an identity from which the hero of the *Aeneid* must disentangle himself to progress, clearly meaningful in the context of Jason's departure from Lemnos. Valerius goes back further, in picking up the Ganymede myth which falls before the scope of the *Iliad*, but which figures several times in the *Aeneid*. Notably it picks up on *Aeneid* 5.250–257, where Cloanthus receives a cloak for his prize in the ship

36 Valerius Flaccus was a 1st century Flavian poet working on his *Argonautica* possibly prior to Statius, possibly around the same time as his *Thebaid* was being composed. Valerius' poem, usually agreed to have been cut short by his death, deals with the voyage of Jason to retrieve the Golden Fleece and draws both on Augustan poetic tradition and on the Homeric and Hellenistic epics, particularly Apollonius' version of the myth. On the dating of the text see Syme 1929; Taylor 1994; on allusion in the text see Hardie 1989, 1993, 83–87; Stover 2003; Zissos 1999 and 2002 on allusion to Homer and Hershkowitz 1998 on the *Argonautica's* relationship with the *Aeneid* of Vergil.

37 Hershkowitz 1998, 141–143 comments on the way in which this scene reworks Dido's imploring of Aeneas and his abandonment of her, and points to the depiction of Hypsipyle's father on the cloak as portraying her *pietas*. She also notes that in giving Jason a sword (2.418) rather than receiving one as Dido had, Hypsipyle follows a different path to the suicidal Dido.

38 Bulloch 2006, 48 notes the intertextuality between Aeneas' encounter with Dido, and Jason's with Hypsipyle in the *Argonautica* of Apollonius. This adds even more resonance to the cloaks of course, highlighting the cyclic nature of epic and, for a post–Catullan Latin audience, for whom the prevalence of Ariadne (again Bulloch 2006) in the context of textile games has deep metapoetic meaning, a cycle which disturbs the natural order of things.

39 As well as potential cosmic associations on which see Bulloch 2006, 58 n. 21; and, more generally: Faber 2008, 212–213.

40 See Carderi 2008, 219–221 on this section, especially on how the Vergilian allusion affects the Apollonian one.

41 Hershkowitz 1998, 142–143 n. 143 reads the depiction of Ganymede on the cloak as signifying Hypsipyle and Jason's mutual rape of each other as 'young virgins' and reminding us that Jason departs from Lemnos 'on Jovian business'.

race, in an intertextual gesture which connects to the competitive poetics which are foregrounded in the games, through the poetically significant nature of the cloak.[42] More to the point the cloak tells stories, not just one but two; in the rush of Jason's departure two more rushed departures are narrated, with the second focusing on a static moment and stalling Jason as he leaves, ekphrasis struggling against heroic impetus.

Valerius, in this scene with Hypsipyle and Jason, confuses the allusive directions which clothing gifts have to offer. He mixes clothing gifts which problematize epic identity with allusions which play Dido and Andromache off each other. In doing so, he points to the way in which the gift of the cloak can confer heroism but also create tension within heroic identity in an epic context. Hypsipyle's speech straight afterwards to Jason highlights a subversive link to Dido. She refers to the "Jason" in her womb, alluding to Dido at *Aen.* 4.327–330, but her gifts to Jason allude to Andromache and so go further than configuring her identity as a "successful Dido" here.[43] On one level she is successful. In establishing epic continuity and in cementing a secure connection to the epic hero, however, she recreates the disparity on display in Vergil's Andromache. The awkward negotiation between Andromache's relationship to her Homeric self and how the gap is bridged into the later literary tradition is revisited. The allusion to Cloanthus points to this as an antagonistic relationship, and in this competition Hypsipyle's relationship with Jason and thus attempt at continuance on this front is thwarted.[44] She hopes that her gift will abide with him (*haesura* "will be about to stay", *Arg.* 2.408), but he makes a gift of what she has woven for him to the dead king Cyzicus later on in the poem (3.340–342), in a reworking of Aeneas' gift to Pallas of cloaks made by Dido for the dead youth on his funeral pyre (*Aen.* 11.72–75).

Internal links: death

When Jason bestows the textiles woven by Hypsipyle in Valerius' *Argonautica* on the slain king Cyzicus, they are said to have been torn from the loom *rapuit telis festina* ("she snatched it hurriedly from the threads" 3.341–342). At this point and not before details are revealed which combine a reference both to its fabrication (2.408– 417) and the gold and burning imagery. These are compressed into the short format familiar from Vergil: *dat pictas auro atque ardentes murice vestes* ("he gives clothes decorated with gold and burning with purple" 3.340). This description is an obvious allusion to Dido's robe, given by Aeneas to Pallas at *Aen.*11.72–75:

> *Tum geminas vestes auroque ostroque rigentis*
> *extulit Aeneas, quas illi laeta laborum*
> *ipsa suis quondam manibus Sidonia Dido*
> *fecerat et tenui telas discreverat auro.*

> Then Aeneas brought out two robes stiff with purple and gold
> Which Sidonian Dido had made for him previously with her own
> hands, rejoicing in the work, and she had separated the threads out
> with delicate gold.

[42] Cf. Lovatt 2005 and Carderi 2008, 220–221 on this.

[43] So Hershkowitz 1998, 144.

[44] Admetus at *Theb.* 6.540–547 also receives a cloak as a prize for coming second in the chariot race in epic games, which plays with competitive weaving as well, as a parallel for Cloanthus.

The burning imagery, gold and purple, evocation of fabrication and the connotations of an ill–fated relationship all tie very well into Polyneices' robe at *Theb.*11.400–2, in a way which forebodes his death. This is the cloak which will serve for identification purposes when Argia is searching his corpse out on the battlefield (*Theb.*12.312–315), and in which he will be wrapped on his pyre. The link to his wife then carries a darker connotation, not only of abandonment but also of death.[45]

The one–line reference to fabrication which captures the funerary context of Hypsipyle's cloak from Valerius, Dido's from Vergil, and the ideas of abandonment and separation through death is an obvious point. The difficulties in identity make it hard for Argia to be identified as anything but an abandoned Hypsispyle or Dido, yet the link between her and the Argive women cements a potential link to Andromache, a figure who is the unseen backdrop for much of the generic play at work in these passages. Polyneices could go in several different directions; the narrator essentially gives him a choice. The text seems about to start a cloak ekphrasis, but is thwarted by chthonic intervention. The fact that this one–line reference to golden and purple threads picks up on a repeated reference throughout the literary tradition both loads it with meaning and counters the idea that the ekphrasis could have been extended, as if the Fury knows the tradition and the narrator's attempt to redirect Polyneices was in fact undermined by literary precedent all along.

The funerary connotations, particularly the pyre, then, add another layer of meaning to the way the cloak 'burns'. Jason's textile gift to Cyzicus comes as part of a sense of competition in the funeral arrangements; the narrator in the *Argonautica* tells us *innumeras nudatis montibus urgent /certatim decorantque pyras et corpora maesti / summa locant* ("they fire up and ornament the innumerable pyres on mountains laid bare and sorrowing; in competition they place the bodies on the highest points", 3.332–4). This competitive aspect brings us back to the competition between Polyneices and Eteocles, one which resonates to the end of the epic with the split pyre of 12.429–36. For Cyzicus receives the robe from Jason to distinguish him as king from the rest, and may therefore be understood as the best, something which resonates with Aeneas' bestowal of his cloak onto Pallas at *Aen.*11.72–75, another important symbolic act of distinction. Again then, with his cloak alluding to the one Jason bestows on Cyzicus, and also to that bestowed on Pallas after death, Polyneices is unaware that he is wearing heroic potential, but equally a potential which spells danger and death.

The purple of the robe is stained by the effusion of blood from Polyneices' body when Argia sees him. She identifies him *in spite* of this: *coniugis ipsa suos noscit miseranda labores / quamquam texta latent suffusaque sanguine maeret / purpura* ("she herself, a woman to be pitied, recognises her own work, belonging to her husband, even though the purple fabric lies hidden and, drenched with blood, is in a sorry state", 12.313–315). We have the *labores* ("works", 313), reference to Argia's part in it, and the purple; now the gold is gone though. The gold has been replaced by blood, which has soaked the garment to such an extent that it has fallen badly out of shape, as one might imagine of a waterlogged item of clothing.[46] Allusion to the Arcadian youth Pallas' death, for whose funeral Aeneas brought out robes so decked out with gold that they were "stiff" (*rigentis* 11.72) highlights an identity issue which perhaps resonates with the identity negotiation of the robes, sending mixed messages about heroism and literary descent. Argia's ability to recognise him

[45] On which see Pollmann 2004,159–160.

[46] Variant readings of *maeret* give *marcet* which would also work with ideas of the contrast between something with shape and something in a more sorry state. The *OLD* (*ad loc.*) gives for *marceo* connotations of rottenness and being soaked, drooping or being withered, being weak, enfeebled, languid or idle. Shackleton Bailey 2003, 273 translates it as 'languishes', and I think the idea of the robe's shape and colour being compromised is an important, but difficult element to render here.

in spite of the blood and loss of golden shine signals that she was his last connection to the world of epic, in that she fights against the tragic depths to which his life has descended.

Polyneices, in his metapoetic cloak of many hues, stands as emblematic for the *Thebaid's* take on epic. With the blood and purple aligned in his cloak, does this suggest that tragedy is an inherent part of epic from which a text like the *Thebaid* cannot escape? Or is it a conscious movement into tragedy which tarnishes the epic shine on the text? Argia's ability to recognise her husband even with the cloak deformed by blood suggests any one searching for epic will find it if they look hard enough, but also suggests, through her craftsmanship of the object by which she recognises him, that the reader has to put in what (s)he reads out of the text. This makes the reader in some way part of the pull back and forth between epic and tragedy which plays out through the course of the text, searching for a lost gleam of epic like Argia on the battlefield.

Conclusion

The gleaming gold, which all of our robes have had as the very basis of their fabrication is something linked powerfully to ideas of epic ekphrasis. This connection, through its light, burning, kingship connotations and perhaps most interestingly foreshadowing of trouble to come, is here brought to a conclusion. We are presented with a robe deformed: as a youth or flower withers in death, so has Polyneices' robe lost its shape and vitality. It has lost the shine, the golden gleam which aligned Jason's robe with Achilles' shield, which designated competitive poetics, heroism and danger. Instead only blood remains, soaking into the purple, dying it a deeper hue.[47] The gaps in understanding between youth and age negotiated throughout the allusive thread of epic robes seem also to find a place in the discovery of Polyneices' corpse. They suggest that not only has he missed the epic connection to the hero who abandons his mistress for an epic quest, but also that he has been somehow misidentified, and even now misses the opportunity to become the hero whose potential is unrealised, but whose status is assured through the actions of someone else, like Pallas or Cyzicus.[48] The gold which connected the robe to the shield, the sun and the stars has all been lost, the connecting thread to live–fast–die–young heroes, or even more calculating heroes like Jason has been severed. The narrator's attempt to connect Polyneices back to his epic association with Argos fails, and Polyneices prefers the garb of his brother, finery which goes unnoticed by the narrator. Lost in the self–destructive return to his origins, to Thebes and the incestuous internal strife of his family, Polyneices has rejected epic and it lies on the battlefield, drenched in the blood of fratricide and tragic degradation.

[47] On purple and blood see Bradley 2009, 190–191.

[48] It is true that in comparison to other members of the Seven, Hopleus and Dymas and indeed the rest of his family, Polyneices in the *Thebaid* has not received as much attention given that he is the one whose exile commences hostilities. Is this a consequence of Statius depriving him of any kind of epic status?

Bibliography

OLD = *Oxford Latin Dictionary.* Oxford.

Alden, M. 2000 *Homer beside Himself, Para–Narratives in the Iliad.* Oxford.

Andersen, O. 1976 Some thoughts on the Shield of Achilles. *Symbolae Osloensis,* 51, 5–18.

Augoustakis, A. 2010 *Motherhood and the Other: Narratives of Kinship in Flavian Epic.* Oxford.

Austin, N. 1975 *Archery at the Dark of the Moon.* Berkeley CA.

Barchiesi, A. 1998 The Statue of Athena at Troy and Carthage. In P. Knox & C. Foss (eds), *Style and Tradition: Studies in Honor of Wendell Clausen.* Stuttgart, 130–140.

Becker, A. 1990 The Shield of Achilles and the Poetics of Homeric Description. *AJPh* 111, 139–153.

Becker, A. S. 1995 *The Shield of Achilles and the Poetics of Ekphrasis.* London.

Belloni, L. 1995 Il manto di Giasone (Apoll. Rhod. *Argon.* 1.721–767). *Aevum Antiquum* 8, 137–155.

Bernstein, N. 2008 *In the Image of the Ancestors: Narratives of Kinship in Flavian Epic.* Toronto.

Bradley, M. 2009 *Colour and Meaning in Ancient Rome.* Cambridge.

Bulloch, A. 2006 Jason's Cloak. *Hermes* 134, 44–68.

Carderi, F. 2008 Le Ekphraseis di Valerio Flacco tra novità e Tradizione. *Hermes* 136, 214–226.

Clauss, J. J. 1993 *The Best of the Argonauts: The Redefinition of the Epic Hero in Book 1 of Apollonius' Argonautica.* Berkeley CA.

Dubel, S. 1997 Ekphrasis et enargeia: la description antique comme parcours. In *Dire l'evidence: philosophie et rhetorique antiques.* Paris, 249–64.

Edwards, M. W. 1991 *The Iliad: A Commentary, Vol. V: books 17–20.* Cambridge.

Faber, R. 2008 The Woven Garment as Literary Metaphor: The "Peplos" in "Ciris" 9–41. In *Roman Dress and the Fabrics of Roman Culture.* Toronto, 205–216.

Farron, S. 1993 *Vergil's Aeneid: a Poem of Grief and Love.* Leiden.

Feeney, D. 1991 *The Gods in Epic.* Oxford.

Ganiban, R. T. 2007 *Statius and Virgil: The Thebaid and the Reinterpretation of the Aeneid.* Cambridge.

Graziosi, B. & Haubold, J. 2010 *Iliad VI: A Commentary.* Cambridge.

Hardie, P. 1986 *Virgil's Aeneid: Cosmos and Imperium.* Oxford.

Hardie, P. 1989 Flavian epicists on Virgil's epic technique. *Ramus* 18, 3–20.

Hardie, P. 1993 *The Epic Successors of Vergil.* Cambridge.

Harrison, S. 1992 The Arms of Capaneus: Statius "Thebaid" 4.165–77. *CQ* 42, 247–52.

Harrison, S. J. 2001 Picturing the Future: the Proleptic Ekphrasis from Homer to Vergil. In: *Texts, Ideas, and the Classics.* Oxford, 70–92.

Hershkowitz, D. 1998 *Valerius Flaccus' Argonautica: Abbreviated Voyages in Silver Latin Epic.* Oxford.

Hill, D. E. 1996 *P. Papini Stati: Thebaidos Libri.* Leiden.

Hubbard, T. 1992 Nature and Art in the Shield of Achilles. *Arion* 2, 16–41.

Hunter, R. 1993 *The Argonautica of Apollonius.* Cambridge.

Juhnke, H. 1972 *Homerisches in römischer Epik flavischer Zeit: Untersuchungen zu Szenennachbildungen und Strukturentsprechungen in Statius', Thebais' und ,Achilleis' und in Silius' ,Punica.* Munich.

Kirk, G. S. 1990 *The Iliad: A Commentary, Vol. II, Books 5–8.* Cambridge.

Legras, L. 1905 *Les Legendes Thébaines en Grece et à Rome. Les Legendes Thébaines dans l'epopée et la tragédie Greques.* Paris.

Lovatt, H. 2005 *Statius and Epic Games: Sport, Politics and Poetics in the Thebaid.* Cambridge.

Lovatt, H. 2013 *The Epic Gaze: Vision, Gender and Narrative.* Cambridge.

Lynn–George, M. 1988 *Epos: Word, Narrative and the Iliad.* London.

McNelis, C. 2007 *Statius' Thebaid and the Poetics of Civil War.* Cambridge.

Moog–Grünewald, M. 2001 Der Sänger im Schild—oder: Über den Grund ekphrastischen Schreibens. In H. J. Drügh & M. Moog–Grünewald (eds), *Behext von Bildern? Ursachen, Funktionen und Perspektiven der textuellen Faszination durch Bilder.* Heidelberg, 1–19.

Mozley, J. 1934 *Valerius Flaccus: Argonautica.* Cambridge MA.

Mynors, R. 1969 *P. Vergili Maronis: Opera.* Oxford.

Nagy, G. 2003 *Homeric Responses.* Austin TX.

Nagy, G. 2009 *Homer the Classic.* Ann Arbor MI.

Nagy, G. 2010 *Homer the Preclassic.* Berkeley CA.

Parkes, R. 2012 *Statius, Thebaid 4.* Cambridge.

Pollmann, K. 2004 *Statius, Thebaid 12: Introduction, Text, and Commentary.* Paderborn.

Race, W. 2009 *Apollonius Rhodius: Argonautica.* Cambridge MA.

Reinhold, M. 1969 On Status Symbols in the Ancient World. *CJ* 64, 300–304.

Reinhold, M. 1970 *History of Purple as a Status Symbol in Antiquity.* Brussels.

Reitz, C. 1996 *Zur Gleichnistechnik des Apollonios von Rhodos.* Frankfurt.

Ripoll, F. 1998 *La morale héroïque dans les épopées latines d'époque flavienne: tradition et innovation.* Leuven.

Scheid, J. & Svenbro, J. 1996 *The Craft of Zeus.* Cambridge MA.

Scully, S. 2003 Reading the Shield of Achilles: Terror, Anger, Delight. *HSPh* 101, 29–47.

Shackleton Bailey, D. R. 2003 *Statius Thebaid, Books 8–12.* Cambridge MA.

Shapiro, H. A. 1980 Jason's Cloak. *TAPhA,* 110, 263–286.

Smolenaars, J. J. L. 1994 *P. Papinius Statius, Thebaid VII: A Commentary.* Leiden.

Sprague Becker, A. 1990 The Shield of Achilles and the Poetics of Homeric description. *AJPh* 111, 139–153.

Sprague Becker, A. 1995 *The Shield of Achilles and the Poetics of Ekphrasis.* London.

Stanley, K. 1993 *The Shield of Homer: Narrative Structure in the Iliad.* Princeton.

Stansbury–O' Donnel, M. D. 1995 Reading Pictorial Narrative: The Law Court Scene of the Shield of Achilles. In J. B. Carter & S. P. Morris (eds), *The Tradition of the Trojan War in Homer and the Epic Cycle.* Baltimore MD, 315–334.

Stover, T. 2003 Conflicting Medea: Genre, Gender, and Allusion in the "Argonautica" of Valerius Flaccus. *CP* 98, 123–147.

Syme, R. 1929 The "Argonautica" of Valerius Flaccus. *CQ* 23, 129–137.

Taplin, O. 1980 The Shield of Achilles within the Iliad. *Greece and Rome,* 27, 1–21.

Taylor, P. R. 1994 Valerius' Flavian "Argonautica". *CQ,* 44, 212–235.

Venini, P. 1970 *P. Papini Stati Thebaidos Liber XI.* Florence.

Vessey, D. 1973 *Statius and the Thebaid.* Cambridge.

Wilcock, M. 1978 *Homer: Iliad I–XII.* London.

Wilcock, M. 1982 *Homer: Iliad XIII–XXIV.* London.

Williams, R. D. 1972 *Publius Papinius Statius Thebaidos Liber X.* Leiden.

Zeitlin, F. 2009 *Under the Sign of the Shield: Semiotics and Aeschylus' Seven Against Thebes.* 2nd ed. Lanham MD.

18. Lexical Expressions of the Texture of Fabrics in Homeric Poetry

*Isabelle Boehm**

This chapter arises from an interest in the representation of textiles in Greek art in a broad sense.[1] It will examine some of the ways the lexicon of Homeric poetry (as an inclusive category encompassing the *Iliad*, the *Odyssey* and the corpus of the *Homeric Hymns*) represents the range of sensorial, material and textural features of fabrics, evokes their symbolism and exploits their narrative potential while often describing them as works of art.

To the viewer, the fabric displays colour and light, and it is characterised by softness or roughness for those who touch it. In addition, textiles either reveal or, on the contrary, dissimulate the identity of the wearer. A fabric is also the product of a *technē* and the result of technical prowess. Textiles constitute invaluable works of art, made to be admired, and were praised in poetry.

The present contribution investigates part of these aspects, namely those found in early Greek hexametric poetry, by studying the lexical expression of sensory perceptions of textiles and by asking what these more or less elaborate evocations could relate to, both symbolically and narratively. Clothing forms an integral part of the physical description of both heroes and gods, and the adjectives used to describe the fabric, whether it be its colour, weight, flexibility or softness, have a particular function. They are, in fact, descriptions which, beyond praising the beauty of the textile, are used to describe the identity of the person wearing it.

The radiance of the material

The textiles evoked by, or playing an important part in the *ekphraseis* of ancient poetry, are in general extraordinary on account of their colour – or, to be more precise, their brightness.[2] When related to textiles, colour adjectives primarily express the sparkle and brightness of the fabric, thereby conveying the sense of its beauty:[3] the semantic and conceptual associations generated by the interplay between a sample of colour–related adjective, and the textiles and garments they describe, will be outlined in the following pages.

* Text and translation of the *Homeric Hymns* are adapted from the most recent Loeb edition: M. L. West 2003. The text of the *Iliad* and the *Odyssey* is the OCT edition of T. W. Allen. The translations of the *Iliad* and *Odyssey* are adapted from the Loeb editions of respectively A.T. Murray 1924–1925 (as revised by W.F. Wyatt) and A.T. Murray 1919 (as revised by G. E. Dimock).

[1] Cf. *e.g.* Aubriot–Sevin 2003, for a study of artistic descriptions in Homer.

[2] For the importance of colour in Homeric poetry, see Rowe 1972.

[3] See *e.g.* Mugler 1964; Irwin 1974.

The adjective πορφύρεος is, for example, associated with the peploi that cover the bones of Hector at his funeral (*Il.* 24.796): πορφυρέοις πέπλοισι καλύψαντες μαλακοῖσιν "covering them over with soft purple robes". There are three areas of semantic and symbolic import that the term πορφύρεος brings with it:

– it is regularly associated with textiles, and often linked to any piece of woven fabric, and, more rarely, to a garment.[4] It is, for example, the colour of Odysseus' χλαῖνα, 'coat' (*Od.* 19.225), which Odysseus describes in detail and which, thereby, assures Penelope of his identity, a passage to which we shall return at the end of this discussion;

– it bears a symbolic association with the domain of death, to which it is sometimes associated (*e.g. Il.* 20.476–477 τὸν δὲ κατ' ὄσσε / ἔλλαβε πορφύρεος θάνατος "and down over his eyes came dark death");

– it conveys the idea of prestige and value by signalling wealth and/or beauty, whether concerning the blankets laid out in Achilles' tent (*Il.* 24.645; *Il.* 9.200) or the tapestries woven by Andromache (*Il.* 22.440–441: ἀλλ' ἥ γ' ἱστὸν ὕφαινε μυχῷ δόμου ὑψηλοῖο / δίπλακα πορφυρέην, ἐν δὲ θρόνα ποικίλ' ἔπασσε "but she was weaving a tapestry in the innermost part of the lofty house, a purple tapestry of double fold, and in it she was weaving flowers of varied hue"). It is also one of the colour adjectives that can be associated to καλός (*Od.* 19.242, καί οἱ ἐγὼ χάλκειον ἄορ καὶ δίπλακα δῶκα / καλὴν πορφυρέην καὶ τερμιόεντα χιτῶνα "I, too, gave him a sword of bronze, and a beautiful purple cloak of double fold, and a fringed tunic").

A further use of πορφύρεος applies to surfaces that have nothing to do with fabric, such as waves, κῦμα / κύματα:[5] while these should be traced back to the semantic sphere of the verb πορφύρω ('to surge'), a connection has been suggested between surging and shining.[6]

The uses of πορφύρεος are comparable to that of another colour adjective, κυάνεος ('dark–blue'). In fact, both qualify the intensity and colour of the depth of the sea, but also that of particular fabrics. The goddess Demeter, when she leaves in search of her daughter, covers her shoulders with a κυάνεος veil. In this case, the function of the adjective is both descriptive and symbolic: it is the colour of the veil that Demeter uses when she sets out to find her daughter (*h.Cer.* 42: κυάνεον κάλυμμα "a dark covering") or, in an exactly parallel context, the material that covers Thetis when she goes to plead with Zeus (*Il.* 24.93–94: Ὣς ἄρα φωνήσασα κάλυμμ' ἕλε δῖα θεάων / κυάνεον, τοῦ δ' οὔ τι μελάντερον ἔπλετο ἔσθος "so saying, the fair goddess took a dark covering, than which was no raiment more black"). These adjectives are applied to textiles not just as descriptive markers, but they seem to add symbolic value to the image: grief and pain for κυάνεος, wealth and high rank for πορφύρεος.[7]

Three further adjectives that express brightness and whiteness will also be briefly considered: ἀργός (or rather its derivatives), λευκός, and φαεινός.

The adjective ἀργεννός ('white', 'shining'), derived from ἀργός ('shining'), is, at least in Homeric poetry, an adjective of colour and of the material – usually wool – which is only associated once with a specific item of clothing – interestingly enough, a linen garment. It is an exceptional and well–known event: Helen exits the palace to meet Priam on the battlements "veiled with shining linen" (*Il.* 3.141, ὀθόνῃσι καλυψαμένη ἀργεννῇσι). What could be the point of the association of a veil, linen, and the shining–white colour, beyond its mere descriptive function? We shall return to ὀθόναι and to this passage below, so I will leave this question open at the moment, and simply

[4] Cf. *e.g.* Telemachus' coat in *Od.* 4.115.

[5] Cf. *Od.* 13.85.

[6] On the semantic relationship between the adjective πορφύρεος and the verb πορφύρω see *DELG* s.v., p. 930.

[7] See Grand–Clément 2004 and 2011 for studies of this term in ancient poetry.

point out that, in the works of Homer, the linen textiles denoted as ὀθόναι seem to be mostly associated with the female sphere of textile workshops (*Od.* 7.107), and with youth and beauty. If we project such an association onto the material adjective ἀργεννός, the occurrence of it at *Il.* 3.141 may refer back to Helen's earlier appearance in Priam's palace, weaving on the loom (3.125–126, where the pattern–woven fabric is described as πορφυρέη). Returning to the image and gesture of veiling, we can observe that, both before and after the *teikhoscopia*, this plays a fundamental role in the relationship between Helen and the goddesses Iris and Aphrodite.[8]

The adjective λευκός, contrary to ἀργεννός, is primarily used for its aesthetic value, for example in *Il.* 24.229: the φάρεα λευκά ("white robes") that Priam brings from his palace treasury to pay the ransom for the body of his son, are part of the sumptuous (περικαλλεῖς) presents that he has selected. The presence of the adjective λευκός in this context might at a closer look indicate the exceptional luminosity of the fabric – a consideration which seems to be suggested by two strands of associations: a) the fact that in the epic corpus this adjective is mostly used for liquids such as water or milk,[9] and b) the fact the semantic value of luminosity emerges clearly in contexts where a comparison is established with snow (*Il.* 10.437) and sunlight (*Il.* 14.185).

An adjective which seems to designate luminosity as its primary semantic domain, φαεινός regularly qualifies pieces of metal, crockery,[10] weaponry such as the quiver of Odysseus' bow,[11] a belt,[12] or even a ceremonial seat.[13] As these occurrences indicate, an object qualified as φαεινός is often a prestigious object. Interestingly, the adjective is used as well in association with a textile, the 'fine robe' (ἑανός), which I propose to see as a veil with which Helen covers herself to leave the battlements of Troy and return to her own quarters inside the palace (*Il.* 3.419–420):

> βῆ δὲ κατασχομένη ἑανῷ ἀργῆτι φαεινῷ
> σιγῇ, πάσας δὲ Τρῳὰς λάθεν ἦρχε δὲ δαίμων.
>
> and she went, wrapping herself in her bright shining fine robe/veil, in silence; and she escaped the notice of the Trojan women; and the goddess led the way.

This is a different situation from the one where Helen leaves the palace to go to the battlements; indeed, that the two scenes should be regarded as sharing a similar pattern of imagery is suggested by the use of two cognate adjectives, ἀργεννός and ἀργής, the one in verse 141, ὀθόνησι καλυψαμένη <u>ἀργεννῇσι,</u> "veiled with shining linen", and the one in verse 419, κατασχομένη ἑανῷ <u>ἀργῆτι</u> φαεινῷ, "wrapping herself in her bright shining fine robe/veil" – the implied difference being that in the latter scene Helen, led by the goddess, escapes detection by means of dazzling light. The association of φαεινός with the luminosity of a textile occurs again in *Iliad* 5, and in the same context, namely that of a fabric which masks and disguises the wearer: Aphrodite's *peplos* is φαεινός, and renders Aeneas invisible. This luminosity of the fabric is a divine feature and, moreover, it plays a specific role in the storyline: it acts as a link between the protecting god and the protected hero.

---

[8] See *e.g. Il.* 3.139–142 (Helen veiling herself as Iris instils in her "sweet longing for her former husband and her city and parents") and *Il.* 3.383–385, when Aphrodite reaches Helen on the bastions and "with her hand laid hold of her perfumed veil, and shook it": I take ἑανός ('fine robe') as indicating a veil. See discussion below.

[9] For a study of the whiteness of milk, see Capponi 2009.

[10] See *e.g.* κρητήρ ('mixing bowl') in *Il.* 3.274 and 15.121.

[11] *Od.* 21.55.

[12] More rarely, the eyes, e.g. *Il.* 13.3.

[13] See *e.g.* θρόνος ('throne') in *Il.* 18.422 and *Od.* 7.169.

The adjective ἀργύφεος[14] 'silver–shining' (*post.* ἀργυφής) is used slightly differently in regard to the matters we are concerned with here: it also refers to a piece of clothing in which a god dresses a hero, but in order to enhance rather than hide him. This is the case of the φᾶρος in which Calypso clothes Odysseus in *Od.* 5.230, or the dress (ἐσθής) with which Athena adorns (κόσμησε) the first woman in Hesiod's *Theogony* (573–574). There is another adjective of the same type, τηλαυγής 'far–beaming', used for the garment of the goddess Selene in the *Homeric Hymn to the Moon* (7–8):[15]

εὖτ' ἂν ἀπ' Ὠκεανοῖο λοεσσαμένη χρόα καλόν,
εἵματα ἑσσαμένη τηλαυγέα δῖα Σελήνη

when from Ocean's stream, where she has bathed her fair body
and put on her far–beaming raiment the lady Selene

In this case, the image of a piece of clothing that shines from afar is 'metonymically' equivalent with the nature deity who wears it.

As the last passage shows, the intense luminosity of these adjectives in the works of Homer is also expressed through comparison with the stars, the moon, and fire. The comparison with ἀστήρ ('star') applies to a particularly ornate and large piece of fabric:[16]

Ἑλένη δὲ παρίστατο φωριαμοῖσιν,
ἔνθ' ἔσαν οἱ πέπλοι παμποίκιλοι, οὓς κάμεν αὐτή.
τῶν ἕν' ἀειραμένη Ἑλένη φέρε, δῖα γυναικῶν,
ὃς κάλλιστος ἔην ποικίλμασιν ἠδὲ μέγιστος,
ἡ δ' ἀρ' ὄπισθε ὣς ἀπέλαμπεν· ἔκειτο δὲ νείατος ἄλλων.

And Helen came to the chests in which were her richly pattern–woven *peploi,* that she herself had made. One of these, Helen, the beautiful woman, lifted out and brought, the one that was most beautiful_in its woven patterns, and the amplest. It shone like a star, and lay beneath all the rest.

The extraordinary beauty of the fabrics is grounded in their ποικιλία ('variegation', here applied to pattern–woven fabrics), and the comparison is clear in the placing of the garment in the palace of Menelaus: it is entirely separate, at the bottom of the chest, glowing in the dark. Finally, this emphasis on the quality of the fabric denotes its marvellous nature: it is the work of Helen and is given as such (v. 126 μνῆμ' Ἑλένης χειρῶν "a remembrance of the hands of Helen") to Telemachus.

In early hexametric poetry, the comparison of the brightness of a fabric to firelight is reserved for a particular piece of clothing in the scene of the seduction of Anchises. The expression φαεινότερον πυρὸς αὐγῆς, "brighter than firelight", is only found twice: in the *Iliad*,[17] and in the *Homeric Hymn to Aphrodite:*[18]

(…) ὁ δὲ σταθμοῖσι λελειμμένος οἶος ἀπ' ἄλλων
πωλεῖτ' ἔνθα καὶ ἔνθα διαπρύσιον κιθαρίζων.
στῆ δ' αὐτοῦ προπάροιθε Διὸς θυγάτηρ Ἀφροδίτη

[14] On the adjective ἄργυφος, see Le Feuvre 2004, 257–264, especially 260. It is not a derivative, but a compound, ἀργ–υφο–, 'in white fabric'. Le Feuvre notes (258–259) that in Homer it is mostly used of fabric, more rarely of sheep's wool. Furthermore, as mentioned by Le Feuvre (262), Eustathius also understands the term in this way in his commentary to *Od.* 5.230 (ἀργύφεον ἤτοι λευκουφὲς φᾶρος μέγα ἕννυτο).

[15] *h.Hom.* 32.7–8.

[16] *Od.* 15.104–108.

[17] *Il.* 18.610, which refers to the corselet of Achilles, made by Hephaestus.

[18] *h.Ven.*, 79–90.

παρθένῳ ἀδμήτῃ μέγεθος καὶ εἶδος ὁμοίη,
μή μιν ταρβήσειεν ἐν ὀφθαλμοῖσι νοήσας.
Ἀγχίσης δ΄ ὁρόων ἐφράζετο θαύμαινέν τε
εἶδός τε μέγεθος καὶ εἵματα σιγαλόεντα.
πέπλον μὲν γὰρ ἕεστο φαεινότερον πυρὸς αὐγῆς,
εἶχε δ΄ ἐπιγναμπτὰς ἕλικας κάλυκάς τε φαεινάς
ὅρμοι δ΄ ἀμφ΄ ἁπαλῇ δειρῇ περικαλλέες ἦσαν
καλοὶ χρύσειοι παμποίκιλοι· ὡς δὲ σελήνη
στήθεσιν ἀμφ΄ ἁπαλοῖσιν ἐλάμπετο, θαῦμα ἰδέσθαι.

(...) while he, left all alone in the steading, was going about this way and that, playing loudly on a lyre. Zeus' daughter Aphrodite stood before him, like an unmarried girl in stature and appearance, so that he should not be afraid when his eyes fell on her. Anchises gazed and took stock of her, wondering at her appearance, her stature, and her shining garments; for she wore a dress brighter than firelight, and she had twisted bracelets and shining ear buds. Round her tender neck there were beautiful necklaces of gold, most elaborate, and about her tender breasts it shone like the moon, a wonder to behold.

It is the context, namely the seduction scene, which imposes the explicit mention of visual perception, through which the hero recognises the goddess and takes fright of her and of his love desire (v. 91: Ἀγχίσην δ' ἔρος εἷλεν "Anchises was seized by desire"). The expression φαεινότερον πυρὸς αὐγῆς is interpreted by Homeric commentators in two ways: either as a bright glare[19] or as shining metal.[20] The present contribution is not concerned with the debates on the interpretations of this comparison, which focus especially on the use of the adjective to describe colour (red–orange) and on the techniques used to ornate fabric; I merely wish to point out that the exceptional nature of this comparison is related to its symbolic value: the brightness of the goddess's clothing and the shine of Achilles' corselet are those of a relatively supple material worn close to the body and notably divine.

The texture of the fabric

My second series of remarks concern another characteristic of the fabric, perceived by looking at it but also by wearing it: the weight and the drape of the clothing on the living body.

In so far as the textiles celebrated in poetry are first and foremost works of art, their quality and beauty derive from their lightness and finesse. Moreover, the textiles and garments qualified as λεπτοί (πέπλοι, φᾶρος and εἵματα) are "finely woven", ἐΰννητοι,[21] and "gracious", χαρίεντα.[22] This is, of course, one of the qualities of the fabric woven by Penelope (*Od.* 2.94–95, στησαμένη μέγαν ἱστὸν ἐνὶ μεγάροισιν ὕφαινε, / λεπτὸν καὶ περίμετρον "she set up in her halls a great web, and fell to weaving – fine of thread was the web and very wide"), and returns, for instance, on the fourth ring on Achilles' shield where young men and young girls dance holding hands (*Il.* 18.595–596):

τῶν δ' αἳ μὲν λεπτὰς ὀθόνας ἔχον, οἳ δὲ χιτῶνας
εἵατ' ἐϋννήτους, ἦκα στίλβοντας ἐλαίῳ·

of these the maidens were clad in fine linen, while the youths wore well–woven tunics faintly glistening with oil.

[19] Joannes Philoponus Phil., *De opificio mundi* (Reichardt, Teubner), 137.14: μείζονος τῆς τοῦ πυρὸς αὐγῆς αἰσθανόμεθα. Choeroboscos (Spengel, Teubner), 253: τὸ πάνυ φωτεινὸν.
[20] Eustathius *ad Il.* 18.610: τὸ δὲ φαεινότερον πυρὸς αὐγῆς εἶναι τὸν θώρακα ἐκ τοῦ λαμπροῦ χρυσοῦ ἔχειν δοκεῖ.
[21] See, *e.g. Od.* 7.96–97 (referred to πέπλοι).
[22] See, *e.g. Od.* 10.543–544, *Il.* 22.510–511.

They dance in costume and the male component is carefully balanced with the female; crowns for the maidens, swords for the youths, as well as ὀθόναι[23] for the young women, and χιτῶνες for the young men. This type of finesse is expressed in a different way in other Homeric passages. Clothing is, in fact, by virtue of its finesse in a symbiotic relationship with its wearer, by making the movements of the body visible. In the *Homeric Hymn to the Sun,* for example, the lightness and the finesse of a garment are evoked in a very particular way:[24]

καλὸν δὲ περὶ χροῒ λάμπεται ἔσθος
λεπτουργὲς πνοιῇ ἀνέμων ἀνέμων, (...)

(...) while about his body glows his fair garment, fine–woven by the blowing winds, (...)

The evocation of lightness, λεπτουργές (a rare term in Homer), is associated to the verb 'to shine', λάμπεται, a radiance in turn linked to movement, as effected by the breeze of winds, πνοίῃ ἀνέμων. The movement caused by the blowing of wind is usually applied to natural elements such as water, trees and wheat: the evocation of such an image here may have been used here because the divinity invoked is Helios, the sun. The fabric that we see moving is generally made of light textiles, thus amplifying the effect of the movements of the body. This feature of the fabric, in turn, permits imperceptible movements to become visible – even the trembling caused by fear or emotion, as in the Homeric scene where a weeping Artemis comes to Zeus (*Il.* 21.506–507):

δακρυόεσσα δὲ πατρὸς ἐφέζετο γούνασι κούρη,
ἀμφὶ δ᾽ ἄρ᾽ ἀμβρόσιος ἑανὸς τρέμε.

and the young girl sat down weeping on her father's knees, while about her the fragrant robe quivered.

The clothing trembles (τρέμε v. 507), shaken by the goddess's body, just like the earth is shaken by the god Poseidon; the clothing moreover trembles under the effects of emotion, precisely like the legs of heroes.[25] The choice of the exceptional association in Homeric language between τρέμω and ἑανός can be explained both poetically and symbolically: it is the emotion of a goddess, and the garment, which is ἀμβρόσιος, "immortal", constitutes the goddess's identity.

As noted before, imagery of light clothing can serve to amplify the movement of the body, as the garment surrounds and is wrapped around a moving deity. Thus, Demeter's *peplos* gets fluttered about (ἐλελίζετο) the goddess's legs, and she gets entangled in it when she rushes to Celeus's house (*h.Cer.* 181–183):

ἡ δ᾽ ἀρ᾽ ὄπισθε φίλον τετιημένη ἦτορ
στεῖχε κατὰ κρῆθεν κεκαλυμμένη, ἀμφὶ δὲ πέπλος
κυάνεος ῥαδινοῖσι θεᾶς ἐλελίζετο ποσσίν.

And she walked behind with sorrowing heart, a veil over her head, while the dark robe fluttered about the goddess's slender calves.

The goddess, veiled from head to toe, disguises her identity, at same time disclosing it through the particular movement of her clothing: here as well, clothing expresses divine emotion.

[23] A kind of thin linen fabric, associated with the female domain of weaving activity; see *e.g. Od.* 7.107, where the servants weave busily and oil drips from the textiles they weave.
[24] *Cf. h.Hom.* 31.13–14.
[25] See *e.g. Il.* 10.390 (Dolon) ὑπὸ δ᾽ ἔτρεμε γυῖα ("his limbs trembled beneath him").

In Homeric poetry, the gesture and imagery of veiling – namely, covering oneself from head to toe in a light garment, the ἑανός – and unveiling (this occurring only rarely) is especially associated with women. On just two occasions, and in exactly the same context, a goddess unveils by removing her clothing: it is, of course, not coincidental that the goddess in question is Athena. She removes her light peplos[26] in order to assume another function, that of a warrior, and puts on (ἐνδύειν) the *chitōn* (χιτών) (*Il.* 8.384–387 =5.733–736):

> αὐτὰρ Ἀθηναίη κούρη Διὸς αἰγιόχοιο
> πέπλον μὲν κατέχευεν ἑανὸν πατρὸς ἐπ᾽ οὔδει
> ποικίλον, ὅν ῥ᾽ αὐτὴ ποιήσατο καὶ κάμε χερσίν,
> ἣ δὲ χιτῶν᾽ ἐνδῦσα Διὸς νεφεληγερέταο
>
> but Athena, daughter of Zeus who bears the aegis, let her light *peplos* fall on her father's floor, richly in–woven (*poikilos*), that she herself had made and her hands had fashioned, and put on the tunic of Zeus the cloud–gatherer.

The choice of the verb καταχεύειν ('pour, shower down') in such a context is exceptional. It serves to reinforce the distinction between the light peplos, ἑανός (here an adjective) and ποικίλος ('richly in–woven'), and the heavier χιτών Athena puts on (ἐνδῦσα), which is associated with "tearful war" (πόλεμον δακρυόεντα, 8.388).[27] With regard to the meaning of καταχεύειν in this passage, it should be noted that the goddess pours down the light peplos on the ground, ἐπ᾽ οὔδει. The use of this verb of movement thus fully conveys the lightness of the garment for at least two reasons:

– in the other scenes of unveiling in Homeric poetry, the verb ἐκδύειν ('take off') is normally used. The reference scene is the famous one where Aphrodite seduces Anchises in the Hymn to Aphrodite, and it is Anchises who undresses the goddess:[28]

> λῦσε δέ οἱ ζώνην ἰδὲ εἵματα σιγαλόεντα
> ἔκδυε καὶ κατέθηκεν ἐπὶ θρόνου ἀργυροήλου
> Ἀγχίσης·
>
> He undid her girdle, and divested her of her gleaming garments and laid them on a silver–riveted chair.

– καταχεύειν is rather used in reference to an exceptional circumstance. Contrary to what its formation might suggest, the verb is not very often used of liquids in Homeric poetry,[29] its usage being rather connected to the domain of natural elements, especially the wind which pours out mist (*Il.* 3.10–11, Εὖτ᾽ ὄρεος κορυφῇσι Νότος κατέχευεν ὀμίχλην / ποιμέσιν οὔ τι φίλην, κλέπτῃ δέ τε νυκτὸς ἀμείνω "Just as when the South Wind pours a mist over the peaks of a mountain, a mist that the shepherd loves not, but that to the robber is better than night")[30] or pours snowflakes in straight lines to form a single, unified, light coloured coat (*Il.* 12.156–158, νιφάδες δ᾽ ὡς πῖπτον ἔραζε, / ἅς τ᾽ ἄνεμος ζαὴς νέφεα σκιόεντα δονήσας / ταρφειὰς κατέχευεν ἐπὶ χθονὶ πουλυβοτείρῃ "from the top of the battlements, pebbles fall in straight lines like the snowflakes and form a cloud which is scattered onto the ground by the wind"): the imagery is straightforward;

[26] On Athena's peplos see Barber 1992.

[27] Interestingly, in the whole scene of the armament of Athena in *Iliad* 5, which follows the passage just discussed, the pattern of associating to each item of the armour a specific verb is maintained (738, ἀμφὶ δ᾽ ἄρ᾽ ὤμοισιν βάλετ᾽ αἰγίδα "about her shoulders she flung the *aegis*"; 743–44, κρατὶ δ᾽ ἐπ᾽ ἀμφίκαλον κυνέην θέτο τετραφάληρον / χρυσείην "and upon her head she set the helmet with two horns and with bosses four, wrought of gold"). Loraux 1989 (esp. 264–270) remains a fundamental reference for Athena's armour.

[28] *h.Ven.*, 164–166. On scenes of seduction in ancient poetry see Hunzinger 1999.

[29] *E.g.* to the oil poured onto the mane of Achilles' horse, *Il.* 23.281–282.

[30] The passage is part of a simile, which establishes a comparison with the dust–cloud rising from the feet of marching warriors (*Il.* 3.13–14, ὣς ἄρα τῶν ὑπὸ ποσσὶ κονίσαλος ὄρνυτ᾽ ἀελλὴς / ἐρχομένων "even in such wise rose the dense dust–cloud from beneath their feet as they went").

– in archaic Greek poetry, the verb is often used metaphorically: examples include grace (χάρις) poured down on Telemachus by Athena (*Od.* 2.12), or, in the sphere of sound, the cicadas spreading out their song (Hesiod, *Op.* 582–584, τέττιξ / δενδρέῳ ἐφεζόμενος λιγυρὴν καταχεύετ' ἀοιδὴν / πυκνὸν ὑπὸ πτερύγων "the chirping cicada, sitting in a tree, incessantly pours out its clear–sounding song from under its wings").[31]

I shall like to end this discussion with a final reference to the finesse of clothing in Homeric poetry, by picking up the image of the clothing of Odysseus (*Od.* 19.225) mentioned above (p. 314). After the description of the cloak (χλαῖνα) shaded in purple and the pattern of the brooch which holds it together, he describes the layers of clothing beneath it (*Od.* 19.232–234):

"τὸν δὲ χιτῶν' ἐνόησα περὶ χροῒ σιγαλόεντα,
οἷόν τε κρομύοιο λοπὸν κάτα ἰσχαλέοιο·
τὼς μὲν ἔην μαλακός, λάμπρος δ' ἦν ἠέλιος ὥς·"

"And I noted the tunic about his body, all shining as is the sheen upon the skin of a dried onion, so sheer it was and soft; and it glistened like the sun."

The garment is stunning and two comparisons are needed to describe it. It comprises three qualities: finesse, softness and light. Finesse is only alluded to through a comparison with onion skin, while its lustre is expressed by two adjectives (σιγαλόεις and μαλακός) and by the comparison with the sun (λαμπρὸς ὥς ἠέλιος). The comparison with onion skin seems to be unique in Homer,[32] as is the adjective ἰσχαλέος,[33] if this is indeed the correct reading. In addition to the commentaries to the passage,[34] this comparison is taken up by Galen[35] in book 10 of *De usu partium* which is dedicated to the eye: the covering of the lens is not only "similar to a dried out onionskin", extremely thin, white, and bright; it is also nearly as transparent as a spider's web.[36]

While a light, and thin, garment is an exceptional marker for its wearer, it is rare that the weight of the garment on the shoulders is mentioned. The weight of a garment is only noted in certain circumstances and it is no coincidence that we find Odysseus struggling with it: the fragrant garments given by the divine Calypso (*Od.* 5.264, θυώδεα εἵματα) are wet and weigh on him in the water (*Od.* 5.321, εἵματα γάρ ῥ' ἐβάρυνε), and he must take them off (ἀποδύς 343), on Ino's advice, in order to save himself. The other uses of the verb ἀποδύειν ('strip off') in Homer converge: the

[31] Similarly, Apollo's lyre spreads a cloud, which gently seals the eyelids of Zeus's eagle (Pind. *Pyth.*, 1.8 κελαινῶπιν δ' ἐπί οἱ νεφέλαν / ἀγκύλῳ κρατί, γλεφάρων ἁδὺ κλάϊ– / στρον, κατέχευας "For you have poured over his curved head a black–hooded cloud, sweet seal for his eyelids").

[32] Its vocalisation –o– in λοπός 'peel' is attested in the most ancient sources whereas the vocalisation –e–, λέπος 'rind, husk', is more recent (like the one used by Galen). This is the only attestation of the term in Homer. Both roots are found in medical and botanical texts. It is important to note that the comparison, indeed exceptional, must obviously be put into perspective: it is clear that the presence of the term is motivated by the relation to the derivative λεπτός 'fine, small' (rare in the literal meaning 'peeled', 'husked': *cf. Il.* 20.497), which, on the contrary, is richly attested in Homer, where it is applied to textiles. For Mycenean *re–po–to,* see the bibliography in *DMic*, vol. 2, *s.v.*

[33] The poetic adjective ἰσχαλέος is associated with ἰσχνός, which means "dry, hard, dried out or weak".

[34] The scholiasts (Dindorf, *s.l.*) refer to the more recent form λέπος, and add τζύπα. They also mention λέπυρον, 'envelope', '(fruit)skin', 'pod/shell'. Eustathius merely points out that it is not only the colour that likens it to an onion (λαμπρὸς δ' ἦν ἠέλιος ὥς. Καὶ οὐ κατὰ χρῶμα δηλαδὴ κρομύου.).

[35] For the use of Homer in Galen, see Moraux 1987. Moraux does not specifically mention the quotation that interests us here. It could, indeed, be perceived as anything between a simple "lexical" reference and a rare and ancient variation of the word. However, it is evident that Galen uses this quote by exploiting Homer's comparison in his anatomical description of the eye.

[36] *De usu partium* 3.787.9-11: ὁ δ' ἴδιος αὐτοῦ χιτὼν οὐ μόνον "οἷός τε κρομύοιο λοπὸς κάτα ἰσχαλέοιο", ἀλλ' ἔτι καὶ τῶν λεπτῶν ἀραχνίων λεπτότερός ἐστι καὶ λευκότερος.

body, or part of the body, of the hero[37] is heavy due to a weapon or an injury, and the verb is used in the context of danger, with death close by, exactly as in our passage.

Conclusions

Like an ornate canvas, the garment itself is not described in Homeric poetry only for its qualities; it is not merely the skilled creation of an artisan. The garment is an integral part of the universe it represents. The terms used to describe it entwine the woven object in the network of everything an artist can create: the garment becomes divine when worn by a divinity, as we have seen. Its exceptional character is consequently evoked by qualifiers which point to a further dimension, as when they link the worn fabric to cosmic realities (the stars on account of their brightness)[38] and natural elements (the snow in the description of a light veil spread out on the ground). Finally, the extraordinary aspect of the divine (and god–given) garment is associated with a specific function: beyond conveying and expressing the power of a deity, which is displayed through the act of clothing (and thus protecting) a hero, the garment worn by a god/goddess reveals his/her identity, even revealing and magnifying emotions: in this sense it is endowed with a dazzling appearance.

[37] Head (*Il.* 8.308), limbs (*Il.* 11.584), arm (*Il.* 20.480).

[38] See Grand–Clément 2009 for ποικίλα garments and their cosmic dimension.

Bibliography

Aubriot–Sevin, D. 2003 Autour des représentations artistiques chez Homère: fausse description, vraie narration: dieu et poète. *Gaia* 7, 135–146.

Bader, F. 1987 La racine de ποικίλος, πικρός. In J.T. Killen, J. L. Melena & J.–P. Olivier (eds), *Studies in Mycenaean and Classical Greek Presented to John Chadwick. Minos* 20–22. Salamanca, 41–60.

Barber, E. J. W. 1992 The Peplos of Athena. In J. Neils (ed.), *Goddess and Polis: the Panathenaic Festival in Ancient Athens*. Princeton, 103–117.

Capponi, M. 2009 Lait blanc, vin rouge et sang noir: l'utilisation stylistique des termes chromatiques. In M. Carastro (ed.), *L'antiquité en couleurs. Catégories, pratiques, représentations*. Grenoble, 85–98.

Clarke, M. 2004 The Semantics of Colour in the early Greek Word–Hoard. In L. Cleland, & K. Stears (eds), *Colour in the Ancient Mediterranean World*. BAR International Series 1267. Oxford, 131–139.

Cleland, L. & Stears, K. 2004 *Colour in the Ancient Mediterranean World*, BAR International Series 1267. Oxford.

DMic. = Aura Jorro, Fr. (ed.) 1985–1993 *Diccionario micenico*. Madrid.

Ferrini, F. 1978 Il problema dei termini di colore nella poesia homerica. *Annali della Facoltà di Lettere e Filosofia. Universita di Macerata* 11, 9–35.

Foster, E. D. 1975 An Administrative Department at Knossos Dealing with Perfumery and Offerings. *Minos* 16, 19–51.

Gernet, L. 1957 Dénomination et perception des couleurs chez les Grecs. In I. Meyerson (ed.), *Problèmes de la couleur. Colloque, Paris, 18–20 mai 1954*. Paris, 313–324.

Grand–Clément, A. 2004 Histoire du paysage sensible des Grecs à l'époque archaïque: Homère, les couleurs et l'exemple de πορφύρεος. *Pallas* 65, 123–143.

Grand–Clément, A. 2009 La bigarrure cosmique. In D. Micalella, E. Berrardi & F. Lisi (ed.), *Poikilia. Variazioni sul tema*. Acireale–Roma, 167–179.

Grand–Clément, A. 2010 Mettre en couleurs, un acte cosmogonique: quand tissus, métaux et chants s'entrelacent. Autour du *poikilon*. In I. Milliat–Pillot (ed.), *Texte du Monde, Monde du texte*. Grenoble, 69–88.

Grand–Clément, A. 2011 Du bon usage du vêtement bariolé. In L. Bodiou, F. Gherchanoc, V. Huet & V. Mehl (eds), *Parures et artifices: le corps exposé dans l'Antiquité*. Paris, 255–273.

Grand–Clément, A. 2011 *La fabrique des couleurs. Histoire du paysage sensible des Grecs anciens (VIIIe–début du Ve siècle avant notre ère)*. Paris.

Handschur, E. 1970 *Die Farb– und Glanzwörter bei Homer und Hesiod, in den homerischen Hymnen und den Fragmenten des epischen Kyklos*. Vienna.

Hunzinger, Chr. 1999 Thaûma. L'étonnement et l'émerveillement dans l'épopée archaïque. PhD thesis.

Irwin, E. 1974 *Colour Terms in Greek Poetry*. Toronto.

Le Feuvre, Cl. 2004 Note sur l'adjectif épique ἄργυφος, ἀργύφεος. *RPh.* 78.2, 257–264.

Loraux, N. 1989 *Les expériences de Tirésias. Le féminin et l'homme grec*. Paris.

Marinatos, Sp. 1967 *Kleidung–, Haar– und Barttracht, Archeologia homerica*, I, A & B. Göttingen.

Maxwell–Stuart, P.–G. 1981 *Studies in Greek Colour Terminology*, vol 1: *Glaukos*. Leiden.

Melena, J. L. 1983 Olive Oil and Other Sorts of Oil. *Minos* 18, 89–123.

Moraux, P. 1987 Homère chez Galien. In J. Servais, T. Hackens & B. Servais–Soyez, *Mélanges de philologie, d'histoire et d'archéologie grecques offerts à Jules Labarbe*. Louvain–la–Neuve, 25–37.

Mugler, C. 1964 *Dictionnaire de la terminologie optique des Grecs*. Paris.

Murray, A. T. 1919 *Homer. Odyssey. Translated by A. T. Murray. Revised by George E. Dimock*. 2 volumes. Cambridge MA.

Murray, A. T. 1924–25 *Homer. Iliad. Translated* by A. T. Murray. Revised by William F. Wyatt. 2 volumes. Cambridge MA.

Rowe, C. 1972 Conceptions of colour and colour symbolism in the ancient world. *Eranos* 41, 327–364.

Shelmerdine, C. W. 1985 *The Perfume Industry of Mycenaean Pylos*. Göteborg.

Shelmerdine, C. W. 1995 Shining and fragrant cloth in Homeric Epic. In Jane B. Carter & Sarah P. Morris (eds), *The Ages of Homer, A Tribute on Emily Townsend Vermeule*. Austin, 99–107.

West, M. L. 2003 *Homeric Hymns. Homeric Apocrypha. Lives of Homer.* Edited and translated by Martin L. West. Cambridge MA.

Wylock, M. 1970 La fabrication des parfums à l'époque mycénienne. *SMEA* 11, 116–133.

19. Textiles and Clothing Imagery in Greek and Latin Literature: Structuring, Ordering and Dissembling

*Giovanni Fanfani, Mary Harlow and Marie-Louise Nosch**

This volume is the result of two Symposia held at the Centre for Textile Research, University of Copenhagen, in 2012 and 2013. The idea of getting together a set of interested scholars to discuss the role of textiles, dress and modes of production in literature grew out of conversations between Mary Harlow and Marie-Louise Nosch which considered the similarities of textile-related tropes across a range of different time frames and contexts. We decided to set up a workshop to discuss these ideas and were quite unprepared for the range of topics which were offered and presented. The discussions were erudite, energetic, informative, and often funny. There was a real exchange of ideas and knowledge: for those with a purely literary background the ramifications of understanding the technique and craft behind metaphorical and other figurative use of textiles, clothing and cloth production has enhanced their work; those whose knowledge was in modes and methods of production were alerted to the ways in which language is influenced by textile techniques at very fundamental levels.

The two Symposia were inspiring days and we thank our colleagues for their time and hard work on the production of the papers that make up this volume. We also apologise to them for taking so long to produce the book, and hope the final production will allow them to forgive us.

Giovanni Fanfani, Mary Harlow, Marie-Louise Nosch
Copenhagen, April 2016

Introduction

The range of textile and clothing imagery in classical literature is immense. The chapters in this volume explore its diverse ramifications by positioning patterns of poetic and literary images within their specific context, as well as by looking at their wider material and cultural implications. What emerges is that an awareness of the craft and technology of weaving and spinning, of the production and consumption of clothing items, and of the social and religious significance of garments, is key to

* We would like to thank the Danish National Research Foundation, the Danish Council for Independent Research and FP7 Marie Curie Actions – COFUND (DFF – 1321-00158), and the University of Leicester for supporting the research for this chapter and the overall project of this volume. Giovanni Fanfani is the author of the sections on definitions, Aristotle, and approaches to metaphor in classical literature; Marie-Louise Nosch and Mary Harlow are the authors of the introduction and conclusions.

appreciating how textile and clothing imagery and metaphors work, their suitability to conceptualize human activities and represent cosmic realities, and their potential to evoke symbolic associations and arouse generic expectations. In brief, the breadth of figurative language that the craft(s) of fabric-making generates invites us to take textiles seriously.[1] This is all the more so when we regard imagery as a conceptual device, and not just an ornament of style. Viewing poetic creation or philosophical inquiry in terms of weaving, or representing emotions in terms of garments, are ways of conceptualizing one process in terms of another. One implication of this, as the contributions in this volume show, is that knowledge of the structure of one process (the 'literal' meaning of the image) gives access to the structure of the other process (the 'figurative' meaning): thus, investigating, for example, the mechanism and features of weaving and spinning technology, their symbolic import as crafts, and the cultural assumptions of clothing, sheds light on the concepts these illustrate, enhancing the interpretation of particular passages and complex images.[2] This afterword positions textile imagery in classical literature within ancient and modern discussions on figurative language; our emphasis is on views of metaphor as a conceptual device, from Aristotle to cognitive linguistics: while this sets the theoretical premises on which several chapters in this volume are grounded in a broader frame, it is further argued that in the specific case of archaic Greek literature different approaches to poetic metaphor may be applied to the analysis of a particular sample of textile and clothing imagery.

The centrality of textiles and clothing in the material life and conceptual universe of the Greeks and Romans is further reflected in literature by the distinctive polysemy of their symbolic and metaphoric associations. Literary textiles seem to bring with them an inherent tension between 'positive' concepts of social cohesion on the one hand, and 'negative' concepts of deception and ambiguity on the other. This represents a recognizable pattern in both the domains of craft (spinning and weaving) and clothing (cloth-making and garments): [3] it thus lends support to the idea of considering these two semantic areas, and their interaction, as a unitary subject in this volume. Spinning imagery, grounded in the concept of continuity (of time and thread), stands as a symbol for both life and death, seen as processes through the myth of the Fates and the conception of cosmologic threads. While invested with similar symbolic connotations as spinning,[4] weaving features a crucial opposition at the very root of its technological principles, *i.e.* division/separation and union/combination:[5] a similar pattern of opposition emerges at the level of poetic imagery,

[1] Fanfani & Harlizius-Klück (this volume) raise this point through a discussion of the 'outsourcing' of textiles from intellectual history, in both ancient and modern authors. See the contributions in Harich-Schwarzbauer 2015 for recent scholarly interest in the significance of textiles in several domains of ancient culture (both material and intellectual) and literature.

[2] A complementary approach, encompassing a shift of focus from the 'figurative' to the 'literal' element of a given image, allows for the possibility to explore textile imagery for what it reveals about ancient weaving technology (and the cognitive patterns this generates: see Fanfani & Harlizius-Klück in this volume, and Harlizius-Klück 2004 for a book-length investigation of the cognitive role of weaving in Plato's conception of *epistēmē*), or for what it suggests about the production of dyed fabrics (see Bakola in this volume).

[3] See Goldhill 1996 (review of Scheid & Svenbro 1996) on the opposition between, and compresence of, normative assumptions and an element of "trickery" and "untrustworthy" in weaving imagery. See Swalec (this volume) on the symbolism of the *chlaina* in Aristophanes' *Lysistrata*. On the multi-faceted (symbolic, economic, anthropological) ambiguity of textiles in the ancient Greek society, culture and literature see Jenkins 1985. *Cf.* Nosch 2014 for the association of weaving and women in ancient literature.

[4] See Bakola in this volume on instances of 'generative symbolism' associated with fabric-making and textile production; Guilleux (this volume) discusses areas of overlapping imagery and shared metaphorical networks between spinning and weaving, grounded on contiguities and analogies at the level of craft and technology.

[5] Famously illustrated by Plato in the *Statesman* 281a, 282b-c, 283a.

where the craft of weaving generates metaphors for both poetic composition and the 'weaving' of wiles.[6] In turn, clothing imagery exploits the peculiar symbolism of garments, which may reveal or conceal the identity of the wearer. Drawing on these oppositions and on further patterns of literary imagery both within and outside the range of material explored in the contributions to this volume, the final section of this afterword presents and briefly discusses categories of textile and clothing imagery, and some of their ramifications.

Definitions

For the scope of the present volume, our notion of 'textile and clothing imagery' includes metaphor, simile, analogy, metonymy, and all literary images referring to the conceptual, symbolic and material features associated with textiles, garments, and the craft of fabric-making.

In the definitions offered by classical rhetoric, metaphor and simile are features of literary language, first systematically investigated by Aristotle in the *Poetics* and the *Rhetoric*:[7] while metaphor (μεταφορά) is defined as the "transference (to a thing) of an alien term" (*Poet.* 1457b), simile (εἰκών) is a metaphor modified by some expression explicitly denoting likeness (*Rhet.* 1406b21-22, where "he was a lion in his attack" gets expanded into "he was *like* a lion in his attack"). A different view of the relationship between the two devices informs the account of later theorists, who see metaphor as an abbreviated simile (Quint. 8.6.8 *In toto autem metaphora brevior est similitudo*): this is also the time when metaphor and simile find their place within an apparatus of *tropes* (τρόποι, *tropi/modi*, single-word deviations from normal linguistic usage) and *figures* (σχήματα, *figurae*, encompassing a large number of rhetorical devices at the level of the sentence), with metaphor listed among tropes (together with metonymy),[8] and simile associated with figures (Cic. *De Or.* 3.205).[9]

Before moving to more modern definitions of metaphor and simile, it is interesting to observe how ancient rhetoric made a sustained use of clothing imagery to illustrate the nature and function of metaphor, and in general to conceptualize the distinction between style and content in the practice of literary prose.[10] Quintilian (*Inst.* 8 *proemium* 20) uses a metaphor when comparing *elocutio* ('style' in the technical terminology of rhetoric) to a variegated cloth: "that translucent and iridescent style of some authors renders effeminate the subject-matter that is cloaked in the verbal costume".[11] Cicero (*De Or.* 3.155-156) provides a historical account of the birth and development

[6] A recognizable pattern of imagery surfacing in both Old Iranian literature and archaic Greek poetry seems to attest to the interaction between the poetological metaphor of weaving and formulations of poetics 'embedded' into the image and semantics of 'poetic thread': the continuity in textile poetic imagery among Indo-European literatures that such a pattern demonstrates suggested us to bring into this volume a discussion of Vedic and Avestan instances of weaving and thread metaphors for poetic composition (see chapter by Andrés-Toledo; see Fanfani & Harlizius-Klück for the emergence of this pattern in archaic Greek poetics).

[7] See below for a more detailed account of metaphor in Aristotle. For simile, some of the relevant Aristotelian passages are *Rhet.* 1406b21-22 (metaphor transformed into simile), 1406b24-25 (simile being poetic and to be used rarely in prose) 1410b18 (simile differing from metaphor by the addition [πρόθεσις] of a word, and also by the fact that "it does not say that this is that").

[8] On metonymy (a deviation based on association or contiguity) and its relation with metaphor (a deviation based on similarity or analogy) in ancient and modern theorists see Silk 2003 and the opening remarks in Nagy 2015, 0§01–0§7.

[9] We draw here on Silk 1996.

[10] See Rosati 1999, 246. For a modern version of the same stance, see Conte 1986, 47. Conte applies the theoretical frame of the figure of speech to literary allusion, with special reference to Virgil: he sees allusion working in the same way as a rhetorical figure, and in the same semantic area.

[11] *illa translucida et versicolor quorundam elocution res ipsas effeminat quae illo verborum habitu vestiuntur.*

of metaphor, which was commonly considered among the *ornamenta* of style, in terms of clothing as *ornamentum* to the body, making use of an extended simile:

> The third mode [of elevating the style through a single-word rhetorical device], that of using words in a metaphorical sense *(transferendi verbi),* is widely prevalent, a mode of which necessity was the parent, compelled by the sterility and narrowness of language; but afterwards delight and pleasure made it frequent; for as dress was first adopted for the sake of keeping off the cold *(frigore depellendi causa),* but in process of time began to be made an ornament of the body and an emblem of dignity *(ad ornatum etiam corporis et dignitiatem),* so the metaphorical use of words *(verbi translatio)* was originally invented on account of their paucity, but became common from the delight which it afforded *(frequentata delectationis)*.[12]

Modern critics generally tend to view both simile and metaphor as figures of speech,[13] in consideration of the fact that the metaphorical sense affects the whole sentence rather than just being associated to a single word (a characteristic of the trope).[14] As a general definition, metaphor implies viewing one thing in terms of another. In 1936 I.A. Richards coined a new terminology for the constitutive elements of metaphor and established a helpful kit of critical tools for the literary and linguistic analysis of imagery, as the fortune of the terms 'tenor', 'vehicle' and 'ground' in later scholarship attests.[15] In a given metaphor (and simile),[16] the 'tenor' is (in Richards' words) the "underlying idea or principal subject which the vehicle or figure means",[17] while the 'ground' describes the conceptual relation between tenor and vehicle.[18] By defining metaphor not as a device of rhetorical ornamentation but as "the omnipresent principle of language",[19] Richards called into question the traditional theory of metaphorical expression as a rhetorical trope, arguing that "fundamentally it is a borrowing between and intercourse of *thoughts*, a transaction between contexts".[20] Such a reorientation of the mechanism of metaphorical expression led critics to regard metaphor as a 'figure of thought', as well as of speech: a new meaning is produced via the interaction of tenor and vehicle, and metaphorical thought is seen as a "distinctive mode of achieving insight".[21] The pervasiveness of metaphor in everyday language is a further element that modern theorists (in areas such as linguistics, philosophy of language and cognitive studies) regard as evidence for seeing metaphor as structuring our thoughts about a large number of abstract concepts. Textile and clothing

[12] Trans. J.S. Watson 1878. On this passage see Martin 2015, 278–279.

[13] Silk 1996, 967 observes that the category of 'imagery' is largely used by modern theorists in relation to metaphor, simile, comparison and informal analogy.

[14] *Cf.* Steiner 1986, 6 on the distinction between 'figure' and 'trope' and Aristotle's confusion of the two categories in his account of metaphor.

[15] Richards 1936, 89–112; see in particular 96–110 for the explanation of the technical terms. See Garner 1990, 5 for the application of Richards' terminology, and thus of the mechanism of metaphor (as rhetorical figure in this case), to the phenomenology of poetic allusion in archaic and classical Greek poetry, building on Conte 1986.

[16] For the terms as applied to a simile see Hutchinson 2012, 277.

[17] Richards 1936, 97.

[18] See Silk 1974, 9 on Richards' term 'ground', meant "to denote the likeness, the feature of features held in common between tenor and vehicle. This 'ground' might seem to be inherently a conceptual, not a verbal, matter".

[19] Richards 1936, 92.

[20] Richards 1936, 94, who in the remainder of the passages argues that "[*T*]*hought* is metaphorical, and proceeds by comparison, and the metaphors of language derive therefrom", thus anticipating, at least partly (by giving logical prominence to metaphor in thought), the assertions of the cognitivists.

[21] Black 1962, 237, in the context of a comparison between metaphor and (scientific) model, one of the differences being that "[A] memorable metaphor has the power to bring two separate domains into cognitive and emotional relation" (236); Black builds on Richards to elaborate his 'interaction' theory of metaphor (38–47).

metaphors, which we consider in this volume, recur in both literary texts of antiquity and in the everyday speech of a number of modern languages, producing systems of images (or 'metaphorical matrixes': see Guilleux's chapter in this volume) in a cross-linguistic and diachronic range.

In the terminology of the cognitivists ('conceptual metaphor theory'), metaphors are conceptual mappings from a 'source domain' to a 'target domain'.[22] Such a framework works as well for similes, seen as "metaphorical comparisons composed of two nominal terms one of which belongs to the target and the other to the source domain".[23] Similes generally observe the principle of directionality which characterizes conceptual metaphors, in that more concrete concepts are mapped onto more abstract ones.[24] This is generally valid for the most prominent sample of literary similes in classical literature, namely Homeric similes. In their narrative extended version, these tend to provide a detailed and imagistic description of a situation familiar to the audience as a means to illustrate events, feelings and concepts:[25] the typicality of Homeric similes has invited critics to analyse them according to thematic classes or families.[26] The phenomenology of literary similes in ancient literature is far wider than we can attempt to describe here,[27] and as a closing example for this section we would like to point to one of the few instances of Homeric similes drawing on the domain of textile activity, *Il.* 23.758-763: here, a detailed description of a woman weaving on the warp-weighted loom is the vehicle image for illustrating the closeness of two heroes engaged in a running race; on the occasion of the funeral games for Patroclus, Odysseus is described as running very close behind Ajax:

ὦκα δ᾽ἔπειτα
ἔκφερ᾽ Ὀϊλιάδης· ἐπὶ δ᾽ ὄρνυτο δῖος Ὀδυσσεὺς
ἄγχι μάλ᾽, ὡς ὅτε τῖς τε γυναικὸς ἐϋζώνοιο
στήθεός ἐστι κανών, ὅν τ᾽ εὖ μάλα χερσὶ τανύσσῃ
πηνίον ἐξέλκουσα παρὲκ μίτον, ἀγχόθι δ᾽ ἴσχει
στήθεος·

Then speedily the son of Oïleus forged to the front, and close after him sped noble Odysseus, close as is the heddle bar to the breast of a fair–belted woman when she deftly draws it in her hands, pulling the spool past the warp, and holds the bar near to her breast, so close behind.[28]

[22] The classic formulation of this theory is in Lakoff & Johnson 1980; a helpful account of conceptual metaphor theory is Kövecses 2002. See below for a more detailed account of the cognitive approach to metaphor.

[23] Shen 2008, 297.

[24] For a recent discussion of similes in terms of cognitive approach to figurative language, see Bark 2011, 18–22, 34–36.

[25] In this regard, Homeric similes cover two of the functions that Silk 1996, 967 and 2003, 126 sees metaphor and simile perform in ancient poetic usage: a) "to make clearer, as through a diagram, usually by appeal to familiar experience", and b) "to exploit the associations, including the contrary associations of the vehicle, beyond any limited point or ground of comparison".

[26] *Cf.* Scott 1974 for a systematic treatment of Homeric similes as a device of oral poetics. Scott divides Homeric similes by creating a typology according to location and subject matters. de Jong 1999, III, 299–382 is a collection of reference scholarship on Homeric similes.

[27] See for instance the recent discussion of Pindar's grammar of comparison, and his use of 'condensed similes' as devices of conceptualization, by Maslov 2015, 156–166.

[28] Translation: Murray rev. by Wyatt 1925 (Loeb) modified after Nosch 2015, 178 n. 71. On aspects of this simile related to textile technology and terminology see Barber 1991, 267 and Edmunds 2012, § 24.

Aristotle on metaphors

As it is well known, Aristotle's highly influential treatment of metaphor in his *Poetics* and *Rhetoric* laid the foundations for the elaboration by later rhetoricians of what is known as the classical theory of metaphor as an ornament of style – a *trope*.[29] The two grounding notions around which the theory is built are substitution (or deviation) and similarity (or analogy), both of which surface prominently in Aristotle's designations of *metaphora* (a term which, as we will see, does not correspond entirely to what we mean by 'metaphor'). Interestingly, however, these two principles do not exhaust the breadth and complexity of Aristotle's account,[30] which features formulations of significant cognitive and conceptual import.

Itself a metaphor, the term μεταφορά 'transference/carrying across' conveys "the basic idea that a term is transferred from its original context to another",[31] as illustrated by the well-known definition in *Poetics* 1457b6 "*metaphora* is the application/transference (to one thing) of a word that belongs to another thing" (μεταφορὰ δέ ἐστιν ὀνόματος ἀλλοτρίου ἐπιφορά).[32] The four-folded classification immediately following such definition ("(a transference) either from genus to species, species to genus, species to species, or by analogy")[33] illustrates with four examples that Aristotle's *metaphora* is a more comprehensive category than that which we call 'metaphor', as it includes different types of transfer.[34] The passage as a whole seems to make clear that the notion of *metaphora* is grounded

[29] An excellent recent overview on the topic, which explores the range of ancient formulations of metaphor as a rhetorical trope, also bringing them in relation to modern views, is offered by Innes 2003. For a semiotic approach to Aristotle's theory of metaphor see Kirby 1997. On the cognitive implications of Aristotle's view of metaphor see Martin 2015, 274–284, who sets it against ancient (the rhetoricians) and modern (Hobbes, Locke) approaches. Lloyd 1996 discusses the use of metaphorical language in Aristotle's account of metaphor in *Rhetoric* and *Poetics*. Silk 2003 focuses on Aristotle's elaboration of a theory of poetic metaphor, and points to some apparent flaws of the theory when seen against actual literary usage. The bibliography on Aristotle's theory of metaphor is massive (Kirby 1997, 518 n.3 provides a helpful orientation): for the scope of the present discussion, Kirby, Martin, Lloyd and Silk represent three particularly interesting approaches to the topic. English translation of *Poetics* and *Rhetoric* are adapted from the most recent Loeb editions (respectively Halliwell 1995 and Freese 1926).

[30] It should be noted that Aristotle's discussion in *Poetics* and *Rhetoric* has a pronounced focus on the use of metaphor in poetry and literary prose/oratory. A different concern and attitude towards metaphor is shown in works like *Topics* and *Metaphysics*: see Lloyd 1987, 183–186, and 1996, 205–222 on Aristotle's critique of the use of metaphor in philosophical demonstration and formal logic.

[31] Innes 2003, 7, who remarks how such "vocabulary of movement, change/exchange, and place/domain" is a recurring one in antiquity for conceptualizing the mechanism of metaphorical transfer, and it reflects the divide between content and style (as ornamentation) fundamental to ancient rhetoric's view of language; interestingly, this distinction is in turn illustrated by rhetoricians in terms of clothing imagery (*cf.* Cic. *Brut.* 262 and pp. 325–326 above).

[32] Translation: Silk 2003 adapted. A more concise translation is offered by Innes 2003, 7: "(metaphor is) the introduction of an alien term". Kirby 1997, 531–533 discusses in particular the semantics of the term ἐπιφορά, built as μεταφορά from the root of the verb φέρω 'carry'.

[33] *Po.* 1457b7–9 ἢ ἀπὸ τοῦ γένος ἐπὶ εἶδος ἢ ἀπὸ τοῦ εἴδους ἐπὶ τὸ γένος ἤ ἀπὸ τοῦ εἴδους ἐπὶ εἶδος ἤ κατὰ τὸ ἀνάλογον. The two latter categories cover usages that we normally associate with our idea of metaphor: as an instance of the 'species to species' type, Aristotle quotes Empedocles fr. 138 DK "drawing off the life with bronze" and fr. 143 DK "cutting with slender-edged bronze", observing that "here he has used 'drawing off' for 'cutting' and *vice versa*, as both are kinds of removing" (1457b14–16); the analogical metaphors are explained as "cases when *b* is to *a* as *d* is to *c*: one will then speak of *d* instead of *b*, or *b* instead of *d*", and illustrated by the famous example "the wine bowl is to Dionysus as the shield to Ares: so one will call the wine bowl 'Dionysus' shield' and the shield 'Ares' wine bowl'" (1457b16–23). On this metaphor see Di Campobianco in this volume.

[34] Silk 2003, 116–117, who observes how the first two illustrative examples provided by Aristotle ("from genus to species, from species to genus") includes a 'normal usage' (or a 'dead metaphor') in Hom. *Od.* 1.185 where ἕστηκεν 'stands' is used of a ship lying at anchor, and metonymy (or, more precisely, *hyperbole*) in Hom. *Il.* 2.272 where "ten thousand (deeds)" is used instead of "many".

on the underlying principle of 'substitution' (*i.e.* using one term *instead* of another); this, in turn, presupposes the view that a term has a single, standard meaning, and that any other usage is therefore non–standard, deviant. This aspect emerges as early as *Poetics* 1457b1-2, the first occurrence of the term μεταφορά in Aristotle, where it figures in a list of word categories which, in relation to poetic language, seems to set a standard/ordinary word (κύριον ὄνομα) in opposition to a number of non-standard usages: "every word is either a standard term, or a loan-word (γλῶττα), or a *metaphora*, or an epithet (κόσμος),[35] neologism, lengthening, contraction, or modification".[36] The suitability of metaphor to prose is spelled out in a passage of *Rhetoric* (1404b32-35) which refers back to the previous one in *Poetics* and positions μεταφορά alongside τὸ κύριον ('the standard') and τὸ οἰκεῖον ('the appropriate'), on the grounds that the three word-types "are alone to be employed in the style (λέξις) of prose; for all use *metaphorai* in conversation (διαλέγονται), as well as appropriate and standard words". The observation that metaphor is a feature of ordinary speech, which "may be seen as actually prescient of the cognitive approach",[37] surfaces again at the end of the discussion on *metaphora* in the *Poetics* (1459a10-14), where it is preceded by a couple of further significant assertions about what it is implied, cognitively, in one's capacity to use metaphors (1459a5-7): such a skill (τὸ μεταφορικὸν), labelled as "much the greatest asset" (πολὺ δὲ μέγιστον), is the only one that cannot be learnt from another, and "is a sign of natural gifts (εὐφυΐα): because to use metaphor well (τὸ εὖ μεταφέρειν) is to discern similarities (τὸ τὸ ὅμοιον θεορεῖν)".[38] The ability to perceive likeness (on the part of both the poet crafting *metaphorai* and the audience decoding them) positions metaphorical language and thinking within the sphere of cognition – a reflection developed in a number of passages in the *Rhetoric*, especially in relation to 'learning' (μάθησις). Given that "all words that make us learn something (μανθάνειν) are the most pleasant (ἥδιστα)" (1410b14), metaphor achieves this effect at full scale (μάλιστα): recognizing the relationship between species and genus that metaphor effects is thus generating learning (μάθησις) and the acquisition of knowledge (γνῶσις, 1410b15). A further linguistic and conceptual function that Aristotle sees *metaphora* performing, beyond the realm of rhetorical and poetic ornamentation, is related to its capacity to fill a lexical gap by "giving names to things that have none" (1405a36 μεταφέρειν τὰ ἀνώνυμα ὠνομασμένως).[39] The metaphorical nature of the technical terminology for metaphor itself may be regarded as an obvious application of this feature, which in turn seems to adumbrate Hans Blumenberg's view of metaphor as a technological device,[40] and the idea, grounded in the conceptual metaphor theory, that there are

[35] For this interpretation of *kosmos*, which normally indicates 'ornament (of style)', see Martin 2015, 280–282, and *passim*.

[36] A further passage in the *Poetics* (1458a22-23) locates *metaphora* within a list of 'exotic' (ξενικόν) usages, with the explicit remark that 'exotic' is "anything that diverges from the standard" (πᾶν τὸ παρὰ τὸ κύριον). A 'parallel' passage in *Rhetoric*, when the reference is to prose, asserts that "it is *metaphora* above all that gives clarity (τὸ σαφές), pleasure (τὸ ἡδύ), and foreign air (τὸ ξενικόν)" (1405a8-9).

[37] So, aptly, Kirby 1997, 539, who considers Lakoff's objection to the 'classical' theory of metaphor (*i.e.* the fact that it distinguishes between 'figurative' language and every-day usage) and remarks that "while it is true that Aristotle has a notion of *kuria* or ordinary terms, he by no means restricted metaphor to poetic or extraordinary contexts".

[38] The perception of likeness is elsewhere regarded by Aristotle (*Top.* 108b7-14) as fundamental to dialectic and logic: see Kirby 1997, 536–537. See Martin 2015, 282–283 on Aristotle's treatment of metaphor "as a *cognitive* tool, an aid to instruction".

[39] This seems to amount to a conception of the nature of metaphor as, originally, a 'necessary' device for generating new terms: such a view is then developed by Cicero (*De Oratore* 3.155) into an 'evolutionary' theory of metaphor: "necessity was the parent [of metaphor], compelled by the sterility and narrowness of language; but afterwards delight and pleasure made it frequent".

[40] On Blumenberg's conception of metaphor see chapter by Fanfani & Harlizius-Klück in this volume.

cases when the *only* linguistic expression of a concept is metaphorical.[41] The creation of such 'name-giving' metaphors should rest on the same guiding principle of analogy ("by deriving the metaphor from what shares the same genus [ἐκ τῶν συγγενῶν] and the same species [ἐκ τῶν ὁμοειδῶν]" 1405a37) that defines the prominent status of *metaphora* within the array of ornamental devices and deviant usages available to the poet (*Poetics* 1459a5-8). The notion of an *a priori* underlying analogy between the two members of a metaphor has been questioned by modern theorists,[42] and Michael Silk has shown that, in relation to ancient poetic usage, analogy is "rarely of central importance",[43] whereas vividness and immediacy, features that Aristotle sees as "bringing things before the eyes" (*Rhet.* 1411b24), are more prominent in poetic metaphors.

Aristotle's wide-ranging discussion of *metaphora*, fundamental to the elaboration of the classical theory of metaphor by later rhetoricians, represents an unavoidable frame of reference for modern approaches to the topic, some of which it seems to have anticipated in various ways. Considering classical and contemporary views of metaphor, and their suitability as theoretical frameworks for the interpretation of imagery in ancient literature, particularly in reference to textile metaphors, will be our concern in the next section.

A short overview on theories and approaches to metaphor in classical literature. The case of textile and clothing metaphors

Any discussion of textile and clothing imagery in classical literature is bound to engage with the prominent role of metaphor in expressing the imagistic and conceptual potential of the craft(s) and technology of fabric- and cloth-making to generate ways of thinking about realities, processes and experiences that fall into the sphere of the unseen and abstract – and as such cannot be directly thought of or represented. At the outset of a lucid account of ancient and modern views on metaphor which introduces her investigation of the device in Pindar's victory odes,[44] Deborah Steiner presents the semantic domain of crafts (*technai*) as a significant source of terminology for the functioning of the world, with weaving and spinning singled out for their generative force in "describing the working of such intangibles as fate and fortune".[45] Indeed, the case of the weaving, spinning and garment metaphors of archaic Greek poetry provide us with a unique vantage point to reflect on how different (and often perceived as competing) approaches to the nature and functioning of metaphor can be (and have been) fruitfully applied to certain patterns of imagery. One of the reasons for

[41] See Cairns 2014, 6 who argues that certain concepts are "inherently metaphorical" *i.e.* "metaphorically constructed" (3 with n.3)

[42] See Ricoeur 1978, 86 for the view that it is metaphor that creates the similarity, rather than expressing an underlying likeness.

[43] Silk 2003, 126; for Silk, analogy should be regarded as the logical basis of metaphorical expression, "irrespective of whether that logical basis is central or peripheral to the instance in question" (124).

[44] Steiner 1986, 1–17. Though the theory of conceptual mappings (which was still in its very infancy at the time) does not feature in Steiner's overview of modern linguistics' approach to metaphorical language, the author shows full awareness of the cognitive potential of metaphor.

[45] Steiner 1986, 1; see 52–65 for a treatment of craftsmanship as vehicle for song/poetry in Pindar's victory odes. On the richness of the metaphorical networks activated by the semantic fields of spinning and weaving in a cross-linguistic and diachronic perspective, see Guilleux' chapter in this volume. From a different perspective see Lloyd 1966, 192–193 and 1987, 174–176 on the extent to which it is possible to draw the distinction between the literal and the metaphorical in the Homeric account of the working of fate as spinning, weaving and binding.

this might be the special status of archaic Greek poetry in terms of 'literacy': factors like orality and formulaic language (in epic), pragmatics and performance context (in lyric) suggest ways of production and reception of imagery (and metaphor in particular) that can be analysed along the lines of the classical theory of tropes as well as of modern approaches to language and cognition.[46] To borrow a recent formulation of the issue at stake, what interests us here is a particular aspect of the broader "dilemma of cognitive universalism and historicism in the study of figuration".[47] Typically, this will invite considerations on broader aspects of metaphor theory[48] – among which is the crucial question of the relation between metaphor in literature and in everyday language – [49] as assumptions regarding the origin and the structure of metaphorical thinking are going to ground the interpretation of poetic imagery.[50] We shall briefly consider a few different approaches to metaphor and to poetic metaphor in particular, and see how these can be applied to samples of textile imagery in archaic Greek poetry.

A first possible divide, at the risk of oversimplifying, can be drawn between views of metaphor as a linguistic phenomenon, encompassing varying degrees of engagement with the conceptual cargo of the image, and an approach that sees lexical metaphors as projections and reflections of a cognitive process, a cross-domain mapping of concepts. In turn, the two focuses generate different views about the nature and function of poetic metaphor.

Metaphor as poetic device: the 'hymnos *of song' in context*

At one end of the spectrum, based on language-oriented analysis and grounded in the classical theory of metaphor as a trope, we may position Silk's 'interactional' approach to archaic Greek poetic imagery.[51] He explores the varied and subtle ways through which the relationship between tenor and vehicle (and the linguistic elements that effect this relation and that represent the 'interaction') is exploited at the level of poetic diction and style in authors such as Pindar and Aeschylus, whose fondness for complex textures of imagery and elaborate metaphors was highly regarded already in antiquity. Crucial to such an approach, which sees imagery as deviant "from the terminological norm",[52] is the theoretical apparatus of criteria according to which a reconstruction of categories such as 'normal usage', 'live metaphors', 'dead metaphors', and 'cliché' can be attempted. The distribution and spread of the usage of a term throughout ancient authors will often decide whether an image is a metaphor or

[46] Pindar's metaphors (especially in the *epinikia*) represent a case in point: for different approaches see Silk 1974, *passim*; Steiner 1986, 18–27; Bonifazi 2001, 76–87; Hutchinson 2012; Maslov 2015, 117–177.

[47] Maslov 2015, 177, stating that such a dilemma "cannot be resolved an aprioristic terms. Any such debate must be conducted with reference to the particulars of the given literary-historical episode": this is what Maslov aims to do in his exploration of the function of metaphor as an emergent conceptual device in Pindar (117–177).

[48] In the last thirty years, research into metaphor within a vast array of academic fields and disciplines, ranging wider than just rhetoric, linguistics, philosophy and cognitive studies, has seen an explosion of scholarly work from both a theoretical and an experimental, applied approach: for the purpose of the present discussion refer to Gibbs 2008 and Cairns 2014, 2–8 with further bibliography; see also Ortony 1993.

[49] See on this topic Semino & Steen 2008, esp. 244.

[50] See the chapters by Guilleux, Cairns, Fanfani & Harlizius-Klück, Papadopoulou and Johncock in this volume for marked theoretical premises to the analysis of particular patterns of metaphorical language in poetry.

[51] Silk 1974. See in particular pp. 6–14 for the decision to adopt Richards' categories *tenor* and *vehicle*, "reinterpreted as matters of words" (9), at the service of a 'terminological' approach to the analysis of poetic imagery.

[52] Silk 1974, 27. As for the classical theory of tropes, the concept of 'deviant usage' presupposes an aesthetic statement on the part of both the modern critic and ancient audience; see Silk 2003, 123: "[T]he essence of deviation is that it is *perceptible* deviation."

not.[53] The consistent application of these criteria leads Silk to shed light on the meaning and context of a large sample of imagery, and on the relationship between conventional and creative diction in a historical perspective. Interestingly enough, he takes the spread (Homer, Bacchylides and Pindar) of the metaphor of 'weaving a hymn/song' (ὑφάνας ὕμνον in Bacch. 5.9; other occurrences: Bacch. 19.8, Pi. fr. 179 S-M; 955 *PMG*)[54] as evidence for arguing that the use of *hyphanein* in these instances should not be considered a cliché, but rather a live metaphor making use of the device of 'alliterative linking' – thus proposing an explanation that is alternative to the *communis opinio* that 'weaving a *hymnos*' is a word-play on the seemingly common etymology between *hyphainein* and *hymnos*.[55]

The metaphor of 'weaving the poem/song' may indeed provide us with a helpful imagetype for illustrating how approaches to metaphorical expressions that are attentive to the literary, historical, and performative specificities of an image in its immediate context can enhance our perception of the functioning of metaphor: they can thus be seen as integrating views that account for the cross–linguistic and universal significance of certain patterns of imagery or conceptual metaphors.

A concern for the development of imagery within the history of literary genres informs the 'historical poetics' approach. A recent work by Boris Maslov traces the emerging conceptual (and cognitive) import of metaphor in Pindar, where for the first time in Greek literature metaphor is used as a device of conceptualization and of abstract thought – thus effecting a detachment from the 'mythopoetic' use of images in personifications and genealogical metaphors in previous authors.[56] In the case of the 'weaving a *hymnos*' metaphor, considerations of the semantics of the term *hymnos* as "choral cult song" particularly suited to epinician performance[57] should be linked to a further core feature of the genre, namely the authority of the choral voice to utter metapoetic statements.[58] Thus, the weaving of the *hymnos* on the part of the choral *persona* is likely to be referring to the performative and pragmatic context of the execution of the poem. While the pragmatic approach to metaphor sees it as a matter of 'utterance meaning' rather than 'sentence meaning',[59] the peculiarity of the epinician performance in terms of pragmatics rests on its being 'occasional', embedded in the specific circumstances of the event: on the one hand, the conversion of the event into choral song is effected through the presence of poetry as a favourite metaphorical target (as in the 'weaving a *hymnos*' image);[60] on the other, material objects (or places) included in the performance become important elements at the level of poetic expression. This is the case, for instance, of *stephanos* ('crown') in phrases like 'I interlace a crown of song', especially given the presence of garlands

[53] The criterion of 'distribution' encompasses three sub-criteria establishing the quality (*i.e.* in which authors and texts the word appears), the quantity (how often it appears) and the period of usage of the given word (Silk 1974, 34). For a similar historical concern for tracing the originality of a poetic image, but in the different perspective of detecting a *série métaphorique*, cf. Taillardat 1965, 15–24.

[54] Silk 1974, 180–181 considers part of the spread of this metaphor also the Homeric instance of *Il.* 3.212 which refers to the weaving of a speech (μύθους ὕφαινον).

[55] See Fanfani & Harlizius-Klück in this volume.

[56] Maslov 2015, 117–177, who argues that "Pindar's poetics inaugurates the conceptualization of poetic image", and this can be considered as "one particularly consequential episode in the history of poetic language" (119).

[57] Maslov 2015, 288, who does not treat the weaving metaphor for song in his discussion of *hymnos*. See pp. 294–307 for a comprehensive investigation into the generic archaeology of *hymnos*; as for the hypothesis of a common Indo-European poetics as a single poetic tradition, which he rejects, Maslov submits that (299) "the Proto-Indo-European metapoetic apparatus survived as part of the genre of cult (choral) song whose likely designation in Greek was *hymnos*".

[58] *Cf.* Maslov 2015, drawing on D'Alessio 1994.

[59] Morgan 1993, 124–125 (referring to Searle 1993). See Bark 2011, 18 on the area of *discourse-pragmatics* in current metaphor research, and Bonifazi 2001, 53–54.

[60] See Hutchinson 2012, 279–280.

crowning the *victor* athlete and the dancers.[61] Turning back to the 'weaving a *hymnos*' metaphor, the chorus may be here hinting at the nexus between weaving and chorality that is a recurrent motif in archaic Greek poetry, and that may have been reflected in performance through the rhythmic and dancing patterns of the song, conceptually associated to weaving patterns.[62]

Universal and *specific: conceptual metaphors reconsidered*

A different focus on textile and clothing imagery is offered by approaches which look for semantic networks of images in a cross-linguistical and diachronic perspective ('metaphorical matrixes'), or for the cognitive mechanism of broad conceptual mappings generating metaphorical systems ('conceptual metaphor theory'). As Nicole Guilleux points out in her chapter, the two approaches reflect an epistemological convergence: both explore (semantic and conceptual) associations organized in larger systems of metaphors that pervade both everyday language and poetic imagery, and in both approaches the recourse to concrete referents for expressing (and constructing) abstract concepts, processes or activities is rooted in human and bodily experience and then projected onto the target domain through metaphor.[63]

The cognitive theory of metaphor, famously formulated by George Lakoff and Mark Johnson, is grounded on the notion that metaphor is a universal feature of language, "pervasive in everyday life",[64] and that it is generated by a mapping between two conceptual domains, with a direction that goes from the more concrete to the more abstract domain.[65] A fundamental import of the theory is that linguistic metaphor is basically the reflection of the cognitive process at the conceptual level. A further crucial consequence of the theory is that many complex concepts are constructed metaphorically. The three constituents of a given metaphor are the 'target domain' (an abstract concept to be described metaphorically), the 'source domain' (a relatively concrete concept which will be mapped onto the target), and the 'mapping' (the cognitive process that project the source onto the target).[66] Any metaphorical mapping consists of "slots in the source-domain schema, which get mapped onto slots in the target domain", together with relations, properties, and knowledge.[67] The conventional representation of the mapping uses small capitals, *e.g.* LIFE IS A JOURNEY. When it comes to poetic metaphors, the cognitive theory sees them as using the same mechanisms of every-day, basic conceptual metaphors, but extending, elaborating and combining them in novel ways.[68] An interesting instance of

[61] *Cf.* Bonifazi 2001, 79, who points out that Pindar's metaphors describe often processes and actions. See pp. 81–83 for further interesting features of Pindaric metaphors in relation to the pragmatics of the performance, among which the necessity for the poet to draw on his audience's competence (*sophia*) in order for the image to be interpreted correctly.

[62] *Cf.* Steiner 1986, 54.

[63] For the semantic approach to metaphorical networks, in the context of the imagery system of Aristophanes, cf. the fundamental Taillardat 1965. The foundational work for the cognitive approach to metaphor theory is Lakoff & Johnson 1980. For poetic metaphors in a cognitive perspective see Lakoff & Turner 1989.

[64] Lakoff & Johnson 1980, 3.

[65] See Bark 2011, 7: "[M]etaphor then is not a lexical ornament designed to elaborate a pre-existing concept, but a cognitive relationship *between* concepts."

[66] Bark 2011, 10.

[67] Lakoff & Turner 1989, 63. Interestingly, the notion that features of the source domain are being mapped onto corresponding features of the target domain is an integral part of the 'interaction' view of metaphor proposed by Max Black; see Black 1993, 28: "[T]he maker of a metaphorical statement selects, emphasizes, suppresses, and organizes features of the primary subject by applying to it statements isomorphic with the members of the secondary subject's implicative complex."

[68] Lakoff & Turner 1989, 67. Hutchinson 2012, though not endorsing the theory of conceptual metaphor, adopts the view that poetic metaphors derive from basic "underlying equivalences" (277) that they develop in various and complex

weaving technology providing the source domain for a complex cosmological concept, namely the structure of atomic compound, is discussed by Matthew Johncock in this volume for Lucretius' *De Rerum Natura*; this is suggestive of two elements related to the function and nature of weaving: a) it must have been a familiar and widely accessible technology, if Lucretius decided to exploit its conceptual import for his didactic purpose, and b) it must have been a complex enough technology to be capable of describing and creating, through analogy and conceptual mapping, the invisible structure of atomic compound.[69] As a corollary of this, it could be noted that the technology of the warp-weighted loom, with its constraints and peculiar features, also seems to play a role in structuring the concept of atomic compound in Lucretius' account: we see in this case the suitability of the conceptual metaphor model to be applied to historical and context-specific cases of metaphorical use. The cognitive theory of metaphorical mapping is often associated with a claim of universalism, in that it seeks to position individual metaphors within larger systems of conceptual mappings, and to trace the peculiarity and novelty of literary imagery to broader patterns of conventional metaphors in every-day language and thinking. In fact, as Douglas Cairns illustrates in his chapter about the role of garment metaphors in conceptualizing emotions in ancient Greek, the universal cognitive mechanism of the formation of abstract concepts from concrete experiences needs to be investigated in relation to culture-specific factors – the linguistic and symbolic 'codices' of a given culture, the dynamics of the interaction between the body and the environment – in order for a metaphorical system to be accessed and understood. He shows in particular how ritual and non-verbal gestures associated with the performance of emotions (*e.g.* grief and the ritual of mourning, implying the covering of the dead body and of the mourners' heads) interact with the physical and bodily symptoms in generating concepts of emotions through metonymy and metaphor.

The weaving metaphor as 'myth' in Scheid and Svenbro

One further approach to textile imagery in classical literature, though not primarily concerned with the application of a specific theoretical model to its source material, deserves a mention at this point. In the introductory remarks to their influential investigation of the weaving metaphor in Greek and Roman cultures, John Scheid and Jesper Svenbro argue in favour of a terminological shift from 'metaphor' to 'myth', in order to account for some specific features of the pervasive "metaphor of weaving and fabric" within the Greco-Roman civilization.[70] They observe how a distinctive trait of the weaving metaphor – indeed the main reason for proposing to categorize it as 'myth' – is that it is "a shared one" rather than "an individual creation". More specifically, it is "a figure of thought used by an entire civilization, repeated, modified, and resurrected over time without ever becoming

ways: he discusses Pindar's favourite targets, poetry and victory, showing the range of variations in the 'fame is light' image. To refer back one last time to the 'weaving a *hymnos*' metaphor, this would normally be investigated as vehicle within the category of 'craftsmansip imagery' (see *e.g.* Steiner 1986, 52–65): the emphasis on targets ('poetry', in this case), which in the case of epinician poetry are particularly rich in attracting and generating conceptual images, shows the potential of the conceptual metaphor theory for analysis of genre-specific patterns of imagery.

[69] On the notion of 'gestalt metaphor', which Johncock applies the sustained employment and the complex structure of the weaving in Lucretius' poem, Lakoff & Johnson 1980, 77 observe that it is a means of "structuring one experience in terms of another", by way of "superimposing the multidimensional structure" of one concept "upon the corresponding structure" of the other concept (81).

[70] Scheid & Svenbro 1996, 2–4 (also for the quotations that follow); the authors' stance in relation to this shift of category is clearly reflected in the subtitle of their book, "Myths of weaving and fabric". Their focus is on the craft aspects of textile production, of spinning and weaving, and less on clothing metaphors (despite chapter names such as *Peplos* and *Khlaina*).

fixed or dead". While Scheid and Svenbro make a good case for associating these features to their nuanced concept of 'generative mythology' as encompassing stories, images, and ritual, it could be observed that, in fact, the notion of 'conceptual metaphor' (as we have outlined it earlier in this section) may as well account for most of those same elements. Indeed, one of the 'myths' of weaving explored by Scheid and Svenbro is the image of the 'fabric of the city', which is representative of the fundamental function of myth as "a kind of commonly shared *idée fixe* by means of which the members of a culture constantly try to explore and organize reality":[71] in providing a conceptual image (fabric) through which a civilization makes sense of, and construct its view of another domain (civic life), the myth of the 'fabric of the city' is again invested with a function ascribable as well to conceptual metaphors. One of the best known applications of the 'fabric of the city' image in classical literature, the paradigm (*paradeigma*) of weaving in Plato's *Statesman*, is interpreted by Scheid and Svenbro as an instance of 'political weaving' grounded on the concept of the interlacing (*sumplokē*) of opposites (the two virtues of *andreia* and *sophrosyne*, 'courage' and 'moderation', that needs to be woven together by the art of good government, *politikē*) as reflecting the crossing of warp and weft in the fabric.[72] Integral to such a normative connotation of the metaphor of the 'fabric of the city' is a further dimension of weaving as *paradeigma* for statesmanship: in the dialogue, the opposition on which the technology of weaving is based, namely the principles of combining and dividing, is presented by Plato as the key concept the key concept grounding the sustained structural analogy illustrating the nature of kingship. As a device of investigation of the underlying analogies between two domains, *paradeigma* is thus invested by Plato with an important cognitive function.[73]

Conclusions: understanding textile expressions over two millennia?

The aim of the original two workshops and the papers that have made up this volume has been to investigate how the imagery of textiles (through simile, analogy, metaphor) worked in ancient literature. The collection has identified a range of interlinked connections that have woven together in a complex fabric of text. This complexity should not, and indeed, cannot be unravelled but in order to guide the less-textile-dress-initiated we have identified a series of categories, all of which are covered in one aspect or another by the chapters in this volume.

Language, terminologies, concepts and categories

In dealing with ancient – or dead – languages, we rely heavily on dictionaries where one term equals another term. In research into textile terminologies, the greatest effort is put into understanding textile, garment and textile tool terms in their contexts and translating them correctly so that both the technology and the concept which the term encompasses are expressed. However, a too flat-footed or over-simplified use of dictionaries tends to lose the multiple layers of meaning, and particularly the metaphorical sense of textile and garment terms. While dictionaries and glossaries are useful sources of information, the fluidity of textile and garment metaphors and expressions must be sought in context and in the author's choice of terms and concepts.

[71] Scheid & Svenbro 1996, 3.

[72] Scheid & Svenbro 1996, 22–31.

[73] The relationship between παράδειγμα ('model') and εἰκών ('image', 'comparison' and even 'metaphor') in Plato is discussed by Pender 2003 in terms of the cognitive import of the two concepts. On the function of weaving in Plato's *Statesman* see Harlizius-Klück 2004.

In recent years, a series of systematic surveys and analyses of textile terminologies have been undertaken. These have focussed on textile terms as part of general or specialised vocabularies; on etymologies and evolution; on verbal and non-verbal expressions; and on the ontologies of specialised languages.[74] The impetus behind the workshops that created this volume was a desire to focus beyond the literal meanings of textile terms, and to understand their symbolic and figurative connotations. From the large body of textile and garment imagery in classical literature we identified three distinct semantic and conceptual areas, each containing a series of inter-linked, and sometimes, oppositional, sub-categories:

I Textile craft: spinning, weaving, stitching

- Spinning as metaphor of life and death, destiny, and time
- Spinning and weaving as metaphor for speech and narrative genres; spinning and weaving imagery and their (semantic and metaphorical) associations with the performance and composition of song and poetry; mechanical and functional affinities between weaving implements and musical instruments
- Spinning and weaving as producing ruses and wiles, cheating and plotting
- Weaving (and spinning) as a connecting force, symbolizing social and ritual cohesion (conceptual images in this realm of imagery: society is a fabric of people, weaving as metaphor for intercourse, *symplokein*)
- Weaving and spinning as technological devices illustrating processes of cosmological creation

II The moral values of/associated with textiles and textile producers

- Spinning and weaving as the litmus test for female virtue
- Spinning and weaving as markers of a lack of masculinity

III Garments as markers of identity and emotional expression

- Clothing as metaphor of identity, personification and characterisation of the wearer
- Clothing as expression of the wearer's emotions
- Clothing as metaphor for concealing and revealing the wearer's identity
- Clothing as a vehicle of illness, disease.

These categories could undoubtedly be further extended and finessed, and there are many ways in which they can be blurred and nuanced. We would refer readers to chapters in the current volume to continue their own analysis. Creating lists has the danger of inferring a finite collection: this is not our aim, and even a cursory reading of the volume will demonstrate the diversity of textile and garment imagery.

There remain a few points worth reiterating. The erudite, male Greek or Roman author had a bodily and visual experience of garments and textile production, he was rarely as withdrawn from it as we are in the modern world. One of the effects of this disengagement with the processes of production is our inability to understand the ramifications of what we are reading in ancient texts. This issue is amplified once we stop reading texts in their original languages and refer to translations, particularly when modern translators no longer possess or recognize the technical terms for textiles and garments of antiquity. The problem is compounded by readers who are also no longer familiar with ancient textile techniques, technical textile vocabularies or garment construction.[75]

[74] Del Freo, Nosch, Rougemont 2010; Michel & Nosch 2010a; Michel & Nosch 2010b; Nosch 2012; Lervad, Dury, Nosch 2012; Lervad, Flemestad, Weilgaard Christensen 2016.

[75] Harlow 2016.

And should they be? This is one of the dilemmas that translators face. It might be correct to describe a women at her warp-weighted loom, but to the modern reader of epics, this probably does not make sense or even blurs the message of role and status of the elite woman, as intended by the epic narrator. Perhaps the old and erroneous translations of a lady embroidering at her loom conveys better the notion of leisure, luxury and domesticity to the modern reader? On the other hand, the weaving woman conveys the idea of value of textile labour and the female contribution to a household in a pre-industrial economy. Thus, the choice of how to translate textile words will modify the interpretation of a scene.[76] Textile expressions can be viewed as useful literary tools and cognitive devices stemming from a concrete materiality and visual experience, and used to express and construct more abstract concepts. Many scholars, however, seem to consider textile expressions as archaisms and detached from a concrete memory or experience. They have become frozen metaphors of the past, and may not even work well as metaphors anymore since they refer to clothing or techniques, which are forgotten and unknown to a (modern) reader.

We have here tried to discern certain threads in Greek and Latin literature which express bonds, relationships, birth and life; cosmic phenomena are viewed as bindings and interlacements of elements; garment metaphors express identity, concealing and revealing both emotions and character traits. A significant dimension of our research has been, and continues to be, to explore the role of textile technology in the mental universes of the past, whether in cult, ritual, mythology, metaphors, political rhetoric, poetry or the language of the sciences. Phrases such as urban tissues, the fabric of the universe, the outskirts of the city, the common thread, the time warp, the world wide web – all these terms belong to a figurative and metaphorical language still of today, and make it blindingly apparent that textile expressions are not merely stylistic tools but rooted in cognitive and experiential realities in the past and the present. They inform us of technical terms and textile practice in daily life in antiquity, and thus have a strong didactic and rhetorical value in ancient literature which we need to recognise in order to understand past cultures and present experiences.

[76] Wace 1948; Droß-Krüpe, K. and A. Paetz gen. Schieck 2014, 207–235; Harlow 2016.

Bibliography

Barber, E. 1991 *Prehistoric Textiles: The Development of Cloth in the Neolithic and Bronze Ages with a special reference to the Aegean*. Princeton NJ.

Black, M. 1962 *Models and Metaphors. Studies in Language and Philosophy*. Ithaca NY.

Black, M. 1993 More about metaphor. In A. Ortony (ed.), *Metaphor and Thought*. Second edition. Cambridge, 14–41.

Bonifazi, A. 2001 *Mescolare un Cratere di Canti. Pragmatica della Poesia Epinicia in Pindaro*. Alessandria.

Bork, H. 2011 Wind, Wave, and Generative Metaphor in Greek. All Theses and Dissertations (ETDs). Paper 463.

Boys-Stones, G. R. 2003 (ed.) *Metaphor, Allegory and the Classical Tradition. Ancient Thought and Modern Revision*. Oxford.

Cairns, D. 2014 Ψυχή, Θυμός and Metaphor in Homer and Plato. *Études platoniciennes* 11: 1–83 (on-line). URL http://etudesplatoniciennes.revues.org/566.

D'Alessio, G. B. 1994 First-person problems in Pindar. *BICS* 39, 117–139.

Del Freo, M., Nosch, M.-L., and Rougemont, F. 2010, The Terminology of Textiles in the Linear B Tablets, including Some Considerations on Linear A Logograms and Abbreviations. In Michel & Nosch 2010b, 338–373.

Droß-Krüpe, K. and Paetz gen. Schieck A. 2014 Unravelling the threads of ancient embroidery: a compilation of written sources and archaeologically preserved textiles. In M. Harlow & M.-L. Nosch (eds) 2014 *Greek and Roman Textiles: An Interdisciplinary Anthology*. Oxford, 207–235.

Edmunds, S. T. 2012 Picturing Homeric Weaving. In *Donum natalicum digitaliter confectum Gregorio Nagy septuagenario a discipulis collegis familiaribus oblatum*. http://chs.harvard.edu/CHS/article/display/4365. Center for Hellenic Studies.

Gibbs, R. W. 2008 (ed.) *The Cambridge Handbook of Metaphor and Thought*. Cambridge.

Goldhill, S. 1996 Review of Scheid & Svenbro 1996. *Bryn Mawr Classical Review* 96.8.3

Harich-Schwarzbauer, H. 2015 (ed.), *Weben und Gewebe in der Antike. Materialität–Repräsentation–Episteme–Metapoetik*. Oxford.

Harlizius-Klück, E. 2004 *Weberei als episteme und die Genese der deduktiven Mathematik: in vier Umschweifen entwickelt aus Platons Dialog Politikos*. Berlin.

Harlow, M. 2016 Literary Representations. In M. Harlow (ed.), *A Cultural History of Dress and Fashion, vol. 1 Antiquity*. London.

Hutchinson, G. O. 2012 Image and world in epinician poetry. In C. Carey, P. Agócs & R. Rawles (eds) *Reading the Victory Ode*. Cambridge, 277–302.

Innes, D. 2003 Metaphor, simile, and allegory as ornaments of style. In G. Boys-Stones (ed.), *Metaphor, Allegory, and the Classical Tradition: Ancient Thought and Modern Revisions*. Oxford, 7–27.

Jenkins, I. D. 1985 The ambiguity of Greek textiles. *Arethusa* 18, 109–132.

Jong, I. J. F. de 1999 (ed.) *Homer. Critical Assessments*. 4 vols. New York.

Kirby, J. T. 1997 Aristotle on Metaphor. *American Journal of Philosophy* 118, 517–574.

Kövecses, Z. 2002 *Metaphor. A Practical Introduction*. Oxford.

Lakoff, G. & M. Johnson 1980 *Metaphors We Live By*. Chicago IL.

Lakoff, G. & Turner, M. 1989 *More than Cool Reason: A Field Guide to Poetic Metaphor*. Chicago IL.

Lervad, S., Dury, P., Nosch, M.-L. 2011 Verbal and Non-Verbal configurations of textiles: a diachronic study. In *Conférence TOTh 2011. Terminologie & Ontologie : Théories et applications, Annecy, 26–27 mai 2011*. 201–220.

Lervad, S., Flemestad, P. and Weilgaard Christensen, L. 2016 *Verbal and Nonverbal Representation in Terminology. Proceedings of the TOTh Workshop 2013* – Copenhagen – 8 November 2013 / Actes de la Journée d'étude TOTh 2013 – Copenhague – 8 novembre 2013. Copenhagen.

Lloyd, G. E. R. 1987 *The Revolutions of Wisdom*. Studies in the Claims and Practices of Ancient Greek Science. Berkeley CA.

Lloyd, G. E. R. 1996 *Aristotelian explorations*. Cambridge.

Martin, R. P. 2015 Against Ornament: O.M. Freidenberg's Concept of Metaphor in Ancient and Modern Contexts. In I. Kliger & B. Maslov (eds) *Persistent Forms: Explorations in Historical Poetics*. New York, 274–313.

Maslov, B. 2015 *Pindar and the Emergence of Literature*. Cambridge.

Michel, C. & Nosch, M.-L. 2010a Textile Terminologies, in C. Michel & M.-L. Nosch (eds), *Textile Terminologies in the Ancient Near East and Mediterranean from the Third to the First Millennia BC*. Oxford, ix–xix.

Michel, C. & Nosch, M.-L. (eds) 2010b, *Textile Terminologies in the Ancient Near East and Mediterranean from the Third to the First Millennia BC*. Oxford.

Nagy, G. 2015 *Masterpieces of Metonymy*. Washington DC.

Nosch, M.–L. 2012 The Textile Logograms in the Linear B Tablets: Les idéogrammes archéologiques des textiles. In P. Carlier, Ch. De Lamberterie, M. Egetmeyer, N. Guilleux, F. Rougemont, J. Zurbach (eds), *Études mycéniennes 2010. Actes du XIIIe colloque international sur les textes égéens, Sèvres, Paris, Nanterre, 20–23 septembre 2010*. Pisa-Rome, 305–346.

Nosch M.-L. 2014 Voicing the Loom: Women, Weaving and Plotting. In D. Nakassis, J. Gulizio & S. A. James (eds) *KE-RA-ME-JA: Studies Presented to Cynthia W. Shelmerdine*. Philadelphia, 91–102.

Nosch, M.-L. 2015 The Wool Age: Textile Traditions and Textile innovations. In F. Ruppenstein & J. Weilhartner (eds.), *Tradition and Innovation in the Mycenaean Palatial Polities*. Veröffentlichungen der Abteilung für Ägäische und Mykenische Forschungen, Österreichische Akademie der Wissenschaften, Philosophisch-historische Klasse, Denkschriften, 487. Band. Mykenische Studien 34. Wien, 167–201.

Ortony, A. 1993 Metaphor, Language and Thought. In A. Ortony (ed.) *Metaphor and Thought*. 2nd edition. Cambridge, 1–18.

Pender, E. E. 2003 Plato on Metaphors and Models. In G. Boys-Stones (ed.), *Metaphor, Allegory, and the Classical Tradition: Ancient Thought and Modern Revisions* Oxford, 55–82.

Richard, I.A. 1936 *The Philosophy or Rhetoric*. New York-London.

Rosati, G. 1999 Form in Motion: Weaving the Text in the *Metamorphoses*. In P. Hardie, A. Barchiesi & S. Hinds (eds) *Ovidian Transformations. Essays on the Metamorphoses and its Reception*. Cambridge, 240–253.

Ricoeur, P. 1978 *The Rule of Metaphor.* London.

Scheid J. & Svenbro J. 1996 *The Craft of Zeus. Myths of Weaving and Fabric*. Cambridge MA-London.

Scott, W. C. 1974 *The Oral Nature of the Homeric Simile*. Leiden.

Shen, Y. 2008 Metaphor and poetic figures. In R.W. Gibbs (ed.) *The Cambridge Handbook of Metaphor and Thought*. Cambridge 295–308.

Semino, E. & Steen, G. 2008 Metaphor in Literature. In Gibbs 2008a, 232–246.

Silk, M. S. 1974 *Interaction in Poetic Imagery. With Special Reference to Early Greek Poetry*. Cambridge.

Silk, M. S. 1996 Metaphor and Simile. In *OCD*[3] 966–968.

Silk, M. S. 2003 Metaphor and Metonymy: Aristotle, Jakobson, Ricoeur, and Others. In G. Boys-Stones (ed.) *Metaphor, Allegory, and the Classical Tradition: Ancient Thought and Modern Revisions*. Oxford, 115–147.

Steiner, D. 1986 *The Crown of Song. Metaphor in Pindar*. London.

Taillardat, J. 1965 *Les images d'Aristophane*. Paris.

Wace, A. J. B. 1948 Weaving or embroidery? *AJA* 52, 51–55.

Watson, J. S. 1878 *Cicero. On Oratory and Orators*. New York.

List of Contributors

Miguel Ángel Andrés-Toledo has been research assistant at the Department of Classical Philology and Indo-European Linguistics of the University of Salamanca, Spain (2005–2009), where he obtained the title of Doctor Europeus with his PhD "Vīdēvdād 10–12: Critical Edition, Translation and Commentary of the Avestan and Pahlavi Texts." He has also been post-doc research assistant at the Institut für Iranistik of the Freie Universität Berlin, Germany (2010 – 2013) and associate professor and Marie Curie Fellow at The Danish National Research Foundation's Center for Textile Research of the University of Copenhagen, Denmark (2013–2015), where he carried out the project "Zoroastrian Text(ile)s: Regulations, Symbolism, Identity." Since January 2016 he is working at the Hebrew University of Jerusalem, Israel. His research interests are Old and Middle Iranian languages and literatures, Old Indian languages and literatures, Zoroastrianism, Avestan and Pahlavi manuscripts and textual criticism, Iranian and Indian lexicography, and Indo–Iranian and Indo-European linguistics.

Carmen Alfaro Giner is an ancient historian, and senior lecturer at the Department of Ancient History in the University of Valencia. She works on ancient textiles, dress colour and symbolism (social and religious). She is editor of the *Purpureae Vestes* series and of the SEMA volumes (Seminario de Estudios de la Mujer en la Antigüedad), among which the monographs *Vestimenta e Identitad* (2011), *Luxury and Dress* (2013), and *Tiarae, Diadems and Headdresses in the Ancient Mediterranean Cultures. Symbolism and Technology* (2014). In 2007 she was guest professor at the Danish National Research Foundation's Centre for Textile Research (CTR) at the University of Copenhagen. Co-organizer of the DressID Project (Clothing and Identities. New perspectives on textiles in the Roman Empire: 2007–2012). In winter 2014–2015 Prof. Alfaro was invited as fellow at the Internationale Kolleg für Kulturtechnikforschung und Medienphilosophie (IKKM) in the Bauhaus University of Weimar (Germany).

Emmanuela Bakola is assistant professor in Greek Language and Literature at the University of Warwick. Previously, she held fellowships at University College and King's College London. She has published a monograph (*Cratinus and the Art of Comedy*, OUP 2010) and several articles and chapters which explore the relationship of ancient comedy to other genres. Her current project uses cultural anthropology and theatre space theory and argues that Aeschylean dramaturgy, imagery, stage action, and engagement with cult and ritual show that Aeschylean theatre is profoundly preoccupied with the human relationship to the earth and its resources. Her monograph on this topic is entitled *The Erinyes and the wealth of the earth.*

Isabelle Boehm teaches Ancient Greek and Linguistics at the University of Lyon and is a researcher at the Laboratoire HiSoMA (UMR 5189, Maison de l'Orient et de la Méditerranée). Her scholarly interests include semantics and technical vocabularies in Ancient Greek, with a focus on technical texts and specialized glossaries, in particular the Greek medical vocabulary of the body and sensory perceptions.

Douglas Cairns is the professor of Classics in the University of Edinburgh. His recent publications include *Bacchylides: Five Epinician Odes* (2010), *Tragedy and Archaic Greek Thought* (2013), *Vision and Viewing in Ancient Greece* (*Helios* Supplement 2013, with Sue Blundell and Nancy Rabinowitz), *Defining Greek Narrative* (2014, with Ruth Scodel), *Emotions between Greece and Rome* (2015, with Laurel Fulkerson), *Sophocles: Antigone* (2016), *Greek Laughter and Tears: Antiquity and After* (2016, with Margaret Alexiou), and *Emotions in the Classical World: Methods, Approaches, and Directions* (2016, with Damien Nelis).

Thomas R.P. Coward recently completed his PhD at King's College London on Pindar and Greek lyric poetry. He studied at Durham University, Università di Bologna, and at King's College London. His interests are in literary history, Greek poetry and music, ancient scholarship, fragmentary literary texts. He is co-editing a volume on interactions between Greek lyric and tragedy, and has forthcoming articles on Greek lyric poetry and drama, *scholia*, and literary fragments. He plans to develop his work on ancient scholarship, poetry, and fragmentary literary texts in a study on Rhodian intellectual and cultural history.

Andrea Doyle is associate professor of Classics in the Department of Greek and Latin Studies at the University of Johannesburg. She received her DLit et Phil from the University of Johannesburg in 2008. Her research combines women in classical myth and the reception of classical myth from Late Antiquity to the Middle Ages in text and image. She is currently working on the topic of 'Weaving Women in Myth: Reception in Art and Literature'.

Giovanni Fanfani is a classical philologist and postdoctoral researcher at the Danish National Foundation's Centre for Textile Research (CTR), University of Copenhagen. He received his PhD at the University of Bologna and his research focuses on intertextuality and Euripidean tragedy. His current project, for which he has been awarded a Mobilex grant (2014–2016) from the Danish Council for Independent Research, investigates the role and function of textile imagery in archaic Greek poetry and poetics.

Judith Fletcher is a professor in the Department of History at Wilfrid Laurier University, Waterloo Ontario. She is a classical philologist who works mainly in the area of Greek epic poetry and drama. She has published three monographs including *Performing Oaths in Classical Greek Drama* (Cambridge 2012), over two dozen refereed articles and book chapters, and as many encyclopaedia articles, short pieces, and reviews. In addition she also researches and teaches in the area of Classical Reception, with a book forthcoming on the Underworld motif in contemporary fiction.

Danielle Frisby lectures in Latin and Greek Language and literature at the University of Manchester, UK. She works on the epic tradition, with particular interests in intertextuality, identity and animal studies. She is currently preparing a monograph on Statian intertextuality with Homeric Epic and beginning work on a project which studies the metaphorical and symbolic functions of horses and horsemanship in Latin Epic of the Imperial period.

Ann Glennie is a PhD candidate studying Classical Archaeology in the Department of Classics at Florida State University. She also received an MA in Classical Archaeology from Florida State University in 2013. Her research interests include the archaeology of pre-Roman and Roman Italy, hydraulic engineering, anthropological perspectives on the management of water systems, and weaving metaphor in late Republican to early Imperial Latin poetry. Since 2013 she has served as the Director of Collections and a Trench Supervisor for the Cosa Excavations.

Nicole Guilleux is a senior lecturer in Ancient Greek and Linguistics at the University of Caen–Normandie. Her scholarly interests are mainly focussed on Mycenaean Greek, Homeric epos, Greek drama and lexicology applied to Latin and Greek. She is also involved in the updating of the *Dictionnaire étymologique de la langue grecque* by Pierre Chantraine and of the *Dictionnaire étymologique de la langue latine* by Alfred Ernout and Antoine Meillet. Her research combines linguistic, historical and literary studies, and relies on methodologies derived from these fields.

Ellen Harlizius-Klück is principal investigator in the ERC Consolidator Grant Project 'PENELOPE: A Study of Weaving as Technical Mode of Existence'. In 2014–2015 she was senior research fellow in the international cooperation project between the Danish National Research Foundation's Centre for Textile Research, University of Copenhagen, and the Department of Ancient History, University of Hannover, funded by the Humboldt Foundation. She was also international investigator in the *Weaving Codes – Coding Weaves* Project, conducted by Alex McLean (University of Leeds) and funded by the Arts & Humanities Research Council (UK). In 2012 and 2013 she was Marie Curie Research Fellow of the Gerda-Henkel Foundation in CTR. After being guest professor in Textile Studies at the University of Osnabrück in 2002–2006, she was scholar in residence at the Deutsches Museum in Munich. In her PhD thesis, she combined mathematics, philosophy and textile studies. Since then she focuses on the question if and in which way ancient textile technology has contributed to the advent of mathematics and sciences in ancient Greece.

Mary Harlow is an ancient historian, senior lecturer at the School of Archaeology and Ancient History, University of Leicester. In 2011–2013 she was guest professor at the Danish National Research Foundation's Centre for Textile Research (CTR) at the University of Copenhagen. She works on Roman dress and the Roman life course. Her research combines literary studies, iconography and archaeology and methodologies derived from history, anthropology and sociology. She is publishing the Cambridge Key Themes volume on Roman dress.

Matthew Johncock (RHUL) recently received his PhD from Royal Holloway, University of London, on the topic of Metaphor and Argumentation in Lucretius. His broader research interests, predominantly in Latin literature, include metaphor, the Latin philosophical language, and the relationship between author and reader. Matthew has co-edited with Ed Sanders *Emotion and*

Persuasion in Classical Antiquity, a collection of papers from a 2013 colloquium of the same name, for publication in April 2016. He is currently the teacher of Classics at Wellington College.

Camillo Neri teaches Greek Language and Literature at the University of Bologna. Director of Dipartimento di Filologia Classica e Medioevale of the University of Bologna (2009–2011) and co-founder of Dipartimento di Filologia Classica e Italianistica (2011), he is a member of the international team Année Philologique (he coordinates the group of Bologna), of the journal *Eikasmós. Quaderni Bolognesi di Filologia Classica* and of the Centro Studi, La permanenza del Classico. He is interested in, and has published extensively on, Greek poetry (expecially Greek women poets: Erinna, Sappho), historiography and philosophy, biblical exegesis, humanities and computing. He is the author of *Erinna. Testimonianze e frammenti* (Bologna 2003), *Lirici greci* (Roma 2011), and his current project is a monograph on Sappho.

Marie-Louise Nosch is a historian and director of the Danish National Research Foundation's Centre for Textile Research (CTR) at the University of Copenhagen and the National Museum of Denmark. She is a professor in ancient history. She was awarded her PhD by the University of Salzburg in 2000 with a thesis on Mycenaean textile administration in Linear B but has subsequently merged Linear B studies with experimental archaeology and textile tool studies; as director of the CTR, she has launched research programmes combining history, archaeology and natural sciences.

Magdalena Öhrman, University of Wales Trinity Saint David, is a senior lecturer in Classics. She is an expert in Latin poetry. She is the author of *Varying Virtue. Mythological Paragons of Wifely Virtues in Roman Elegy*, Lund 2008, and she has published on love, emotions and textiles in Latin poetry. Among her recent works are 'Listening for *licia*: A Reconsideration of Latin *licia* as Heddle Leashes', in *Textile Terminologies from the Orient to the Mediterranean and Europe 1000 BC – AD 1000*, eds S. Gaspa, C. Michel, and M.-L. Nosch, Ancient Textile Series, Oxford (forthcoming); 'Fake Farewells: The Elegiac Cast of Ov. *Trist*. 1.3', in *Latinet i tiden. En festskrift till Hans Aili* (Stockholm University 2014), eds E. Andersson, E. Kihlman, and M. Plaza, 427–438; 'Adding An Audience: Notes on Mart. 11.104', *BICS* 56, 2013, 117–121. In 2016–2018 she is a Marie Skłodowska Curie fellow at CTR, University of Copenhagen.

Maria Papadopoulou earned a BA in Classics from the University of Athens and a MA in Linguistics from the University of Lancaster, UK. She completed her PhD in Classical Studies at the University of Athens on the semantics of colour terms in Hellenistic poetry. She has carried out research on historical lexicography and has published on textile terminology and Hellenistic Egypt. She is currently a Marie Skłodowska Curie Post-doctoral Fellow at the Centre for Textile Research, University of Copenhagen. Her project focuses on the foundation of Alexandria and its impact on textile production and trade in Hellenistic Egypt. Her research interests include textile studies, colour studies, history of urban spaces and gender.

Luca Sansone di Campobianco completed his PhD at the University of Bologna, with a thesis on conceptual history focusing on the analysis of a new concept of immortality in Archaic Sparta and of the languages used to convey it (2011 *Linguaggi di Immortalità nella Sparta antica*. Bologna).

Currently he is a post-doctoral research fellow at the University of KwaZulu-Natal (SA), where he is researching the metaphorical language developed in Pre–Socratic philosophy to account for the interactions between the body and the soul.

Jennifer Swalec is a PhD candidate in Classics at Brown University. Her dissertation draws upon archaeological, literary, epigraphic, and iconographic sources to investigate the role of textiles in ancient Greek religion from the perspective of contemporary dress theory. Her research focuses on Greek religion, history, epigraphy, and drama; textile production and the social function of dress; and sex and gender in the ancient world. She was the 2015–2016 Edward Capps Fellow at the American School of Classical Studies at Athens.

Index of Passages

This index is selective, focussing on the passages discussed in the main body of the text (page numbers in bold).

I. Ancient texts

II. Late Antique and Medieval Texts